22nd VIRGINIA INFANTRY

TERRY LOWRY

22nd VIRGINIA INFANTRY

TERRY LOWRY

quarrier press

Charleston, West Virginia

quarrier press

Charleston, West Virginia

Originally published in 1988 by H. E. Howard, Inc.
Reprinted with permission.

ISBN 13: 978-1-942294-56-6
ISBN 10: 1-942294-56-5

Front cover photo of George S. Patton courtesy of Virginia Military Institute Archives

Back cover photo of George S. Patton courtesy of Wenham Museum

Distributed by:

WEST VIRGINIA BOOK COMPANY

West Virginia Book Company
1125 Central Ave. Charleston, WV 25302
www.wvbookco.com

DEDICATION

This Volume Is Dedicated To John Pardue

—Boyhood Friend—

You Left Us Much Too Soon

TABLE OF CONTENTS

ACKNOWLEDGEMENTS

This volume of the Virginia Regimental Histories Series would not have been complete without the outstanding contributions of Brenda Williams, Newport, Virginia, who aided with the roster and also contributed a wealth of information on Craig County soldiers. Also providing invaluable assistance was Robert Driver, Brownsburg, Virginia, who continually guided me to names and sources, helping to make the roster more complete.

Other individuals who provided meritorious assistance included John and Pam Scott, Roanoke, Virginia; Jack Dickinson, Barboursville, West Virginia; John Dyer, New Milford, Connecticut, who kindly loaned his files on the 22nd Virginia Infantry; Gary Walker, Roanoke, Virginia; Scott Hutchison, Powhatan, Virginia, for information on Alleghany County soldiers; Randy Rice, Superintendent, Carnifex Ferry Battlefield State Park, Summersville, West Virginia, who also helped type the roster; Mike Smith, Superintendent, Droop Mountain Battlefield State Park, Hillsboro, West Virginia; Roland Hamilton, Charleston, West Virginia; John Arbogast, White Sulphur Springs, West Virginia; Noble Wyatt, Nitro, West Virginia; Stan Cohen, Missoula, Montana, for technical advice; William Wintz, St. Albans, West Virginia; William Price, Charleston, West Virginia; Larry Legge, Barboursville, West Virginia; and James 'Slim' Combs, Charleston, West Virginia.

Various institutions greatly aided this project including the entire staff of the West Virginia State Archives, Charleston, West Virginia; Howson W. Cole, Virginia Historical Society, Richmond, Virginia; Museum of the Confederacy, Richmond, Virginia; Virginia State Library, Richmond, Virginia; Diane B. Jacob, Virginia Military Institute, Lexington, Virginia; West Virginia University, Morgantown, West Virginia; National Archives, Washington, D.C.; Library of Congress, Washington, D.C.; Russ Pritchard, War Library and Museum, Philadelphia, Pennsylvania; Michael Winey and Richard J. Sommers, United States Army Military History Institute, Carlisle Barracks, Pennsylvania; Rutherford B. Hayes Presidential Center, Fremont, Ohio; Ohio Historical Society, Columbus, Ohio; Tim McKinney and Bob Beckelheimer, Fayette County Historical Society, Ansted, West Virginia; Nicholas County Historical Society, Summersville, West Virginia; Greenbrier County Historical Society, Lewisburg, West Virginia; Monroe County Historical Society, Union, West Virginia.

Special thanks to the many people who contributed information on their ancestors who served with the 22nd Virginia Infantry: Larry White, Clendenin, West Virginia; Tim McKinney, Kanawha Falls, West Virginia; Josephine Halcomb, Ansted, West Virginia; Helen Garner, Beckley, West Virginia; Fuller and Lillian McClung, Alderson, West Virginia; Dorothy Atkins, Nitro, West Virginia; Elaine Richmond, Craigsville, West Virginia;

Mabel Moore, Nallen, West Virginia; Marion McQuade, Oak Hill, West Virginia; Mrs. E. C. Settle, South Charleston, West Virginia; Connie Graves, La Jolla, California; Isaac Noyes Smith, Jr. Charleston, West Virginia, June Lindner, Charleston, West Virginia; Jeffrey Halcomb, Charleston, West Virginia; Kevin Ruffner, Alexandria, Virginia; Theodore Bailey, Pineville, West Virginia; Jack Eads, Mechanicsburg, Pennsylvania; Sandra Rabell, Cross Lanes, West Virginia; Guy Lanham, Charleston, West Virginia; Myrtle Flint, Lookout, West Virginia; William Bess, Melbourne, Florida; Russell Turley, St. Albans, West Virginia; Barbara Bess, Orlando, Florida; and Rembrandt and John Morgan of Morgan's Landing, West Virginia. Of special note in this category are Ruth Ellen Patton Totten, for the letters of Colonel George S. Patton; and Lee Mays, Ripley, West Virginia, for first-hand knowledge of his father, James H. Mays, who served in the 22nd regiment.

Although space limitations excluded a number of excellent pictures, I extend thanks to the many people who sent photographs of their ancestors who served in the 22nd Virginia, including Lucy Echols Huffman, Nellie Stebar Echols, Myrtle Meredith Huffman, Charlotte Hentz, Roscue Huffman, Russell Caldwell, Kenneth Mills, Ruth Hale, Waynard Caldwell, Mayme McCluer, Phillip Lee and Louise Lee, and the Craig County United Daughters of the Confederacy. I am especially grateful to Jane Johnston, New Castle, Virginia, for assisting with the photographs.

Noteworthy contributions also came from Milton Murray II, Mathews, Virginia; Gary Bays, South Charleston, West Virginia; Cleo Hogan, Memphis, Tennesse; Marjorie Jarvis, South Charleston, West Virginia, for typing the manuscript; Richard Armstrong, Millboro, Virginia; Pat Trout, Roanoke, Virginia; Don Haynes, Saltville, Virginia; Gary and Cheryl Baker, South Charleston, West Virginia; Judy Bibbee, Circulation Department, Charleston Newspapers, for leniency with my work schedule so I could complete this project; Jackie Martin, Charleston, West Virginia; Sam Pittman, Charleston, West Virginia; and Christine Beckelheimer, Southern Cross Chapter, U.D.C., Oak Hill, West Virginia.

Finally, special thanks go to Robert K. Krick, Fredericksburg, Virginia, for use of his extensive files; the late Robert Chapman, Huntington, West Virginia, for continued inspiration; Dr. Stephen Cassis, Charleston, West Virginia, who made the entire book possible by restoring my eyesight; the late Charles Lowry, my father; my brother Tim and his family, Nancy, Chuck, and David; and especially my mother, Ruth Ann Lowry, for supporting and encouraging me throughout this project despite the many inconveniences it caused her. Lastly, to Harold E. Howard, for affording me the opportunity to record the legacy of "The Fighting 22nd."

Terry Lowry
South Charleston
West Virginia

INTRODUCTION

It has been thirty-five years (1988) since I released my second book, *22nd Virginia Infantry.* Upon reflection I now realize that, with the home digital age of publishing in its infancy, I wrote the original manuscript on my fathers manual Underwood typewriter, and such items as White Out and correction paper were crude and not very useful. Every time I made a mistake or failed to include something I just removed the paper and started over. In hindsight I find it an absolute miracle I could do this, which I also had done on my first book and my next four books following *22nd Virginia Infantry.* This method of writing complicated matters as authors who contributed to the Virginia Regimental Histories were instructed that once they submitted their original manuscript it would not be returned to the respective authors for review. Such a process culminated in numerous mistakes of varying degrees. I was particularly stunned when I read in the published version the "Field of Lost Shoes" at the battle of New Market became the "Field of Last Shots" in the final version. For years I have had to endure good-natured teasing from friends for this misnomer, which would have been corrected if I had seen the manuscript after the publisher reviewed it.

I found some relief from the mistakes when a second edition was published, and I was permitted to add the corrections as well as some new material. Although the first edition of *22nd Virginia Infantry* may become some sort of collector's item due to the errors, it is with pride we reprint the second edition, with no changes other than this intro and the covers.

The Virginia Regimental Histories series went out of business in 2012, at which time publisher Harold Howard officially relinquished reprint rights to each respective author in the series. Howard, who passed away in 2017, once told me the *22nd Virginia Infantry,* was one of the largest selling volumes in the Virginia Regimental Histories series, and when it sold out and went out of print, it commanded large sums on the rare books market, sometimes as high as $1,000.00. My original intent was to revise and completely rewrite the book after Harold Howard returned the publishing rights, but in early 2022 I suffered a near fatal heart attack which resulted in a severe loss of vision and being unable to efficiently read or write. I am thankful and honored that Bill Clements of West Virginia Book Company and Quarrier Press has stepped up and made this book once again available to all.

Terry Lowry
2023

CHAPTER I

The Kanawha Boys

When the United States erupted into Civil War during the early months of 1861 the Kanawha Valley of western Virginia immediately became a vital and strategic area to both combatants. Situated on the western edge of Appalachia, the Kanawha Valley is formed by the (Great) Kanawha River flowing from the quaint little community of Gauley Bridge, where the confluence of the New and Gauley rivers form its headwaters, to the equally rural town of Point Pleasant, where it empties into the Ohio River. Covering a total distance of some 90 miles, this region became one of the most openly contested areas of sentiment at the outbreak of the war. Truly, the phrase "brother against brother, father against son" applied to the Kanawha Valley, and it was in such a turbulent environment that the 22nd Virginia Infantry of the Confederate army was formed.

Undoubtedly, salt was the primary factor attracting both North and South to the Kanawha Valley. Salt was essential for meat preservation, a necessity in keeping a mobile army well fed, and the valley boasted numerous rich salt deposits. Additionally, the area had a surplus of forest and agricultural products, with particularly exceptional farmland from west of Charleston to the lower end of the valley at Point Pleasant. Being able to produce hay, grain, and fodder abundantly, as well as salt, the resources of the Kanawha Valley could easily provide the essential supplies for armies of men and horses.

The Kanawha Valley was also a direct transportation and communications link between the Ohio River Valley and Richmond. It stood as a sort of buffer zone between North and South, running rife with highly divided political sentiment; many in the same families engaged in sharp disagreements. As a result, a large number of men remained highly uncertain as to their loyalties, providing a good recruiter with much potential manpower. That the fiercely independent mountain man of the valley was held in high esteem, particularly by the Confederacy, was clearly indicated in the April 26, 1861, edition of the *Kanawha Valley Star,* which alluded to the excellent ability of the mountaineer with guns. It was men such as these mountain and farm boys, raw and undisciplined but skilled with horses and weaponry, who enlisted in the Virginia state volunteers and militia in the Kanawha Valley. The result was the formation of three loosely-knit Confederate infantry regiments in the valley shortly after the outbreak of hostilities. These were the 1st Kanawha, which would evolve into the 22nd Virginia Infantry; the 2nd Kanawha, which became the 36th

Virginia Infantry; and the 3rd Kanawha, or 60th Virginia Infantry.

As the Kanawha Valley lay entirely on Virginia soil, it saw the immediate formation of a number of state and militia volunteers, organized for defense against any invaders or aggressors. In fact, the actual nucleus of the Southern army in the Kanawha Valley, as well as the 22nd Virginia in particular, originated through the foresight of Captain George Smith Patton, grandfather of General George S. Patton of World War II fame. Patton, an 1852 graduate of the Virginia Military Institute, had spent time studying law in his father's Richmond office. During that time he had often viewed nearby the drilling of the Richmond Light Infantry Blues, Virginia's oldest and among its most distinguished military organizations.

In 1856 Patton and his new bride moved to Charleston in the Kanawha Valley where he set up a law practice with Thomas Lee Broun. In his new profession he served as Commissioner of Chancery to Kanawha Circuit Court and Kanawha County Court. Ironically, Patton and Broun both served on the board of directors of the Coal River Navigation Company while William Starke Rosecrans, a future military opponent, was company president.

Upon arrival in Charleston, principal economic center and largest town in the Kanawha Valley, Patton purchased the "Elm Grove" estate and quickly became a popular citizen, nicknamed "Frenchy" due to his pointed goatee. He was arrogant, a smart dresser, and displayed classic chivalry toward ladies, making him a dashing, romantic figure. Patton undoubtedly anticipated the forthcoming conflict and formed a militia company in 1856, dubbing it the Kanawha Minutemen. Their formation took place at Brooks Hall (also known as the "Assembly Room") in Charleston, and about a week later they elected officers. The name soon changed to the Kanawha Rifles, and finally , in November of 1859, it changed permanently to the Kanawha Riflemen.

Charleston's newspaper, the *Kanawha Valley Star,* reported of the Riflemen: "This Company is already large and rapidly increasing. It is holding drills regularly." In fact, according to a veteran of the unit, the earliest organization was a "drilling club," something of which Patton was a master. But another statement in the by-laws hinted at their seriousness:

> No provision is made for the failure of a member of this company to attend in time of war or tumult, for it is naturally supposed that every member will be present. However, should any member fail to answer roll call under such circumstances, and not show good reason therefor, he shall be dealt with as a deserter from the army.

The Kanawha Riflemen, with Patton as captain, obviously was patterned after the Richmond Light Infantry Blues, and was composed of the socially prominent and wealthy men of the valley, including some 20 lawyers. The unit was privately financed. Patton, a harsh disciplinarian, drilled the men extensively in the old Mercer Military School lot, in Joel Ruffner's meadow (present Ruffner Park, where a commemorative marker to the Riflemen now rests), and in Brooks Hall. Because they were impressive in drill, the Riflemen were often invited to appear at parades, balls, and social functions, earning a reputation that they could dance as well as, and maybe better, than they could fight. Their decimated ranks at the end of the war put that belief to rest forever.

When the Riflemen appeared at such social affairs, Patton had the guard mounted. As one of them said, "I recall a July 4th celebration and picnic at Ruffner grove, in Slab Town as we used to call the eastern end of the city, and the Riflemen turned out en masse. I was one of the mounted guards, marching back and forth for hours in the hot sun, in my new uniform with my rifle on my shoulder — proudly guarding the fried chicken and lemonade."

This life of leisure turned serious in 1859 when there was talk of sending the Riflemen to Harpers Ferry to help quiet the John Brown affair, but this never materialized. However, in the autumn of 1860 another incident almost erupted into violence when the Riflemen were invited to attend the Agricultural Fair at Point Pleasant. Two uniformed members took occasion to visit friends at nearby Gallipolis, Ohio, and soon found themselves jeered and harrassed by the strongly pro-Union townsfolk. They were quickly rescued by a small detachment of the Riflemen. According to Noyes Rand, one of the unwelcome visitors (and a future adjutant of the 22nd Virginia), had it not been for the discipline under which the Riflemen had been so well trained by Patton, tempers would have flared and "hostilities would have begun then & there in Gallipolis instead of next Spring at Ft. Sumter."

The Kanawha Riflemen included a fluctuating membership of some 75 to 100 men, although member Joseph A. Brown said there were 120. The Riflemen had an excellent brass band led by an English cornetist. They were best known for their gaudy uniforms which Patton had personally designed for them. Slightly varying descriptions survive, but the uniform basically consisted of brass buttons stamped with the Virginia coat of arms on a uniform of "dark [olive] green broadcloth, matching overcoats, broad gold [some veterans claimed black] stripes down the pantaloons, a fancy headgear consisting of black hats with ostrich feathers dangling from the wide brim with a gold KR on the front of the cap, the entire headress covered with oilcloth in bad weather, and white [cotton] Berlin gloves." The "coats were long, with epaulets of gold braid . . . and a short shoulder cape graced the coat." Patton also required the men to blacken their shoes regularly. This uniform was undoubtedly patterned after that of the Richmond Light Infantry Blues. When the Riflemen later mustered into the Confederate army as Company H, 22nd Virginia Infantry, authorities ordered them to dispose of the flashy uniforms. The new more modest outfit featured "light blue jackets and dark gray trousers with yellow trimmings." Some men still wore the original uniform in the early campaigns of 1861.

Weaponry of the Riflemen consisted of a various assortment of guns, including "a short Neiss Rifle, with bayonet, cartridge box and scabbard." Others carried Belgian rifles or a Harpers Ferry Rifle which had a bore of about .45, an iron ramrod, fired with percussion caps, and had a grease box located in the stock. Joseph A. Brown said that they were armed with the noted Mississippi Rifle. One Rifleman reflected:

> Although there is no record of it, I think Patton must have offered the services of the company as a whole to the Governor very shortly after we organized. For had we not been part of the Virginia state militia, and under the command of old Colonel Hogue of Winfield, commandant of this district, we could not have secured rifles from the state. We were inordinately proud of our appearance. After we had learned to drill we never lost an opportunity to turn out, especially after the rifles had come from Richmond. I recall they were locked in Brooks Hall under guard and none of us were allowed to look at those precious rifles at first. I managed to get a peep at them, however, by calling through the door that Colonel Hogue had come and wanted to inspect the guns. When the door was opened I walked in.

The proud ladies of Charleston bought the Kanawha Riflemen a flag, which had the regular 13 stripes and the Virginia coat of arms. With the outbreak of the war they converted the stripes to the Confederate colors and left the coat of arms.

So it was that Patton, the transplanted Richmond lawyer, full of daring and dash and only 24 years of age in 1861, prepared the Kanawha Valley of western Virginia for its defense well before the war began. Patton was not alone. A number of other other militia groups organized in the valley prior to the war, the most prominent being the Kanawha (or Charleston) Sharpshooters. This company was composed of many Kanawha River boatmen and was organized in 1859 by Captain John Sterling Swann, a Charleston lawyer. He was assisted by Joel Ruffner, Jr., 1st Lieutenant John R. Taylor, 2nd Lieutenant Charles Ufferman, 3rd Lieutenant Dr. A. E. Summers, and Sergeant James Venable. Although some sources claim that the Sharpshooters wore gray uniforms, according to the minutes of their meeting on March 12, 1860, they adopted a uniform of Navy blue flannel material. It included a high-collared frock coat with a single row of nine buttons, a short shoulder cape, and pants with two green cords at the seams. It is known that in 1860 the Sharpshooters received permission to drill in a field north of Charleston belonging to a Mr. Clarkson, and that later they were often found practicing maneuvers in a field belonging to the Welch family. The Sharpshooters often worked in conjunction with the Kanawha Riflemen. Once, when both companies attended the funeral of a kinsman of General Winfield Scott, the *Kanawha Valley Star* said: "The two companies were handsomely uniformed and equipped, presenting an imposing appearance. Captain Swann's Company has exhibited it's herosim frequently in the capacity of a fire brigade, battling at all hours of the night in heat and cold." The Charleston Sharpshooters would serve only briefly with the 22nd Virginia Infantry, being mustered into service for one year on June 30, 1861, to date from April 25. On July 25, they transferred to become Company K, 59th Virginia Infantry.

From these two groups, and a number of others from Kanawha and surrounding counties, the 1st Kanawha Regiment was organized on April 25, 1861. The exact composition of the regiment at that time is difficult to ascertain, but when brought to completion it apparently was constituted as follows. Many of these company designations would change when the 1st Kanawha became the 22nd Virginia Infantry, probably in August of 1861, and again when the regiment reorganized on May 1, 1862.

1st Kanawha Regiment

May - August 1861

Border Rifles (Putnam County)
Charleston Sharpshooters (Kanawha County)
Kanawha Rangers (Kanawha County) — cavalry
Bruce Rifles (Greenbrier County)
Mountain Cove Guards (Fayette County)
Nicholas Blues (Nicholas County)
Elk River Tigers (Kanawha County)
Kanawha Riflemen (Kanawha County)
Boone Company (Boone County)
Kanawha Artillery (Kanawha County) — artillery

On April 29, 1861, Virginia authorities commissioned John McCausland a lieutenant colonel of state volunteers and placed him in charge of the situation in the Kanawha Valley. Having been raised and educated in the valley he had a familiarity with the land and its people. Additionally, he had a military background as an 1857 graduate (with first honors) of V.M.I., later serving as an assistant professor of mathematics at that institution. He organized the 1st Rockbridge Artillery and was elected captain, but turned down the command in order to return to the embroiled Kanawha Valley.

McCausland was to raise no more than ten companies in the Kanawha region, and the membership of each was not to exceed 82. He was told that his policy was to be strictly defensive and that the companies of Patton and Swann, as well as two from Putnam County, were already holding themselves in readiness for him.

When he arrived in the valley McCausland immediately set up training camps at strategic locations, including Charleston and Buffalo, but the strongly divided sentiment within the area hindered his recruitment effort, so much so that at one point he proposed a draft, a move that greatly infuriated the militia. In order to bolster his efforts Richmond authorities brought in Christopher Quarles Tompkins, who on May 3 was commissioned colonel of Virginia volunteer forces in the Kanawha Valley, superseding McCausland, who was to assist him.

Tompkins, then 47 years of age, was a highly experienced military man and a well known resident and business man of the Kanawha Valley. He owned an elaborate summer estate three miles east of Gauley Bridge known as "Gauley Mount." Tompkins was an 1836 graduate of West Point and had served in the United States Army artillery until resigning from the service in 1841. Then he continued in the Virginia militia until 1853. Tompkins was not in full agreement with the Southern cause, but sensing the war would be of a short duration, he went ahead and accepted the command in full earnest and soon was appointed as the first colonel of the 22nd Virginia.

On May 8, Patton's Kanawha Riflemen were mustered into the service of the state and

> . . . escorted by the Sharpshooters, Capt. Swann, with flying banners and martial music, the tears and sobs and smiles and waving of handkerchiefs of the fair, the encouragement of the sterner sex and the blessings of all, they marched to the Methodist Camp Ground, which they speedily put in order and designated Camp Lee.

Two days later the *Kanawha Valley Star* printed orders for the Kanawha Riflemen to assemble for dress parade and inspection, and also attempted to boost enlistment by running McCausland's advertisement ordering the volunteer companies of Mason, Jackson, and Putnam counties to assemble at Buffalo. Those from Cabell, Wayne, and Logan were to gather at Barboursville; those from Kanawha, Boone, Wyoming, Raleigh, Fayette, Nicholas, and Clay were to meet at Charleston. Each company captain was ordered to see that every man was provided with the essentials required of a soldier. The volunteers did come, though not in the numbers anticipated, but enough turned out to form numerous companies, many of which became attached to the 22nd Virginia. Most were dreadfully lacking in the required materials, particularly uniforms and weapons, but their spirits were high and they quickly began to gel into a competent fighting unit of western Virginia mountaineers.

During the early morning hours of May 11 the Kanawha Riflemen and Charleston Sharpshooters boarded the steamer *Julia Maffitt* and sailed west on the Kanawha River to the camp at Buffalo, where they assisted the Buffalo Blues from Putnam County. In a letter written from Camp Buffalo on May 15, Kanawha Rifleman Thomas Lee Broun, Patton's former law partner, reported that the company then consisted of 90 men and that their exercise at the camp was very severe with constant drilling and other chores to keep them occupied. McCausland was in charge of the camp. On May 16, he wrote to the Superintendent of V.M.I. requesting supplies and expressing his concern that the war might actually begin at Buffalo.

On May 22 the Putnam County Border Riflemen of Captain Andrew Russell Barbee arrived in camp and were mustered into service. Barbee, a practicing phyiscian from Poca Bottom, claimed that the entire company, including himself, opposed secession but decided to remain loyal to Virginia. One source said that in the summer of 1861 the Border Rifles wore gray uniforms. Barbee claimed that the company consisted of "163 bear and deer hunters, most of whom were six-footers." According to John Morgan, an early member of the group, the company was comprised of 160 men, about half of them from Putnam County and the other half from Kanawha. Putnam was privileged to choose the two highest positions, captain and 1st lieutenant. Barbee defeated Morgan for Captain by one vote. He said that in the voting the two candidates stood side by side and the soldiers lined up behind the man of their choice. Morgan's brother Jim was unanimously elected 1st lieutenant, and Kanawha County claimed the right to the next highest offices, 2nd lieutenant, quartermaster, and orderly sergeant; then back to Putnam County for the next highest, 2nd sergeant, to which John Morgan was elected. Morgan said he was a very tall man, 2nd sergeant on the extreme right, and was the first man called in after the officers to take the oath to the state of Virginia and to take the physical examination.

Also on May 22, McCausland's scouts spotted two companies of the 21st Ohio Volunteer Infantry operating near Gallipolis, the first Federal troops to arrive for participation in the impending and inevitable invasion of the Kanawha Valley. Tompkins visited Buffalo the next day and reported his total strength there as 350 men comprising five companies. On the 25th Swann's Sharpshooters were mustered into state service. Realizing his manpower shortage and the precarious geographical position of Buffalo, Tompkins raced to Kanawha Falls on May 27 and sent an urgent appeal for assistance to Richmond. Ironically, the remainder of the 21st Ohio arrived at Gallipolis the same day.

Additional recruitment efforts continued throughout the Kanawha

Valley. On May 30 Tompkins issued a persuasive appeal to the men of Kanawha and Virginia. The Nicholas Blues enlisted in the First Kanawha the same day, but Tompkins had already come to the realization that the highly divided sentiment in the area would result in few recruits to defend the valley. Fearing for the safety of his men, Tompkins ordered them to board two steamers at Buffalo on May 29 and return to Charleston, where they quartered in Brooks Hall and performed obligatory guard duty. Captain Patton was against leaving Buffalo but his objections apparently failed to deter Tompkins.

To the end of May the Virginia volunteers had retained full possession of the Kanawha Valley with a force numbering between 340 and 500. Most were ill-equipped farmers and mountain boys, with outdated weapons and meager supplies, few uniforms, and sparse training. With men such as Tompkins, Patton, McCausland, and cavalryman Albert Gallatin Jenkins leading them, however, they were a force to be taken seriously by the enemy.

CHAPTER II

Into The Fire

On June 3 the Kanawha volunteers were ordered to Camp Tompkins, about 14 miles west of Charleston near the mouth of Coal River, where life consisted of seemingly endless drilling, guard-marching, dress parade every afternoon at 4 p.m., and other such typical camp life. It once got so boring that the men requested a day off to go fishing. The first three companies to arrive in camp were the Kanawha Riflemen, the (Putnam County) Border Rifles, and one cavalry unit known as the Sandy Rangers. Yet within days the ranks swelled until there were seven infantry companies present, plus four of cavalry, and some artillery. Among those present who would make up the early ranks of the 22nd Virginia were the Kanawha Riflemen, the Charleston Sharpshooters, the Putnam Border Rifles, the Fayetteville Rifles, the Elk River Tigers, the Kanawha Rangers, and the Kanawha Artillery. Typical of these companies was the Fayetteville Rifles, which originated as early as May 10 when the 149th Regiment Virginia Militia assembled in reponse to a call for volunteers in a field near Fayetteville. Speeches were made by Christopher Q. Tompkins, afterwards colonel of the 22nd Virginia, and Buehring, H. Jones, soon to be colonel of the 60th Virginia Infantry. Only ten men stepped forward to volunteer but within the next few days many others joined, bringing the total strength of the company to 76. Their average height was 5'9½". Member William F. Bahlmann said the company took all the men on Laurel Creek in Fayette County except two, and that there were only ten married men in the company.

The County Court appropriated $5000 to clothe the Fayetteville Rifles and another company that was forming at Mountain Cove on the opposite side of New River. A merchant tailor cut the cloth for their uniforms and the area ladies gathered at the only church in Fayetteville in order to sew them. The "jackets were made of blue flannel with yellow stripes across the breast like the European hussars, the pants were of dark gray jeans with a wide yellow stripe down each leg." Yellow chevrons were put on officers' sleeves. In the place of knapsacks they had long and narrow bags called pokes which fastened to each shoulder. This colorful uniform later provoked the enemy to call them the "yellowjackets."

The company was organized on May 23. Despite having no drill master, they drilled in drill-squad every morning. On June 3 they were ordered to depart for Charleston, which they promptly did, bidding farewell to family and friends. Looking at the citizens that day William

Bahlmann thought to himself, "Take your last look at these men, you will never see them all again," which proved tragically correct. The first day's march, approximately 21 miles, took them to Cannelton (Montgomery) where they divided up and shared meals with various area families. Captain Robert Augustus ('Gus') Bailey was laid up at Mr. Montgomery's house that night, his feet sore from having worn ill-fitting French boots on the march. Apparently Bailey was not highly thought of by Bahlmann who, upon observing Bailey's predicament, said, "If I should say that I felt sorry for him, I would tell an untruth." During a late roll call that night, Jim Arthur and Pat Murray fainted in the ranks.

The following day the Fayetteville Rifles arrived in Charleston and quartered in the Yankee Sawmill. A few days later they marched to a store in town in order to receive rifles. A near riot erupted among the company as an attempt was made to give them smoothbore muskets. Some already had old mountain rifles and squirrel guns which they had brought with them; this effort to arm them with smoothbores insulted them. They claimed that they had been promised rifles, had volunteered as riflemen, and had "Fayetteville Rifles" on their flag. Bahlmann, the 1st sergeant, accepted his gun, but 2nd Sergeant Joel H. Abbott refused, causing tempers to flare. Finally, Colonel Tompkins made a speech and a compromise was effected: the boys would be trained in the rifle manual of arms and would receive rifles as soon as the government could provide them. Bahlmann said they eventually did get the rifles — "on the battlefield from the enemy."

The company was mustered in June 6 and after a few days to master their crude muskets, boarded a steamer on the Kanawha River. The steamer took the men to the mouth of Coal River, then they marched the remainder of the distance to Camp Tompkins.

In consequence of Colonel Tompkins' constant movements throughout the valley on official business, Captain Patton was placed in charge of Camp Tompkins and the Kanawha Riflemen provided drillmasters. V.M.I. Cadet John Koontz Thompson would arrive in camp on July 3, enter the Putnam Border Rifles, and become Drill Master and Officer In Charge at the camp. The Valcoulon estate on the property served as a headquarters and the men gave nicknames to the rooms they quartered in, such as "The Aristocratic Seven," "The Filthy Four," and "The Thieving Three." On June 4, while modestly requesting a promotion, Patton gave his strength at Camp Tompkins as "about 850 men, which will be largely increased in the next few days."

Fortunately for Patton and Tompkins, Federal General George B. McClellan did not wish his army to appear as the aggressor and delayed an attack on the valley. This gave time for the Richmond authorities to send the hot-tempered and militarily untrained Brigadier General Henry A.

Wise, an ex-governor of Virginia, to the Kanawha Valley to take charge. Although he was sick at the time, Wise jumped at the opportunity and arrived in Charleston on June 26 accompanied by a number of troops, including Captain Jacob N. Taylor's Bruce Rifles from Greenbrier County, picked up en route at White Sulphur Springs. Taylor's men would serve briefly with the 22nd Virginia before being transferred to the Wise Legion.

After a few days in Charleston, Wise established Camp Two-Mile at the western outskirts of town on the Littlepage farm. Companies associated with the 22nd Virginia arrived and trained there, including Taylor's company, which had marched 112 miles. The Rocky Point Grays, according to original member James H. Mays, had enlisted at Rocky Point in Monroe County and included 100 young men. After spending their first few days in training at Rocky Point, the Grays set out for Charleston. They spent their first night at Palestine and the second at Blue Sulphur Springs, both in Greenbrier County. The Grays spent the next few days crossing the Sewell and Gauley mountain ranges until they arrived at Gauley Bridge. There they were issued their first rifles, old rusty flintlocks of which many had obviously seen service in the American Revolution. From Gauley Bridge they traveled down the Kanawha River to the mouth of Kelly's Creek, where they saw their first Confederate flag, a red, white, and blue banner bedecked with stars. A steamboat came up the river from Charleston to pick them up and took them to Bibbee's Mill near the mouth of Elk River. There they took a slight physical examination; so slight, in fact, that only one man failed, and he because he had a wooden leg. Then the company moved to Camp Two-Mile, where the men went on dress parade for the first time.

McCausland was in charge at Camp Two-Mile and reported the volunteers there as being "very rugged, ill-equipped, most without uniforms or weapons" but insisted that they had the potential to develop into first-rate soldiers. General Wise did not help matters, as he quickly became "erratic, impulsive, and thoroughly unpredictable," not trusting anyone and soon having the jails of the valley full of suspected traitors. On June 29 he reported 30 companies present for duty, including the state volunteers and his own Wise Legion. He feared Charleston might be attacked from the north and sent an 800-man force, including Patton's Kanawha Riflemen, to Ripley. This detachment departed on the night of the 29th and marched 40 miles, only to find that the enemy had abandoned Ripley.

While yet at Ripley on July 1, the 22nd Virginia Infantry (First Kanawha) was accepted into Confederate service and Patton led a contingent comprised of the Richmond Light Infantry Blues, and possibly the Kanawha Riflemen, three miles west of Ripley en route to Cottageville. A small skirmish ensued with some home guards at Antill (now Conrad) Hill.

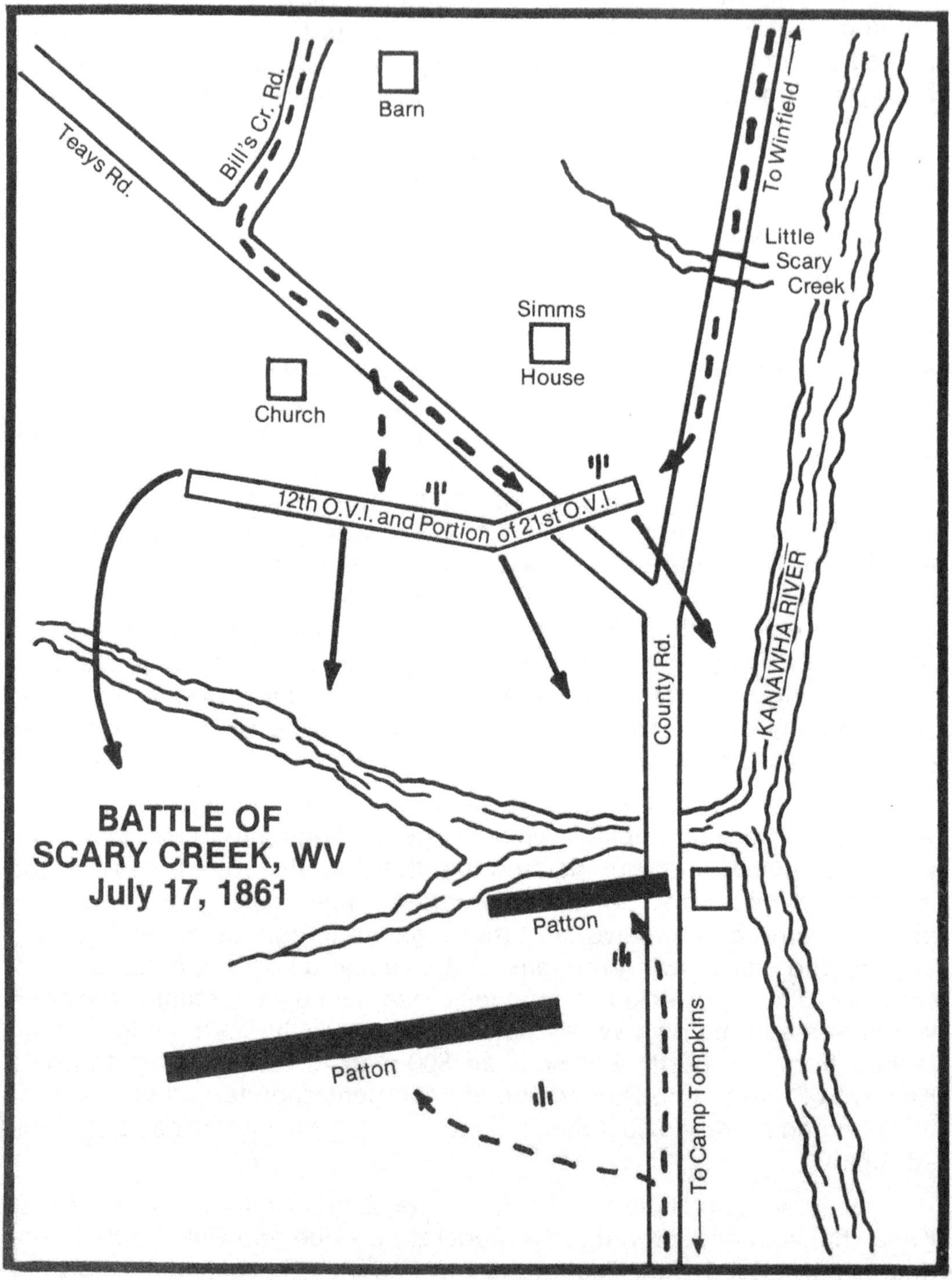
Teays Rd.
Bill's Cr. Rd.
Barn
To Winfield
Little Scary Creek
Simms House
Church
12th O.V.I. and Portion of 21st O.V.I.
County Rd.
KANAWHA RIVER
BATTLE OF SCARY CREEK, WV July 17, 1861
Patton
Patton
To Camp Tompkins

This was Patton's baptism of fire, and the Riflemen's, too, if they were indeed present. There were no casualties and the entire movement to Ripley proved to be nothing more than a lot of useless marching, countermarching, and false alarms. But Patton did manage to pick up the Jackson County Border Rifles of Captain William H. Lipscomb (a Ripley lawyer), who would soon join the 22nd Virginia. By July 10 the entire force was back at Camp Two-Mile. Joab Smith of the Border Rifles described the foray as a "tedious journey through the rain . . . spent another night at Scofford's, or rather from two in the morning until daylight . . . lay down on the wet ground, where I was nearly devoured by fleas . . . rose at daylight . . . after two hours of restless slumber . . . without breakfast reached Ripley . . . ." Kanawha Rifleman William C. Reynolds described the return march as "traveled 24 miles and slept in the dog Kennel [at Sissonsville] . . . minus dinner and supper," and returned to Charleston the next day in a drenching rain.

Wise had launched a separate attack north of Charleston at Glenville on July 7, the very same day that Patton was made a lieutenant colonel; but the 17th Ohio Volunteer Infantry held off the 160 rebels and sent them hurrying back to Charleston the following day.

On July 8, Wise's "Army of the Kanawha" consisted of 2,863 men, although many were not yet present for duty. The 1st and 2nd Kanawha reported that they had 68 officers and 1,332 men with an aggregate present of 1,482. Wise's soldiers had few uniforms and weapons, no knapsacks, tents, blankets, and so forth, and were led "by incompetent officers, resulting in a near total lack of discipline and military training." Therefore, Wise felt that little could be expected of them if brought into contact with the enemy.

Wise's assessment proved entirely untrue when the Federals, under Brigadier General Jacob D. Cox, began a three-pronged invasion of the valley on July 11. The first inevitable clash came at Barboursville, at the mouth of Mud River, on July 14 when the 2nd Kentucky Volunteer Infantry attacked some 600 to 900 rebels. After briefly being pinned down by a galling fire the Federals rallied and launched a bayonet charge, driving away their foe in complete disorganization. On that same day a small force of the 21st Ohio, attached to Cox's main force moving up the Kanawha River, struck the Kanawha Riflemen, the Putnam Border Rifles, and the Kanawha Artillery at the mouth of Scary Creek, only a mile or so west of Camp Tompkins. The action only lasted long enough for the Federals to test the Confederate strength. They then fell back to the main encampment at Poca with no damage to either army.

Private James H. Mays of the Rocky Point Grays (Co. F, 22nd Virginia) recalled that the two clashes on the 14th brought the troops at Camp Two-Mile to full alert and the men were ordered to Scary Creek to give assistance. Before the soldiers boarded a steamer, the chaplain held a prayer meeting and there was much pentenance, and probably some conversions, among men anticipating their first brush with death. "Some fellows were feeling so weak and sick they had to be led onto the boat. The conversations en route to battle of these men of peace from the farms of Virginia were very amusing, as well as pathetic."

The Rocky Point Grays were not needed that day. Three days later, however, on July 17, some 900 men under the command of Patton, primarily members of the 22nd and 36th Virginia regiments, were attacked by General Cox at the mouth of Scary Creek. The first contact came early in the day. Cox's advance, led by Lieutenant Colonel Carr B. White, skirmished with Captain Andrew Russell Barbee's Putnam Border Rifles, who were posted in a house situated at the mouth of Little Scary Creek about a half-mile west of Scary Creek. The Federals, having probed the rebel line, withdrew to their camp at Poca and assembled a larger attacking force under the command of Colonel John W. Lowe of the 12th Ohio Volunteer Infantry. He took with him about 1,200 men, including his own 12th Ohio, portions of the 21st Ohio, and some cavalry and artillery.

The skirmish at Little Scary Creek alerted Patton at Camp Tompkins to the danger and he immediately rushed troops to Scary Creek, deploying them strategically, mostly on the eastern bank of the creek. The Kanawha Riflemen, under Lieutenant Nicholas Fitzhugh, were placed on the extreme left as skirmishers. Also with Patton were Barbee's men, the Fayetteville Rifles, two field pieces comprising the Kanawha Artillery, an infantry company from Wheeling, and assorted cavalry under the command of Captain Albert G. Jenkins. Patton asked John K. Thompson where he felt the artillery should be posted.

The battle opened at about 2 p.m. but after some 5 hours of heavy fighting, with the outcome often in doubt, the badly outnumbered Confederates defeated the well-trained and equipped Federals. But the victory came at a high cost. While Patton was attempting to rally his men during the affair, his horse had become unruly and the officer was hit in the left shoulder by a bullet. The wound proved so severe that he had to be removed from the action and Captain Jenkins assumed command. Other casualties among the 22nd Virginia were Lieutenant James C. Welch of the Kanawha Artillery, who was killed when an enemy cannonball nearly decapitated him. Lieutenant Welch probably was the first battle casualty of the 22nd Virginia. Also killed were Privates Charles Blake and William C. Fellers of the Fayetteville Rifles. Other than Patton the wounded included Captain Barbee; Sergeant John T. Dudding of Company A, who was struck by a cannonball yet survived; Privates Warren Jones and Jonathan Weaver of the Fayetteville Rifles; Beverly O. Markham of Company A, who was wounded by a musketball; and Private Madison Craigo of the same company, who received a severe musketball wound of the thigh and hip. The total Confederate casualties at Scary Creek were some 4 killed, 10 wounded, and 2 captured. Cox lost 15 killed, 11 wounded, and 7 captured, including a number of high-ranking officers.

The Charleston Sharpshooters arrived at the battlefield but their participation was minimal. The Rocky Point Grays were on a reconnaissance at the time of the clash, heard the sounds of battle, and rushed toward Scary Creek, but arrived too late to participate. The following day James H. Mays of the Grays helped to bury the dead Federals at Scary Creek. One of the dead officers had a love letter on his person which Captain Barbee returned to the lady, praising her lover's bravery in the fight. Most of the Confederates had returned to Camp Tompkins by 8 a.m. on July 18, having been at Scary Creek for some 19 hours with but little to eat or drink. Cox's army was only slightly bloodied and would soon reorganize and renew its drive on the Kanawha Valley, but for the time being the 22nd Virginia, as well as all the rebels who fought at Scary Creek, basked in their first true victory over the enemy.

Despite success at Scary Creek the Confederates were compelled to abandon the Kanawha Valley. Federal victories in the northern section of western Virginia by McClellan had opened the way for Union soldiers to advance from the north against Gauley Bridge, thereby possibly cutting off Wise's avenue of retreat and trapping the rebels in the valley. General Wise realized that he was about to be caught in a deadly pincer and ordered a "retrograde movement" out of the valley beginning on July 19. The retreat went relatively well until July 25 when some 700 men under Colonel Tompkins, primarily members of the 22nd Virginia, were nearly captured by General Cox's Federals. With all of the Confederates' supplies and some of their men aboard the *Julia Maffitt,* Cox's artillery opened fire on the boat from Tyler Mountain. After sustaining a hit near present-day Dunbar, the boat went to the opposite side of the river near the mouth of Davis Creek, where Kanawha Rifleman Phil Doddridge, the boat's pilot, set it ablaze so it would not fall into enemy hands. Almost all of the supplies of the 22nd Virginia went down with the boat. The troops, though, escaped on the south side of the river.

Moving on to Charleston the hometown boys bid their families and friends farewell. The "weeping and moaning accompanying these partings were audible to the soldiers" on the opposite side of Kanawha River. By evening much of Charleston was in flames and the Confederates were at Clifton (Pratt). During the day the Charleston Sharpshooters had transferred out of the 22nd Virginia to Company K, 59th Virginia, taking their approximate 101 men.

On July 26 Wise was at Gauley Bridge and Cox was in competent pursuit. To hinder the Federals the rebels burned the wooden bridge spanning the mouth of Gauley River and retreated over Gauley Mountain to Miller's Ferry, then through Raleigh, Mercer, and Monroe counties, arriving at Bunger's Mill near Lewisburg August 1. Cox ended his advance by taking possession of Gauley Bridge, as well as the entire Kanawha Valley, on July 29.

The "retrogade movement" from the valley proved costly to Wise. Many of the mountaineers had enlisted under the impression that they were only obligated to defend their homes and would not be required to leave the valley. As a result, many of them deserted along the way. To make matters worse, many of the officers were against the retreat and furloughed large bulks of their men while in retreat. Some companies arrived at Bunger's Mill with only a few men. Company I of the 22nd Virginia, from Boone County, deserted en masse on the night of July 25, although they all reportedly returned of their own accord on August 22. Wise could count only 3,500 effectives when he reached Bunger's Mill, and that was probably an overestimate. He claimed that 300 to 500 of the state volunteers had deserted, but his own Legion was intact. The worst damage to the 22nd Virginia was the loss of Patton, who was too seriously wounded at Scary Creek to be moved. Therefore, he was left with the Federals under an odd exchange agreement, and would not be able to rejoin the regiment until the following year. The stark reality of the entire situation became apparent when Wise said that during the occupation of the Kanawha Valley his troops barely had enough ammunition to have fired one round at the enemy, a fact only he and Colonel Tompkins knew.

The army moved on to White Sulphur Springs, where Wise and Tompkins desired to rest and refit the troops, which had been plagued by the constant marching and counter-marching, desertions, disease, and battle casualties while in the Kanawha Valley. First, the artillery and cavalry companies were transferred or dissolved. The 22nd Virginia was created from the 1st Kanawha, although company organizations at that time would change once more at the official reorganization of the regiment the following year. The Fayetteville Rifles, originally assigned to the 36th Virginia Infantry, traded with another company and became a unit of the 22nd Virginia, and received gray shirts and jackets to replace their "yellowjackets." The Rocky Point Grays performed their first battalion drill and had their flintlock muskets converted to percussion.

The rest was cut short when Wise's old political opponent, also an ex-governor of Virginia, General John B. Floyd, arrived and took charge. Floyd, a military incompetent as well as Wise's senior in rank, desired to retake the Kanawha Valley immediately. He was not the least bit concerned about resting the exhausted mountaineers. Wise was in full disagreement, but on August 20 the two quarrelsome generals began a drive back to the valley. They had to leave behind numerous soldiers in the hospital at White Sulphur Springs. A member of Company A of the 22nd Virginia remarked that more than half of the regiment remained behind with measles and typhoid.

On August 21 General Robert E. Lee ordered both the 22nd and 36th Virginia regiments to join Floyd's column. Both left White Sulphur Springs at 10:30 a.m., marching to Lewisburg and then spending the night at Bunger's Mill after covering a "very hot and disagreeable 13 miles." That night the 22nd Virginia received tents for the first time.

Both regiments camped at the foot of Big Sewell Mountain on August 22 and at Locust Lane August 23, joining Floyd at Carnifex Ferry in Nicholas County on the 24th. Two days later, on August 26, Floyd's entire army completely surprised and routed Colonel Erasutus B. Tyler's 7th Ohio Volunteer Infantry at Kessler's Cross Lanes, just north of Carnifex Ferry. The Federals were eating breakfast when attacked and put up a feeble resistance for about forty minutes. A number of companies of the 22nd Virginia participated, including Company A (Putnam Border Rifles), which was armed with Mississippi and Derringer rifles, "well equipped, clothed, and in good cheer." They double-quicked to the field, took position on a hill near the enemy, and after a partial fire the already retreating Yankees were soon out of sight. Later in the evening, with Tompkins at the head of the column, they captured a squad of Federals in the wooded mountainside. Col. Tompkins' kind treatment of the prisoners earned him the respect of the Ohio soldiers.

Company F of the 22nd was on the left wing during the fight and assisted in the capture of a captain and 14 prisoners. Company H, being on the left wing, could not participate before the enemy retreated. Companies B and D also participated as did Company G (Wyoming Riflemen) which was led into battle by 1st Lieutenant Floyd McDonald, filling in for officers Erastus Duncan and James A. Cook, who were ill. Member Theodore F. Bailey witnessed McDonald in the fight "with his arm bandaged, having received a sabre cut, riding his horse among the troops, encouraging them to 'Give 'em Hell.' "

Following the victory, in which he only took slight casualties, Floyd fell back to Carnifex Ferry and established Camp Gauley on the high mountain bluffs overlooking the mouth of Meadow River where it empties into the Gauley. The 22nd Virginia was mustered and inspected there by Col. Tompkins on Aug. 31 and went on dress parade on September 4. Two days later the regiment was recalled to Wise's camp near Hawks Nest as he was expecting to be attacked by Cox. The alarm proved false and the 22nd began to march back to Carnifex Ferry on September 9, camping that night on the Sunday Road in a meadow near the Reynolds place.

Federal General William S. Rosecrans, having taken over from McClellan, had launched a drive against Floyd with three brigades via Sutton

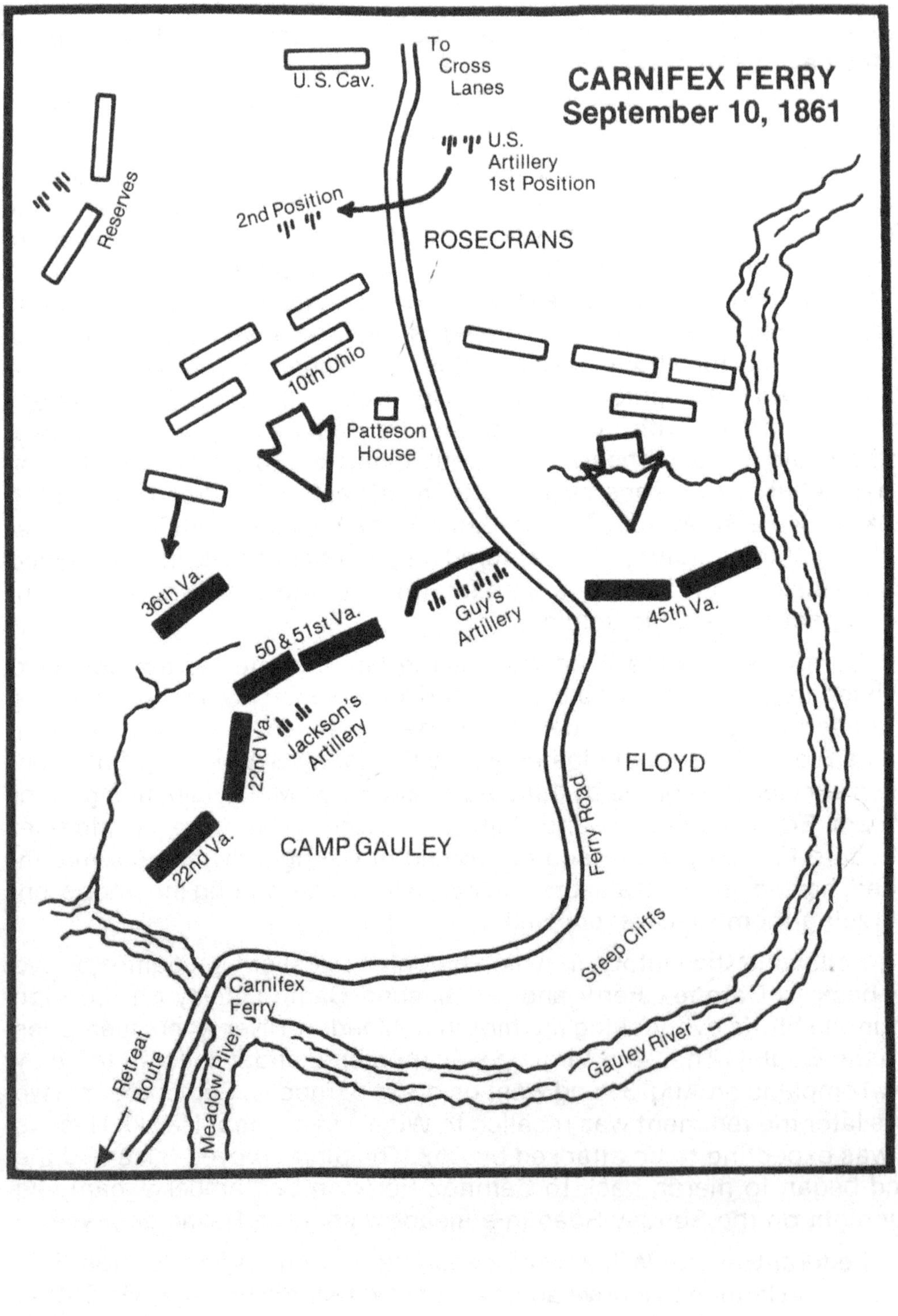
CARNIFEX FERRY
September 10, 1861
U. S. Cav.
To
Cross
Lanes
U.S.
Artillery
1st Position
2nd Position
Reserves
ROSECRANS
10th Ohio
Patteson
House
36th Va.
50 & 51st Va.
Guy's
Artillery
45th Va.
22nd Va.
Jackson's
Artillery
FLOYD
22nd Va.
CAMP GAULEY
Ferry Road
Steep Cliffs
Carnifex
Ferry
Retreat
Route
Meadow River
Gauley River

and Summersville. With him were perhaps as many as 9000 soldiers, mostly from Ohio. By September 9, these Federals were within striking distance of Floyd's fortifications at Camp Gauley. With less than 400 effectives the 22nd Virginia arrived at Carnifex Ferry at about 2 p.m. on September 10, just in time to be sent to defend Floyd's left flank. Altogether Floyd had about 2000 men, including Colonel John McCausland's 36th Virginia, Colonel Henry Heth's 45th Virginia, Colonel Alexander Welch Reynolds' 50th Virginia, Colonel Gabriel C. Wharton's 51st Virginia, and assorted detachments of artillery, cavalry, and home guards.

Shortly after 3 p.m. on September 10, Rosecrans launched the first of numerous assaults on Floyd's defensive line. The 22nd Virginia, in the thick brush on the left flank, probably had the safest place on the field, as all they had to contend with were some artillery shells that flew over their heads. Some of the men pried them out of the ground as souvenirs. Adjutant John K. Thompson was slightly injured when a musketball bounced off his head, his first of six wounds incurred during the war. Private Levi Tincher, Jr., of Company K ran out of ammunition and Colonel Tompkins reportedly told him to throw rocks at the enemy.

Floyd held his position against Rosecrans until nightfall, having inflicted severe punishment on the Yankees. Colonel John W. Lowe of the 12th Ohio was killed and Colonel William H. Lytle of the 10th Ohio was seriously wounded leading a charge on Floyd's breastworks. But Floyd realized that he could not repel a full-fledged attack the next day so he opted to retreat that night, moving his troops down the treacherous ferry road in the darkness and to safety by daylight. Much to the disgust of Colonel Tompkins and Major Isaac Noyes Smith, the 22nd and 36th Virginia regiments were made the rear guard in the retreat, exposing them to great danger, but they managed to get across the river and destroy the ferry so Rosecrans could not pursue. In his haste to get away quickly, though, Floyd had left behind many supplies as well as his own sick and wounded. As a result Rosecrans captured a number of men at the breastworks on September 11, including at least three members of the 22nd Virginia. But Floyd had escaped, with no men in his own ranks killed and only some 20, including himself, slightly wounded. Rosecrans, on the other hand, had more than 150 killed, wounded, or missing.

Wise joined a disoriented Floyd at Dogwood Gap on September 13 and both retreated east to Big Sewell Mountain, reaching there on the 17th. Wise appreciated the defensive possibilities of the position and remained there with his men. Floyd disagreed and moved his men, including the 22nd Virginia, eastward to Meadow Bluff, reaching there the same day. Floyd felt his position was the strongest and had the troops dig extensive earthworks. Private William C. Reynolds of Company H of the

22nd said that they worked on the fortifications on the right flank of the Meadow River on September 19 and that his company had only one axe and two butcher knives with which to dig.

The constant bickering between Floyd and Wise caused the Confederate authorities much headache. Even when the whining ex-governors combined their forces they had far fewer men than Rosecrans, so their refusal to work together divided the numerical strength of the army drastically. In an effort to resolve this situation General Robert E. Lee arrived at Meadow Bluff on September 21 and met with Floyd. That officer did his best to persuade Lee to order Wise back from Sewell Mountain, pointing out that the impending winter would make the roads on Sewell Mountain impassable, and that if the Federals were defeated at Meadow Bluff they could not retreat over the mountains. Lee took this under consideration, checked Wise's position, and assessed it the strongest of the two points. He ordered Floyd's men up from Meadow Bluff. On September 24 Floyd returned to the mountain with his troops. The following day the 22nd Virginia built breastworks at Camp Defiance on the eastern summit of Sewell Mountain, accidentally wounding two of their sentinels by being overly alert.

Rosecrans and Cox managed to get on another peak of Sewell Mountain with only a mile-wide gorge separating them from Lee, but incessant rains began to fall and for almost two weeks the opposing armies sat facing each other. They could not mount an attack due to the weather and rugged, mountainous terrain. Lee sent Wise back to Richmond on September 29 and by October 5 the Federals decided to withdraw back to Gauley Bridge, thereby ending the stalemate. The two weeks on Sewell Mountain proved disastrous to both armies as sickness became widespread. Floyd complained that more men were lost here by illness than by combat in the battle of Manassas. Private Joab Smith, Company B of the 22nd Virginia, was put in charge of a ward at White Sulphur Springs on September 29 and said it contained 60 men who were suffering from various maladies, "being in filth and destitute of every necessity of life, and all imploring relief," which Smith could not give. In fact, disease took Smith's own life there in late October.

Floyd decided to leave Sewell Mountain on October 11 and pursue Rosecrans, taking a southerly direction by way of Packs Ferry. By October 16 he was within a days march of Red Sulphur Springs, camping on the south side of New River at Richmond's Ferry. Colonel Tompkins left the 22nd Virginia at this time to go to Lewisburg to meet Mrs. Tompkins, whom Rosecrans had permitted to pass through Union lines. This left Major Isaac Noyes Smith in command of the regiment.

Hoping to go into winter quarters at Logan Court House, Floyd moved in that direction, but on October 23 he discovered that Rosecrans was camped at and near Gauley Bridge, accessible to attack from Cotton Hill, south of New River. Floyd immediately set up camp east of Cotton Hill and just north of Fayetteville, naming it Camp Dickinson (also referred to as Camp Dickerson), and moved his artillery and some infantry to the heights overlooking Gauley Bridge and "Gauley Mount."

Many of Floyd's officers did not approve of the position and signed a petition requesting that the camp be moved to a safer location. Floyd refused and began moving his artillery into position to open upon the enemy. This created a delicate situation as Colonel Tompkins' home, "Gauley Mount," was then being used as a camp by Rosecrans. An attack on the estate might destroy Tompkins' property and endanger his family. Tompkins had already been in much disagreement with Floyd, who had earlier refused him a pass to go to Richmond to clear up the situation with his family and property, so Tompkins soon afterward resigned in disgust. This move left the 22nd Virginia without a colonel, as Patton was yet an un-exchanged prisoner, so Major Isaac N. Smith found himself in temporary command — a most difficult position for him since Floyd continually accused him of being a Unionist. Smith said he was like "a prisoner in command." Actually, Colonel McCausland of the 36th Virginia led the 22nd during much of this period. Private William C. Reynolds wrote on October 24 that nine companies of the 22nd Virginia were then on Cotton Hill "under the command of McCausland."

Those nine companies of the 22nd Virginia were sent as an advance guard on Cotton Hill to Huddleston's farm, situated alongside the Fayette, Giles, and Kanawha Turnpike. These men were led by Major Smith, who detested the position, which was in a deep hollow surrounded on all sides by commanding hills and within earshot of the Yankee camps at Kanawha Falls and Gauley Bridge. McCausland liked the position and even desired to get closer to the enemy. Smith felt that his men should camp on top of Cotton Hill. His fears were justified on October 31 when a group of Federals opened a heavy musketry fire upon them from the hill in front of the house. Caught off guard, the soldiers of the 22nd fired wildly in all directions until Major Smith managed to get out of the house and moved to a stone wall located at the rear of the structure. With bullets flying all about him he attempted to rally the men as he stood on the wall and yelled for the men to follow him and charge the enemy. A minie ball hit at his feet and threw dirt in the face of "young Reynolds." After a brief hesitation the men obeyed Smith and the Yankees fled, leaving behind one of their men dead and another mortally wounded. The 22nd Virginia reported one of their men shot in the foot and another scratched on the hand. McCausland arrived at Huddleston's later that day and moved the men to the top of Cotton Hill.

Floyd's artillery on Cotton Hill began to shell the Federal camps at

Gauley Bridge and "Gauley Mount" on November 1, catching Rosecrans and Cox slightly unaware. But the shells did little damage and for almost a week the Federal command tolerated the nuisance, often returning the fire from their own guns on the hills above Gauley Bridge and "Gauley Mount." Rosecrans eventually tired of the ineffectual artillery duel and on November 10 set in motion a three-pronged movement intended to drive Floyd off Cotton Hill and cut his avenue of retreat. The 22nd Virginia was camped at this time at the foot of Cotton Hill, where, Major Smith complained, the entire regiment was "unfit for duty for want of clothes." Smith asserted that only 202 officers and men were available for duty. He attributed the small number to Floyd's wrongful abuse of the regiment, continually assigning it all of the difficult and demeaning tasks, resulting in much sickness and desertion. Smith himself, fed up with Floyd's accusations, had resigned. On that same day William A. Jackson, an unpopular man who had been a coal operator from Charleston, was made lieutenant colonel of the 22nd regiment.

Rosecrans' movements resulted in a series of small skirmishes, the first coming at Blake's farm on Cotton Hill on November 10-11. Floyd was well aware of Rosecrans' intentions and ordered the troops off Cotton Hill. The 22nd Virginia received marching orders at 1 a.m. on November 11 and by 6 a.m. was on the road to Camp Dickinson. When the regiment arrived there, the men camped around the house. Federal shells from across New River began to fall around them, so the men built earthworks on the heights north of camp. The 22nd Virginia remained all night under arms.

The next day the Federal shelling became more intense, some rounds falling no more than 20 steps from the 22nd's tents, and almost injuring some of the regiment's horses. Realizing that a Federal force was about to cut them off south of Fayetteville, the Confederates retreated rapidly, falling back via McCoy's Mill to Piney Creek. The 22nd camped at Piney Creek for three or four days before moving to Beaver Creek and then to Princeton. This retreat was plagued by wintry cold, rain, sleet, and snow, and followed horrendous mountain roads. The men had been ordered by McCausland, whom Major Smith charged was not sober, to burn their tents and discard unnecessary items. Quartermaster Sam Miller refused to obey the order, tossing aside nothing more than tent poles and three or four tents. As a result, the 22nd Virginia was the only regiment to complete the retreat with its equipment fairly intact. Additionally, the 22nd and 36th regiments had been made the rear guard in the retreat and were able to pick up many items discarded or partially destroyed by Floyd's other regiments while in hurried flight. Major Smith mentioned that the 22nd Virginia picked up at least 100 percussion muskets.

The Federals ended their pursuit following a skirmish at McCoy's Mill, having gained firm control of the Kanawha Valley and its wealth of salt, lead, and coal. Floyd was safe but the retreat from Cotton Hill had been physically devastating. Many of his soldiers were only half-clothed and a large number in the 22nd Virginia were without shoes, resulting in "lacerated and bleeding" feet. Quartermaster Miller described the movement by saying: "I doubt any Quartermaster in any service has gone through in a four month period as many retreats as I have . . . and Cotton Hill was the worst." Major Smith was more emphatic: "I believe in the utterly destitute condition of the 22nd Virginia and the retreat from Cotton Hill, I have seen the sufferings of Valley Forge realized & with it have witnessed a comparative cheerfulness & determined fortitude, with a hearty readiness for action, so as no other men under similar circumstances have ever before been surpassed."

Floyd had been told to winter at Piney Creek, or even Meadow Bluff, but due to lack of forage he moved on to Princeton and to Peterstown on November 29, where he also entertained the idea of winter quarters. But he soon fell back to Dublin Depot, arriving there around December 9. Floyd's Brigade, which consisted of the 22nd Virginia under the command of Lieutenant Colonel Jackson, was inspected at Newbern Depot on December 14. The men were listed as in a poor physical state, "having suffered much hardship and exposure in the constant mountain campaigning in western Virginia." Disease, desertions, resignations, and combat casualties had greatly reduced the ranks, and those who had remained were "poorly equipped, half-clothed, undernourished, and unpaid." But their spirits were still high and they had performed against tremendous odds without a murmur. The camp at Newbern seemed to Private Joseph A. Brown to be "humble and rough."

On December 16, Floyd was ordered to take his brigade to Kentucky to join Albert Sidney Johnston, but he was to leave one regiment of his choice at Lewisburg to protect Greenbrier and Monroe counties. The 22nd Virginia was selected for this assignment, primarily because it was not in any condition to make the trip to Kentucky. Floyd's troops left western Virginia shortly after Christmas Day and on December 31 the 22nd Virginia Infantry embarked on the Virginia and Tennessee Railroad at Newbern Depot and rode to Lynchburg via Salem and Liberty (Bedford). The men remained in the rail cars that night as all the town hotels were full.

New Years Day was spent at Lynchburg, with the regiment "quarreling among themselves." Many of the boys scattered to the various residences to party. By 10 a.m. on January 2 they had boarded the Orange and Alexandria Railroad, and they reached Charlottesville by 6 p.m. Near there Orderly Sergeant George P. Hedrick of Company K was caught by telegraph wires and thrown from the top of a box car. His leg and arm were amputated and he died from shock.

Switching to the Virginia and Tennessee railroad the 22nd Virginia headed west and arrived at Jackson's River Depot at 9 a.m. on January 3, going into camp as rain and sleet fell. On January 7 the boys marched west and arrived at White Sulphur Springs the following day. Colonel Alexander Welch Reynolds, commanding at Lewisburg, reported that Lt. Col. Jackson's command had been reduced to about 325 effectives due to sickness and other causes, and that their "weakness invites attack." Private William C. Reynolds, Company H, said that there were then 170 men sick in the hospital at White Sulphur Springs, although at one time earlier there had been 1,600.

Finally, on January 12, the 22nd Virginia Infantry moved west of White Sulphur Springs to winter quarters at Lewisburg. There, according to William H. Bahlmann, they had "altogether . . . a not unpleasant time." The officers boarded in private houses, while many of the soldiers became acquainted with young ladies of the town.

The first year of the war had been one of sheer determination and endurance for the mountaineers of the 22nd Virginia. Not only had they had to contend with incompatible and incompetent generals, but the command structure within their own regiment had been shattered. The regiment was led by at least five different officers: Tompkins, Patton, Smith, Jackson, and McCausland. As the 1st Kanawha they had enlisted more than 1,003 men at the opening of the war; but desertions, transfers, resignations, battle casualties, and discharges had left the regiment with no more than 570 to 650 men, and at least half of that figure was not available for duty.

CHAPTER III

Reorganization To Disorganization

In March of 1862 General Henry Heth commanded all Confederate forces in the New River region, and the 22nd Virginia was made a part of his "Army of New River." With his headquarters at Lewisburg, Heth could muster about 1,400 men and four field pieces to contest any Federal advance or threat against the Virginia and Tennessee Railroad and his supply center at Dublin. He had been ordered to retire if attacked but to put up a formidable resistance if possible. To accomplish his goal he strategically placed 1000 men along a line extending from Packs Ferry in Monroe County to Flat Top Mountain, just north of Princeton. The remaining 400, all from the 22nd Virginia, Heth sent to Brushy Ridge on Muddy Creek Mountain, west of Lewisburg. There, alongside the James River and Kanawha Turnpike, the 22nd Virginia constructed extensive earthworks and fortifications. On March 11 Colonel William W. Finney, commanding the District of Gauley at Lewisburg, informed Heth that he had the entire 22nd regiment at work upon the fortifications, "but the men work reluctantly and so little is done."

Meanwhile, George S. Patton, still an unexchanged prisoner, wrote from Culpepper that he would at present remain with his family. He had been offered a teaching job at V.M.I. if unable to return to service. Patton said that he would learn to use his uninjured arm if necessary to come back to the regiment. He closed by saying that he hated Yanks. By late March Patton's long-delayed exchange came through and he moved to Muddy Creek Mountain to assume his rightful place as colonel of the 22nd Virginia. On April 8 he reported the strength of the regiment there as 424 rank and file present, and about 140 on furlough or sick, making the total strength of the regiment 560 not including those without leave. Patton reported that the 22nd's arms were in excellent order and the men looked clean and neat, although most were dreading any future marching. Reenlistment was going well although the enticement of guerrilla and cavalry companies impeded the effort. More than 25 men had enlisted in the regiment for the first time while the regiment camped at Brushy Ridge. Patton closed by saying that he did not feel Lewisburg was threatened, but if it were he would not have sufficient numbers to meet the enemy. On April 20 he signed a supply request for 375 men.

The official reorganization of the regiment took place on the wet grounds of White Sulphur Springs on May 1. The 22nd, being from within the boundaries of present-day West Virginia, was able to get few recruits, primarily receiving men from the counties of Greenbrier, Alleghany, and Craig. Since all new men enjoyed the privilege of choosing whichever company they preferred, Company K found itself with an abundance of men from Craig. William Bahlmann described these new conscripts as being first-class soldiers "who took war seriously, but the volunteers of 1861 hardly regarded them as equals." Somewhere between 317 and 347 men enlisted in the 22nd Virginia at White Sulphur Springs, although a portion of that figure were reenlisting.

Following the reorganization, in which Patton was confirmed as colonel and Captain Andrew R. Barbee of Company A was made lieutenant colonel, the 22nd left White Sulphur Springs and marched 17 miles east to Callahan's. It then moved on to Buchanan in Botetourt County, where the town ladies placed flowers in the rifle muzzles of Company A. From Buchanan the men marched to Bonsacks, where they arrived hungry and exhausted. At this point Company K was sent to New River Bridge (Radford) to guard the rail crossing over New River. On May 5 the 22nd was at Jackson's River and then moved south to help guard strategic points at Narrows and Rocky Gap.

In early May General Jacob D. Cox made a move to strike the Virginia and Tennessee rail line, advancing southward by way of Flat Top Mountain and Princeton. His Federal force was composed of two brigades under the commands of Eliakim P. Scammon and Augustus Moor. Working in conjunction with Cox was a third brigade under Colonel George Crook, who moved east to Lewisburg, and a fourth brigade led by Colonel Joseph A. J. Lightburn, which was to hold and defend the Kanawha Valley.

Cox's advance, Colonel Rutherford B. Hayes' 23rd Ohio, quickly moved south and took possession of Giles Court House on May 6, encountering little resistance. The bulk of the Federal force with Cox made headquarters at and near Princeton. Aware of this threat, Heth divided his army into two brigades with which to strike Hayes. The First Brigade, led by Colonel Walter H. Jenifer, contained the 45th Virginia Infantry with 650 men; one company of the 8th Virginia Cavalry comprising 80 men; and Otey's Battery of 40 men with one 6-pounder and one 12-pounder. The 2nd Brigade, under charge of Colonel John McCausland, had the 22nd Virginia, one company of the 8th Virginia Cavalry, a reserve of fragments of three companies, Chapman's Battery, Lowry's Battery, and two mountain howitzers.

The 22nd Virginia was at the Narrows May 9. The regiment, along with both brigades, moved to Camp Shannon just north of Cloyd's Mountain, where the soldiers received battle orders in the evening and began the advance to Giles Court House between 10:00 and 10:30 p.m. At about 4:00 a.m. on May 10, Heth's men encountered Federal pickets about two-

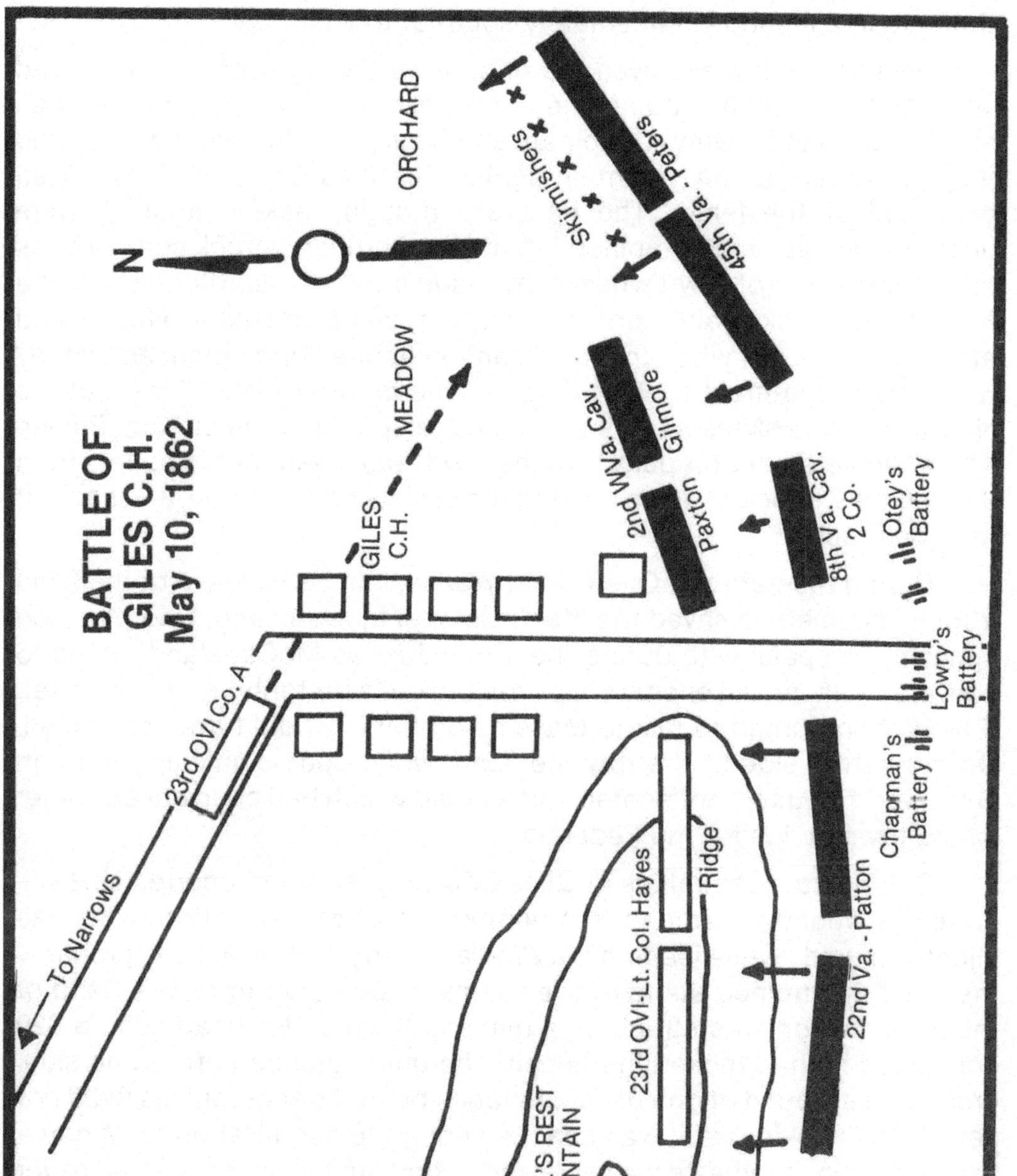
BATTLE OF
GILES C.H.
May 10, 1862
N
ORCHARD
MEADOW
GILES
C.H.
Skirmishers
45th Va. - Peters
2nd W.Va. Cav.
Gilmore
Paxton
8th Va. Cav.
2 Co.
Otey's
Battery
Lowry's
Battery
Chapman's
Battery
22nd Va. - Patton
23rd OVI Co. A
To Narrows
23rd OVI-Lt. Col. Hayes
Ridge
ANGEL'S REST
MOUNTAIN

and-a-half miles south of Giles Court House and drove them in. When only one-half mile south of town, Heth found the 23rd Ohio behind a fence along a ridge running from the main road and up Angels Rest Mountain. Paxton's 2nd (West) Virginia Cavalry was defending the opposite side of the road in a meadow immediately south of town.

Jenifer quickly deployed his brigade to the right of the main road while McCausland moved his men to the left. The battle opened with an artillery barrage by Otey, Chapman, and Lowry. The left wing of the 22nd Virginia moved up the mountain against the 23rd Ohio, which was well protected by the fence. The Federals, though, lacking artillery, were unable to repulse the combined infantry and artillery attack and fell back to the north side of town where they made another dispirited stand. The Federals fell back again, skirmishing as they went, and made a final stand at the Narrows. But when the rebel cannon opened upon them again they were forced to retreat to East River. At the mouth of Wolf Creek, at the Narrows, the Yankees set fire to the bridge spanning the stream. Private Albert Singleton of Company H of the 22nd "ran a gauntlet of bullets from the Yankees, knocked down the burning brands of the bridge, and saved it from destruction."

During the battle at Giles Court House, James H. Mays of the 22nd Virginia claimed he saved the life of Colonel McCausland, whom he had occasion to speak with during the affair. Just as McCausland turned to converse with him an enemy bullet "ripped out a buttonhole" in his jacket. If he had not turned to talk to Mays he probably would have been killed. On the lighter side of the fray the 22nd, while double-quicking through Giles Court House, confiscated some bread which had been baked earlier by the town ladies for the Federals.

Confederate casualties at Giles Court House were reported as 2 killed and 4 wounded. Among this number was Colonel Patton, who was slightly injured "while leading the 22nd and carrying the last and probably the most determined stand of the enemy." According to Noyes Rand of the 22nd, Patton was struck by a minie ball which fortunately hit a $20 gold piece he had tucked in his belt. The bullet glanced off to one side, producing a wound of ghastly appearance but not dangerous as it did not penetrate the skin. Yet it was serious enough to send Patton back home to recover; while there he was shocked to find out his exchange had never been formally completed. In respect for military tradition, as well as to avoid being shot if captured again, Patton refused to violate his parole and remained at home. Eventually he was exchanged for a Colonel Wood, but in the confusion at the time Colonel Jesse S. Norton of the 21st Ohio was relieved of his command on July 4 to prevent his possible execution if caught by rebels.

At Giles Court House the rebels claimed that the enemy loss was 20 killed and 50 wounded. The Southerners soon afterward set in motion a plan to dislodge Cox's main force at Princeton. In charge of this was General Humphrey Marshall, with some 2,195 men plus a battery. Marshall directed Colonel Gabriel C. Wharton and Heth to assist him. Heth's job was to strike the enemy advance at the mouth of East River and then join Marshall at Princeton. On May 16 and 17 Marshall fought and defeated the Federals at Princeton, but Heth never managed to connect with him. The 22nd Virginia was yet at the Narrows on May 15, where the regiment was reported in good condition. A nearby skirmish recently had taken place in which "one of the new Kanawha Riflemen was killed" and four of the enemy were captured and several "crippled." The following day, as the fighting at Princeton raged, the 22nd remained at the Narrows skirmishing with the enemy who remained at the mouth of East River. Heth, who had been ordered to attack those particular Federals was surprised to find the enemy camp vacated when he arrived there on May 17, but he did confiscate some abandoned baggage and transportation. This prompted him to move within four or five miles of Princeton, at which point he received unreliable information that Marshall had been repulsed. Heth never entered the town, and therefore never got into the fight at Princeton. The 22nd Virginia may not even have participated as a whole with him in this movement as the record book of Company H states that they were at the Narrows from May 16 to the 20th.

While Heth had been occupied at Giles Court House and Princeton, Colonel George Crook's brigade of Federals had advanced to Lewisburg, arriving there May 12. Facing only slight opposition, he continued eastward to Jackson's River Depot and the Virginia and Tennessee Railroad. Upon hearing of Heth's success at Giles Court House, Crook feared being cut off from the rear and immediately retraced his steps to Lewisburg, going into camp on the western edge of town. With him were between 1,200 and 1,300 foot soldiers and more than 300 cavalrymen. Included in this figure were his own 36th Ohio Infantry, the 44th Ohio Infantry, and the 2nd (West) Virginia Cavalry.

Gaining knowledge of Crook's position, Heth and his "Army of New River" sensed an excellent opportunity to completely surprise the troops at Lewisburg and then to continue west in an effort to regain the valuable salt works of the Kanawha Valley. Departing the Narrows, he arrived at Union in Monroe County on the afternoon of May 21. His army, composed of many raw and undisciplined soldiers, appeared "much run down." The men quartered in the various public buildings and 30 to 40 of the sick and wounded remained there when Heth resumed his march north on May 22 to the vicinity of Greenbrier Bridge (Caldwell).

In the early morning hours of May 23, Heth's advance captured Crook's pickets at Greenbrier Bridge, just east of Lewisburg. This opened

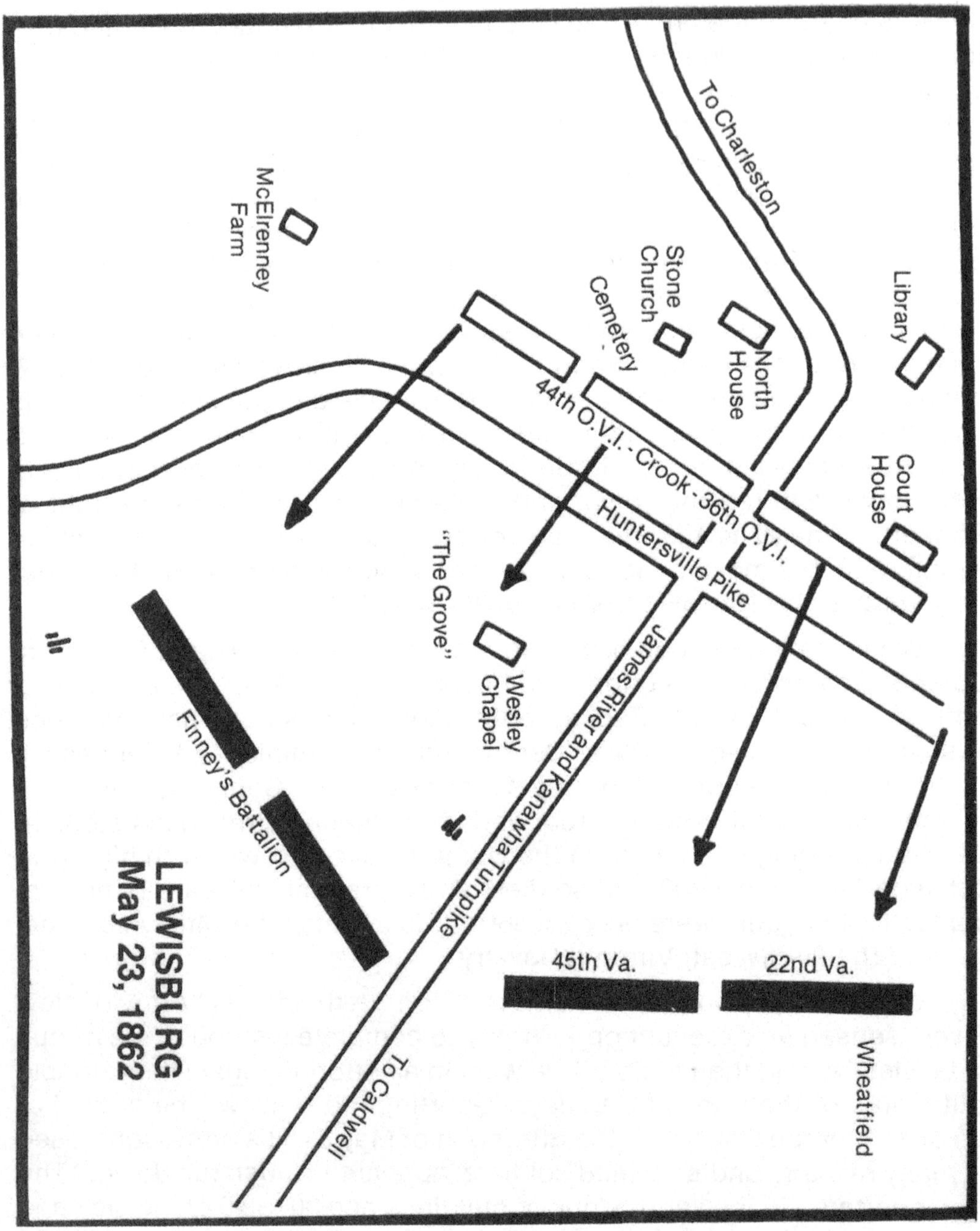
LEWISBURG
May 23, 1862
To Charleston
Library
Court House
North House
Stone Church
Cemetery
McElrenney Farm
44th O.V.I. - Crook - 36th O.V.I.
Huntersville Pike
"The Grove"
Wesley Chapel
James River and Kanawha Turnpike
To Caldwell
Finney's Battalion
45th Va.
22nd Va.
Wheatfield

the way for Heth's surprise attack. Many in the ranks of the 22nd Virginia, which had never actually been defeated in open combat, were overconfident of success. William Bahlmann recalled putting on his old clothes for the battle with the thought that he would soon drive the enemy from Lewisburg and then return to town to have breakfast with the ladies. Private John McMullen of Company H "plucked a sprig of evergreen and pinned it to the lapel of his coat," quoting Scott's poems "Charge, Chester, Charge" and "On, Stanley."

At about 5:00 a.m. the rebels arrived on the western edge of Lewisburg "fatigued, hungry, and somewhat demoralized," probably a description of the conscripts as opposed to the veterans. By this time Heth's force totaled about 2000 foot soldiers, about 100 cavalry, three batteries and two mountain howitzers. Included in this number was Lieutenant Coloinel William W. Finney's Battalion, which was a hodge-podge of men from the 189th (Craig County) Militia and two companies of the 51st Virginia, and would evolve into the 26th (Edgar's) Battalion Virginia Infantry. Heth also had the 45th Virginia, the 22nd Virginia (bolstered in strength by the timely arrival the previous night of one-half of Company K, back from guarding New River Bridge), and Lieutenant Colonel Alphonso F. Cook's Battalion of dismounted men from the 8th Virginia Cavalry. He held the troopers in reserve.

Heth struck in force, his artillery actually sending some shells into the town. One of the misdirected missiles killed Private Charles Chewning of Company H of the 22nd. Finney's Battalion moved through a wheatfield to some oak timber to cover the left. The 45th Virginia was placed in the center, protected by houses, and the 22nd in a wheatfield on the right. Private Joseph A. Brown of Company H said a lieutenant colonel (probably Barbee) yelled, "Come on boys, we've got them by the umbellicus, come on." Brown and Private Henry D. McFarland of the same company pushed forward and took position near the corner of a stable, where a few steps away they saw Henry Radford killed by an enemy bullet.

Crook was caught partially unaware while many of his men were still at breakfast, but they quickly rallied and charged through the town and up the hill at Heth's line. Leading the assault north of the turnpike was Lieutenant Colonel Melvin Clarke with his 36th Ohio; their withering fire stunned the 22nd Virginia. William Bahlmann said it seemed "as if the other side had a granary full of bullets and was throwing them at us with a scoop shovel." Men were "falling like ten pins in a bowling alley."

South of the turnpike Colonel Samuel A. Gilbert and his 44th Ohio advanced up the hill against Finney's Battalion. When the Ohio troops reached the crest of the hill the green conscripts under Finney broke into complete disorganization and ran. This movement left the other rebel units, particularly the 22nd, open to a severe enfilading fire. Companies B and K of the 22nd, on the left flank, got the worst of it. Private George Caldwell of Company C said, "the balls flew like hail . . . you ought to have heard the balls whiz past us." Company K saw a number of its men go down, including Bud Sandige with a bullet in the heart, Miles Johnson struck in the temple, Sergeant Major Claudius Blake and Pat Murray shot in the thigh, Bill Taylor through the shoulder, and William Bahlman, in his old clothes, wounded in the right arm and hand and captured. George McKnight of Company G was shot through both thighs and afterward died of consumption. Captain John K. Thompson, Company A, had his right eye shot out and was captured. Captain Ben Chase of Company B was shot through the heart and fell dead on his face, while Captain James H. Shumate of Company F was wounded so severely in the knee that he had to retire from service.

Finney, under a severe fire, unsuccessfully tried to rally his men as Heth's entire line disintegrated and fled in chaos. Finney was taken prisoner and the jubilant Ohio boys captured the rebel artillery position and seized a cannon reportedly left over from the Revolution which Heth had pressed into service. What had at first appeared to be a rebel victory had quickly deteriorated into a devastating route which Heth called a "most disgraceful retreat." Heth fell back via Greenbrier Bridge, the soldiers tossing aside arms and accoutrements between Lewisburg and that point, and burning the wooden bridge to hinder pursuit. Then they turned southward along Monroe Draft and did not halt until reaching Union, where the 22nd Virginia was in camp on May 24.

Casualties at Lewisburg for the Federal army were 13 killed, 53 wounded, and 7 captured. But they took 4 artillery pieces and approximately 300 stands of rebel arms. Heth suffered much more heavily, losing 38 dead, 66 wounded, and 93 captured. The 22nd Virginia was the hardest hit, going into battle with 395 men and losing 149, or 37%. Included in this number were at least 22 killed, 24 wounded, and 66 captured, including Adjutant Noyes Rand. Company K alone lost 21 out of 36, with 16 of those being either killed or wounded. Almost all of those taken prisoner were sent to Camp Chase and exchanged in August by way of Vicksburg.

Despite such tragic losses, there was one rather light story about the battle told by Private Joseph V. Rollins of Company A. He said that while posted behind a fence waiting for the battle to open a comrade lowered his pants for a bowel movement. While so engaged the enemy broke through the line and the rebels fled. The surprised soldier took off running with his pants yet around his ankles and Rollins paused to look at him and laugh. The laughter caused Rollins' mouth to open and a bullet passed through one jaw and out the other, and he later grew a beard to hide the scar. Had he not opened his mouth to laugh the wound could have been deadly, or at least knocked his teeth out.

The 22nd Virginia was well aware that the disaster at Lewisburg was caused by the panic-stricken conscripts of Finney's (later Edgar's) Battalion. Until the battle at White Sulphur Springs the following year, in which Edgar's boys repulsed numerous Federal assaults, the two units could not march together as they would continually get into fights. Many, including George Caldwell of the 22nd, blamed the debacle on Heth, saying that he was drunk during the battle and used poor judgement.

General Heth was at Union on May 27 and heard rumors of a Federal approach, prompting him to assemble his army in full battle array on the north side of town. It proved to be a false alarm, though, as the success of "Stonewall" Jackson in the Shenandoah Valley had caused Crook to retire his army to Meadow Bluff, fearing that the rebels might try to cut him off from the rear as Heth had attempted.

The 22nd Virginia spent the remainder of May and the first part of June in camp at Salt Sulphur Springs in Monroe County. Before the summer ended the regiment had also camped at Union and Red Sulphur Springs. On June 2 George Caldwell of the 22nd wrote from Salt Sulphur that "Heth is a mean man. He said he would march his men to Craig County and kill the deserters and us, too. If I get a chance I will split the old rascal wide open." But good news came shortly after June 9 when Colonel Patton's confusing exchange was finally completed and he returned to the regiment at Camp McCausland in Monroe county.

Colonel Crook made another attempt to attack Heth, this time north of Union on June 24, but Heth refused to give battle and retired over Peters Mountain. This lack of aggressiveness on Heth's part infuriated the local citizenry, who wanted Heth removed, an attitude apparently also shared by the soldiers. Such changes began to take place and on August 16 General John Echols took charge of the First and Third Brigades at the Narrows. He held a grand reivew on the heights overlooking New River and the men, who by now had come to detest Heth, found a new optimism under the well-liked Echols and felt that they could defeat any force thrown against them. Late in the month General William Wing Loring took command of the Army of Southwestern Virginia.

On August 20 the 22nd Virginia was encamped at Camp Royal Oak at Salt Sulphur, where Loring released a list of deserters from the regiment. James H. Mays recalled two deserters who had been found guilty and were sentenced to be shot. But when the day for their execution arrived they were taken to the middle of camp, their hands tied to a post, their shirts ripped off their backs, and given a severe flogging. Their "screaming was piteous" and undoubtedly unnerved the Virginia officer in charge, but it was a fate noticeably better than death.

In early September Loring realized that conditions had become ideal for possibly retaking the salt works of the Kanawha Valley, or as the boys of the 22nd joked, it was a chance for "the soldiers to salt their corn meal mush." The Federal government had abolished the Mountain Department as early as June 27 and on August 14 General Cox took most of his seasoned Northern veterans out of the Kanawha Valley to the Potomac front. This left behind only seven infantry regiments, one of cavalry, and several "incomplete local organizations" to defend the Kanawha area. These troops were placed under the command of Colonel Joseph A. J. Lightburn, who also kept outposts at Flat Top Mountain, Meadow Bluff, Summersville, and Wyoming Court House. Adding to the probability of success for Loring was a successful two-week raid into western Virginia and Ohio by General Albert G. Jenkins and 600 horsemen, which had begun on August 4 and saw Jenkins pass through sections of the Kanawha Valley virtually unmolested.

Lightburn had officially taken command of the Kanawha District on August 17. Serving under him were the 34th, 37th, 44th, and 47th Ohio infantry regiments; the 4th, 8th, and 9th (West) Virginia infantry, and the 2nd (West) Virginia cavalry. He posted the 34th and 37th Ohio, under Colonel Edward Siber, at Raleigh Court House (Beckley). The 44th and 47th Ohio along with two cavalry companies, all led by Colonel S. A. Gilbert, were stationed at Camp Ewing near present-day Ansted. Major R. S. Curtis took two companies of the 2nd (West) Virginia Cavalry to hold Summersville, while the remainder of the troops were placed at strategic locations between Gauley Bridge and Charleston. Additionally, Lightburn had eight mountain howitzers, three rifled pieces and three smoothbores, all manned by an infantry detail. In reality, this force was more than enough to defend the Kanawha Valley against Loring, but Lightburn's overly cautious attitude would hinder any successful defense.

The 22nd Virginia was in camp at Red Sulphur Springs on September 1 and General Loring left the Narrows for the Kanawha Valley on September 6. He had with him approximately 5000 men, including the 22nd (picked up at Red Sulphur en route the same day), 36th, 45th, 50th, 51st, and 63rd Virginia infantry regiments; the 23rd, 26th, and 30th Virginia infantry battalions, and Otey's, Stamps', and Lowry's batteries. Moving via Princeton and Flat Top, the formidable army trudged north toward Raleigh Court House but found that the Federals of Colonel Siber had learned of his advance and fallen back to the fortifications at Fayetteville, a strong position but accessible to attack from wooded hills on the right flank. Other Federal outposts had been called in closer to Charleston and the quartermaster and commissary stores at Gauley Bridge were ordered shipped to Charleston.

An area secessionist alerted Siber on the night of September 9 that Loring intended to attack him at Fayetteville, prompting him to send for reinforcements and post reconnoitering parties. This proved of no avail when at 5 a.m. on September 10 Loring's army appeared about four and one-half miles south of Fayetteville and deployed for an assault. Patton's 22nd Virginia was placed in Wharton's Brigade, which was ordered to make an effort to go left of the turnpike, take a 7 to 13 mile circuitous march, surround the enemy, and keep them from retreating while the main force attacked in front. While this movement was in progress the battle opened with Loring's men making numerous frontal assaults against the well-fortified Yankees of Siber. The artillery from both sides wrought havoc, causing the rebel infantry to struggle for every inch of ground.

By 2:15 p.m., after having traveled over a rough road and woods, Wharton's Brigade arrived at the designated position, but found it not to be as it had been described. Despite this they attacked the enemy in the rear and the 22nd Virginia was thrown farther to the right on a spur which commanded the position of the Yankee long-range guns. Major Robert A. Bailey took a detachment of the regiment and went deeper to the right, where he got closer to the guns. Late in the evening he made three successive attempts to take the artillery position. Bailey succeeded in the final effort, but could not hold on due to another Federal gun emplacement which commanded the point.

Company B of the 22nd reported being exposed to a severe fire of shot and shell which they were unable to return with muskets. Private James Smith, a conscript in Company G, fired his musket once and "the musketeers were ordered to fall back to a lying down position behind the hill." Private James W. Rust of Company A was the only man in the regiment killed; among the wounded was Captain Richard Q. Laidley of Company H.

Nearly all of Loring's men were engaged in the fight but at nightfall Colonel Siber and his men continued stubbornly to resist the rebels. Ironically, at sunset General John S. Williams of the rebel army suddenly, and for some unknown reason, withdrew his regiments, including the 22nd Virginia, from the Federal right flank. This opened the road for Siber to retreat to Gauley Bridge. At 1 or 2 o'clock in the morning of September 11, after the arrival of reinforcements and destroying the commissary stores, Siber fell back toward Gauley Bridge.

Learning of Siber's retreat on the morning of September 11, Loring sent all men under Williams, Wharton, and General John Echols in hot pursuit. They caught up with Siber on a ridge near the top of Cotton Hill but were immediately repulsed by the Federal artillery. Siber moved on to Gauley Bridge and found that the position had already been evacuated by Lightburn, who was unable to remove most of the commissary stores and put them to the torch, including the ammunition magazine at Zoll's Hollow. Finding Lightburn had also destroyed the crossing at Montgomery's Ferry below Kanawha Falls, Siber was forced to retire toward Charleston on the south side of the Kanawha River, skirmishing the entire way and crossing the Kanawha by boat at Camp Piatt on a rainy September 12.

While briefly in camp at Dickinson's farm, just north of Fayetteville, Colonel John McCausland had been given charge of Echols' Brigade, Echols being sick, and was ordered to pursue the enemy to Charleston. The brigade was composed of the 22nd and 36th Virginia, Otey's and Lowry's batteries, and Major Salyer's cavalry. On September 13, Siber arrived at Charleston and joined with Lightburn. The Federals made preparations to meet Loring, setting up a thin defensive line on the west side of the mouth of Elk River and destroying the bridge spanning that body of water.

By sending a number of good swimmers across the Kanawha at Montgomery's Ferry, including "Cabell and Brown of the 22nd," McCausland was able to retrieve the undamaged ferry boats and move his men across the river. Arriving at a point just east of Charleston, McCausland ran into Federal skirmishers who gradually fell back to the main line at the mouth of the Elk. McCausland sent in the other brigades, as well as his own, and the 22nd moved to hold the left. For them it was a homecoming. In recognition of that fact the Kanawha Riflemen, the Charleston boys, were placed in the advance skirmish line, which was long and extended from the river to the hills. They swept through Charleston, driving the enemy back, but were hampered by their many friends and relatives who ran out to greet them.

Resistance was fierce so McCausland made an effort to get some men across the Elk farther upstream and also posted his artillery on the various hills overlooking Lightburn's position. A battery rattled up Cemetery Hill and Thomas Jeffries, a young Charleston lad, saw John Donaldson and Monroe Quarrier of Company H of the 22nd helping to bring the guns up. They stopped for a few minutes to shake hands and then took the pieces to the top of the hill where the water reservoir was later built. The Federals were unable to withstand the atillery and musket fire of the rebels and retreated in near-complete disorganization by way of the Ripley Road, eventually making their way to safety at Point Pleasant. With the Elk River bridge destroyed, Loring was unable to pursue but took possession of Charleston, holding it as well as the vital salt works of the Kanawha Valley until early October. For many hometown men in the 22nd and 36th Virginia this would be the last time they would see home before the end of the war, provided they survived the conflict.

In the Kanawha Valley campaign of 1862 Loring's Army of the Kanawha, also called the Army of Southwestern Virginia, lost a number of men, most at the Fayetteville action, with 18 killed and 89 wounded. The

22nd's contribution to the total was 1 killed and 9 wounded. For the Federals Lightburn reported 25 killed, 95 wounded, and 190 missing. His superiors considered the entire affair, excepting Siber's stand at Fayetteville, a disgrace.

Two groups of men from the 22nd Virginia missed participating in the campaign. First was Company F, which was placed under the command of Lieutenant Colonel Barbee and detached on September 5, one day prior to Loring's advance. The men of Company F went to Palestine in Greenbrier County for five days and then marched by way of Packs Ferry and Fayetteville to Gauley Bridge, which took another five days. After camping at Gauley Bridge for five days they were ordered to scout Clay and Nicholas counties for six days. Returning by way of Gauley Bridge they captured six of the enemy and fell back to Lewisburg. Then they marched back to Gauley Bridge and camped nine days.

The second group of members who failed to participate in the campaign were those who had been captured at the battle of Lewisburg back in May. In August the formal prisoner of war exchange had taken place and numerous rebel soldiers were sent to Vicksburg for exchange. William Bahlmann of the 22nd was put in charge of 379 Virginia prisoners of war, including 65 from the 22nd and 9 from Company K. The exchange took place at Vicksburg on September 10, the same day as the Fayetteville battle. The contingent with Bahlmann was at Chattanooga, en route home, while the 22nd Virginia was in Charleston.

While at Charleston Company H of the regiment camped where the Baptist Temple was later built. At this camp their meals consisted of a big cup of bean soup, a hunk of bread baked in a Dutch oven, some fried bacon, and plenty of sop, a treat made by stirring flour into the bacon grease. One day William F. Goshorn, a young Charleston youth, sat down in the sop and was run out of camp. Also while in Charleston Captain Laidley and Adjutant Noyes Rand were married, sending their wives to the safety of Lewisburg the following day.

Remaining at Charleston, Loring ignored General Lee's order to lead a foray into the Monongalia Valley and to destroy bridges and tunnels of the Baltimore and Ohio Railroad. He did nothing more than send 1,500 troopers under General Jenkins to inflict the damage. On October 8 Loring began to extricate his troops from Charleston and move east to Lewisburg. This complete disregard for Lee's order ended all possibilities of breaking up the Reorganized Government of Virginia at Wheeling, as well as failing to cause any pertinent damage to the rail system. Confederate authorities were appalled and demanded that Loring return to Richmond and turn over his command to Gen. John Echols on October 16.

Despite Loring's failure the reoccupation of Charleston had accomplished much positive, including use of the salt works, capturing a wealth of Federal supplies, visiting family and friends, and a strong recruitment. The 22nd Virginia alone gained approximately 75 men, although a few deserted during the retreat from Charleston.

At Lewisburg Echols quickly reversed the Army of the Kanawha and began the return trip to Charleston, often in snow and sleet. But by mid-October General Cox's Federal veterans had returned from the Potomac front and Echols had to fall back by way of Cotton Hill and Fayetteville, where the 22nd Virginia was encamped on October 11. Echols then moved on to occupy the old Confederate positions at Princeton and the Narrows. Lieutenant Colonel Barbee and Company F covered the retreating army and rejoined the 22nd at Packs Ferry on November 2. The 22nd had occupied Beckley on October 31 and by early November was camped at Centreville in Monroe County, and could still be found there on the 16th of the month. On December 1 the regiment was reported near Union and on the 31st it was at Camp Sam Miller in Monroe County. This camp was named after the 22nd's quartermaster, who in March of the following year would be elected to serve in the Confederate Congress. The regiment was yet at Camp Sam Miller on January 10, 1863. On January 1 a complaint was filed to General John S. Williams at Salt Sulphur Springs by Oliver Bierne. He protested that the 22nd Virginia had been destroying wood on his property. Williams advised him to point out to the boys the best possible location for chopping wood.

The regiment finally went into winter quarters at Lewisburg. There, William Bahlmann reported, they "had nothing to complain of." An outermost picket post one-and-a-half miles outside of town was kept up all winter.

Some 316 men had been lost from the ranks during 1862, including 23 killed in action, 34 wounded, 61 captured in battle and 43 others captured, 95 desertions, 18 dead from disease, 28 discharged for medical or other reasons, and 9 transferred or resigned. Based on William Bahlmann's casualty figures for Lewisburg the numbers here would be much higher. Despite these losses many of the wounded quickly returned to the ranks and by the end of the year many of the prisoners of war lost at Lewisburg and elsewhere were back with the regiment. Some 640 men had enlisted in the regiment during the year, but the above loss figures quickly reduced that number.

CHAPTER IV

Glorious Victory — Terrible Defeat

The winter of 1862-63 proved uneventful for the 22nd Virginia. In February William Bahlmann commanded the outer picket post at Lewisburg with 16 men and reported that they ate cold bread and beef but lacked coffee. He also said that the temperature once dipped to 15 degrees below zero during the month. During the winter Albert Singleton of Company K got into a gambling scrap with a local farmer who hit him in the face with an iron, leaving a bad scar on his nose.

Coming out of winter quarters in April the 22nd Virginia found itself detached to support General William E. "Grumble" Jones in a raid on northern (West) Virginia in conjunction with Brigadier General John D. Imboden. Actually, Imboden had conceived the plan but its execution was entrusted to Jones. The raid was intended to destroy the Baltimore and Ohio Railroad between Oakland and Grafton, to defeat numerous scattered detachments of Federals, to enlist manpower for the Confederacy, and to threaten the Unionist government of Virginia based at Wheeling. William Bahlmann simplified the effort by saying it was an attempt to destroy the bridges on the Baltimore and Ohio Railroad.

In order to accomplish this daring feat Jones had 2,100 men at his disposal while Imboden had 3,365 in his command. Imboden's force consisted of the 22nd, 25th, 31st, and 62nd Virginia infantry regiments, the 18th and 19th regiments and 37th battalion of Virginia cavalry, and McClanahan's Battery of six pieces. Approximately 700 of the cavalry were mounted while the dismounted remainder were to acquire horses by capture or purchase along the march. This long line of infantry and dismounted cavalry would hinder rapid movement.

Leaving Lewisburg on April 13, the 22nd Virginia marched for three days, arriving at Greenbrier Bridge (Marlinton). After making camp there for four days the regiment continued on to Hightown, arriving there on April 20 and joining Imboden at that point. Imboden had left his camp at Shenandoah Mountain and moved to Hightown via the Staunton and Parkersburg Turnpike. The first four days of the march were plagued by a steady downpour that turned the roads into a sea of mud, causing difficult crossing of rain — swollen streams. James H. Mays of Company F recalled wading across Greenbrier River in Pocahontas County when it was full of mush ice from the spring thaw. He said the soldiers' feet, "which were but poorly clad at best, were almost frozen" before they got across. But by the evening of April 23 the raiders had managed to reach Tygarts Valley near Huttonsville, having covered some 70 miles.

A clash with Federals was nearly inevitable. It came on April 24 when Imboden moved against some 900 Yankees at Beverly under the command of Colonel George R. Latham of the 2nd West Virginia Infantry. Latham had with him a portion of the 2nd and 8th (West) Virginia Infantry, two cavalry companies, and one section of Ewing's Battery armed with two cannon. In the clash Company E of the 22nd Virginia went out as skirmishers while Company K was placed next to the flag. Caught in a ravine between the two opposing batteries, the 22nd was constantly exposed to artillery fire, and there was fear that they would be struck by one of their own shells. In the ravine William Bahlmann ordered his men to fix bayonets. Some took pieces of paper and wrapped them around the muzzles of their guns to make the bayonets fit more securely. A local farmer got caught between the batteries attempting to drive his cattle out of danger.

Latham's vastly outnumbered men put up valiant resistance until late in the evening and were finally forced to fall back to Philippi, but not before setting fire to the government stores. The 22nd reported only one man, Private John Smith of Company A, killed in the battle. The victorious Confederates moved into Beverly, where Colonel Patton found an order from General Robert Schenk for the Federals to defend the Baltimore and Ohio Railroad. While at Beverly, William Bahlman reported that the regimental provision wagons were 20 miles to the rear. The men had no bread or meat to consume so some of them smoked cigars and ate candy in order to stifle their appetites. This somewhat contradicted the account by James H. Mays, who said that at Beverly the regiment "captured as much government provisions and cattle" as it could eat.

When Imboden learned that Philippi was held by a brigade under Colonel James A. Mulligan, he abandoned plans for an attack and moved on to Buckhannon. The Confederates evidently arrived there on the morning of the 29th, although some records of the 22nd Virginia indicate that they may have reached Buckhannon as early as April 26.

While Imobden had been moving toward Buckhannon, General Jones had implemented his part of the raid, departing Lacey Springs in Rockingham County and passing through Oakland, Rowlesburg, Kingwood, Morgantown, Fairmont, Phillippi, and on to Buckhannon. He had a number of encounters with Yankee forces along the route but still managed to cause severe destruction of government property, particularly the railroad, and then joined with Imboden at Buckhannon. On the morning of May 3 the combined forces moved on Weston but found no opposition, the Union defenders having once again evacuated the area.

During the brief occupation of Weston foraging parties went out and various cavalry units scoured the countryside. These detachments stumbled into some small clashes with Federals at numerous points, including a skirmish at the mouth of Simpson Creek with the 3rd West Virginia Cavalry on April 30, and with a force at Jane Lew on May 5. But Imboden's worst problems came from within his own ranks as some 200 men, mostly from Dunn's 37th Battalion Virginia Cavalry, deserted between Beverly and Weston in retaliation against an order prohibiting the dismounted cavalry from seizing horses from citizens for private use. Further reductions had resulted from the large number of sick and exhausted men left at various points along the route of the raid, as well as many men who had been sent to drive cattle herds to the east. By the time Imboden got to Weston his command numbered only 2,300 men.

Realizing that Clarksburg was being strongly reinforced by Yankees, the two generals abandoned plans to capture the town and agreed to separate again and engage in a hit-and-run campaign. Jones left Weston on May 6 with the intent of hitting Parkersburg; finding it was too heavily defended, he struck at Cairo and the Burning Springs oil district. Imboden, again slowed to a near halt by inclement weather, also left Weston on May 6 but only covered 14 miles in three days in an effort to reach Sutton. His advance cavalry reached the outskirts of Summersville on the evening of May 12 and engaged in a brief skirmish with the retreating Federal wagon train, capturing 23 men, 28 supply wagons, and 168 mules with harness. This proved an opportune catch, as Imboden's men had been allowed only a half-pound of meal per day since leaving Beverly.

Imboden occupied Summersville on the morning of the 13th and was joined there that evening by General Jones with part of his command. A planned foray into the Kanawha Valley by both was abandoned due to the exhausted condition of the soldiers. After resting a few days in the friendly confines of Summersville the two began their return to the Shenandoah base camps. Imboden moved by crossing the Gauley River, then the Cherry River, and entered Greenbrier County by way of Cold Knob Mountain, thence on to Warm Springs. Jones returned by crossing the Gauley River at Hughes Ferry and entered Lewisburg via the Wilderness Road and the James River and Kanawha Turnpike. By their own request the 22nd Virginia and 37th Battalion Cavalry came back with the contingent under Jones and remained at Lewisburg. The raid officially ended May 22.

The Jones - Imboden raid was considered highly successful by its participants. Jones covered 700 miles; killed 25 to 30 of the enemy and possibly wounded three times that amount; took nearly 700 prisoners with small arms, one cannon, and two trains; burned 16 railroad bridges and wrecked one railroad tunnel; destroyed 150,000 barrels of oil; and captured about 1,000 head of cattle and 1,200 horses. His losses were minimal, with 10 killed, 42 wounded, and not more than 15 missing.

General Imboden claimed that his troops covered 400 miles; captured more than $100,000 worth of horses, mules, wagons, and arms; put to the torch some six to eight railroad bridges west of Fairmont; forced the enemy to burn three turnpike bridges; and made it feasible for government agents to purchase and move to safety more than 3,100 cattle. Imboden's losses were less than Jones, with only 2 killed and 14 captured.

While the raid certainly did severe damage to the Federals, its negative aspect was overlooked. It was carried out during the spring thaw while many of the roads were still packed with ice and snow, or were deep quagmires of mud. The soldiers were gone 35 to 37 days and marched 24 of them. More than 40 men of the 22nd Virginia marched barefooted. Food was often scarce and it was said that the trail fo the 22nd "could be traced by the birch bark pulled off the trees" for the men to eat. The regiment began the raid with only 7 days rations, and later got 2 or 3 days rations from another brigade. William Bahlman said that at one time they marched 2 days on 1½ crackers, crossed a mountain 22 miles from foot to foot, and slept in the rain at night. At least 67 members of the 22nd regiment had been left sick at various points along the raid, as well as in its aftermath, although most all would return in time for the ensuing campaigns. Once during the raid Company A found three kegs of whiskey and took it to Coloniel Patton, who decided to jigger it out to the companies alphabetically. This infuriated Company K, which would be last in line, and they refused to accept it. Bahlmann summed up the raid saying that it failed because "the Confederate artillery could not hit the supports of the bridges."

During the summer of 1863 the 22nd Virginia remained in the vicinity of Greenbrier County, mostly at Lewisburg, and for a brief time at Meadow Bluff, where they built a small camp town out of chestnut timber and lined it with streets on nearly level ground. The marshy country caused much fever and ague within the ranks. Company K was sent to guard the Narrows of New River and had a rather pleasant time residing in log cabins and exchanging their bacon with farmers for string beans and roasted ears. They also built fish traps in Wolf Creek, although they did have to serve guard duty every other day. In June the 22nd Virginia was encamped at Camp Jackson.

On July 6 Federal Gen. E. P. Scammon, from his post at Charleston, reported that the 22nd Virginia was then on the north side of New River near Lewisburg and consisted of 400 men. On the first of the month Colonel John McCausland of the 36th Virginia had been ordered to scout Fayetteville for the feasibility of an attack, but once there he found the Federal fortifications had been reinforced. McCausland requested that the 22nd Virginia come to his assistance at Piney but they failed to arrive before the Federals began to move against him. He fell back to Flat Top Mountain on the night of July 14 and continued on to Princeton, where the 36th regiment dug earthworks which proved useless as the Federals flanked the position and moved toward Wytheville. McCausland's cavalry gave pursuit but General Scammon's Federals escaped to the west due to what McCausland said was incompetence on the part of one of General John S. Williams' staff. The 22nd Virginia remained at Camp Feamster near Lewisburg.

The Jones - Imboden raid earlier in the year had had a severe impact on Federal authorities and brought about the succession of General William W. Averell to command of Federal troops in western Virginia. Averell, an aggressive soldier and proponent of mounted infantry, strongly felt that the key to success in western Virginia was possession of the Allegheny Mountain passes. To do this he would have to gain control of the Greenbrier Valley, the last major Confederate stronghold in western Virginia, and the location of Lewisburg, the "war home" of the 22nd Virginia.

Utilizing Beverly as a base camp Averell assembled a force of 1,300 men and advanced from Winchester on August 5 into the northwestern Virginia counties of Hardy, Pendleton, Highland, Bath, Pocahontas, and then on to Greenbrier in August. By August 22 he managed to drive Colonel William L. Jackson's rebels out of Pocahontas County to Warm Springs, burn Camp Northwest near Huntersville, completely destroy the salt works near Franklin and later at Covington, and compel the quarterly court at Monterey to abscond.

Confederate authorities were highly aroused by Averell's actions. Anticipating that he planned to strike Confederate positions near Staunton, or the Virginia and Tennessee Railroad, they attempted to intercept him. Major General Samuel Jones, commanding the Department of Western Virginia and East Tennessee, tried to cut Averell off by sending an infantry regiment, believed to have been the 22nd Virginia, to Little Levels in Pocahontas County, along with four cavalry companies sent to Marlin's Bottom (Marlinton). Employing the Army of Southwest Virginia Jones ordered Colonel George S. Patton, commanding the First Brigade in the absence of the often-ill General John Echols, to move on the Anthony's Creek Road to Warm Springs. Jones also made an effort to get the brigades of General Albert G. Jenkins and Colonel Gabriel C. Wharton to come to Warm Springs.

By August 25 Averell had managed to elude all pursuers and clear all potential threats to his rear. He moved toward Lewisburg in an effort to seize the state law library located there, which Federal authorities believed belonged to the new state of West Virginia. His column, four miles in length, passed Callahan's, about 13 miles east of White Sulphur Springs, and continued to move west on the James River and Kanawha Turnpike. Leading the column were two companies each from the 2nd and 11th West Virginia Mounted Infantry, serving as an advance guard under Captain Paul Von Koenig. Next in line came the 2nd West Virginia Mounted Infantry; the 8th West Virginia Mounted Infantry; Major Thomas Gibson's Independent Cavalry Battalion; Captain Chatham T. Ewing's Battery G, 1st West Virginia Light Artillery; the 14th Pennsylvania Cavalry; and the 3rd West Virginia Mounted Infantry.

To counter Averell's boys, Colonel Patton had at his disposal some 1,900 men, including his own 22nd Virginia and the 45th Virginia Infantry; the 26th Battalion Virginia Infantry; the 8th Virginia Cavalry; and Captain George B. Chapman's Battery of four guns. At nearby Greenbrier Bridge (Caldwell) he had the 23rd Battalion Virginia Infantry and the 37th Battalion Virginia Cavalry, which would play a crucial role in the impending battle.

On August 25 General Jones met up with Colonel Patton between Huntersville and Warm Springs on the Anthony's Creek Road. The next day between 9 and 9:30 a.m., after a march of nearly 24 consecutive hours, Patton's force unexpectedly met Avereil's about one-and-a-half miles east of White Sulphur Springs at the intersection of the Anthony's Creek Road with the James River and Kanawha Turnpike. Edgar's Battalion (26th Virginia), made up of some 182 men from the vicinity, made the initial contact with Averell. These boys, many of whom had caused the tragedy at Lewisburg the previous year, held Averell at bay until the remainder of the army could get to the field. Colonel Patton quickly ordered Edgar's men to form a line, facing southeasterly, across the junction of the two roads, and to deploy a company of skirmishers to both left and right. From there they were to advance to the Henry Miller house located between the battle lines. The skirmishers, under Captain Edmund S. Read, opened a severe fire upon Averell's advance units.

Chapman's guns were brought up and placed on a knoll to the left of the Anthony's Creek Road and Patton ordered up the 22nd Virginia, entering the battle 500 strong (Company K was not present, remaining on guard at the Narrows), and placed the men to the left and in front of Chapman. He told them to hold the center of the line at all costs. Companies A, B, E, G and H were detached under Lieutenant Colonel Andrew R. Barbee and deployed as skirmishers on a ridge about 400 yards in advance of and stretching some distance to the left of the four other companies of the regiment, which were under the command of Major Robert A. Bailey.

Edgar took notice of the Yankee horsemen gathering for an assault and had his men tear down a fence and use it as a barricade. His battalion continued to hold back a number of Federal thrusts on his line. Patton, standing on a small hill to the right of the turnpike, observed Edgar's valiant defense and was heard to remark, "Glorious! Hurrah for the gallant Edgar!" Quickly he directed a volunteer from his staff, acting assistant adjutant general Noyes Rand, to relay his congratulations to

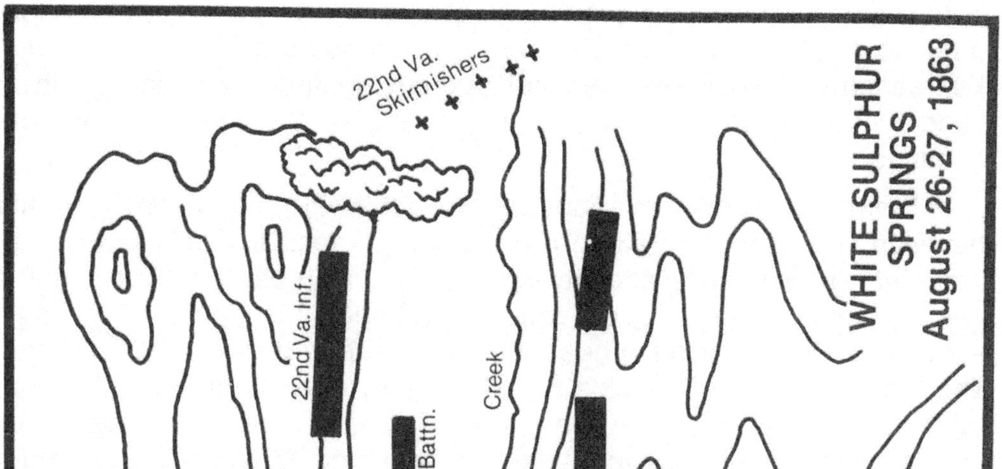
WHITE SULPHUR
SPRINGS
August 26-27, 1863
22nd Va.
Skirmishers
22nd Va. Inf.
23rd Va. Battn.
Creek
Miller
House
26th Va. Battn.
H.Q.
To White Sulphur
45th Va.
Inf.
37th Bttn.
Va. Cav.
Oley

Edgar. But Major D. S. Hounshell insisted that he be allowed to give the message, to which Rand consented.

Chapman's guns fired rapidly at the enemy while the 22nd Virginia remained in line of battle to the left of the artillery, the 45th Virginia to the right, and Edgar holding the advance position. In retaliation to Chapman the Federal guns of Ewing were placed in the first available position alongside the James River and Kanawha Turnpike and blazed away. The 2nd and 8th West Virginia Mounted Infantry dismounted and ran to the hillsides to the left and right of Ewing, where they could give him good support. At about this time one of Ewing's guns burst and the remaining five were moved to within 500 yards of the rebel lines. Captain Koenig advanced the infantry and was killed in one of the numerous Federal assaults.

From this point until about 2:30 p.m. the battle remained a contest between sharpshooters along the opposing lines. Barbee's skirmishers had met hot resistance and were gradually driven back to the main line. Averell sent in fresh troops and strengthened his position, posing a serious threat to Patton, whose men were running low on ammunition. The timely arrival of Derrick's Battalion (23rd Virginia) and the 37th Battalion troopers from nearby Greenbrier Bridge, as well as the nearly exhausted ammunition of Averell and the failure of General E. P. Scammon to reach the Federal forces via the Kanawha Valley, turned the odds in Patton's favor.

Companies H, E, G, B, and part of A, of the 22nd took position on the left of the line near Derrick's Battalion, while the remainder of Company A fell back to the companies under Major Bailey. During this movement Lieutenant Colonel Barbee received a serious wound of the arm and was beaten in the hip with a rifle barrel. The wound proved so severe it would take him out of active combat for the remainder of the war, although he would stay with the regiment until the battle at New Market the following year.

Averell sent another cavalry charge against Major Bailey, then Edgar, and finally Patton's right, but all efforts failed and he lost heavily. By sunset Averell had attacked all points in Patton's line, and yet made one more attempt before dark, but the Confederates would not yield an inch. Following nine straight hours of heavy fighting the opponents ceased fire and Patton reported, "Sentinels were posted in front of the lines, and the two forces lay down to rest less than 300 yards apart."

At daybreak on August 27 the battle resumed, but both armies had nearly used up their ammunition. Averell mounted one more attack on the Confederates, but according to Patton it was "with so little spirit it was evident that the enemy had lost confidence." In fact, Averell had already come to the realization that General Scammon was not going to reinforce him and had earlier prepared his men to retire from the field. They did so in an orderly fashion at about 10:30 a.m. and gradually fell back to Callahan's, blockading the road to hinder pursuit by elements of the 8th Virginia Cavalry. Efforts by "Mudwall" Jackson to cut him off at Covington also failed and Averell made it safely back to Beverly on August 31. The battle, known by a number of names but primarily as White Sulphur Springs or Dry Creek, was over and Patton could claim his most outstanding victory of the war, proof he could command an army in the field.

The two-day battle had proved costly. Averell lost 26 men killed, 125 wounded, and 67 captured, for a total loss of 218. Patton's army counted 20 dead, 129 wounded, and 18 missing, for a total of 167. More than one-half of the Confederate casualties were sustained by the 22nd Virginia, which lost at least 12 killed, 62 wounded, and 9 captured. Henry D. McFarland of Company H was moved to write his mother and complain "it seems as if we always have to bear the brunt of battle in every fight (in which) we are engaged." One of the most tragic losses was 2nd Lieutenant John Gay Carr of Company H who was killed on the field, shot through the brain. The 22nd held a special service for him at Lewisburg soon afterward.

Despite the heavy losses Patton's star rose. Following the victory he moved his men to Lewisburg where he wrote his report of the battle on August 31. He stated that the men were under fire for 12 hours for the first day and 5 the following, and claimed that they captured 20 prisoners and one field officer. He praised Lieutenant Edward T. Jackson of Company E, acting adjutant, for prompt execution of orders; Sergeant Major Monroe Quarrier for keeping the men at their posts; and Captain John K. Thompson of Company A, acting field officer, for assisting Major Bailey by his cool and conspicuous gallantry. Thompson was also wounded in the battle, being shot in the face. Others contributed as well, including V.M.I. Cadet Levi Welch, an original member of the regiment who had resigned in 1861 due to ill health and to attend school. He was home on leave and visited with the regiment during the battle, where he served as an aide-de-camp to Patton. Most distinguished of the men of the 22nd Virginia in the battle was Albert Singleton of Company H. During the affair Albert and "Huffman" were on skirmish line behind trees, and Albert would step out from behind the trees and draw the Yankee fire. At the same time, Huffman, who was a dead shot, would step from behind his tree and picked off the Yankee. When William Bahlmann asked Albert why he took such a risk he replied, "The Yankee's couldn't shoot."

The fall of 1863 was spent by many of the 22nd Virginia working on the farms in the Lewisburg area, cutting corn and other such typical chores. The regiment remained in camp near town, recruiting, drilling, holding courts-martial, and taking time to enjoy the thrill of their victory at White Sulphur Springs. Their enjoyment would be cut short. Averell's failure did not deter either him or General Benjamin F. Kelley from their effort to drive the Confederates out of the Greenbrier Valley region, as well as to continue in their attempt to destroy the Virginia and Tennessee Railroad. Once again utilizing Beverly as a base, Averell planned to take 4000 to 5000 veteran soldiers and attempt to trap the Confederate army at or near Lewisburg. In conjunction with him a 970 to 1,700 man column under General Alfred N. Duffie would move east from Charleston on the James River and Kanawha Turnpike and secure any enemy retreat route from Lewisburg.

Averell had at his command Lieutenant Colonel Alexander Scott's 2nd West Virginia Mounted Infantry; Colonel John H. Oley's 8th West Virginia Mounted Infantry; the 14th Pennsylvania Cavalry of Colonel James Schoonmaker; Gibson's Independent Cavalry Battalion; the 1st West Virginia Light Artillery, Batteries B and G (Keeper and Ewing, respectively); Colonel Augutus Moor's 28th Ohio Volunteer Infantry; and the 10th West Virginia Mounted Infantry of Colonel Thomas M. Harris. To supplement this group Duffie had with him the 2nd West Virginia Cavalry; the 34th Ohio Mounted Infantry; and Seth J. Simmond's Kentucky Battery. Joining Duffie along the route, at Tyree's on Sewell Mountain, would be the 12th Ohio Volunteer Infantry and the 91st Ohio Volunteer Infantry, both commanded by Colonel Carr B. White.

Averell was ordered to move on October 26 and Duffie left Charleston on the morning of November 3. Moving by way of Camp Bartow and Greenbank, Averell was at Huntersville by noon of November 4. At that point he gained knowledge of William P. Thompson's 600-man force of "Mudwall" Jackson's command then at Marlin's Bottom, and learned that Jackson himself was falling back to Mill Point, just north of Hillsboro. Unsuccessful efforts were made to intercept both Confederate columns, but the rebels managed to join forces at Mill Point, where Averell had a brief clash with them on November 5. After a short exchange of artillery and musket fire the vastly outnumbered Confederates fell back south of Hillsboro and took up a strong defensive position on the crest of Droop Mountain.

Hoping to give Duffie enough time to get to Lewisburg, Averell intentionally moved slowly and hesitated striking Jackson on Droop Mountain, settling into headquarters at Hillsboro. But by so stalling he gave Jackson time to be reinforced from Lewisburg by the brigade of General John Echols. Samuel D. Edmonds of the 22nd Virginia recalled departing Lewisburg on November 5, marching north past Frankford, and camping that night on the Greenbrier River, only to be rousted out of sleep at 2 a.m. to go to Droop. Other members of the regiment said that the 22nd Virginia, along with the rest of Echols' Brigade, were awakened on November 6 at 2:30 a.m. and prepared to march from Lewisburg to Jackson's support. It

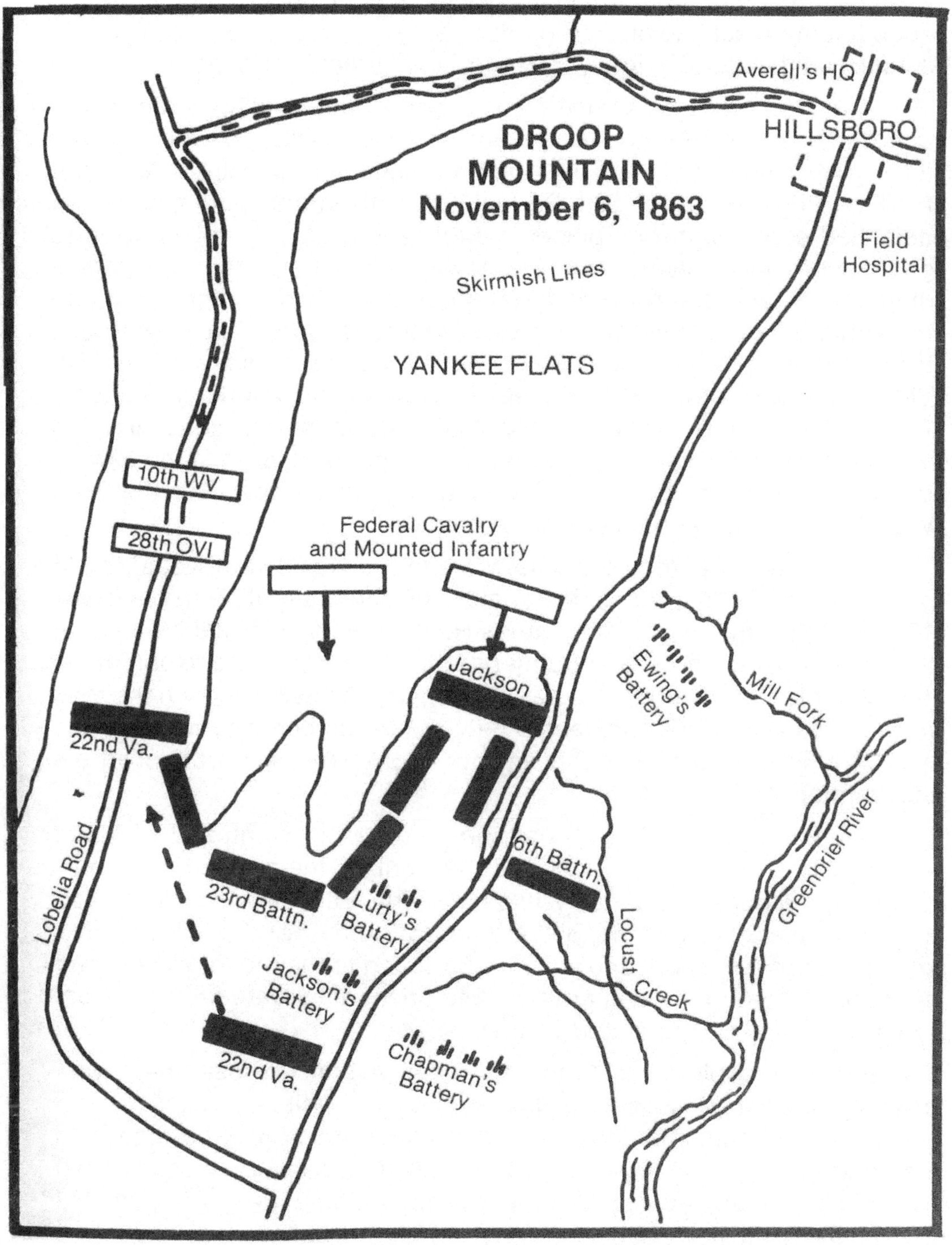
DROOP
MOUNTAIN
November 6, 1863
Averell's HQ
HILLSBORO
Field
Hospital
Skirmish Lines
YANKEE FLATS
10th WV
28th OVI
Federal Cavalry
and Mounted Infantry
Jackson
Ewing's
Battery
Mill Fork
22nd Va.
Lobelia Road
6th Battn.
Greenbrier River
23rd Battn.
Lurty's
Battery
Locust
Creek
Jackson's
Battery
22nd Va.
Chapman's
Battery

was a gloomy day and Colonel Patton sensed a trap. He made it clear that he was against leaving Lewisburg, but being a well disciplined officer he obeyed orders and led his regiment out of town. William Bahlmann had to be called from a court-martial proceeding in town; along the way he acquired some biscuits with red preserves. He shared them with Captain John Donaldson of Company H, whom he met along the road.

Arriving at Jackson's position on Droop Mountain at approximately 9 a.m. Echols placed Patton in command of the First Brigade, which included his own 22nd Virginia and the 23rd Battalion Virginia Infantry. Also with the First Brigade was the 26th Battalion Virginia Infantry, but it had been detached earlier with one piece of artillery and sent 12 miles away to guard another approach, thereby removing the battalion from the impending action. The entire Confederate army which would fight at Droop Mountain barely amounted to 1,700. It included in addition to Patton's Brigade the 14th Virginia Cavalry led by Colonel Cochran; the 19th Virginia Cavalry of Colonel William P. Thompson; Colonel William W. Arnett's 20th Virginia Cavalry; Chapman's Battery of four guns; and two guns with Jackson's cavalry. Colonel Jackson took charge of the cavalry brigade while Major Robert A. Bailey assumed command of the 22nd Virginia in lieu of Patton leading the brigade.

Three other Confederate units were in the vicinity but would fail to participate, including the 26th Battalion on the alternate approach; the 16th Virginia Cavalry of Major James H. Nounan, an original member of the 22nd Virginia, who was at Bunger's Mill to block Duffie; and part of the 14th Virginia Cavalry under Captain Strain, serving as pickets on the road to Nicholas County. Echols feared he was outnumbered by a two-to-one margin, and possibly as much as four-to-one, and was correct in his assumption.

Chapman and Jackson's batteries, under the command of Major William McLaughlin, were placed to the front on the crest of the mountain, near the guns of Captain William S. Lurty. Echols then ordered Patton to put the 22nd Virginia, 550 strong, to the rear of the artillery on a knob to act as a support, and the 23rd Battalion to the rear of his line and the turnpike road. The rebel artillery then opened a severe fire on the advancing Federal troops at and near Hillsboro.

Echols redeployed his line to meet Averell, whose men were assaulting Droop Mountain via Hillsboro in a semi-circle. Echols managed to withstand the initial attacks, particularly the artillery barrage from Ewing and Keeper which was directed at his right flank. He also repulsed numerous dismounted cavalry probes along his line. But by midday the Confederate left flank began to falter. William Bahlmann said that the 22nd was standing deployed in the woods when Adjutant Noyes Rand came by and took two volunteers from each company to cover another road. Bahlmann later heard that they barely got under fire.

The critical point of the battle came at about 2:30 p.m. when the 28th Ohio, the 10th West Virginia Mounted Infantry, and a part of the 14th Pennsylvania Cavalry completed a long, circuitous march and hit the Confederate left flank along Lobelia Road, ironically the only point at which Echols had failed to post any men. This inexcusable mistake would cost him the battle.

In order to meet this new threat General Echols ordered Patton to take command on the left flank of the 19th Virginia Cavalry. Companies A, E, and I of the 22nd, under Captain John K. Thompson, went with Patton at a rush and formed under a galling fire. Two of the companies and one-half of the other went to the right of the 19th Cavalry in order to sustain the 23rd Battalion, while the remaining half-company went to the left of the 19th. In this movement Captain Thompson and his three companies managed to stall the Federal attack and actually drove the enemy back a small distance. He had deployed his line parallel and about 20 yards from his opponent in this effort. But it cost him dearly as the three companies went into this action 125 strong and lost 9 killed, 30 wounded, and 12 missing. Among the injured was Captain Thompson who received his fourth wound of the war. being struck in the neck. Patton remained attempting to rally and encourage the men and was "as usual, conspicous on the field in the thickest of the fight."

Patton came to Colonel Thompson of the 19th Cavalry and had not been with him more than three or four minutes when word came from Echols to fall back. Despite the valiant efforts of Captain Thompson and Patton, the Federal breakthrough on the rebel left proved to be too much and the entire rebel line began to break into complete disorganization, with panic-stricken soldiers fleeing in all directions. Two Federal guns also also moved up the road on the right and opened on the remainder of the 22nd Virginia, posted to the left of McLaughlin's position. William Bahlmann recalled the scene of confusion: "We were standing in the woods and some men of another brigade broke right through our company, starting to the rear. My men didn't move at all. They, like me, had been sure that those men would not stand. I must do them the justice to say that their officers quickly rallied them. One of them was shot, probably by his own men. I saw him turn pale and just then I caught something. An old-fashioned ounce minie ball struck me in the left leg, and I fell to my knees." Gus Tyree of Company C and Erastus Starkes and Napoleon Baker, both of Company K, attempted to carry Bahlmann to the rear but were unable to do so and placed him in a small depression in the ground, where he remained until captured.

Private Samuel D. Edmonds of Company B stated that his company and Company H had been placed in front of the 24-pounder. In a short time heavy fighting erupted across the hollow to his left. As the cannon was turned to fire in that direction Company B, equipped with their Austrian rifles, and Company H with their Enfields, were ordered to meet the Yankee thrust just as "Mudwall" Jackson's soldiers fell back in a disorderly retreat. A stand was made on the middle bench of the mountain on the right flank as Edmonds went to the top of the mountain to seek help. There he requested assistance from Echols, who reportedly told him, "Yes, boys, we better go and do all we can, but we are about whipped." Edmonds returned to the middle bench to give aid but was quickly driven back to the crest, where he found Echols gone and three officers rallying the troops. In an attempt to save John Bennett of Company B, who had been shot in the hip, the 22nd's colorbearer passed the flag to Major Bailey. Then the troops fell back to the turnpike and stopped briefly until the Yankees came bearing down on them again.

Major Bailey, waving the colors in a futile attempt to rally the men, fell mortally wounded. James H. Mays of Company F had just run past Bailey and was ordered by him to stop and help, but sensing that Bailey was doomed, continued his flight. Lieutenant William S. McClanahan of Company A fell dead from an enemy bullet and Captains Richard Q. Laidley and George S. Chilton received serious wounds while "gallantly leading and encouraging their men." John H. McDevitte of Company G managed to save a flag and was later promoted for his brave action. Others in the 22nd regiment went down as well, such as Pete Stribling of Company K who was shot in the chest, the ball entering his left side and exiting to the right of his spine, contributing to his death months later. Mason Helms of the same company went into the battle sick and was shot through the head and body and had a finger shot off, dying of the wounds the following day. Arch Young had a bullet graze his head, literally creating a wound from front to rear of his scalp.

The completely routed Confederates ran through the woods in all directions and not until Patton reached Frankford, just north of Lewisburg, was he able to collect a nucleus of the command. Averell was confident Duffie would cut Echols off at Lewisburg and did not press pursuit. But at Frankford the rebels learned of Duffie's approach and hurried on to and through Lewisburg at about 3 or 4 a.m. on November 7, their rear guard leaving town just as Duffie arrived. In the confusion, all the knapsacks of the 22nd Virginia were left at Lewisburg and captured by Duffie. William Ludwig of the 34th Ohio said "the rebels ran leaving knapsacks, ammunition, and burning some of their artillery carriages. We also captured the colors of the 22nd Virginia — a brag regiment."

Echols hurried on to Union and Duffie's cavalrymen moved southward but gave up pursuit near Second Creek due to obstructions placed in the road. James H. Mays said the 22nd retreated through Greenbrier and Monroe counties to Peters Mountain, where after a brief stay they moved to Anthony's Creek and thence to Dolan's Tavern, where they camped a few days. He said that the regiment then moved down Anthony's Creek to Dry Creek and then to the summit of Kate's Mountain and camped. Mays also said that during this disorganized period many of the regiment dodged into local houses for a few days until the unit finally made camp at Salt Sulphur and reorganized. He said "the boys kept coming in from Droop for more than a month." His description of the post-Droop movements is a bit confusing, and possibly not entirely correct as his memory may have failed him. Samuel D. Edmonds said that he did run from Droop and that Jim Starcher was with him. Jim had no shoes and only socks on his feet. Edmonds recalled that Starcher's feet bled on the frozen ground. They finally found the 22nd at Stoney Creek near Salt Pond Mountain, and from there they returned to Lewisburg.

Averell had avenged his loss at White Sulphur Springs against many of the same men who had earlier defeated him. In his own ranks he counted 30 killed, 98 wounded, and one missing. Echols, on the other hand, was devastated. His total loss was 275 killed, wounded, and missing. Once again the 22nd Virginia suffered the worst, with 113 killed, wounded, and missing. Due to the disorganization following this defeat the Compiled Service Records on Droop are very confusing. They show that the 22nd lost 16 killed, 22 to 35 wounded, and 52 captured, but actual figures were probably much higher. In addition to Laidley and Chilton, the wounded included Captain John P. Donaldson of Company H and Captain Martin L. Smiley of Company K. Captain James McNeil of Company D and Captain John Toney of Company I were captured, as was the wounded Captain Thompson of Company A. William Bahlmann said that the 22nd lost 15 officers out of 24, or 61%, including two captured and 13 wounded, two mortally. Companies A and K had no officers left at all. James H. Mays wrote that the 22nd only numbered 300 after the fight. The wounded were taken to the Josiah Beard house at 8 p.m. after the battle, were treated for a few weeks, and then were released after the Federals departed, Duffie and Averell reaching Beverly on November 12 and 17, respectively.

One group of men from the 22nd Virginia who did not participate in the battle included Lieutenant Robert S. Campbell and Lieutenant George Cofer, who took 35 men of their own Company D and went on a scout of Nicholas County starting on November 4. Two days later, on the same day as the Droop Mountain fight, they attacked a small party of Federals at Summersville under the command of Capt. Riley Ramsey, capturing four men, four horses, several guns, and blankets and overcoats. Upon gaining knowledge of the disaster at Droop Mountain the same day, they decided to parole Ramsey's men and remained in Webster County until Averell left Greenbrier County. Then they rejoined the regiment at Camp Beauregard on November 17.

From Lewisburg on December 7 Lieutenant Coloniel Barbee, having returned from a trip East to recover from his White Sulphur Springs wound, wrote that his injury had just about healed but his hip still caused him pain. "I can not just yet bear the pressure of my belt and sword or pistol," he wrote. "My right arm is some stiff about the elbow joint." He also mentioned that while in the east he had been with Lee in his recent fights and had been within 15 miles of Alexandria. On the 10th Barbee said that the wounded from White Sulphur Springs and Droop Mountain were doing extremely well due primarily to his own technique of stopping long continual application of water to wounds.

As much physical damage as General Averell had inflicted at Droop Mountain, has victory proved rather hollow as within weeks he was gone from Greenbrier County and the Confederates had returned to all of their old positions. Yet he was not ready to stop, and in December began a raid on Salem. Numerous Confederate units unsuccessfully attempted to intercept him at various points. Lieutenant Coloniel Barbee said that on December 12 the 22nd Virginia was issued three day's rations and began the march to catch Averell. Echols, McCausland, and Jones were at Pickaway Plains, just north of Union, on December 14 where Echols, whose command included the 22nd Virginia, was ordered to block the passages over Sweet Springs Mountain. Fighting winter cold and inclement weather, Echols accomplished his task at Sweet Springs and his works at the mountain top were considered formidable. But on December 16 he reported his men "poorly clad and often without shoes, a condition which seriously tested the morale of the mountaineers." Barbee, who claimed he was in command of the 22nd regiment on this campaign, said it was very cold, windy, and sleeting, and that they camped three days on top of Sweet Springs Mountain, the "coldest place in Virginia."

On the 18th Jones joined Echols on the mountain but Averell had managed to elude all pursuers and by December 22 Echols was back at White Sulphur Springs. The only recorded loss of the 22nd in this affair was Private Saunders Gray of Company K who was wounded at either Sweet Springs or Potts Mountain on December 20. Barbee said it was the most severe campaign the 22nd had known.

Once again Lieutenants Campbell and Cofer of Company D missed this campaign. On December 8 they took a captain and 22 men and went on a scout of Big Sewell Mountain, where on December 12 they skirmished with enemy pickets, probably advance elements of the 91st Ohio Volunteer Infantry, and captured six, wounding two of them, as well as taking two horses, several guns and blankets. Private Martin V. B. Powell of Company D was killed in the fight. They remained in the enemy rear attacking pickets and tearing down bridges until the Federals retreated, then rejoined the regiment at Sweet Springs.

December proved to be a bad month for desertion from the regiment. Some sort of skirmishing took place in the vicinity of Meadow Bluff and Lewisburg on December 12 and possibly also on December 23. Quite a few men "deserted in the face of the enemy" on these days. The 23rd Ohio reported the "rebels scattered and the Federals destroyed their Lewisburg barracks." At Charleston Andrew Stairwalt of the 23rd Ohio reported that 15 members of the 22nd Virginia gave themselves up on December 19 and the following day 8 more came in.

Echols' Brigade, consisting of the 22nd Virginia and the 23rd and 26th Battalions, went into winter quarters about four miles east of Lewisburg near present Caldwell. On December 23, according to Barbee, they began constructing quarters but they had nothing to eat but bread and beef. Even so, Barbee stipulated that he was in good cheer and would rather have nothing but bread, or even starve, "than yield one pot or kettle to the enemy." "This war may last always," he wrote, "but we will never be subjugated." On Christmas Eve Barbee noted that the officers were in Lewisburg drinking egg nog with the ladies.

The year 1863 had been a difficult one for the 22nd Virginia. Although the regiment had enlisted approximately 121 men during the year, it had lost about 335 men, including at least 30 killed in action, 106 wounded, 57 captured in battle and 28 others captured, 47 desertions, 32 discharges, 17 who had transferred or resigned, and 19 who died of disease or other causes.

CHAPTER V

1864 — The Year Of Human Suffering

Winter quarters near Caldwell were located at the head of Monroe Draft, a long hollow coming into the Greenbrier River from Monroe County. The soldiers often called the area Camp Gauley or Camp Bunches Hill. Many of the men were very religious, such as Colonel Patton, who never missed a camp service. These pious men built a church in camp and spent several days fasting — not too difficult a task at that stage of the war — and praying for the future success of the army. Breastworks and fortifications were built in the breaks of the river and artillery pieces were posted in emplacements on the top of Bunches Hill, strategically arranged to command all approaches. Eventually the brigade built a town theater and, according to Samuel D. Edmonds, lived in relative quiet that winter, remaining there until May when the weather cleared.

In January six men of the 22nd obtained furloughs and attempted to return to the Kanawha Valley to visit relatives and friends. This party consisted of Sergeant Henry W. Rand, Privates Andrew Matt Donnally, Andrew Van Donnally, Sam Young, John Duling, all of Company H, and Private James Van Bibber of Company D. Near the mouth of Big Sycamore Creek in Clay County they were captured by Federal guerrillas and taken to a home near the mouth of Bell Creek, just north of Gauley Bridge. There they managed to overpower their captors, wounding one of the Federals, and returned to Camp Gauley with their former guards as prisoners.

During the same month Major General Sam Jones, commanding the Department of Western Virginia, with headquarters at Dublin, reported that Echols' Brigade was then composed of the 22nd Virginia, the 23rd and 26th Virginia battalions, two companies of partisan rangers, and Chapman's artillery, making a total of 68 officers and 1,193 men, with an aggregate present of 1,416 and an aggregate present and absent of 2,293. Lieutenant Colonel Barbee said at that time that his commission paid him only $170 and that of his brother-in-law, Captain John K. Thompson was $130, sums which he said could hardly "feed, clothe, and support" them. As a result he sought additional work on area farms to supplement his income.

On February 15 the Confederate Congress approved a joint resolution of thanks to the men of the 22nd Virginia for their service. But many had already become weary of the war. From Charleston on February 20 Andrew Stairwalt of the 23rd Ohio reported that 50 members of the 22nd regiment had surrendered that day and were anxious to take the oath of allegiance. They told him the army was suffering from a severe lack of provisions and clothing and that they had "grown tired of the Confederacy."

General Jones, in his report on the organization of troops in the Department of Western Virginia on the 29th of the month, stated that Lieutenant Colonel Barbee was then in command of the 22nd Virginia and Patton was in charge of Echols' Brigade. On April 9 some 900 members of the brigade were reportedly seen at Union and on the 26th General Alfred N. Duffie captured some members of the 22nd regiment "on the road to Lewisburg." Two days later Echols made reference to three desertions from the regiment. The deteriorating situation began to improve on April 30 when General John C. Breckinridge replaced Jones as commander of the Department of Western Virginia. The following day he said that Echols had 107 officers and 1,339 men, with that same amount present and an aggregate present and absent of 2,182.

Echols' Brigade was inspected at Camp Gauley on May 6. The men were found to be well-equipped and armed, and all had been supplied with a full complement of clothing. The brigade was in a "thorough state of discipline and efficiency, and their military bearing and appearance was soldierly and imposing." Apparently the setbacks of the previous year were fading from memory. The brigade numbered 2,150 men, with a full complement of officers.

On the very same day that it was inspected, the brigade began a march toward Staunton in the Shenandoah Valley to help defend against the invasion of 6,500 Federal troops under Major General Franz Sigel. The first day's march took them 17 miles to "Hughes' on the turnpike" and the next day Echols, again ill with heart problems, led them another 20 miles, passing Covington and going into camp on the "Karnes plantation." On May 8 they reached Jackson's River Depot where they met up with Breckinridge. He had the brigade form in a square and informed them that he had been sent from Tennessee to take charge and he understood the mountaineers were "good soldiers who would not hide behind a hill, stream, or rock in a battle." Then they proceeded another 10 miles to the Covington-Lexington Pike and went into camp at Cow Pasture Bridge. The next day they came within two miles of Goshen and camped near a small stream, then moved past Goshen on May 10 and camped near Crawford Springs near Little Calf Pasture River. The roads on this entire march were still muddy from the spring thaw and Echols' men and horses were nearly exhausted by the rapid pace. Although in desperate need of grain, and unable to locate forage, they pressed on over the mountain roads and came within sight of Staunton on May 11, then remained in camp on the 12th. At Staunton the boys of the 22nd attended services at the Presbyterian Church where Reverend Wilson, father of future U.S. President Woodrow Wilson spoke.

Along with the arrival of Wharton's Brigade, Breckinridge had accomplished the nearly impossible task of gathering together an army of about 4,500 to block Sigel. On May 13 Breckinridge moved north and the 22nd camped at Mt. Crawford on the banks of Cook Creek. Private John A. Maston of Company B recalled that "William Ankrum and George Gillispie deserted last night." For some reason, perhaps the fear of impending battle, a large number of the regiment, particularly those from Craig County, deserted on the 13th.

By May 14 the brigade had passed through Harrisonburg and Breckinridge reached Lacey Springs, 10 miles from New Market. There he met with General John D. Imboden, who was then in charge of the rebel force defending against Sigel at New Market. After discussing strategy Imboden returned to New Market. There he found that Colonel George H. Smith, Patton's cousin and the leader of the 62nd Virginia, had deployed his regiment to contain Sigel, and by nightfall had done so.

Breckinridge arrived at New Market on the morning of May 15 and at about daylight his men came up "weary, wet, and muddy, and halted for a rest." With him were the brigades of Echols and General Gabriel C. Wharton, all "veteran troops, and equal to any in the Confederate army." After studying the terrain Breckinridge took command of the field and said, "We can attack and whip them here, and I will do it." First he attempted to bait Sigel into attacking his line, which was protected by a rail barricade. Sigel refused the lure and fell back north of New Market to Bushong's Hill. Rain poured steadily, turning much of the battlefield into a sloppy mudhole. On the wooded hill Sigel awaited Breckinridge's attack.

Sigel was not to be disappointed. Breckinridge threw in most of his troops, eventually even sending in the V.M.I. Cadets to help some of the veteran regiments which had become stalled by heavy Union fire. The 22nd Virginia, going into the battle 580 strong, and the 23rd Battalion, were sent to straddle the Valley Pike, the bulk of the regiment on the eastern side holding the right flank, along with Chapman's guns and the Berkeley Cadets. Sigel defended well against Breckinridge's left and center for some time but eventually gave way. On the left flank his cavalry, under General Julius Stahel, charged Echols' Brigade. Echols, dressed in civilian clothing throughout the day, exercised so little command that many of the soldiers actually felt he was not on the field and Patton was in charge. Patton did indeed move his 22nd Virginia out against the cavalry charge and acted "with distinguished gallantry."

The Federal horsemen attempted to strike the center of the 23rd Battalion of Lieutenant Colonel Clarence Derrick. Quickly Patton and Derrick improvised and had the "men on Derrick's left and center jam shoulder to shoulder into small squares, so they would be able to fire outward in all directions." Derrick's right would then wheel inward in order to develop

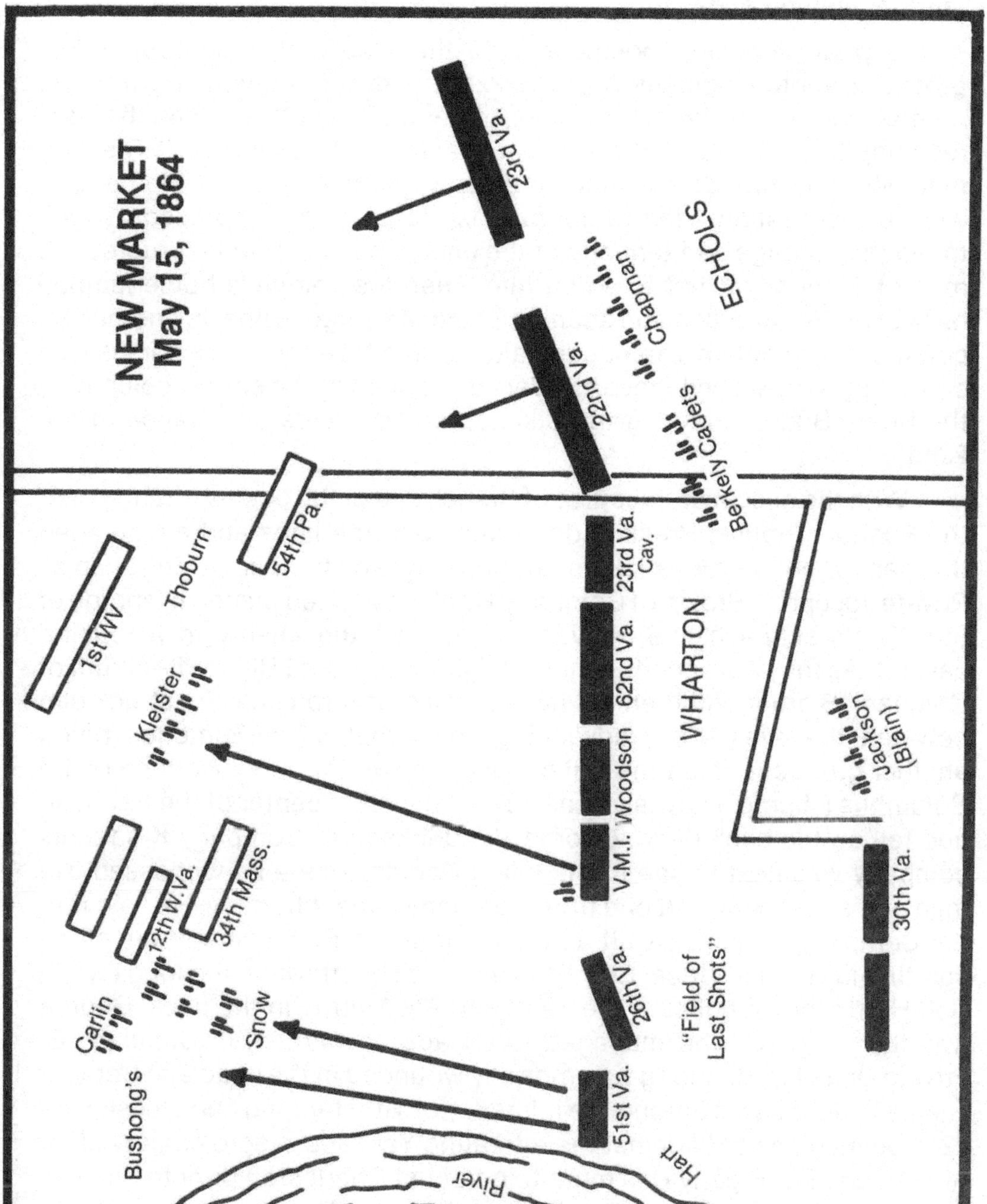
NEW MARKET
May 15, 1864
23rd Va.
22nd Va.
Chapman
ECHOLS
Berkely Cadets
54th Pa.
23rd Va. Cav.
1st WV
Thoburn
62nd Va.
WHARTON
Kleister
Woodson
Jackson (Blain)
V.M.I.
34th Mass.
12th W.Va.
30th Va.
26th Va.
Carlin
Snow
"Field of Last Shots"
Bushong's
51st Va.
Hart
River

flank fire against the horsemen. The 22nd Virginia "wheeled about in the opposite direction," causing the Federal troopers to ride into "a deadly corridor with no end."

Captain John K. Thompson said that due to the contour of the ground it fell to Company A of the 22nd to stop the enemy cavalry. He ordered his men to aim at the horses instead of the riders, which they did, resulting in a barricade of dead and "panic stricken animals." The only man who got through the tangled mess was the colonel of the cavalry, who unsuccessfully tried to get his men to press on. Thompson so admired the courage and tenacity of the enemy colonel that he ordered the men of Company A not to shoot him. Then the colonel's horse jumped back over the barricade and scampered back to Sigel's line. In another account of this incident Thompson said that the fallen mass of horses was caused by a rebel shell that exploded at the head of the enemy column on the Stone Bridge crossing the pike, about 200 yards in advance of the 22nd.

With the successful repulse of Stahel, and of the enemy infantry near the Bushong house, Breckinridge rode along the lines and encouraged the men to "put on your stickers (bayonets) and go for them." Men such as Private Joseph A. Brown of Company H, who had been placed in charge of the 22nd's skirmish line, moved out against the enemy in a rousing assault. As they advanced Samuel D. Edmonds heard Richard Ankrum of Company B grunt and then get up, a small musketball having struck him between the eyes. After advancing somewhat more Edmonds heard another grunt and this time Ankrum went down for good. He then heard Columbus ("Lump") Lewis yell as he was hit in the center of the forehead and fell on his back dead. George Hendrickson of Company K said his company was next in line to the V.M.I. Cadets, where he witnessed five men of his company and one from Company E meet tragedy. He said that the Company E man was hit and fell across his feet, some of his blood spurting in George's face. Billy Franklin was shot through the lungs while Bob Harris took a bullet in the heart and Pat Murray in the neck. George was the only one who managed to escape unscathed. In Company B, Private Daniel H. Bushong was mortally wounded in the head and Captain James C. Johns was among the injured. Notwithstanding such losses the 22nd pushed onward against the retreating Yankees. Negro cook William Armistead of Company H, a mulatto barber at Charleston prior to the war, got so caught up in the excitement that he picked up an abandoned rifle and joined in the fight.

Captain Thompson said that during the route of the enemy a shell exploded in the color guard of the Cadets and they immediately reformed as if on dress parade. The men of the 22nd, though under a heavy fire, paused in their advance momentarily and gave the Cadets a "glorious cheer, which was taken up by the entire line of battle." Adjutant Noyes Rand said that after repulsing the cavalry charge the 22nd Virginia "crossed the Pike in a long diagonal line, and when the pursuit ended near Rude's Hill, we were away to the right toward Smith's Creek."

Sigel managed to escape with the loss of 831 killed, wounded, and missing, while Breckinridge achieved a glorious victory at a cost of 577 killed, wounded, and missing. The 22nd Virginia paid with 6 killed and, according to John Maston, as many as 35 wounded. Two original members of the 22nd regiment, Oliver Perry Evans and Levi Welch, performed well at the battle as members of the V.M.I. Cadets. Evans, nicknamed "Big," was 2nd Sergeant of Company B of the Cadets and served as colorbearer because the regular flag man was on furlough. He was noted for his bravery in rallying the corps. Welch, who had originally been in Company H of the 22nd, was a 2nd Lieutenant in Company B of the Cadets and was detached to command one of the 3-inch rifles of the cadet artillery.

Samuel D. Edmonds recalled that after the smoke had cleared he saw a man from Company H standing over a fallen comrade from the same company. When Edmonds inquired as to what was wrong the man standing said the dead man was his brother and that he had removed much of his own new clothing to put on the corpse. As the saddened man stood shivering in the rain Edmonds gave him his blanket for warmth.

On May 16 the 22nd Virginia camped on Rude's Hill, in the rain, and helped to bury the dead. Lieutenant Colonel Barbee, unable to continue in the field due to his wounds of the previous year, left the regiment to serve in the Confederate Medical Department and was replaced by Major John Clairborne McDonald. At 5 a.m. on May 17 the regiment marched 25 miles in the rain and camped at Big Spring in Rockingham County. En route they passed "The Meadows" where Colonel Patton's wife and children were staying. The next day they came within two miles of Staunton and camped on a muddy wooded hillside where they were visited by a young Woodrow Wilson, then a lad of 7 or 8 years.

In the meantime Breckinridge had been called to the eastern front to bolster Lee's defenses north of Richmond, so on May 19 he moved his army to Staunton and placed the men on the railroad flat cars en route, via Charlottesville, to a point east of Hanover Junction. Companies B, G, and H of the 22nd waited until the following day to board the train. On that same day Breckinridge arrived at Hanover Junction and the 22nd Virginia sat on its arms all night in anticipation of a fight. None came and the next day the regiment camped in an old field at the junction where the water was very bad and the heat seemed unbearable.

On May 22 the regiment moved about one mile from Hanover Junction and camped beside Ewell's Corps. Private John Maston said: "we are drawing one-fourth pound of bacon and one pound of crackers a day rations and are camped near the South Anna River." He believed at that time that the 22nd had also been temporarily attached to Johnson's Brigade. Two days later the boys crossed the South Anna River at Taylorsville and moved toward Hanover Depot, which they arrived near on May 25 and heard cannonading. The next day the regiment listened to a "pretty smart skirmish all along the lines." During the 26th the men received a tobacco ration.

Up again at 8 a.m. on May 27, the regiment moved down the railroad past Taylorsville to the rose-covered grounds of Henry Clay's birthplace at Ashland. Departing there at 4 p.m. the 22nd left the railroad, took the Old Turnpike Road, and camped 8 miles from Richmond in a pine thicket in Henrico County. More marching followed on May 28 as the weary men moved on the road to Atlee's Station, where they were ordered to stand in line in the breastworks at night at Totopotomoy Creek. Remaining in the breastworks the following day, the 22nd was attacked at about 2 p.m. by Federal skirmishers.

While still int he trenches at Gardner's Farm at Totopotomoy Heights on May 30, the fighting became hot for the 22nd Virginia. The Federal artillery opened a furious assault and Private Joseph A. Brown had a shell land at his feet and fail to explode. He kicked it out of the trench with his feet and as it rolled down the embankment it exploded. Private Daniel Harkins of Company H suffered a severe side wound and a number of other men became casualties. The barrage continued on the 31st as bullets "whipped over" the boys. Sergeant Heath J. Miller of Company E was killed and Lieutenant Samuel Counts of Company A was slightly wounded in the head.

Company B of the 22nd went out on picket duty as skirmishers on the morning of June 1 and Breckinridge and General William Mahone drove the enemy back. Sergeant James H. Mays said that he, a lieutenant, and nine others were ordered to hold a position at all costs, but were overrun by superior numbers. They feigned death and let the Yankees pass over them, and then rejoined the regiment. The "shattered" 22nd Virginia got back together and was ordered to Cold Harbor. Mays said that many of the new conscripts in the regiment had not had a single day of drilling; one of them was shot by the enemy and another deserted and fought against the 22nd at Cold Harbor the next day.

During the march to Cold Harbor on June 2 the boys of the 22nd passed through Mechanicsville where the commissary put well-deserved bacon on their bayonets; they would eat that night. As they passed Lee's headquarters they were cheered by the Army of Northern Virginia as the "heroes of New Market" and went into camp a short distance beyond. John Maston said that their poorly built earthworks had been crudely erected by cavalrymen. The 22nd was posted on Lee's right in a marshy area known as Boatswain's Swamp. There the regiment entrenched in earthen breastworks dug by bayonets and bare hands.

Fighting at Cold Harbor became full-scale on June 3 as General Ulysses S. Grant made three successive assaults on the rebel works, resulting in much hand-to-hand combat. The Federals broke over the 26th (Edgar's) Battalion and got in the rear of the 22nd, capturing 20 of Company H, including Captain John P. Donaldson. All of Company B were captured but were soon afterward recaptured. Among the killed were 2nd Lieutenants Alanson Arnold of Company H and Allan C. Rowan of Company C. Captain John K. Thompson was shot, a bullet entering his left lung. He would later claim that he recovered from the wound due to the personal care and attention given to him by Dr. Andrew Russell Barbee, former lieutenant colonel of the 22nd. James Mays said there was but little time for rifle fire as the enemy was too close for there to be time for reloading. As he put it, to do so would be "sheer suicide as any man who reloaded was immediately thrust through with half a dozen enemy bayonets or had his skull crushed by a gun barrel." Mays was also wounded, a minie ball striking his left hand, taking him out of action with the 22nd for the rest of the war. He was sent to Chimborazo Hospital at Richmond. Mays looked at his reflection in a mirror there and found himself so "uncouth and unkempt" that he scarcely knew himself. His hair was uncut and uncombed, his face unshaven, his clothes torn and ragged and generally filthy. Overall, he was somewhat typical of the physical condition of the rebel army involved in trench warfare.

Breckinridge's men were slowly being driven back when Finegan's Florida battalion, situated on a height to the left, observed the Federal breakthrough and came up and drove the enemy back, regaining the position. Colonel Patton was again noted for his gallantry in this action. Finegan's move put an end to the battle. With the 22nd remaining in the trenches in the evening, Joseph Brown went behind the lines with some prisoners he had captured. He ran into William Armistead, who had ready a big kettle of boiled beef and a bag of bread for the 22nd. As a courier could not be found to relay the goods to the men in the trenches, Brown and William's 19 year old son, "Strong," were impressed for the duty. But when they got near the scene of carnage "Strong" panicked, knocked Brown out, and took off running. When Brown later asked William if he had seen "Strong." The Negro cook replied that his son had run by "with his toes barely touching the ground in the high places." During the evening Anderson's Division came up and relieved the 22nd, which fell back toward Gaines' Mill.

The Federals continued to press the rebels on June 4. The 22nd was back in the trenches. Joseph Brown said musketry prevailed the entire day, making "every inch of ground a spot of peril." The flag of the 22nd Virginia was "pierced and torn to pieces by flying bullets, the flagstaff was struck 75 times." Breckinridge was also injured that day as an exploding shell caused him to fall from his horse. But the worst of the fighting at Cold Harbor was over and Lee was able to claim a victory. The fighting at and around Cold Harbor had seriously depleted the ranks of the 22nd Virginia, with 5 killed, 33 wounded, and 32 captured. Fortunately, on June 5 and June 6 the boys got to remain in camp at Gaines' Mill and rest.

Their relaxation was only of brief duration. Federal General David Hunter had been cutting a path of destruction through the Shenandoah Valley, crushing "Grumble" Jones at Piedmont on June 5, burning the Virginia Military Institute, and then threatening General John D. Imboden near Lynchburg. Breckinridge's men were ordered to return to the valley and assist Imboden in protecting the vital rail center of Lynchburg. The 22nd departed Gaines' Mill on the night of June 7 and marched into Richmond where the men boarded the train and took it to the South Anna River, getting off there and marching to Butler's farm. The next day they were back on the train and rode to Blue Ridge Tunnel and camped on the west side near a spring. Here the regiment anticipated cutting off Hunter's advance. Yet on June 8 Hunter had not arrived and the boys went into camp on the western slope of the tunnel.

Waking the following morning the 22nd remained in the breastworks on Blue Ridge about a mile from the Tunnel, in view of Waynesboro, on a ridge to the left of the road. One more day's march took the regiment to Waynesboro and by June 14 were camped 28 miles from Lynchburg. Hunter nad discreetly moved around Rockfish Gap. He thereby avoided the 22nd Virginia, which was compelled to fall back to Lynchburg, arriving there at about 11 p.m. on June 16. Breckinridge, still in bed recovering from his injury, placed General D. H. Hill, in town on furlough, in charge and ordered him to deploy all available troops for the city's defense. The 22nd Virginia was placed on Graveyard Hill in the cemetery. Company B hid behind the brick walls of the cemetery, where they were utilized as skirmishers. Private John Maston spent a most uncomfortable night on top of a grave behind a tombstone. Later that night a portion of General Jubal Early's 2nd Corps arrived from Richmond and "Old Jube" took command. By blowing train whistles and constantly moving troops around, the soldiers and citizens fooled Hunter into believing that they had superior numbers, making him hesitant to launch a general assault.

On June 18 the 22nd Virginia moved forward about a mile and erected breastworks. Breckinridge's Division, composed of Patton's (Echols') Brigade, Colonel Augustus Forsberg's and Colonel Smith's Brigade, was placed on the Confederate right, with General Stephen Dodson Ramsuer to their left and General McCausland on their right. The 22nd was south of Forest Road and northeast of Blackwater Creek. Hunter made various probes of the line, the most serious by his cavalry, which was checked on Forest Road by part of Breckinridge's infantry and

McCausland's cavalry. Lacking success, Hunter concluded that Early's position was too strong and began a retreat back into West Virginia. In the two days of intermittent fighting at Lynchburg the 22nd Virginia lost 2 killed and possibly 5 wounded. The two dead were Privates John A. Lemons of Company F and Allen J. Lypes of Company C, while Sergeant George A. Thorne of Company D was among the wounded.

Early's total losses were slight and on the night of June 18 the remainder of his 2nd Corps arrived, bringing his total strength to about 8,000. Early planned to attack Hunter the following morning but at midnight the Confederates discovered that Hunter was in retreat. Ramseur's Division pursued and caught up with Hunter's rear at Liberty (Bedford), 20 miles from Lynchburg. The skirmish took place just prior to nightfall and resulted in Hunter's rear guard being driven through the town. Early later brought up infantry and resumed pursuit on the morning of the 20th, but Hunter had managed to place his artillery strategically so as to command Buford's Gap and stall Early. Pressing forward on June 21, Early discovered that Hunter had turned off from Salem to Lewisburg, although McCausland's cavalry managed to attack his column at Hanging Rock that day, capturing 10 cannon. Then Hunter got into the mountain ranges and Early deemed further pursuit useless.

Early camped four miles from Botetourt Springs on June 22 and decided to rest his men. They "had marched sixty miles in three day's pursuit" and many had not had anything to eat for two days except for some bacon at Liberty. Early also mentioned that nearly half of his men and officers were barefoot, or nearly so. But the primary reason for his halt, which included the 22nd Virginia, was to let the men recuperate and prepare for a march upon Washington, D.C. When Lee had sent Early to Lynchburg he told him to drive off Hunter and and then move down the Valley, cross the Potomac River near Leesburg or Harpers Ferry, and threaten Washington, D.C. He was also informed that Breckinridge would cooperate with him in the plan. It was a bold idea intended to lure some of the forces of Grant from the Richmond front, thereby taking pressure off Lee's Army of Northern Virginia.

On June 23 Early continued his march north and the 22nd Virginia camped on Mill Creek four miles from Buchanan. The regiment moved 14 miles farther from Buchanan on the Lexington Road the next day, during which the 22nd reported, "losing a great many men by breaking down and desertion." Breckinridge's Division, temporarily under command of General Gabriel C. Wharton, was to march in the rear so as to protect Early's wagon train. Many of Early's veterans nicknamed them "Breckinridge's wagon-guard." As Early's long column passed Lexington on June 25 the men removed hats and reversed arms as they passed the grave of "Stonewall" Jackson, and then went into camp 6 miles outside of town.

By the 26th Early was at Staunton in advance of his troops, who arrived the next day. At that point he said he had 2,000 cavalry and 10,000 foot soldiers, including Breckinridge's Division. In addition, Gordon's Division was assigned to Breckinridge, which meant that Early possibly had as many as 20,000 men. With five days' rations in their wagons and two in their haversacks the Southerners moved north and on June 30 camped near the New Market battlefield. Finally reaching Winchester on July 2, Early received orders from Lee to wait there until all was ready for a safe crossing of the Potomac. General John D. Imboden was sent to destroy a bridge over the South Branch of the Potomac near Martinsburg.

Confederate cavalry skirmished with the enemy at Bunker Hill the next day as Breckinridge approached Martinsburg and encountered some feeble resistance from men under General Franz Sigel, who quickly fell back and crossed the Potomac at Shepherdstown. On July 4 Breckinridge was at Duffield Depot, the 22nd Virginia at Bunker Hill, and Early entered Martinsburg, where his army captured many spoils of war and broke up a Fourth of July celebration. The first division to enter town took advantage of the "spoils" and was celebrating "when the rear division was marched through the place, the men casting envious glances at those of their comrades who had been fortunate enough to be in front that day."

Early crossed the Potomac into Maryland on July 5 and camped near Sharpsburg, remaining there part of the following day, when Ramseur's Division also crossed the river. The Valley Army reached Rohrersville on the 7th and Breckinridge skirmished at Maryland Heights overlooking Harpers Ferry. That same day much-needed shoes arrived for the weary army and the chronically ill General Echols returned to duty. Orders had already been issued for the rebel soldiers to avoid destruction of Northern property. The men found this command difficult to obey, having just marched through the Shenandoah Valley and witnessed the horrible vandalism inflicted by Hunter. For the most part, though, they did as told. Breckinridge moved his Virginians through Fox's Gap at South Mountain on July 8, camping near Middletown, where Federal troopers attacked their rear but were quickly dispersed.

While Early had been progressing northward, General Lew Wallace had been gathering together a hodge-podge of home guards and regular Federal troops to block his passage. He knew that his force was not strong enough to stop Early but he hoped to at least delay him and give Grant time to send some reinforcements to Washington, D.C. As Wallace maneuvered, Early's Southern horsemen pushed the Yankees out of Frederick on July 9 and the fiesty general demanded $200,000 from the city in retribution for Hunter's depredations in the Valley. The citizens stalled paying, hoping that Early would soon meet defeat. Outside of town at Monocacy Junction Early ran into Wallace's rapidly assembled defense.

In the battle, Breckinridge's Division was posted east of Ballenger Creek, to the rear of the B & O railroad. In that location Patton's Brigade held the left, Smith the center, and Wharton the right. The division was primarily held as a reserve and the 22nd Virginia was ordered to guard the train. As a result the 22nd was not engaged in the battle and only one man in the regiment is known to have been wounded in the fight, that being Sergeant James M. Ayers. Wallace put up stubborn resistance to Early but was forced back, having gained valuable time for the protection of Washington, D.C. Early went back and collected the money from the citizens of Frederick and sent his cavalry toward Baltimore to cut telegraph wires and destroy rail lines. This contingent, led by General Bradley T. Johnson, was to continue eastward and attempt reaching the prisoner of war camp at Point Lookout to free the thousands of Confederate prisoners confined there. Early was clearly on the loose. Washington, D.C., seemed to be at his mercy and the citizens were in a state of panic. Grant had come to realize that Early was a crafty opponent and belatedly sent troops to the Federal capitol.

Covered with dust and thirsty from the blazing sun, Early's men trudged into Silver Spring on July 11, where they could see Washington, D.C., and Fort Stephens near Rock Creek. Wharton's Division, including the 22nd, was situated on the left of the road between Monocacy and Washington, D.C., with Patton's Brigade on the right, Forsberg in the center, and Smith on the left. Early made slight probes against the enemy defenses but learned during the night that two Federal divisions were en route to reinforce the city. He realized the position was too strong, and even if he did take it he could not hold it for long. In any case, he had accomplished his main purpose, which was to throw the United States capitol into chaos and draw some troops from Grant, thereby decreasing pressure on Lee. As Joseph A. Brown of the 22nd Virginia put it: "we had no intention of capturing the city due to limited numbers and our sadly deficient equipment."

Not wishing to waste foolishly numerous Southern lives on a futile effort, Early began to withdraw to Virginia on the night of July 12, passing through Rockville and Poolesville, and crossing the Potomac at White's Ford. Although many of the soldiers crossed on ferry boats the men of the 22nd Virginia reportedly took off all of their clothes, held them above their heads, and waded across. Early had also recalled his cavalry and abandoned the proposed assault on Point Lookout. On the 14th and 15th of the month he rested his men at Leesburg.

Another event affecting the 22nd Virginia took place many miles away on July 15 when a train transporting Confederate prisoners to Elmira prison camp in New York crashed at Shohola, Pennsylvania. Numerous rebel prisoners and Northern guards were killed or injured. From the ranks of the 22nd, all captured in the fighting at Cold Harbor, Creed Parks, John M. Patrick, and William A. Jones were killed; Albert R. Singleton had both legs broken and John W. Mathews received a serious cut on the temple.

On July 16 Early moved toward Snicker's Gap in the Blue Ridge Mountains, skirmishing en route with enemy cavalry at Purcellville. In this encounter Wharton's Division held the area south of Purcellville at a road intersection. Patton's Brigade took the center while Smith's defended the left and Forsberg's the right. The enemy did capture part of Early's train with some of the "spoils from Maryland" before being driven off. With the way cleared Early marched on to Snicker's Gap on the 17th, crossed the mountains and the Shenandoah River, and returned to the Valley. Joseph Thoburn's Division of Federals attempted to cross Snicker's Gap on July 18 but was repulsed by Early. Echols drove off Federals who crossed at Parker's Ford, or Cool Spring. In this combined combat at Castleman's Ferry Wharton's Division was in the center, placing Patton on the right, Smith in the center, and Forsberg on the left. To the left of Wharton was Rodes and on Wharton's right was Gordon. Early's forces retired to Middletown and then Strasburg.

Crook and Averell, both old opponents of the 22nd Virginia, attacked Ramsuer's Division at Stephenson Depot, just north of Winchester, on July 20. Ramseur's men struggled valiantly but were overwhelmed and broke and ran. Left behind were some 400 Southern casualties, a loss Early could ill-afford. The enemy lost 214. At the very same time there was a fight at Berryville. Such aggressiveness on the part of the Federals seemed to indicate that they were anxious to get into a serious fray with Early. Taking this into account, Early moved to Fisher's Hill on July 21 and had the men dig rifle pits and earthworks, then remain in line of battle. The Federals arrived, took a concerned look at the imposing enemy position, and went into camp near Kernstown. The rebels at Fisher's Hill, including the 22nd Virginia, received rations of beef and bread, but they found that the beef was uncooked and contaminated with larvae. Despite the horrid prospects of digesting such a meal, the men were so hungry that they went ahead and forced it down. Obviously, the spirit and morale of such men must have still been quite high to have endured so much.

Early was firmly established at Strasburg on July 22, and Patton was listed in command of the First Brigade, Wharton's Division. Early had grown weary of being chased about the valley. His army was quite capable of striking a serious blow and did just that on July 24 when, like a wounded animal, it turned and struck the Federals at Kernstown. General George Crook had been warned off Early's approach but had falsely assumed it was only a small reconnaissance party. He sent a detachment to block them but when they made contact with the rebels they proved to be the entire division of General John B. Gordon. Early used three divisions in the fight, having Gordon demonstrate in the front while Breckinridge's Division struck the flank. Breckinridge put a skirmish line to the front and had the three brigades of his division, Patton's, Smith's, and Forsberg's, take a long, circuitous route around some hills and open on the Union left flank. The bluecoats were completely surprised and retreated in a disorganized mass. Only nightfall kept Early from actually eliminating the entire force. Early pursued the fleeing Yankees to Middletown, where he went into camp. In the clash at Kernstown the 22nd Virginia suffered at least 6 wounded, including 1st Lieutenant Edward T. Jackson of Company E and Privates Henry Wilson and Francis M. Cox, both of Company H.

On the morning of July 25 Early moved through Winchester, the men helping themselves to captured Federal supplies, and then marched on to Bunker Hill where they made camp for the night. The 22nd Virginia apparently did not share in the plundered Union supplies as John Maston wrote in his diary that night that the men were "upset about having no rations." The following day, Maston said, the boys were "much run down" as they marched on to Martinsburg, reaching there on the 27th. On the 28th the regiment was at the Baltimore and Ohio railroad, where they destroyed several miles of track, or as it was called, "Early finishing 'Stonewall' Jackson's contract on B. & O. railroad." The 22nd Virginia camped at Darkesville on July 31.

Early's cavalry had also been quite active, with General John McCausland taking his horsemen into Pennsylvania to Chambersburg, where he demanded payment from the citizens of $100,000 in gold or $500,000 in paper money to help atone for Hunter's acts in the Valley. The citizenry refused, however, and McCausland burned Chambersburg to the ground on the 31st.

After a brief respite to rest his exhausted warriors, Early was on the march again on August 4, passing through Shepherdstown, crossing the Potomac the next day and skirmishing with the enemy near Sharpsburg. He then advanced to Boonsboro and "feigned an invasion of the north," then turned to Williamsport on August 6 where he recrossed the Potomac River back into Virginia. Early's cavalry suffered a serious blow on August 7 when the Federals surprised McCausland and Johnson at Moorefield and defeated them, while Early moved by Darkesville to Bunker Hill.

The most serious threat to Early began on August 9 when General Philip Sheridan took command of the Federal troops in the Valley. Lee and Early refused to take Sheridan seriously, assuming he would follow in the footsteps of his predecessors and chase Early about in a non-aggressive style. Sheridan actually had no such intention. "Little Phil" was extremely aggressive and had the full blessings of Grant. With that combination he proved to be a deadly entity in the Valley.

Smith's Brigade skirmished on the Berryville Road on August 10 and then Early moved east of Winchester where Wharton's Division, comprised of Smith's and Patton's brigades, took up position in the evening. The 22nd Virginia reported being "within 4 miles of Winchester." On this day, Private Henry Fisher of Company I was shot for desertion. In the evening of August 11 Early repulsed some Federals between Millwood and Winchester, having placed Wharton's Division on the left, Patton's Brigade holding the right of the division. On Patton's right was Gordon at the double toll gate. Fighting continued at Cedar Creek on August 12, where Wharton's Division went west of the pike, south of Winchester, Patton in the center, Smith on the right, and Gordon east of the turnpike. Private William Henderson Lucas of Company C of the 22nd Virginia was killed by a bullet in the left thigh. Afterward the regiment moved to within a mile of Strasburg and marched to Fisher's Hill. At that strong location Early hoped to lure Sheridan into combat on August 13. He deployed Gordon's Division west of the pike and Wharton's on the eastern side of the road. Patton's Brigade protected the left of Wharton's Division, alongside the pike, while Smith held the far right. Sheridan refused to take the bait, though, and fell back to Berryville on the 14th without a fight.

Skirmishing continued on the 15th as Early followed Sheridan, and Smith's Brigade dispersed a Federal outpost. The next day the 22nd Virginia was still at Fisher's Hill skirmishing and Wharton's Division moved near Hupp's Hill. Much heavier fighting occurred on August 17 as the Southerners moved on the main road in the direction of Winchester, where Early defeated Sheridan's rear guard. Having passed Kernstown, Wharton's Division was halted and Forsberg's Brigade was sent to the left of the road to support the cavalry. The entire line advanced and steadily drove the enemy to "within a mile of the fortified heights southwest of Winchester." Then Breckinridge, with a white linen duster over his uniform, cooly directed the attack at about sundown, placing Forsberg's Brigade on the left in advance and Smith's and Patton's on the right in echelon. Combined with Gordon's, Rodes', and Ramseurs' divisions, the rebels drove the Federals from their position on Bower's Hill.

Early continued to press the retreating Federal column and marched toward Harpers Ferry, arriving near Winchester on the 18th where "the dead were buried and preparations made to advance." Arriving at Bunker Hill on August 19 the exhausted gray-clad warriors went into camp until the 21st, when they trudged past Smithfield toward Charleston, the cautious Federals remaining at nearby Cameron Depot. The next day the two opponents skirmished at Halltown, and General Echols left for Dublin, having been appointed to command of the District of Southwestern Virginia. This move left Patton exclusively in command of Echols' former brigade, although in essence he had led the brigade almost from its inception.

As part of Wharton's Division the Patton Brigade participated in a battle at Charleston on August 23 along with Gordon, Ramseur, and Rodes. More serious fighting developed near Leetown on August 25, where Forsberg's Brigade was sent to the front and moved down the road near Kearneysville, encountering General Torbert's Federal cavalry. The 51st Virginia went into the fight first, followed by the 45th Virginia and the 30th Battalion Sharpshooters. As the battle became intense Patton's and Smith's brigades advanced on the left, Patton on the right of the position, just south of Shepherdstown. General Terry's Brigade of Gordon's Division came up and gave support, as did Ramseur and Rodes. The Yankees put up a spirited resistance but could not hold and fell back toward Charleston, quickly pursued by Early. General Gordon was wounded in the fighting at Leetown and Private William M. Mason of Company A of the 22nd Virginia was killed.

Later in the evening Wharton's Division caught up with the fleeing Yankees near Shepherdstown, where Patton supported the left of the Division. Then Early's men advanced to Sharpsburg and camped for the night. They moved back through Leetown to Bunker Hill on August 26, where they remained in camp the following day. Soon afterward Early ran into some Federals at Smithfield and marched back to Winchester. On the 29th Wharton's Division skirmished at Payne's Chapel, near Opequon Creek, driving the enemy through Smithfield to Charleston. At this point Early decided to rest his men and went into camp at Bunker Hill, stationing detachments at points between Winchester and Berryville.

In the Bunker Hill camp on August 31 Echols' (Patton's) Brigade of Wharton's Division reported an aggregate of 503, a definite sign of depleting membership yet a small indication that Early's rebels were still numerically capable of undertaking a major engagement. But that possibility soon dimmed when General Lee ordered General Richard H. Anderson's Division to leave the Valley and move to his aid at Petersburg. Lee had falsely assumed that Sheridan would continue to display a lack of aggressiveness, having been in command of the Federal army in the Valley for nearly a month and having failed to offer any substantial resistance. Lee felt that Early would be relatively safe and Anderson could help him to cope with Grant. In reality, Sheridan had every intention of crushing Early, but had been wise to the wily general's cat-and-mouse games. Lee's ill-timed move left Early with fewer than 15,000 effectives to compete against Sheridan's army of nearly 40,000, having recently been reinforced by troops from Grant. Sheridan's golden opportunity was about to arrive.

On September 2 Early moved to Brucetown, six miles from Winchester, and the following day, before leaving the Valley, Anderson was struck at Berryville. Early came to his assistance and Wharton's Division

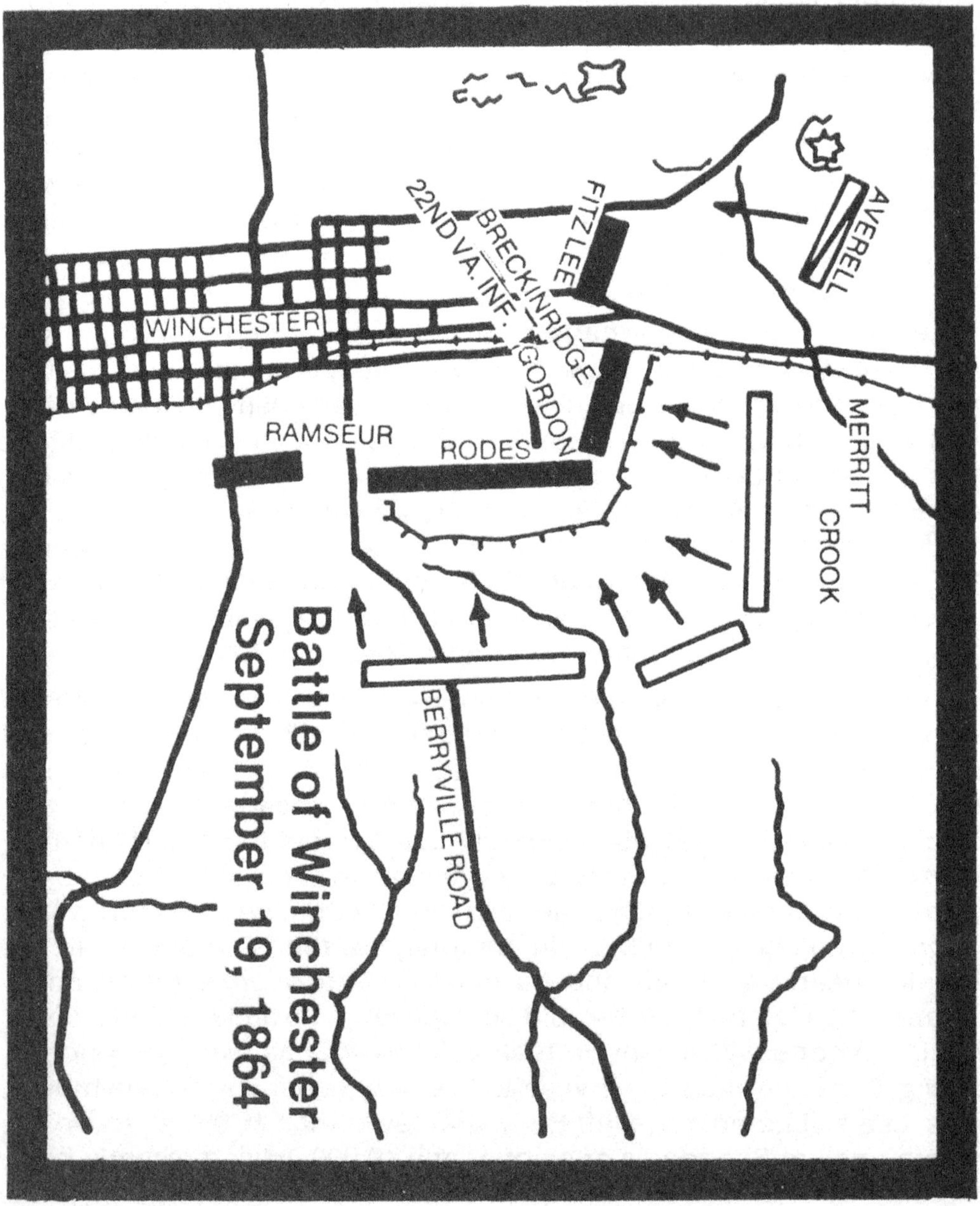

Battle of Winchester
September 19, 1864
WINCHESTER
RAMSEUR
RODES
GORDON
22ND VA. INF.
BRECKINRIDGE
FITZ LEE
AVERELL
MERRITT
CROOK
BERRYVILLE ROAD

went into the action with Patton on the left, Wharton (Forsberg) in the center, and Smith on the right. Rodes' Division held the far left while Ramseur's defended the right.

Between September 5 and 9 Early fell back across Opequon Creek in a move designed to draw out the Federals, but Sheridan, ever wise to such tactics, could not be lured out. On September 15 Wharton's Division, comprising some 1,700 men, was at and near Stephenson's Depot, just north of Winchester, and the 22nd Virginia went on picket one mile beyond Brucetown, returning to camp at night. The following day the regiment remained in camp.

Early remained confident that Sheridan was the picture image of his predecessors and would not attack him. With that dangerous thought in mind on the morning of September 17 he detached the divisions of Rodes and Gordon and sent them toward Martinsburg, critically reducing his numbers at Winchester. Patton's Brigade was sent to picket the road leading to Jordan's White Sulphur Springs, and on the 18th moved one mile back in the direction of Winchester.

Sheridan was quick to seize the moment. At 3 a.m. on September 19 he sent a division of cavalry, backed by 35,000 infantrymen, across Opequon Creek north and east of Winchester, and drove Early's pickets and defensive line back to Ramseur's Division, the only Southern infantry in town. Ramseur fought stubbornly and held the enemy in check, giving Early time to recall Rodes and Gordon. By noon the two divisions had returned to Winchester to bolster Early's line. Patton's Brigade had been near Stephenson's Depot when the battle opened and supported Fitz Lee's cavalry. His brigade held the extreme left of the infantry line, fighting in retrograde movement, maintaining "fire, oblique left, oblique right, and front, futilely attempting to halt" the mounted enemy brigades.

At 1 p.m. Sheridan moved against Early's entire line with his cavalry south of town, the Sixth Corps in the middle, and the Nineteenth Corps to the north. The attack briefly succeeded until Gordon came up and forced some of the Federals to withdraw. The Federal line was further broken up as Rodes' Division advanced into a gap between the Sixth and Nineteenth Corps. With support, Rodes forced back General James Rickett's Division on the Sixth Corps' right and General Cuvier Grover's Division on the Nineteenth Corps' left. This movement caused the entire Federal line to balk, although Rodes was killed in the action.

Sheridan quickly regained the advantage by bringing up a reserve, General David Russell's Division of the Sixth Corps, which pounded the Confederate left flank, forcing the defenders to fall back and reestablishing the Yankee line. Russell lost his life in the movement, though, shot first in the chest and followed up a few minutes later by an artillery fragment through the heart.

A short lull followed, then Sheridan brought up General George Crook's Eighth Corps, the seasoned Army of West Virginia and old opponents of the 22nd Virginia, which he had held in reserve. At 4:30 p.m. they crossed Red Bud Run and plowed into the rebel left flank, breaking Gordon's Division. Yankee horsemen supported Crook and rode into Breckinridge's Division in "a whirlwind of galloping horses and swinging sabers." Initially Early met this attack with a galling fire, particularly by the 51st Virginia, seriously thinning Crook's numbers; but the enemy attack was too overwhelming and the left flank, where the 22nd Virginia was posted, collapsed, enabling Sheridan nearly to surround Early.

The entire southern army fell back in chaos through the streets of Winchester and headed south toward Fisher's Hill. Colonel Patton, riding his horse through a narrow street in town, halted and attempted to rally his men. As he stood up in the saddle waving his sword he received a mortal wound in the leg. William Bahlmann would later say that Patton was struck with a one-fourth piece of shrapnel which passed through both hips. Command of Patton's Brigade devolved upon Lieutenant Colonel John Clairborne McDonald, while Captain John K. Thompson, still suffering from his Cold Harbor wound, managed to lead the remnants of the 22nd Virginia from the field, although he soon afterward returned to the hospital.

At Winchester Early lost one-third of his army, suffering some 4,000 to 4,500 casualties. Wharton's Division had the heaviest loss. The 22nd Virginia, which reportedly entered the fray with 386, lost more than 217, including at least 10 killed, 18 wounded, and 189 captured. Private James Harvey Leach of Company F said that his unit went into the fight with 22 men and lost 18 killed and captured. Patton's brigade organization was almost completely broken up, the unit being left without an officer above the rank of Captain. The wounded Colonel Patton was carried through the streets of Winchester to the house of a relative, John J. Williams, located on the south side of Piccadilly Street between Main and Braddock. Federal surgeons took note of his badly mangled leg and insisted it be amputated, as he was fast losing blood. But Patton sternly refused and pulled his pistol from his holster and placed it on the night table as a deterent to the surgeons. (Why the Federals permitted Patton to remain armed has never been made clear.) Soon afterward Patton's father was called to Winchester in an effort to convince his son to have the leg, which had turned gangrenous, amputated. He reportedly said, "Why don't you let them cut your leg off, Colonel?" to which George replied, "What would I look like stumping through life on a wooden leg?"

Bleeding profusely, Patton grew weak and pale as his father sat in vigil at his bedside. Weeping, the elder Patton asked his son, "What will your mother do? What will your mother do?" But it was all to no avail and

Colonel Patton passed away on September 25. He was buried at Mount Hebron Cemetery, near the spot where he fell, in a grave later also occupied by his brother, Colonel Waller Tazewell Patton, who had been killed in action the previous year leading the 7th Virginia Infantry at Gettysburg. Patton left behind a nine-year-old son, George Smith Patton II, who would give birth to General George S. Patton III, "Old Blood and Guts" of World War II fame. Since Patton's death stories have circulated that he was promoted to Brigadier General prior to the battle of Winchester, but died before his commission ever reached him. There is some sketchy evidence to support this claim but no valid documentation has ever been found, although he was recommended for the position a number of times during the war. In fact, George's brother, John M. Patton, actually attempted to convince authorities to give George the well-deserved commission, but George was against his brother doing this and preferred for his actions on the field of battle to speak for themselves in gaining any promotion.

Early felt that he lost at Winchester due to the great superiority of the Federal cavalry over his own horsemen. He claimed he had already defeated Sheridan's infantry, which was basically true, and could have inflicted more severe damage on the enemy foot soldiers, but Sheridan's troopers were much too strong for him. Sheridan did lose more men in the fight, some 5,000, but he could easily spare the loss whereas Early could not. Despite this, Sheridan won the battle and on September 20 Early was in the earthworks at Fisher's Hill, the only spot he felt strong enough to make a stand. Reduced to 10,000 men he formed his battle line with Wharton's Division on the right. The 22nd Virginia remained with Wharton, Breckinridge having that day left for command of the Department of Southwest Virginia. Patton's Brigade was immediately to the right of the Valley Pike, with Wharton to its right and Smith to the right. Early's line continued west across the road, posting Ramseur's Division and Pegram's and the cavalry on the left flank.

Sheridan had already moved in front of Fisher's Hill but was hesitant to attack, well aware of the difficulty in making a frontal assault on this familiar piece of terrain. After consultation with his staff Sheridan decided that Crook would move against the weak rebel left, the same tactic that had worked so well at Winchester. Early on the morning of September 22 he launched a frontal assault on Fisher's Hill and the combat raged until about 4 p.m. when Crook reached his position on the Confederate left. Crook encountered some feeble resistance from the "undermanned and demoralized" soldiers of General Lunsford L. Lomax, who had left their positions in the entrenchments. Lomax requested that his former position be filled by the infantry but the charging Federals caught sight of the gap in the line and rushed into the hole, causing the Confederate left flank to turn. Wharton's Division, including the approximate 186 men of the 22nd

Virginia, was ordered from the right in an effort to support the faltering left, but they were unable to reach the position in time and Early's army fled from the field in complete disorganization, discarding knapsacks, weapons, and accoutrements. Sheridan had completely routed Early, with his own loss less than 500.

Early's Army of the Valley was left in a state of shock, having suffered between 1,200 and 1,500 casualties, including much artillery. The 22nd Virginia, being on the right, did not suffer as much as the men on the left, but they did lose 6 men captured, and John Maston said that in the combined fighting at Winchester and Fisher's Hill Patton's Brigade lost 3 flags, 14 men killed, 70 wounded, 46 commissioned officers captured, and 374 total taken prisoner. Another member of the brigade said that 45 men of the 22nd Virginia were missing in action at Fisher's Hill, which if true meant that the regiment was down to approximately 140 men.

The Confederates retired through Woodstock to Narrow Passage and stopped just north of Mt. Jackson on the 23rd. Wharton's Division deployed west of the Valley Pike, with Wharton's (Forsberg's) Brigade on the left, Smith's in the center, and Patton's on the right next to the Pike. General John Pegram, replacing Rodes, was situated just east of the road. In the day's skirmishing Private John Calvin Lucas, from Company C of the 22nd, was wounded in the thigh, having already been wounded in the hand at the Winchester battle.

On September 24 Wharton's Division was at Rude's Hill, Patton's Brigade holding the center. A stand was made against the Federals and the right flank managed to hold, then Early fell back through New Market to Tenth Legion under intermittent fire from pursuers. Gordon's and Wharton's Divisions were west of the Pike and Pegram and Ramseur on the eastern edge. The next day the Southern army passed through Port Republic, where Early received news that General Kershaw's Division would be sent to him as a reinforcement. At that time Early made it clear that his army had suffered heavily, and most were exhausted or without shoes, yet he promised that when Kershaw connected with him he would do his best to stop Sheridan. But he was not optimistic about the result of such an encounter.

Early was at Brown's Gap on the 26th, Patton's Brigade holding the center, east of the Pike. Kershaw arrived from across the Blue Ridge and on September 27 Early drove an enemy force from Weyer's Cave as the 36th Virginia moved to Port Republic. On the 28th Early moved through Waynesboro to attack the Federal cavalry at Rockfish Gap. Wharton's Division remained behind to guard the trains at Waynesboro.

In camp near Waynesboro on September 29 Patton's Brigade of Wharton's Division was inspected by Wood Bouldin, Jr., who reported that the brigade had opened the year with 2,150 men but was now down to

266 with an aggregate of 275. Lieutenant Colonel McDonald was listed as commanding the brigade, while Captain Henry B. Dickinson of Company C was at the helm of the 22nd Virginia, which had only 111 men report for inspection out of an aggregate of 158. Bouldin attributed the severe losses to the previous five months casualties of war. He also said that the brigade had been asked at almost every camp to burn rails, with but few axes to accomplish the chore. On a bright note Bouldin ended the report by saying that the present officers were mostly "efficient and anxious" and the men were armed with .58 caliber Enfields and Springfields.

The 22nd Virginia spent the first four days of October camped to the right of Mt. Sidney and on the 3rd heard for the first time of the death of Colonel Patton. The news produced much sadness in the camp. On the 5th Early advanced toward the Federals at Harrisonburg, but they discreetly withdrew and the following day Early's men remained in the Federal camps at Harrisonburg. On October 7 Early was camped one mile south of New Market on the Shenandoah River, and remained in camp the next two days. His cavalry, under Lomax and Thomas L. Rosser, was badly beaten by Federal cavalry at Tom's Brook on the 9th, and the troopers fell back and joined Early at New Market.

General Early again advanced against the Federal camp at Cedar Creek at 6 a.m. on October 13. The enemy advanced to meet him and found Early, including Wharton's Division, on Hupp's Hill. The Yankees were driven back and lost several men. Between October 13 and October 16 Early remained at Fisher's Hill and Hupp's Hill near Strasburg. With Early's defeats at Winchester and Fisher's Hill, Sheridan had convinced himself that Early was no longer a serious threat. Indeed, the farthest thing from Sheridan's mind was that Early would be audacious enough to turn and strike him. What he didn't know was that Early could not have cared less that he was outnumbered three-to-one, that his army supplies were nearly exhausted, and that the morale and spirit of the men was at an all time low. Early had been biding his time, regrouping his men, waiting to catch Sheridan unaware, and when the Federals settled into camp at Cedar Creek the time was right.

Setting his plan into motion on the night of October 18, Early sent Wharton's Division toward Cedar Creek. The bulk of his army moved farther east toward Bowman's Mill and Ford. By 5 a.m. on October 19 Wharton's Division and the Confederate artillery were on Hupp's Hill ready to pounce on the unsuspecting Federals nearby. The division lined up left of the Valley Pike, actually just west of Hupp's Hill, Patton's Brigade on the left, Smith's on the right, and Wharton's (Forsberg's) directly behind Patton's. At about that time Gordon's and Kershaw's Divisions attacked the Federal camp at Cedar Creek with a ferocity never before witnessed. The Northern soldiers of the Sixth, Eighth, and Nineteenth Corps, many yet asleep, were caught completely unprepared and were swept from the field by the surging rebel line. Wharton's boys moved up and the panic-stricken Yankees were driven back through Middletown, where they managed to organize a weak line of defense.

Within a matter of hours Early's vastly outnumbered troops had nearly routed Sheridan's 30,000 Federals. Many of Early's men, half-clothed and starving, fell out of the ranks to plunder the wealth of the captured Federal camps. Early also delayed the second assault, feeling secure that he had already won a brilliant victory. Unfortunately, these incidents gave General George W. Getty time to partially rally the Federals, who were struck again by Wharton and fell back. By that time, Sheridan, who had been away on business, arrived on the field and his presence restored the vigor and courage of his soldiers. He got his men back into the fight, causing the Confederate line to stall and then fall back around the Belle Grove house. At that point General Ramseur fell mortally wounded. Wharton's Division was hit by enemy cavalry as Smith's Brigade held the left, Patton's the center, and Payne's cavalry defended the right.

Early's men could not hold and the line disintegrated, the rebels fleeing back to New Market. Early had almost pulled off the most surprising victory of the war, but due to his own caution and the rummaging of the enemy camps, his opponent was able to seize the initiative and finally reduce the Valley Army to an ineffective force. Early took full blame for the disaster and even offered to turn his command over to another, but his gesture was refused by the authorities.

In the fighting at Cedar Creek the 22nd Virginia lost at least 8 killed, 8 wounded, and 9 captured, losses the regiment could ill-afford. Among the wounded was Captain Henry B. Dickinson of Company C. Private Thomas Oliver Surber from Company H had a bullet strike his rifle, knocking him down and lodging in his haversack.

Early's depleted ranks remained in the vicinity of New Market. There on October 29 the 22nd Virginia reported 120 present for inspection out of an aggregate of 136. McDonald still led the brigade while Captain James C. Johns of Company B commanded the regiment.

No longer a serious threat to Sheridan, Early spent the remainder of 1864 marching about the Valley and camping at various locations virtually unmolested. The whereabouts of the 22nd Virginia during this period are sketchy at best, but it is assumed that they remained with Wharton's Division and participated in all its movements. It is known that the 36th Virginia from Smith's Brigade left camp on November 10 and within two days had marched to within 10 miles of Winchester, where the men spent the entire night in line of battle. Returning the next day they went into camp at Edinburg on November 26. The 22nd Virginia was reported camped near New Market on November 28, as well as the next two days.

At the end of the month, the brigade consisted of 223 men present out of an aggregate of 237. McDonald was still in charge and Captain James A. Cook of Company G led the regiment, although he reportedly resigned November 11.

In early December the Division marched to Fishersville, east of Staunton, to go into winter quarters, but on December 20 Wharton was sent 30 miles down the Valley toward Harrisonburg in order to catch some troopers under General George A. Custer. Before they arrived Confederate cavalry had already driven Custer off and the men marched back to Fishersville in snow and sleet, arriving in camp on December 23. The 22nd Virginia probably failed to participate in this movement as the regiment reported from Fishersville on December 21 that the brigade then consisted of "425 present and temporarily absent, possibly increased by Spring to 600."

Soon after the Harrisonburg expedition arrived back in camp the men boarded a train and went to Charlottesville to intercept the enemy; but again, before they arrived, Confederate cavalry had driven the Federals off and the rebels returned to winter quarters at Fishersville.

Losses in the 22nd Virginia in 1864 were tremendous. Only some 99 men enlisted yet more than 33 were killed in action; between 100 and 131 were wounded; at least 236 were captured in battle; another 76 were captured elsewhere; more than 130 deserted; 22 died of disease; 6 were discharged; and 9 transferred or resigned. This would account for a total loss of no less than 610 to 640 men, nearly the full strength of the regiment at the beginning of the year.

CHAPTER VI

1865 — Too Little, Too Late

The bloodied and battered remnants of the 22nd Virginia spent the first part of the winter encampment of 1864-65 at a point between Fishersville and Waynesboro, about two or three miies east of the latter, to the right of the railroad. Private Sampson Smith of Company G wrote a letter from there on December 29, explaining that the regiment had not yet gone into winter quarters but the men were then cutting logs in preparation. He also said that it was difficult to work as the snow was then about two or three feet deep and it was extremely cold. The 50th Virginia Infantry was camped about two or three hundred yards away and the prior day, while cutting lumber, a tree fell on a member of that regiment and killed him. The 22nd Virginia had had a similar experience the previous night when a tree the soldiers cut fell across a tent, killing one man and seriously injuring another.

Sampson also complained of the scarcity of food, stating that the men only drew a pound of flour some days, or a pound of corn meal with bran in it, or a pound of beef. They could buy whatever else they could find, which usually amounted to very little or nothing. Sampson's attention was focused primarily on returning home, going so far as to ask the recipient of his letter to try and get him out of the army so he could come home and make shoes. Although Smith had been in the army a short time he had not yet been in a battle, but he had marched incessantly, and encouraged the letter's recipient to stay out of the army himself.

On December 31 officers of Wharton's Division reported Echols' Brigade was then led by Lieutenant Colonel McDonald, while Captain Montraville P. Roach of Company G commanded the 22nd Virginia. The brigade adjutant, Lieutenant James French Patton, a brother of Colonel Patton, said 425 members of Echols' (Patton's) Brigade were present, and that the 22nd Virginia had an aggregate of 262 with 191 present for duty.

Typical winter camp life occupied the men of the 22nd Virginia at Fishersville during the first half of January. On the 15th of the month Federal authorities reported members of the 22nd Virginia had been seen in the vicinity of Lewisburg recruiting. Despite such reports, which may or may not have been true, on January 18 Echols' Brigade, including the 22nd Virginia, was detached from Wharton's Division and ordered to move to Dublin and serve under Echols, who was then in command of the Department of Southwest Virginia and East Tennessee. In a letter written from Lewisburg on January 24, Mason Mathews wrote: "Edgar recently passed through Staunton on the way to Jackson Depot, and I understand to Greenbrier, and Echols' Brigade goes to Dublin in that section." In February Edgar's Battalion is known to have been in camp at Saltville.

The 22nd Virginia again went into winter quarters, this time at Red Sulphur Springs in Monroe County. McDonald reported from that location that he had 230 men and 15 officers who were "absent with leave and prisoner of war." He claimed that he reported them as such by order of General Early and would continue to do so until further ordered. William Bahlmann, who had become an Enrollment Officer at Lewisburg due to his severe wound at Droop Mountain, reported visiting the regiment once while they were camped at Red Sulphur Springs. He said it numbered only 200 men. They were preparing to begin their fifth campaign. On February 28 Echols' Brigade, still composed of the 22nd Virginia and the 23rd and 26th Virginia battalions, recorded 475 effectives present, with McDonald in command. On that same day, Echols, with his headquarters at Wytheville, reported a total aggregate present in his brigade as 662. That made it the largest number of men in the entire Department of Southwest Virginia and East Tennessee. The cavalry brigade under Lieutenant Colonel Vincent A. Witcher, numbering 215, was the second largest.

Having been detached to join Echols, the 22nd Virginia escaped the debacle at Waynesboro, where on March 2 General Sheridan struck Early's army, including Wharton's Division, and nearly captured the entire force. Generals Early and Wharton both barely managed to escape personal capture, but Early's army had been almost completely destroyed.

On March 6 Federal authorities reported the 22nd Virginia, along with Edgar's and Derrick's battalions, at Union and grossly overestimated their numbers at 800. The following day all troops under Echols and General H. H. Walker were ordered to Lynchburg to aid Lomax, who was expecting an attack from Sheridan. Early and Wharton had reunited and Early sent Wharton to Lynchburg to report to Echols. Lynchburg was not attacked. On March 15 a massive prisoner exchange took place which soon helped to return a number to the ranks of the 22nd Virginia, but materials were not available to equip those who did return, so it became the old story of "too little, too late."

To complicate matters for the thinned ranks of the Department of Southwest Virginia and East Tennessee another enemy raid through southwest Virginia and North Carolina by General George Stoneman began on March 23 and continued through April 23. Its purpose was to "destroy property more than armies." The raiders entered southwest Virginia via Morristown, Tennessee, and struck first at Boone, North Carolina. With barely enough men to hold a town, much less an entire department, Echols had the unpleasant task of defending against Stoneman. The exact movements and amount of participation in the campaign by the 22nd Virginia is little known.

On March 25 Echols' Brigade was at Dublin and reported 425 present and temporarily absent, with 247 paroled prisoners about to return, which would bring the total present and absent to 800. From Abingdon on March 30 the 22nd Virginia listed an aggregate of 392 and said it had no Major. Also that day Echols replaced Early in command and began plans to move east toward Lynchburg to unite with General Lee. Lieutenant Colonel Mcdonald stated that the 22nd Virginia had been ordered to Saltville to intercept Stoneman but arrived after he had already retreated. The regiment then was ordered back toward Richmond, marching most of the way as the majority of the railroad lines had been destroyed.

Echols began the movement to unite with the Army of Northern Virginia on April 2, but disillusionment in the ranks was highly evident. Basil Duke remembered that Echols' men were veterans who understood all too clearly the end was near for the Confederacy. Duke said "they were not altogether hopeless, and were still resolute although their old enthusiam was utterly gone." Along the march the citizens continued to show their support and helped to bolster morale.

A portion of Stoneman's raiders, some 500 men from his 3rd Brigade, attacked Wytheville on April 6 and were surprised when the rebels "attacked with infantry and cavalry." The Yankees were driven off with 35 killed and wounded and some captured. The exact rebel force involved in this clash is not known but it is possible that the 22nd Virginia may have participated. After the war Captain John K. Thompson recalled that after he returned to the regiment in January he was involved in "a skirmish at Wytheville."

Echols reached Christiansburg on April 10 and the 22nd Virginia went into camp alongside the Roanoke River on the east side of town. Later that night confirmed reports arrived on Lee's surrender the previous day at Appomattox. According to Basil Duke, "the effect of the news upon the infantry was to cause an entire disorganization." He said they gathered in large groups and many tossed their weapons aside and departed, while "those who remained lost all sense of discipline." Lieutenant Colonel McDonald and his staff decided to move back to Christiansburg and hold a council of war to see if they could take their command over to General Kirby Smith's. After consultation they decided to disband and furlough the boys. Echols had also called a council of war and announced his plan to link up with General Joseph E. Johnston in North Carolina. The infantry officers balked at that idea and told Echols their men would not go and it would be "useless to attempt to make them." In a compromise Echols furloughed the infantry for 60 days in the frail hope that at the expiration of that time the war would still be in progress and many would be encouraged to return. As a result, Echols had only his cavalry and a few die-hard infantrymen when he left to seek Johnston.

What little remained of the 22nd Virginia was probably with the 36th Virginia Infantry at Christiansburg on April 12 when that regiment disbanded. The 22nd most probably was disbanded at that time as well, although one member of Company F said that the regiment disbanded April 15. Thomas Oliver Surber of Company H said the last march of the 22nd Virginia "began near the Tennessee state line and ended near Salem when the boys heard of Lee's surrender and disbanded."

When the 22nd Virginia ceased to exist the members were free to find their own way home as best they could. Company B, the Jackson County Border Rifles, reportedly had not been home in four years. Most of the regiment's veterans returned to the new state of West Virginia or the southwest counties of Virginia and attempted to resume the careers they had left behind when they had become soldiers. Lawyers, though, of whom there had been many in the 22nd Virginia, were not permitted to practice law for the immediate years following the war, prompting many to leave the state. As the years passed many went into politics, business, medicine, and engineering, but most had been farmers and remained so.

For those who did not survive the war there were many sad memories for their families. Private James A. Rust, who was killed in action at Fayetteville in September 1862, had gone with his aunt for a stroll at the beginning of the war and picked out the spot where he wished to be buried if killed, and there he now rests. Private John Flemin Reynolds died of dysentery while incarcerated in a northern prison camp, leaving one child and a pregnant widow. The Yankees had taken her slaves away when they passed through Craig County and she had difficulty surviving without their help. Colonel Patton's widow married his cousin, Colonel George Smith of the 62nd Virginia, and they moved to California. Private William S. Rose arrived home from the war one week after his wife had died and Private Meinhart Wehrle was hanged during the war by Federal soldiers for killing a Yankee who attempted to rape his wife.

Possibly the most representative example of the attitude of the 22nd Virginia both during and after the war was that of Captain John K. Thompson. Wounded in action six times, considered "brave to the point of recklessness," he entered his farm house at Poca, West Virginia, one day during the postwar years and noticed a four-inch-square Confederate flag placed over a picture. He tore it down, broke it into bits, and informed those present that "while he had fought under that flag, henceforth it must never be displayed" and it never was in his house again. Despite this he would often relate proudly his war stories and "show off" his war wounds by sticking his finger in the bullet holes.

Many of the regiment's survivors attended veterans' reunions. One such group of men, surviving officers of the First Kanawha, posed for a photo taken at a Charleston, West Virginia, drugstore in late 1865. On June 3, 1922, a monument was erected to the Kanawha Riflemen on Kanawha Boulevard in Charleston. General John McCausland, still surviving at that time, recalled fondly many of his comrades whose names graced the memorial.

Other tributes have been paid to the regiment, including one which transpired during the war when three soldiers, none belonging to the 22nd Virginia, literally saved Lewisburg from being burned by the Yankees. According to Lieutenant S. W. N. Feamster, one of the participants, he and two other Confederates were in Lewisburg on business not too long after the 22nd Virginia had achieved a nearby victory, most probably White Sulphur Springs. They noticed a Federal squad approaching the town with intent to burn. The three rebels mounted up and charged at the head of the blue column yelling, "Forward 22nd" in a ruse to scare off the Yankees. It worked, as the enemy thought Feamster and his friends were leading the "feared 22nd" and they fell back to camp at Frankford. To keep this from being written off as the wild imagination of an aging veteran, Feamster supplied names of numerous living eyewitnesses who could verify his story.

There was also Kanawha Rifleman Joseph A. Brown, purportedly the first Confederate veteran to be buried with full military honors at Arlington, who said he was "shot four times, captured four times, and was scared all the time." Squire Hamilton, whose father fought with Company C of the regiment at Carnifex Ferry and White Sulphur Springs, said his dad told him the regiment was known as "The Fighting 22nd." The regiment's record, incomplete as it is, justifies the name.

A STATISTICAL RECORD

An exact statistical record of the 22nd Virginia Infantry would be impossible to present as records of the regiment are extremely incomplete due to various factors, including the confusion surrounding the original structure of the 1st Kanawha Regiment Infantry and it's evolvement into the 22nd Virginia Infantry; poor bookkeeping on the part of the Confederacy; the destruction of much Confederate paperwork in the Richmond evacuation fire of 1865; and the nearly complete lack of material for late 1864 and early 1865, caused by the regiment being nearly destroyed in the 1864 campaigns. But by combining the Compiled Service Records with various memoirs and personal narratives, one can get a statistical, although not entirely accurate, record of the regiment's service for 1861-65. Undoubtedly all figures presented here are much lower than actual returns, and the exact figures for the final months of the war may never be known.

Taking these flaws into consideration it can be concluded that the 22nd Virginia consisted of more than 2,473 men who served at various times during the war. For the most part the men who made up the ranks were rugged mountaineers from the hills of western and southwestern Virginia. The vast majority were farmers although there was a large number of lawyers, particularly in Company H, which boasted more than 20 members of the Kanawha County Bar. These men came primarily from the counties of present day West Virginia, including Kanawha, Putnam, Nicholas, Clay, Roane, Boone, Wyoming, Jackson, Fayette, Greenbrier and Monroe. There was also a healthy complement of men from the southwest Virginia counties of Craig, Allegheny, Botetourt, and Giles.

The regiment was undoubtedly at its numerical strongest in 1861, when it enlisted some 1,003 men. That figure is misleading, however, as by the end of the year the number was down to about 427 due to various losses, including the transfer of two entire companies in July (Charleston Sharpshooters with 101 men and Bruce Rifles with 69) which eliminated 170 soldiers from the ranks. The removal and dissemination of the artillery and cavalry companies also took away a large, but undetermined amount. General Henry A. Wise's retreat from Charleston in July contributed heavily to desertions (at least 82 are shown on the records to have deserted and only 18 of those are known to have returned). Disease also wreaked havoc, with 101 men reported left sick in the hospital at White Sulphur Springs in August; of that number records indicate that 63 eventually returned to duty. At least seven men are on record as having died of disease in 1861 although that figure is a gross misrepresentation as many others were obviously sick and dying in the hospitals at White Sulphur Springs and Pearisburg, typhoid and measles being the primary culprits. Adding to these losses were three killed in action, nine wounded in battle, eighteen captured, an undetermined number missing in action, eight medical discharges, twenty either transferred or resigned, and fourteen deserted other than in Wise's retreat. By the end of 1861 the regiment probably had an aggregate of about 427 men, most lacking uniforms, weapons, and essential equipment.

During 1862 the 22nd Virginia enlisted 640 but lost 316, the largest contributor to the deficit being battle losses (118) and desertions (95). Enlistments were down in 1863 with 121 and losses were severe, numbering about 335. Again, battle losses (193) and desertion (47) accounted for the largest quantity. Battle casualties for Droop Mountain (November 6, 1863) are difficult to determine as the regiment was in a severe state of disorganization and confusion after the loss, so that many casualties probably were not recorded. As service records were kept extremely well in the aftermath of victory at White Sulphur Springs three months earlier, it can be assumed that the soldiers listed as wounded in action on the December 1863 roll, with no mention of location, are attributable to Droop Mountain.

The worst year for the regiment was undeniably 1864, when, in a five month span of fighting, the regiment's brigade organization was nearly destroyed. Only some 99 men enlisted yet the regiment lost between 657 and 688, including 369 to 400 in battle alone, followed by 131 desertions. Battle losses are difficult to determine as a number of men are listed as wounded in action on the late 1864 roll, but with no mention of location. Additionally, the number of wounded at the battle of New Market (May 15, 1864) as found in the usually very accurate diary of Private John Maston far exceeds the amount in the CSR.

Unfortunately, records for 1865 are almost non-existent for the 22nd Virginia. It is known that numerous members deserted and surrendered in the final months of the war but there is little statistical record of this, although there are 59 recorded "desertions" and at least 30 who surrendered but did not serve prison terms.

It should be noted that in all figures listed herein the wounded who were captured are not separated from the non-wounded who were caught, so these figures may overlap, but the discrepancy in numbers is minimal. Also, one must remember that many of those wounded recuperated and returned to the ranks later in the year. But they were usually away from the regiment long enough to be considered a temporary loss, and many others' wounds were too serious to return at all.

In addition to all of the above losses the 22nd Virginia had at least 42 men die in northern prison camps including nine at Camp Chase, thirteen at Pt. Lookout, ten at Ft. Delaware, eight at Elmira, and one each at Johnson's Island and Rock Island. Also lost were three killed and two severely injured in 1864 as a train carrying POW's to Elmira wrecked at

Shohola, Pennsylvania. There was also one man killed while riding a train in 1861; two known to have been executed for desertion; about ten accidentally wounded; and one hung by Federal troops for killing at Yankee who attempted to rape his wife.

In summary, out of the some 2,473 men who enlisted in the 22nd Virginia Infantry, more than 87 were killed in action, between 251 and 282 were wounded in battle, 358 were captured in battle while another 168 were captured in non-battle situations, 331 deserted, 77 were discharged for medical or other reasons, 42 died in POW camps, and a small number of others were either missing in action or injured by accidental causes.

CASUALTIES

Battles/Campaigns	KIA	WIA	POW	MIA	TOTAL
Antill Hill (Co. H) 7/1/61	0	0	0	0	0
Scary Creek 7/17/61	3	7	1	0	11
Cross Lanes 8/26/61	0	1	0	0	1
Carnifex Ferry 9/10/61	0	1	3	0	4
Sewell Mountain - Cotton Hill Oct.-Nov. 1861	0	2	0	0	2
Giles Court House 5/10/62	0	1	0	0	1
Lewisburg 5/23/62	22	24	63	3	110(149
Kanawha Valley Campaign Sept.-Oct. 1862	1	9	0	0	10
Beverly 4/24/63	1	0	0	0	1
White Sulphur Springs 8/26-27/63	12	62	8	1	83
Droop Mountain 11/6/63	16	22	49	3	90(113)
Sweet Springs 12/20/63	0	1	0	0	1
Other 1863	1	21	0	0	22
New Market 5/15/64	6	8(35)	0	0	14(41

Cold Harbor Campaign 5/20/64-6/4/64	5	33	32	0	70
Lynchburg 6/17-18/64	2	1(5)	0	0	3(7)
Monocacy 7/9/64	0	1	0	0	1
Kernstown 7/24/64	0	6	0	0	6
Winchester 9/19/64	10	18	189	0	217
Fisher's Hill 9/22/64	0	0	6	(45)	6(39)
Cedar Creek 10/19/64	8	8	9	0	25
Other 1864	2	25	0	0	27
TOTAL	**87**	**251(282)**	**360**	**7(52)**	**703(831)**

Note: Figures given are those based upon information on the Compiled Service Records. Numbers in parenthesis are those given by an actual participant in the 22nd Virginia, which accounts for the different sum totals.

Most of the names for "Other 1863" came from the 12/30/64 roll and are probably results of the Droop Mountain fight although some may represent casualties from the battle at White Sulphur Springs.

Most of the names for "Otner 1864" were taken from records in which no battle was specified, and the figure for the Cold Harbor Campaign is a sum total for the actions from 5/20/64 to 6/4/64 but the majority are for the 6/1-3/64 period.

STATISTICAL RECORD

	1861	1862	1863	1864	1865	TOTAL
Enlisted	1,003	640	121	99	0	1,863
KIA	3	23	30	33	0	87
WIA	11	34	106	100-131	0	251-282
POW (battle)	4	61	57	236	0	358
POW (non-battle)	14	43	28	76	12	173
MIA (undetermined)	Unknown	3	4	45	0	52
Desertions	179	95	47	130	59	331
Died of disease	7	18	19	22	0	66
Discharged	8	28	32	6	0	74
Transfer/Resigned	190	9	17	9	0	225
Other (sick)	101	—	—	—	—	101
TOTAL LOSSES	517	316	335	657-688	71	1717-1749

Note: All data is based primarily upon the CSR and does not incorporate postwar and other sources, which accounts for the large discrepancy in men actually enlisted and those on the CSR. All figures were undeniably much higher.

Col. George S. Patton

Courtesy: Virginia State Archives

Col. Christopher Q. Tompkins

Lt. Col. Andrew R. Barbee

Lt. Col. John Clairborne McDonald

Reunion photo taken in a Charleston, WV drugstore 1865. Back (left to right): Thomas Broun, Thomas Smith, Sam Miller, Thomas Fife, Nicholas Fitzhugh. Front (left to right): Joseph Watkins, William Quarrier, Richard Laidley, John Swann. All except Fife served in the 22nd Va.

Courtesy: Isaac Noyes Smith, Jr.

Major Isaac Noyes Smith

Jackson Burwell Huddleston, Co. K

Lt. John P. Donaldson, Co. H

William F. Bahlmann

Pvt. William T. Thayer, Co. H

Lt. Henry D. Ruffner, Co. H

Sgt. Henry Bradford, Co. H

Capt. John K. Thompson, Co. A

Corp. Archibald McHenry Caldwell, Co. C (right) with brothers John and Robert of the 28th Virginia

Capt. William Tyree, Co. C

Capt. Thomas B. Swann, Co. E

Capt. Lewis F. Watts, Co. F

Courtesy: Helen Garner, Beckley, WV

Capt. John P. Hale, Kanawha Artillery

Pvt. Andrew Jackson Reynolds, Co. C

Pvt. James Hector Huffman, Co. H

James H. Mays, ca. 1860, Co. F

Henry D. McFarland, Co. H

George Craft

Pvt. David M. Beard, Co. K

Pvt. Lloyd Caldwell, Co. C

Henry E. Simms, Co. A

Jeffrey Steele

Courtesy: WV State Archives

Thomas L. Broun, Co. H

Pvt. Jonathan Calvin Lucas, Co. H

Pvt. Hiram Hines Anderson, Co. H

Pvt. Hugh Caldwell Wright, Co. K

Isaac Dotson, Co. D

Lt. James C. Welch, Kanawha Artillery

Jonathan Rundle, Co. H

John 'Dusty' Lugar, Co. K

Phil Doddridge, Co. H

David B. Moore

Courtesy: Fuller McClung, Alderson, WV

Adoniram J. McClung

Courtesy: WV State Archives

Overcoat of Levi Welch, Kanawha Rifleman (Co. H)

From Allan Pinkerton's *Spy of the Rebellion*

Pryce Lewis, Pinkerton spy, toasts Lt. Col. George S. Patton at the Valcoulon Mansion, Camp Tompkins. Sam Bridgeman, Lewis' assistant, looks on.

APPENDIX

FIELD AND STAFF OFFICERS OF THE 22ND VIRGINIA INFANTRY

1861 - 1865

Christopher Quarles Tompkins — Colonel
George Smith Patton — Lieutenant Colonel / Colonel
Andrew Russell Barbee — Lieutenant Colonel
William A. Jackson — Lieutenant Colonel
John Clairborne McDonald — Lieutenant Colonel
Robert Augustus Bailey — Major
P. H. Moore — Major
Isaac Noyes Smith — Major
John M. Hendrickson — Adjutant
F. D. Irving — Adjutant
James French Patton — Adjutant
Noyes Rand — Adjutant
David L. Ruffner — Adjutant
Thomas Smith — Adjutant
John K. Thompson — Adjutant
George Burnsides — Quartermaster
Ottmar L. C. Hinesman — Quartermaster
Joseph M. Brown — Asst. Quartermaster
Samuel A. Miller — Asst. Quartermaster
Thomas R. Roberts — Asst. Quartermaster
John Dryden — Quartermaster Sgt.
George W. Huddleston — Quartermaster Sgt.
Joseph B. Roberts — Quartermaster Sgt.
J. S. F. Smith — Asst. Quartermaster Sgt.
Monroe Quarrier — Sgt. Major
Elijah D. Dotson — Ensign
William Leigh Barksdale — Surgeon
Gabriel C. McDonald — Surgeon
George McDonald — Surgeon
Joseph F. Watkins — Surgeon
Alfred Spicer Patrick — Asst. Surgeon / Surgeon
S. R. Swann — Asst. Surgeon
James A. Templeton — Asst. Surgeon
John C. Ruby — Asst. Commissary Sgt.
William Fisher — Chaplain
G. S. McCutcheon — Chaplain
Samuel Rodgers — Chaplain

Charles M. Smith — Chaplain
Thompson L. Smith — Chaplain
William Campbell — Hospital Steward
Benjamin D. Rust — Ordnance Sgt.
William Armistead — Chief Cook (Negro)
Thomas J. Fleshman — Fife Major
Gilson Hambrick — Fifer
Elijah Miller — Fifer
Morris H. Reese — Fifer
George W. Huddleston — Drum Major
John P. Phillips — Drummer
William Stone — Drummer
William H. Thompson — Drummer
Isadore Portier — Bugler
Lindsey Bradley — Musician
Xavier Brandsteller — Musician
Spencer Bright — Musician
Sandiforth M. Griffin — Musician
Jonathan Calvin Lee — Musician
Thomas O. Lemon — Musician
John A. Reese — Musician
Andrew J. Reynolds — Musician
William W. Tyree — Musician
George L. Wood — Musician
Marshall Wood — Musician

Company Organizations And Officers Of The 22nd Virginia Infantry

Organized about April 25, 1861, this regiment was originally known as the 1st Kanawha Regiment from May to August, 1861. In early August the regiment was organized as the 22nd Virginia Infantry, and company organizations apparently were as follows:

22nd Virginia Infantry

(1st Organization - August 1861)

Company A: Border Rifles (Putnam County)
Company B: Mountain Cove Guards (Fayette County)
Company C: Fayetteville Rifles (Fayette County)
Company D: Nicholas Blues (Nicholas County)
Company E: Elk River Tigers (Kanawha County)
Company F: Border Rifles (Jackson County)
Company G: Rocky Point Grays (Monroe County)
Company H: Wyoming Riflemen (Wyoming County)
Company I: Kanawha Riflemen (Kanawha County)
Company K: Boone Company (Boone County)

The regiment was accepted into Confederate service on July 1, 1861 (although some veterans claimed that they were officially mustered into service on August 31, 1861), and the regimental organization was finally brought to completion on May 1, 1862, as follows:

Company A - "Border Rifles"

This company, known as the Border Rifles, was primarily from Kanawha County, and was reportedly organized before the war. The company enlisted for one year's service at Mouth of Poca (River) in Putnam County on May 22, 1861, and reorganized May 1, 1862, at White Sulphur Springs.

Captains: Andrew Russell Barbee, John Koontz Thompson.

Lieutenants: James B. Morgan, George B. Rust, William S. McClanahan, Thomas A. Bryan, Thomas L. Mason, Alex G. Handley. William H. Bailey, Samuel Franklin Counts.

Company B - "Border Rifles"

Better known as the Jackson County Border Rifles, this company was composed almost entirely of men from Jackson County and was first led by a young Ripley lawyer, William H. Lipscomb. Formerly called Company F, this group enlisted at Jackson Court House (Ripley) June 1, 1861, for one year. On regimental morning reports for July 7-8 they are listed as an "incomplete" company of the unorganized 3rd Kanawha Regiment, serving under the command of Colonel Christopher Q. Tompkins. In August 1861 they were reported as a company of the 36th Virginia Infantry but evidently transferred to the 22nd Virginia in the same month. The company was reorganized at White Sulphur Springs May 1, 1862. It is believed that this company was "picked up" by General Henry A. Wise during his foray to Ripley in July 1861.

Captains: William H. Lipscomb, Benjamin Chase, James C. Johns.

Lieutenants: Jules L. Progler, Charles N. Austin, Samuel Hunter Austin, William Crow, Abram S. Parks, David McCoy, Ottmar L. C. Hinesman.

Company C - "Mountain Cove Guards"

Formerly called Company B, this unit is believed to have enlisted June 5, 1861, and was composed primarily of men from the Mountain Cove district or community of Fayette County. Records indicate that a number of recruits had served with the 142nd Virginia Militia. The regimental roster indicates that the company was organized June 6, 1861, and reorganized May 1, 1862, at White Sulphur Springs, during which time many recruits from the 189th (Craig County) Virginia Militia enlisted. William F. Bahlmann of the Fayetteville Rifles stated that the Fayette County Court appropriated $5,000 to clothe his company and another which was forming at Mountain Cove on the opposite side of New River. It would appear that he was referring to the Mountain Cove Guards.

Captains: William Tyree, Henry B. Dickinson, T. A. Nutter.

Lieutenants: Samuel Stone Dews, Thomas J. Masterson, John M. Wood, E. C. Peters, Woodson A. Tyree, Allan C. Rowan, Henry S. Troutt, J. R. McNutt.

Company D - "Nicholas Blues"

On May 1, 1861, a secessionist flag was raised at Nicholas Court House (Summersville) at which time Captain Winston Shelton raised a 100-man company known as the Nicholas Blues. The company enlisted May 31, 1861, at Nicholas Court House, under Captain Newman, for one year's service. They reorganized June 6, 1861, under Captain Shelton, and were formally reorganized May 1, 1862, at White Sulphur Springs.

Captains: Winston Shelton, James M. McNeil.

Lieutenants: Robert S. Campbell, Michael B. McClung, John F. Campbell, George N. Cofer, Edward J. Rutledge.

Company E - "Elk River Tigers"

The Elk River Tigers was composed primarily of men from Kanawha, Clay, and Roane counties, and was formerly known as Captain Thomas Belt Swann's Company, 80th Regiment (Kanawha County) Virginia Militia. This company enlisted for one year's service June 8, 1861, at Kanawha Court House (Charleston). Charleston lawyer Thomas Belt Swann, a man who opposed secession but remained loyal to Virginia, who was the brother of Captain John S. Swann of the Charleston Sharpshooters, was elected captain. The company was reorganized May 1, 1862, at White Sulphur Springs.

Captains: Thomas Belt Swann, George Steptoe Chilton.

Lieutenants: John Connell, Marshall Triplett, Joel Ruffner, William M. Triplett, David Monroe, Edward T. Jackson, M. G. Woods, Thomas Miller, George Triplett.

Company F - "Rocky Point Grays"

Formerly called Company G, the Rocky Point Grays were mostly from Allegheny and Monroe counties. They enlisted June 11, 1861, at Kanawha Court House (Charleston) and then enlisted and organized for one year's service on July 4, 1861. On the regimental reports for July 7-8, 1861, they were reported as a company of the unorganized 3rd Kanawha Regiment, under the command of Colonel Christopher Q. Tompkins. They were assigned to the 22nd Virginia Infantry on August 13, 1861, and reorganized May 1, 1862, at White Sulphur Springs.

Captains: Lewis Franklin Watts, John Campbell, James H. Shumate.

Lieutenants: S. G. Waite, George T. Nicol, J. W. McCormack, Ballard Preston Shumate, James O. Miller, James Campbell.

Company G - "Wyoming Riflemen"

This group, formerly called Company H, was from Wyoming County and was mustered into State service for one year June 28, 1861, then enlisted in Confederate service for one year on July 3, 1861, at Kanawha Court House (Charleston). On the regimental morning reports for July 7-8, 1861, they reported as a company of the unorganized 3rd Kanawha Regiment, Colonel Christopher Q. Tompkins commanding. On the August 31, 1861 roll, 11 men are shown as having enlisted July 7, 1861, all apparently having transferred to this company from Captain William T. Sarver's Company of the 3rd Kanawha on July 25, 1861. Sarver's company is believed to have broken up at about that time.

Captains: Erastus A. Duncan, James A. Cook, Montraville P. Roach.

Lieutenants: Floyd McDonald, Floyd S. Gore, Theodore F. Bailey, John Johnson.

Company H - "Kanawha Riflemen"

This company was originally formed in 1856 as the Kanawha Minutemen, the name was soon afterward changed to the Kanawha Rifles. According to Noyes Rand, Adjutant of the 22nd Virginia and an original member of this company, the group organized in 1858, although other veterans claimed the name changed permanently to the Kanawha Riflemen in November of 1859. The company was composed almost entirely of men from Kanawha County at its inception, including some 20 lawyers of the Kanawha Bar as well as manufacturers, merchants, and college graduates. The Riflemen enlisted for one years service on May 8, 1861, as Company I. Reorganization of the company took place at White Sulphur Springs May 1, 1862.

Captains: George Smith Patton, David L. Ruffner, Richard S. Q. Laidley, John P. Donaldson.

Lieutenants: Andrew Moore, Nicholas Fitzhugh, Henry D. Ruffner, John Gay Carr, Henry W. Rand, John Lewis, Alanson Arnold, Joseph A. Brown.

Company I - "Boone Company"

From Boone County, this company was evidently organized from the 187th Virginia Militia and was formerly called Company K (1st) when first assigned to this regiment and prior to their disappearance. They enlisted for one year's service on June 22, 1861, at Kanawha Court House (Charleston). The entire company disappeared (deserted) on the night of July 25, 1861, during General Henry A. Wise's retrogade movement out of the Kanawha Valley, but returned of their own accord on August 22, 1861. Regimental morning reports for August-October 1861 show Captain Huddleston's cavalry company then serving as Company K and report the "Boone Company" as being "unattached." Huddleston's cavalry was later assigned to Company I, 8th Virginia Cavalry, while the "Boone Company" was assigned as Company I, 22nd Virginia Infantry. The company reorganized May 1, 1862, at White Sulphur Springs.

Captains: William S. Chandler, John P. Toney.

Lieutenants: Daniel Elkins, William Stone, Lorenzo R. Dodd, David J. Huff, Martin F. Simpson, Andrew J. Barrett, John R. Ballard.

Company K (3rd) - "Fayetteville Rifles"

Also known as the Fayette Rifles and the Fayetteville Riflemen, this company evolved from the 142nd Virginia Militia and was first organized May 23, 1861. The men enlisted at Kanawha Court House (Charleston) for one year's service on June 6, 1861, and was originally assigned as Captain Robert A. Bailey's Company, 36th Virginia Infantry. While at White Sulphur Springs in August of 1861 they traded positions with another company and entered the 22nd Virginia Infantry. On May 1, 1862, they were reorganized at White Sulphur Springs. In order to distinguish the three different companies bearing the letter K it should be recognized that Company K (1st), the Boone Company, became Company I, 22nd Virginia Infantry. Company K (2nd) became Company I, 8th Virginia Cavalry; and the Fayetteville Rifles became Company K (3rd) of the 22nd Virginia.

Captains: Robert Augutus Bailey, John Clairborne McDonald, William Frederick Bahlmann, Martin L. Smiley.

Lieutenants: Samuel B. Woods, Isaac Howson Easley, Francis M. Paynter, John M. Hendrickson, James H. Wilson.

The following companies served as original companies of the 1st Kanawha Regiment (22nd Virginia) but were transferred to other Confederate organizations:

Charleston Sharpshooters - Swann's Company

Also known as the Kanawha Sharpshooters and Swann's Company, the Charleston Sharpshooters was organized in December 1859 and was composed of many Kanawha River boatmen. The men, primarily from Kanawha County, often worked in conjunction with the Kanawha Riflemen in performing emergency and ceremonial services throughout the Kanawha Valley. These riflemen were mustered into service on June 30, 1861, for one year's service, to date from April 25, 1861. They were transferred from the 22nd Virginia on July 25, 1861, and mustered into Confederate States service on August 1, 1861, as Company K, 59th Virginia Infantry (2nd Regiment, Wise Legion). By S.O. No. 139, A & IGO, dated June 17, 1862, they were detached from the 59th Virginia and assigned as Company A, 26th Battalion Virginia Infantry. Charleston lawyer John Sterling Swann organized this company and served as its Captain.

Captains: John Sterling Swann.

Lieutenants: Joel Ruffner, Jr., John R. Taylor, Christopher C. Roy, Charles Ufferman, Robert Bibby, Dr. A. E. Summers.

Bruce Rifles - Taylor's Company

Also known as the Greenbrier Rifles, the men of the Bruce Rifles were from Greenbrier County and enlisted there on June 19, 1861, at Frankford for one year's service. After marching 112 miles to Camp Two-Mile at Charleston, they were mustered into state service on June 24 and 25, 1861, although General Henry A. Wise's record book states they were mustered into state service June 26. The riflemen enlisted in Confederate service July 1, 1861, for one year's service and transferred to the Wise Legion (Brigade) July 10, 1861, by order of General Wise and consent of Colonel Tompkins. With the Wise Legion they served as Company F (1st), 46th Virginia Infantry. In August of 1861 they were transferred as Company E, 3rd Regiment, Wise Legion, later becoming Company E, 60th Virginia Infantry.

Captains: Jacob N. Taylor.

Lieutenants: Jonathan Mays, Moses McClintoc (or McClintic).

Captain John Peter Hale's Artillery Company - "Kanawha Artillery"

Along with his commission as lieutenant colonel of all Virginia state forces in the Kanawha Valley on April 29, 1861, John McCausland was directed to "organize a company of artillery." This was immediately accomplished by Dr. John Peter Hale, a wealthy Charleston resident who became the principal financier of the Kanawha Artillery, also known as Hale's Battery. William Taylor Thayer and his brother, Otis A. Thayer, who had established a foundry at Malden (Kanawha Salines) just east of Charleston, made a 10-pounder smooth-bore, iron-cast cannon there and nicknamed it the "Peacemaker." This piece purportedly helped turn the tide at the battle of Scary Creek on July 17, 1861.

William Alexander Quarrier and James Clark Welch, both from prominent Charleston families and original Kanawha Riflemen, served as lieutenants in the artillery. Welch's commission, dated June 3, 1861, listed him as a 2nd Lieutenant, 5th Regiment of Artillery, 22nd Brigade, 5th Division of Virginia Militia. Welch was killed in the affair at Scary Creek, thereby possibly making him the first casualty of the war for the South in the Kanawha Valley, and maybe for the 22nd Virginia (1st Kanawha) as well.

The Kanawha Artillery was one of the original companies of the 22nd Virginia. Most of its members, including Captain Hale, were drawn from the ranks of the Kanawha Riflemen. In early June and July the company was stationed at Camp Tompkins near Coalsmouth (St. Albans) and consisted of two pieces commanded by Welch and Quarrier. In the Scary Creek battle both pieces were engaged. Welch's sustained severe damage and the artillery horses were injured.

In General Orders #1 for the Kanawha Artillery, dated June 26, 1861, at Headquarters, Charleston, Captain Hale appointed the following non-commissioned officers:

Presley Chapman Eastham - 1st Sgt.
Andrew J. Jarrell - 2nd Sgt.
Thomas Mathews - 3rd Sgt.
L. G. Maufin - 4th Sgt.
William D. Shrewsbury - 1st Corp.
John Parlier - 2nd Corp.
William Taylor Thayer - 3rd Corp.
James Brake - 4th Corp.

The Kanawha Artillery reported on the regimental morning records for July 4-8, 1861, and was transferred sometime in July or August of that year, becoming Captain John P. Hale's, afterwards Captain Thomas E. Jackson's Company Virginia Light Artillery (1st organization), which was captured at Ft. Donelson. Some members transferred to the Kanawha Riflemen, since many were original members of that company. Some others reenlisted in 1863 in Captain Thomas E. Jackson's Company Virginia Light Artillery (2nd organization).

Captains: John P. Hale.

Lieutenants: William A. Quarrier, James C. Welch, Charles F. Walker.

Company K (2nd) (cavalry) - "Kanawha Rangers"

Records of this company indicate that the charter members enlisted May 27, 1861, for one year under Colonel Christopher Q. Tompkins, which would imply that this was an original company of this regiment. Charles Irvine Lewis, a prominent physician from Cedar Grove in Kanawha County, appears to have been the organizer of this company as the unit was known as Lewis' Company in it's earliest stages. In the June 4, 1861, edition of the *Kanawha Valley Star* mention is made of a meeting at Fayette Court House on May 21, 1861, at which the Lewis company of cavalry was present. The company was probably attached to the 1st Kanawha from July through August of 1861. The Kanawha Rangers reported as Company K, 22nd Virginia from August through October of 1861, at which time they are believed to have transferred to the 8th Virginia Cavalry. Lewis was listed as a cavalry captain in the Virginia state militia and enlisted in this company May 27, 1861, apparently leading the company in state service while Thomas Huddleston took over the helm in Confederate service. Lewis was elected captain of the company in December 1861 following the death of Huddleston during a raid on Guyandotte in November 1861.

Despite the fact that the Kanawha Rangers, made up of men from Kanawha and Fayette counties, was an original company of the 22nd Virginia, the Compiled Service Records are filed under Company I, 8th Virginia Cavalry. One man is known to have transferred to Company D of the 22nd on July 29, 1861.

Captains: Thomas Huddleston, Charles Irvine Lewis.

Lieutenants: John A. Crockett, A. J. Dickinson, William B. Banister, John W. Smith, G. H. Hanna.

Captain Lewis' Company

Much speculation exists concerning this company. It was reported as an original company of this regiment on the regimental morning reports for July 4 and 8, 1861. It is generally believed that this is the state militia cavalry of Captain Irvine Lewis (not Green as some sources state) and once it entered Confederate service the leadership was taken over by Tom Huddleston. Lewis again commanded the company upon Huddleston's death in November 1861. This company probably became Company K (1st), Kanawha Rangers, an original company of this regiment, which soon afterward became Company I, 8th Virginia Cavalry.

REGIMENTAL COMMANDERS OF THE 22ND VIRGINIA INFANTRY

During the years 1861-65 at least eleven different men commanded the 22nd Virginia Infantry (1st Kanawha) including colonels, lieutenant colonels, majors, and captains:

CHRISTOPHER QUARLES TOMPKINS

Christopher Quarles Tompkins was born ca. 1813 and resided at Mathews, Virginia. He served 16 years in the United States Army, graduating 27th in a class of 49 at West Point in 1836. Tompkins then served in various United States Army artillery units in the Florida Seminole War, at the battle of Locha-Hatchee, and later in the Mexican War, where he commanded William T. Sherman. He resigned from the U.S. Army in 1847, having attained the rank of captain. At that time he continued on as a captain of Virginia militia from 1848 to 1851, followed by a term as lieutenant colonel for two years.

From 1847 to 1855 Tompkins engaged in the iron manufacturing business at Richmond, Virginia, and from 1855 to 1861 he was Mining Agent and Superintendent of Collieries (coal mines) in Kanawha and Fayette counties of (West) Virginia. He was also Chief Agent of the Paint Creek Coal and Iron Manufacturing Company.

Tompkins owned an elaborate estate three miles east of Gauley Bridge, just outside the Kanawha Valley, known as "Gauley Mount." This summer home was a "veritable showplace, with slaves, a splendid library," a 70-foot-long barn, and a vineyard of 800 vines. Tompkins kept a tutor present for his children, and carriages were continually on call so that he and his wife, Ellen Wilkins, and their children could be transported to Richmond whenever they wished.

At the outbreak of the war Tompkins was 47 years old and admittedly not in full agreement with the Southern cause. Despite this, on May 3, 1861, Robert S. Garnett, Adjutant of Virginia forces, commissioned Tompkins a colonel of the Virginia volunteer forces in the Kanawha Valley. He would originally command the 1st Kanawha Regiment, which evolved into the 22nd Virginia Infantry. Along with John McCausland and George S. Patton he helped to organize the volunteers of the Kanawha Valley into a competent fighting army.

Tompkins' effectiveness as a leader began to crumble in August of 1861 following the retreat of General Henry A. Wise from the Kanawha Valley, when Federal forces occupied and made camp on the grounds of "Gauley Mount," and his family was left to their mercy. Colonel Tompkins wrote Federal General Jacob D. Cox and asked that his family and property be treated with due respect. Against the wishes of many of his subordinates, the gentlemanly Cox agreed.

Colonel Tompkins led the 22nd Virginia at the battles of Cross Lanes and Carnifex Ferry, but resigned his commission in early November of 1861 while at Cotton Hill. He made it clear that the reason for his departure was his inability to work with General John B. Floyd, although other possibilities have surfaced including Floyd's decision to shell the Federals at "Gauley Mount," thereby endangering Tompkins' family. He was also accused, most probably by Floyd, of resigning due to his disagreement with the Southern cause. General William S. Rosecrans permitted Mrs. Tompkins and her children to pass through the lines and visit her husband at Richmond on October 14 and then gave her a permanent pass December 3, 1861.

From Richmond, Tompkins aided the war effort through his industrial connections. He was involved with coal mining near Richmond from 1862 until 1877. He apparently had a meeting with General Robert E. Lee in which the possibility of his return to command was discussed, but nothing ever came of it. Taking the oath of allegiance at Richmond on April 29, 1865, he remained there in the coal mining business until his death on May 28, 1877. Tompkins is buried in Hollywood Cemetery at Richmond. He was a half-brother of Captain Sally Tompkins, the noted Confederate nurse.

GEORGE SMITH PATTON

Born June 26, 1833, at Fredericksburg, Virginia, George Smith Patton was the fourth son of John M. Patton. Early in George's youth the family moved to Richmond, where his father practiced law and George was educated. Patton entered the Virginia Military Institute in 1849 and graduated in 1852. The following two years he spent teaching in Richmond, where he also studied law.

In 1856 Patton and his new family moved to Charleston, (West) Virginia where he formed a law partnership with Thomas Lee Broun. Prior to the war Patton served as Commissioner in Chancery to Kanawha Circuit Court and Kanawha County Court. Patton and Broun also served on the board of directors of the Coal River Navigation Company while future Federal General William S. Rosecrans was company president.

Perhaps anticipating the conflict, George S. Patton formed a militia company in 1856 and called it the Kanawha Minutemen, and was designated its captain. The name soon afterward changed to the Kanawha Rifles and in November of 1859 it was permanently changed to the Kanawha Riflemen. Patton enlisted in Confederate service May 8, 1861, as Captain of Company H (Kanawha Riflemen), 22nd Virginia Infantry. On July 7 he was commissioned lieutenant colonel of the regiment and led the victorious Confederate troops at the battle of Scary Creek on July 17, 1861. During the action he received a severe wound in the left shoulder and was left at Charleston as an exchange prisoner following General Henry A. Wise's retreat from the Kanawha Valley later in the month. This oddly-agreed-upon exchange caused much confusion and he was not able to return to the 22nd Virginia until April of 1862.

At the battle of Giles Court House, May 10, 1862, Patton was slightly wounded and had to return home to recover, only to find out that his exchange was yet not formally completed. His exchange was finally completed on May 25. Patton returned to the regiment some time shortly after June 9 with the rank of colonel, a position he had assumed in November of 1861 while still under exchange.

Due to the constant illness and absence of General John Echols the dashing Colonel Patton often commanded Echols' Brigade throughout 1863 and 1864, so much so that it was more often referred to as Patton's Brigade. In this capacity Patton won a sharp victory at White Sulphur Springs on August 26-27, 1863. Patton, who fought gallantly throughout the Shenandoah Valley campaign of 1864, was mortally wounded in the leg and captured at Winchester, Virginia on September 19, 1864, and died there September 25 after sternly refusing to have his leg amputated. He was buried at Mt. Hebron Cemetery at Winchester in a grave with his brother, Colonel Waller Tazewell Patton, who had died the previous year leading the 7th Virginia Infantry at Gettysburg.

Claims have been made since that Patton was promoted to the rank of brigadier general shortly before his death, but that he died before the commission ever reached him. Although he certainly was recommended for that position a number of times throughout the war, as of this date no valid documentation has been uncovered to support this assertion.

Patton's widow married his cousin, Colonel George H. Smith of the 62nd Virginia Infantry, and they moved to California. Colonel Patton's son gave birth to General George S. Patton III, known as old "Blood and Guts" of World War II fame. That officer would often say that his inspiration was his grandfather, the brave colonel of the 22nd Virginia Infantry.

ANDREW RUSSELL BARBEE

Born at Hawsburg, Virginia, December 9, 1827, Andrew Russell Barbee became a tanner and received his early education at Petersburg, Virginia, studying medicine under Dr. John J. Thompson of Luray, Virginia. He continued to study medicine at the University of Pennsylvania from 1848 to 1849 and then the Richmond Medical College from 1849 to 1850. Barbee graduated from the University of Pennsylvania in April 1851 with a medical degree.

Following graduation Barbee moved first to Flint Hill, Virginia, and then to Criglersville, Virginia, in 1852. During that same year he married Margaret A. G. Thompson, the daughter of Dr. Thompson. He then moved to Poca Bottom in (West) Virginia to plant, farm, and practice medicine and surgery. He opposed secession but at the outbreak of the war formed the Putnam County Border Rifles and was made captain. He would later claim that the entire company opposed secession but decided to remain loyal to Virginia.

Barbee enlisted in Company A (Border Rifles), 22nd Virginia Infantry, May 12, 1861, and served as captain. His company was often used as skirmishers and at the battle of Scary Creek on July 17, 1861, he was seriously wounded, a bullet apparently passing through the elbow joint causing the bone to break and severing the ulnar nerve. Barbee's commission was destroyed by fire on the steamer *Julia Maffitt* when Federal artillery struck the boat July 24, 1861, during General Henry A. Wise's retreat from the Kanawha Valley.

Andrew wrote a letter refusing to serve under Lieutenant Colonel William A. Jackson from November 1861 until early 1862. On May 1, 1862, at the regimental reorganization, Barbee was made lieutenant colonel of the 22nd Virginia. The following year, at the battle of White Sulphur Springs, August 26-27, he received another serious wound, this time in the left arm as well as being hit in the right hip by a gun barrel, which was driven into the groin causing a paralysis of the great muscles of the hip and thigh and affecting the leg.

Lieutenant Colonel Barbee went east to recover and participated with Lee in some of his actions. He returned to the 22nd Virginia in time to help nurse their wounds from the Droop Mountain battle and he led the regiment in the unsuccessful attempt to catch Federal General William Averell in December 1863. On February 9, 1864, he was still listed as commanding the regiment, with Colonel Patton in charge of Echols' Brigade. But his wounds had prevented him from participating in active combat duty and on May 16, 1864, following the battle of New Market, he was ordered to report to General Kemper. With his thorough knowledge of medicine and surgery he was assigned to the Confederate Medical Department, and served much of that time at Saltville. In addition to being in charge of the medical facilities there, he was once more placed in field command for one day during the battle of Saltville in late 1864. In a postwar interview Barbee also claimed that he succeeded to command of the 22nd Virginia upon the death of Colonel Patton at Winchester, but records show Lieutenant Colonel John C. McDonald assuming command of the regiment.

Barbee retired from service in April of 1865 and was paroled at Charleston, West Virginia, on May 9. He resided in Buffalo, West Virginia until 1868 when he moved to Point Pleasant, West Virginia, and continued to work as a doctor. He was elected to the West Virginia State Senate in 1880-84 despite the fact that he was ill with blood poisoning caused by an abrasion of his index finger while engaged in a life-saving operation on a patient. This often left Barbee unconscious and quite sick during the election. In 1881 he was appointed to the West Virginia Board of Health. Twice he was nominated for the United States Congress but was defeated. Appointed as president of the United States Examining Board in his area in 1885, as well as presiding surgeon of the board, he was also a member of numerous medical associations. He succumbed to heart disease at Pruntytown, West Virginia in 1903 and is buried at Lone Oak Cemetery at Point Pleasant, West Virginia.

JOHN CLAIRBORNE McDONALD

John Clairborne McDonald was born in Montgomery County, Virginia, March 14, 1831. He lived in Wyoming County, (West) Virginia shortly before the outbreak of hostilities. In January of 1861 he was a member of the Fayette County patrol.

McDonald was an original member of the Fayetteville Rifles (Company K, 22nd Virginia Infantry) and enlisted in Confederate service at Charleston, (West) Virginia June 6, 1861. He was elected to 2nd lieutenant at that time and promoted to captain of Company K on December 24, 1861. On November 12, 1863, he was appointed Major to replace Robert A. Bailey, who was mortally wounded at Droop Mountain. At the battle of New Market, May 15, 1864, he was appointed lieutenant colonel of the 22nd Virginia, replacing Andrew R. Barbee whose previous wounds prevented him from continuing in that capacity.

Lieutenant Colonel McDonald was at the Winchester Hospital for erysipelas July 27-31, 1864, and then took command of the regiment and brigade following Patton's mortal wounding at Winchester on September 19, 1864. During this time he often led Patton's (Echols') Brigade and remained in that capacity until the close of the war. He claimed that he was promoted to colonel upon Patton's death but no documentation of this has been found, although he was recommended for the position by General Gabriel C. Wharton on December 26, 1864, and March 23, 1865.

Paroled at Charleston, West Virginia, May 25, 1865, McDonald took up residence at Hinton, West Virginia. In the early 1880's he became interested in the development of Craig County and moved there. On August 26, 1897, he attended a veterans' reunion at Union, West Virginia and gave his residence as New Castle, Virginia. He is listed as a resident of Craig County in 1898 and died at Craig City December 22, 1914. He is buried in the McDonald Cemetery at Craig City, Virginia near the Virginia Highways shop.

In his memoirs William F. Bahlmann said that McDonald was made captain of Company K on May 1, 1862, although McDonald's service record says he was elected captain of the Company in December 1863.

WILLIAM A. JACKSON

Information is scarce on William A. Jackson. He was born about 1806 in Fredericksburg, Virginia, and in 1860 was a coal dealer at Malden, (West) Virginia. At the outbreak of the war he was in command of the 100-man Home Guard of Charleston, (West) Virginia. He was appointed lieutenant colonel of the 22nd Virginia Infantry by General John B. Floyd on November 7, 1861, following the resignation of Colonel Tompkins and because of the status of Lieutenant Colonel Patton as an exchange prisoner. Prior to Jackson's appointment the regiment was briefly led by Major Isaac N. Smith.

Jackson wrote a letter to Colonel Tompkins on November 19, 1861, and told him that Floyd had appointed him lieutenant colonel; Jackson requested the influence of Tompkins in securing that commission. On December 21, 1861, Jackson was officially appointed lieutenant colonel. Members of the 22nd Virginia indicated that Jackson was not well liked. Captain Andrew R. Barbee said that he was insulted by Jackson and was refused to serve under him. It is known that Colonel John McCausland often led the regiment while Jackson was in command.

At the regimental reorganization on May 1, 1862, Jackson was dropped and replaced as lieutenant colonel by Andrew R. Barbee. He later served as lieutenant colonel of the 4th Virginia State Line and was recommended for commandant at Dublin Depot on October 10, 1863. Jackson also served in the Nitre and Mining Bureau. He had a large family including four sons. Jackson died at his son's home in Pulaski County, Virginia, on January 15, 1875. He is buried in the Fredericksburg Confederate Cemetery.

ISAAC NOYES SMITH

Born at Charleston, (West) Virginia on April 6, 1831, Isaac Noyes Smith was the only son of Benjamin Harrison Smith and Roxalana Noyes Smith. He was an 1849 graduate of Washington College and a lawyer in partnership with his father at Charleston prior to the war. He also served two years in the Virginia legislature. At the outbreak of the war he enlisted in the Kanawha Riflemen (Company H, 22nd Virginia Infantry) and was appointed major in the regiment in August of 1861. Smith tendered his resignation on October 26, 1861, due to his great dislike of General John B. Floyd, who accused him of being loyal to the North since his father was a known strong Unionist. His resignation was accepted on November 6, 1861, although Floyd did not give it to him until the 17th of the month. He left the regiment for Lewisburg the following day and remained there and at other points in Virginia until 1862.

Isaac commanded the regiment during Colonel Tompkins' frequent periods of absence in October and November of 1861 and from Tompkins' resignation until November 10, 1861, at which time Wiliam A. Jackson took command of the 22nd. According to Smith, he was like a "prisoner in command" due to Floyd's accusations against him.

Smith kept a diary from early September 1861 until November 30, 1861, which was confiscated from him by General Jacob D. Cox when captured; it is now in the collection of Oberlin College in Ohio. It was printed in *West Virginia History* in 1966 and gives a detailed account of the regiment's movements during his service. He was given a safe pass through Federal lines on February 6, 1862, and returned to Charleston, where he apparently worked in the salt business.

After the war Smith resided in Charleston and in 1865 formed a law partnership with E. B. Wright and later worked with the law firm of "Smith & McKnight." He also served in the West Virginia House of Delegates, was President of Kanawha Salt Company, and president of a bank. He died October 6, 1883 and is buried in Spring Hill Cemetery, Charleston, West Virginia.

ROBERT AUGUSTUS BAILEY

Nicknamed "Gus," Robert Augustus Bailey was born in 1839, the son of Circuit Judge Edward P. Bailey of Lewisburg, West Virginia. In 1860 Robert was a resident of Fayette County, West Virginia and an attorney. Prior to the war he served as a captain in the 142nd Regiment Virginia Militia, 27th Brigade, 5th Division. He enlisted at Charleston, (West)

Virginia June 6, 1861, as captain of the Fayetteville Riflemen (Company K, 22nd Virginia Infantry). On August 26, 1861, he led a small scouting party on Cotton Hill to study the feasibility of placing artillery on the heights and shelling the Federal camps at and near Gauley Bridge, a plan used by General John B. Floyd in November of that year.

On November 23, 1861, Bailey was promoted to major and often led the 22nd Virginia or segments of it in various battles, receiving particular notice for his bravery at Fayetteville, (West) Virginia, September 10, 1862. On March 1, 1863, he was major in command of the Department at Lewisburg, (West) Virginia and further demonstrated his gallantry while leading a portion of the regiment at White Sulphur Springs on August 26-27, 1863.

William Bahlmann implied in his memoirs that Bailey was not particularly popular, at least not with him, but despite such an allegation Bailey continued to serve the regiment faithfully until mortally wounded at Droop Mountain, West Virginia, November 6, 1863. He had saved the flag and was attempting to rally the men when he was struck by enemy fire. In Federal hands he managed to linger on in the Federal hospital at Hillsboro, West Virginia until November 11. He is buried in the Old Stone Church Cemetery at Lewisburg, West Virginia.

In his Compiled Service Record can be found some vague reference to his possible association as a lieutenant colonel and colonel of the 14th Virginia Cavalry on September 5, 1862, but apparently nothing ever came of this.

HENRY B. DICKINSON

Born circa 1835-36, Henry B. Dickinson was a resident and physician in Fayette County, (West) Virginia in 1860. He enlisted in the Mountain Cove Guards (Company C, 22nd Virginia Infantry), on July 15, 1861. At the reorganization of the regiment, May 1, 1862, he was elected captain of the company. He is listed as in command of the 22nd Virginia on September 30, 1864, with Captain Edmund S. Read of the 26th Virginia Battalion leading the brigade. Dickinson was soon afterward wounded at Cedar Creek, October 19, 1864, and was paroled at Charleston, West Virginia, May 5, 1865. He was 5'6" tall, of fair complexion, light hair, gray eyes, and light whiskers. Dickinson died in 1912 and is buried in the Dickinson Cemetery just north of Fayetteville, West Virginia.

JAMES C. JOHNS

James C. Johns was born circa 1836 and was a resident of Elk Fork, (West) Virginia in 1860. He enlisted as a Private in Company B of the 22nd Virginia Infantry at Ripley, (West) Virginia June 1, 1861. He was elected 1st lieutenant on May 1, 1862, and promoted to captain May 23, 1862, the very same day that he was captured at the battle of Lewisburg.

After spending his prison time at Camp Chase and Johnson's Island he was exchanged on September 1, 1862. At the battle of New Market, May 15, 1864, he was wounded. He was listed in command of the 22nd Virginia on October 31, 1864, with the rank of captain, as Lieutenant Colonel McDonald was commanding Patton's (Echols') Brigade.

Johns was paroled at Charleston, West Virginia, May 10, 1865, and was 5'10" in height, of light complexion, dark hair, blue eyes, and black whiskers.

JAMES A. COOK

James A. Cook was born in 1838 and in 1860 was a store clerk in Wyoming County, (West) Virginia. Shortly before the war he was busy recruiting volunteers at Oceana, (West) Virginia, along with William T. Sarver and Erastus A. Duncan. He enlisted in Company G of the 22nd Virginia Infantry on July 3, 1861, and was elected 2nd lieutenant five days later. On January 24, 1862, he was elected captain of Company G and held that rank until late in 1864.

Cook was in a hospital at Charlottesville, Virginia, for diarrhea from June 12 to 17, 1864, and in a hospital at Danville, Virginia on July 15, 1864. Although his service record states that he resigned at Staunton, Virginia, on November 22, 1864, he was listed as being in command of the regiment with Lieutenant Colonel McDonald leading the brigade as of November 30, 1864.

Cook was a farmer and after the war was known as "Captain Jack" when he managed the Mountain House resort at Oceana. He was still engaged in that business in 1889.

MONTRAVILLE P. ROACH

Born in 1841, Montraville P. Roach was a farmer and resident of Wyoming County, (West) Virginia and in 1860 was a farm laborer there. He enlisted as a corporal in Company G at Charleston, (West) Virginia, July 3, 1861. On January 24, 1862, he was elected 2nd lieutenant and on March 1, 1862, 1st lieutenant.

Roach was apparently promoted to captain of Company G on November 22, 1864, and was listed as in charge of the 22nd Virginia on December 31, 1864, while Lieutenant Colonel McDonald headed Patton's (Echols') Brigade. Roach later surrendered and took the oath at Charleston, West Virginia, May 9, 1865. He was 5'11" tall, of dark complexion, black hair and blue eyes. Other than John C. McDonald he is the last individual known to have commanded the 22nd Virginia in any capacity.

INTRODUCTION TO THE ROSTER OF THE 22ND VIRGINIA INFANTRY

During the years 1861-1865 approximately 2,473 men served in the 22nd Virginia infantry, all contained in the following roster. This membership figure is based primarily upon the Compiled Service Records found in the National Archives, and abbreviated herein as CSR. These Confederate military records were gathered together beginning in 1903 and brought to completion twenty-six years later. The CSR contains information derived from muster rolls, supplemented by pay vouchers, hospital records, prisoner of war records, paroles, appointments, transfers, resignations, court martial charges, personal letters, as well as supply requests. The record for Virginia soldiers is on Microcopy M324, and that of the 22nd Virginia Infantry is found on rolls 647 through 655.

These muster rolls were usually taken every two months and present the status of the soldier on the day the roll was taken, not necessary for the two month time period preceding the roll. To eliminate as much confusion as possible surround this, wherever possible the exact date of the roll is given, not the two month time period covered.

Whenever possible each soldiers data is presented in chronological order although all are preceded by their military rank and regimental company. This information is followed by date and place of birth, academic record, prewar vocation, military record, postwar careers and residence, date and place of death and where buried. Often there is no information beyond what is found in the CSR.

Supplementing the CSR are assorted personal narratives and memoirs, county histories, newspapers, cemetery readings, obituaries, manuscripts, and so forth. Particularly useful in this area is William Bahlmann's various detailed writings as a member of the 22nd Virginia Infantry, and the printed memoir of service by James H. Mays, written by him in 1922.

Due to space limitations a soldier is presumed present at all times unless otherwise noted. An exception to this is if he is absent "sick" for no significant reason and for no substantial amount of time. Following an absence the next time he appears present for roll is recorded. Additionally, the final military status of each soldier is given whenever possible. For many there was no supplemental record found, resulting in the abbreviation NFR for no further record known. Regretfully, the final military status of many in the 22nd Virginia is difficult to determine as the last extensive roll was taken in October 1864, when the regiment was a mere ghost of it's former self. Some sketchy rolls were taken in November and December of 1864 but they provide such little information that they are not justified in being considered the final roll.

Most prisoners of war (POW) were either paroled or required to take an oath of allegiance to the United States, and "took oath" denotes this. The main prison camps members of the 22nd Virginia Infantry spent time in were: Camp Chase, Ohio; Point Lookout, Maryland; Fort Delaware, Delaware; Elmira, New York; Johnson's Island, Ohio; and Rock Island, Illinois. Most were held briefly in an interim prison while it was decided which camp to send them to. Typical of these was the Atheneum at Wheeling, West Virginia. Although these temporary holding places are sometimes mentioned in the CSR they are not included in this roster as the tie spent there was usually never more than a few days.

Physical descriptions are generally taken from POW records, discharges, pardons or paroles. Ages presented are usually how old the man was in the chronological sequence of events. For example, if he was captured at Droop Mountain on November 6, 1863 and his POW record says he was twenty, then that is the age he gave when captured. By applying simple math one can subtract the age from the year it was given and come up with the approximately year of birth.

Locations within West Virginia are cited as such, although it did not become a state until June 20, 1863 and was a part of Virginia prior to that. Points such as Kanawha Court House (Charleston), Nicholas Court House (Summersville), and Jackson Court House (Ripley) are simplified by using the present name, since both names were in use during the war.

It must be noted that numerous soldiers of the 22nd Virginia Infantry deserted during General Henry A. Wise's retreat from the Kanawha Valley in July of 1861. If the man is known to have returned his military record would continue, while a lack of additional information would indicate he probably failed to return to the regiment. A great number of men are listed as being "left sick" at various locations, and again, that is often the last entry for many. If so, it can be assumed they probably did not return to service.

Men who served in companies only briefly associated with the 22nd Virginia Infantry (such as Swann's Company, Taylor's Company, Kanawha Artillery, and the cavalry) and then transferred to other regiments, are not listed in detail. For specifics on these men one should consult the organizations to which they transferred.

Most abbreviations used throughout the roster are standard, and other are self-explanatory. But for a better understanding of the roster, the following abbreviations are used:

Absent without leave — AWOL
Adjutant — Adj.

Aide-de-camp — A.D.C.
Artillery — Art.
Assistant — Asst.
Assistant Adjutant General — A.A.G.
Assistant Commissary of Supply — A.C.S.
Attended — Attd.
Born — B.
Brigade — Brig.
Brigadier General — Brig. Gen.
Buried — Bur.
Captain — Capt.
Captured — Capt.
Cavalry — Cav.
Cemetery — Cem.
Colonel — Col.
Commissary — Comm.
Commissioned — Comm.
Company — Co.
Confederate Veteran — CV
Corporal — Corp.
County — Co.
Court House — C.H.
Court Martial — C.M.
Died — D.
Discharged — Dischd.
Division — Div.
Enlisted — Enl.
Exchanged — Exchd.
First Lieutenant — 1st Lt.
First Sergeant — 1st Sgt.
Fort — Ft.
General — Gen.
General Court Martial — G.C.M.
Hospital — Hosp.
Infantry — Inf.
Judge Advocate — J.A.
Killed in action — KIA
Lieutenant — Lt.
Lieutenant Colonel — Lt. Col.
Major — Maj.
Major General — Maj. Gen.
Missing in action — MIA
Mortally wounded in action — MWIA
Mount or Mountain — Mt.

No further record — NFR
Orderly — Ord.
Ordnance — Ord.
Postwar record — PWR
Prisoner of war — POW
Private — Pvt.
Quarter Master — Q.M.
Reenlisted — Re-enl.
Regiment — Regt.
Released — Relsd.
Resigned — Resgnd.
Route — Rt.
Second Lieutenant — 2nd Lt.
Sergeant — Sgt.
Sergeant Major — Sgt. Maj.
Third Lieutenant — 3rd Lt.
Transferred — Transf.
United Confederate Veterans — UCV
United Daughters of the Confederacy — UDC
United States — U.S.
Virginia — Va.
West Virginia — WV
Wounded in action — WIA

ABBOTT, ANDREW: Pvt. Co. B. Enl. 11/10/62 at Red Sulphur Springs, Monroe Co., WV. Substitute. Det. recruiting in Alleghany Co., Va. Jan.-Feb. 1863 roll. At home sick Nov.-Dec. 1863 roll.

ABBOTT, JOEL HOUGHTON: Sgt. Co. K. B. 11/4/39 Rockingham Co., Va. Res. Fayette Co., WV 1850. Enl. 6/5 or 6/61 at Charleston, Kanawha Co., WV. On special duty training recruits for 19th Brig. Va. Militia Aug. 1861. Apptd. Lt. Col. 19th Brig. Va. Militia 1862. Served as a Capt. on Gen. Heth's staff Apr. 1862. Reportedly promo. to Capt. of Co. H, 8th Va. Cav. CSR says enl. in Co. H, 8th Va. Cav. 5/2/62 in Giles Co., Va. and apptd. 1st Lt. 5/14/62. Served in 8th Va. Cav. until end of war. Postwar res. Beckwith, WV. Farmer, lumbering, livestock business. Married 1885. Res. Charleston, WV 1913. Active in vet. organizations. Attended a vet. reunion at Charleston, WV 1924 and gave address Charleston, WV. An original member of Jeb Stuart UCV Camp, Fayette Co., WV. Historian Shirley Donnelly recalled: "In his last year, I went to see him. Always he was dressed in a suit of Confederate gray and wore a long-snow-white beard." D. 1/2/1926 at Charleston, WV. Bur. Abbott Cem., Fayette Co., WV. Wrote brief war memoir called "A Civil War Narrative" published in *History of Fayette County* 1926.

ABBOTT, J. W.: Pvt. Co. H. Enl. 2/4/63 in Monroe Co., WV. Absent on furlo. on Jan.-Feb. 1863 roll. Present on Mar.-Apr. 1863 roll.

ABBOTT, LENT: Pvt. Co. K. Date and place of enlistment not stated. Bahlmann says he enl. early 1862. Dischd. at Camp McCausland, WV 7/26/62.

ABBOTT, SINCLAIR C.: Co. H. Deserted 5/28/62 on Aug. 1862 broadside. Craig Co., Va. Also served in Co. K, 54th Va. Inf. Bur. Brickey Cem., Rt. 311, near Mountain View Church, Craig Co., Va.

ABBOTT, WILLIAM: Pvt. Co. I. Enl. 5/22/61 or 6/29/61 at Charleston, Kanawha Co., WV. AWOL on 8/26/61 roll. Returned on 10/15/62 roll. Later card says he enl. 5/2/62 at White Sulphur Springs, Greenbrier Co., WV. Serving sentence at Dublin, Va., for GCM on 12/31/62 roll. Returned to co. 4/1/64. Deserted near Staunton, Va. 6/28/64. Postwar res. Craig Co., Va.

ACKER, M. H.: Pvt. Co. C. Date and place of enlistment not stated. Capt. at Winchester, Va. 9/19/64. POW Pt. Lookout, MD 9/24/64.

ACORD, JOHN: Pvt. Co. G. Enl. 7/31/61 at Charleston, Kanawha Co., WV. Furlo. on surg. cert. being "unfit for duty" 8/4/51. Scouting with Capt. Thurmond July 1862. Transf. to Capt. Harrison's Co., (?) 9/30/62. (or transf. to Thurmond's Co. July 1862). Mustered 1st Corp Thurmond's Co. 9/12/62.

ACORD, JOSEPH: Pvt. Co. G. Enl. 7/31/61 at Charleston, Kanawha Co., WV. Scouting with Capt. Thurmond July 1862. Transf. to Capt. Thurmond's Co. 9/30/62.

ACORD, ROBERT: Pvt. Co. G. Res. head of Trough Fork, Wyoming Co., WV. Enl. 7/3/61 at Charleston, Kanawha Co., WV. Age: 23. Det. as a mounted guard at White Sulphur Springs, WV 8/13-14/61. Dischd. 7/3/62. Marr. Michel Oliver.

ACORD, WILLIAM: Co. G. Postwar roster. PWR says enl. May 1861 at Charleston, WV. Age: 18. Res. Wyoming Co., WV Surrendered 1865.

ACRE, ADAM C.: Pvt. Co. B. Res. Jackson Co., WV. Enl. 7/1/61 at Ripley, Jackson Co., WV. In hosp. at White Sulphur Springs, WV on June-Aug. 1861 roll. Det. as a scout for Gen. Loring on 10/62 - 4/63 rolls.

ADAMS, ADAM: Pvt. Co. C. Enl. 5/1/62 at White Sulphur Springs, Greenbrier Co., WV. On furlo. on rolls from Nov.-Dec. 1862 to Mar.-Apr. 1863. Present under arrest on Nov.-Dec. 1863 roll. Under sentence of GCM and not entitled to bounty 4/1/64. In prison on 4/30/64-Oct. 1864 roll. Relsd. 12/1/64.

ADAMS, JEFFERSON: Pvt. Co. C. B. 1825 Montgomery Co., Va. Enl. 5/1/62 at White Sulphur Springs, Greenbrier Co., WV. Enl. in Co. F, 36th Va. Inf. 5/15/62. Present in 36th Va. until deserted 10/19/62. Returned 3/1/63 and placed under arrest. Transf. to 22nd Va. 6/25/63. Under sentence of GCM and not entitled to bounty 4/1/64. In prison on 4/30/64 - 10/31/64 roll. Relsd. from confinement 12/1/64. D. before 1880.

ADAMS, STEPHEN: Grad. Yale 1850. Res. Lynchburg, Va. Studied law with Robert J. Davis. Admitted to bar 1854. Practiced law Raleigh Co., WV. Organized a co. and comm. Capt. Det. and put in command of post at Sulphur Springs, Va. Rejoined regt. at Big Sewell Mountain, WV Aug. 1861. Assigned to 30th Va. Battn. as Capt. Served with that regt. until end of war. Postwar res. Lynchburg, Va. Lawyer. Virginia Legislature 1880. Apptd. Judge of Campbell Co. Court 1896. Still alive in 1899.

ADKINS, ALEX: Pvt. Co. K. Enl. 1862. Postwar roster.

ADKINS, JAMES S.: Pvt. Co. D. Enl. 5/30/61 at Summersville, Nicholas Co., WV. Present on 8/31/61 roll. NFR.

ADKINS, JOHN T.: Pvt. Co. K. Enl. 6/6/61 at Charleston, Kanawha Co., WV. Present on 9/14/61 roll. B. in KY. Res. Fayette Co., WV 1860. Carpenter. Age: 27. PWR says middle initial F.

ADKINS, LEVI: Pvt./Corp. Co. I. B. 2/18/41 at Rock Creek, Boone Co., WV. Res. Boone Co., WV. Enl. 6/20 or 29/61 at Charleston, Kanawha Co., WV. Capt. at Lewisburg, WV 5/23/62. POW Camp Chase, OH. 5/30/62. Exchd. 8/25/62. Sick at Giles C.H., Va. on Jan.-Feb. 1863 roll. Deserted and surrendered at Charleston, WV Mar. 1865. Took oath at Charleston, WV and sent north 3/1/65. Age: 24. Farmer. 6'. Tan complexion, brown hair, blue eyes, light whiskers on chin. Married: Elizabeth Ann Halstead. D. Nov. 1908 at Rock Creek, WV. Bur. Atkins - Elkins Cem. Also spelled name Atkins.

ADKINS, WILLIAM: Pvt. Co. F. Enl. 5/1/62 at White Sulphur Springs, Greenbrier Co., WV. Det. to drive hosp. wagon on Mar.-Apr. 1863 roll. Det. as a teamster for the hosp. at Lewisburg, WV 3/30/63. Appears on a clothing list of Gen. Hosp. at Liberty (Bedford), Va. Jan.-Mar. 1863. Deserted 5/19/64. Appears on a list of prisoners at Christiansburg, Va. as a deserter 1/28/65.

AGEE, FREDERICK M.: Pvt. Co. A. Farm laborer, Charleston, WV 1860. Age: 21. Enl. 5/22 or 25/61 at Poca, Putnam Co., WV. WIA (slight hand) at White Sulphur Springs, WV 8/26-27/63. Deserted and surrendered near Charleston, WV Mar. 1865. Took oath at Charleston, WV and sent north 3/20/65. Age: 26. Farmer. 5'11". Fair complexion, fair hair, blue eyes.

AGEE, HEZEKIAL L.: Pvt. Co. A. Farm laborer, Charleston, WV 1860. Age: 18. Enl. 7/2/61 at Camp Tompkins, Kanawha Co., WV. KIA at Cedar Creek, Va. 10/19/64.

ALB, J. J.: Pvt. Co. E. Enl. 5/1/62 at White Sulphur Springs, Greenbrier Co., WV. Deserted on Sept.-Oct. 1862 roll.

ALBRIGHT, MOSES: Pvt. Co. I. Enl. 6/29/61 at Charleston, Kanawha Co., WV. Capt. in Fayette Co., WV 9/13/61. POW Camp Chase, OH. Exchd. 8/25/62. Returned to regt. from prison on Sept.-Oct. 1862 roll. D. at Montgomery White Sulphur Springs, Va., of periconditis 2/25/63.

ALDERSON, JESSE J.: Pvt. Co. C. Enl. 5/1/62 at White Sulphur Springs, WV. AWOL on Sept.-Oct. 1862 roll. Present on Nov.-Dec. 1862 roll. Capt. and paroled at Charleston. WV 5/10/65. Took oath and sent north. Age: 28. 5'5". Sallow complexion, dark hair, dark eyes. PWR says enl. 1861. Res. Fayette Co., WV 1860.

ALEXANDER, ARTHUR JR.: Pvt. Co. K. Enl. 6/6/61 at Charleston, Kanawha Co., WV. Present on June-Sept. 1861 roll. NFR.

ALEXANDER, J. M.: Pvt. Co. C. Date and place of enlistment not stated. Dischd. at Camp McCausland, WV 7/29/62. Must have re-enlisted as name appears on a list of patients at Chimborazo Hosp. 5/5/63.

ALEXANDER, M.: Pvt. Co. C. Date and place of enlistment not stated. Dischd. at Camp McCausland, WV 7/8/62.

ALFORD, GEORGE WASHINGTON: Pvt. Co. I/Co. K. B. 11/24/37 Cabell Co., WV. Enl. 6/29/61 at Charleston, Kanawha Co., WV. AWOL 7/26/61. PWR says fought at Scary Creek in Co. I but soon after took sick with measles. Nursed in private quarters. When recovered, his co. was gone and he returned home. Postwar res. Garrett's Bend, Lincoln Co., WV. School trustee, farmer.

ALFRED, JOSEPH A.: Pvt. Swann's Co. Enl. 6/9/61 at Charleston, Kanawha Co., WV. Age: 21. Co. transferred to Co. K, 59th Va. Inf. 7/25/61. Brother of G. W. "wounded 17 times in war but survived two balls passing thru body."

ALIFF, HARRISON: Pvt. Co. F. Enl. 1/1/63 at Camp Sam Miller, Monroe Co., WV. Sick at Union, WV hosp. on Jan.-Feb. 1863 roll. AWOL on Mar.-Aprl. 1864 roll. Deserted about 5/2/64.

ALLEN, ALBERT: Pvt. Co. G. Enl. 10/1/62 at Charleston, Kanawha Co., WV. Wded. and in hosp. on Nov.-Dec. 1863 roll. Present on Mar.-Apr. 1864 roll. In Charlottesville, Va. hosp. with debilitas 6/10-11/64. Listed as POW but since escaped and returned on 10/31/64 roll. Appears on a list of rebel deserters surrendered at Charleston, WV May 1865. Took oath at Charleston, WV 5/9/65. Age: 24 or 25. Farmer. 5'3". Fair complexion, brown hair, black eyes. PWR says enl. May 1861 at Charleston, WV. Res. Wyoming Co., WV.

ALLEN, AMON: Pvt. Co. G. Enl. 7/3/61 at Charleston, Kanawha Co., WV. Left sick at White Sulphur Springs, WV 8/20/61. Present on Sept. 1862 roll. Left sick at Lewisburg, WV 4/13/63. Possible died as a POW having POW Record of Death & Interment: G. A. Allen - G - 22nd Va. - D. 10/17/64. Place and cause of death not stated. Res. Wyoming Co., WV

ALLEN, JAMES H.: Pvt. Co. I. Enl. 6/29/61 at Charleston, Kanawha Co., WV. Present on June-Aug. 1861 roll. NFR.

ALLEN, JAMES L.: Co. E. Listed as deserter on Aug. 1862 broadside.

ALLEN, JAMES MURPHY: Pvt. Co. H. B. Kanawha Co., WV. Cooper, Charleston, WV 1860. Age: 35. Enl. 7/9/63 in Monroe Co., WV. Det. as Q.M. Dept., Dublin, Va. on July-Aug. 1863 roll. Appears on a list of employees of Post Commissary, Dublin Depot, Va. 1/30/64 (65?). Employed as a cooper. Deserted Feb. 1864 (65?). Took oath at Charleston, WV. and sent north 2/28/65. Age: 42. Cooper. 6'. Dark complexion, dark hair, brown eyes. Bur. Teays Hill Cem., St. Albans, WV.

ALLEN, JOHN A.: Pvt. Co. G. Enl. 7/3/61 at Charleston, Kanawha Co., WV. Det. as a mounted guard at White Sulphur Springs, WV 8/13/61. Left sick at White Sulphur Springs, WV 8/20/61. Present on Sept. 1862 roll. Capt. at Cold Harbor (Gaines Mill), Va. 6/3/64. POW Pt. Lookout, MD. Transf. to Elmira, NY 7/17/64. Exchd. 9/11/64. Admitted to U.S. Gen. Hosp., Baltimore, MD 9/13/64. D. in Baltimore hosp. 9/17/64. *History of Alleghany Co.* says died at Ft. McHenry and buried in London Park Nat. Cem. Res. Wyoming Co., WV. Age 24 at enl.

ALLEN, WILLIAM: Pvt. Co. I. Enl. 7/10/62 at Union, Monroe Co., WV. Entitled to a $50 bounty. WIA (slight arm) at White Sulphur Springs, WV 8/26-27/63. KIA at Droop Mountain, WV 11/6/63.

ALLEN, WILLIAM H.: Pvt. Co. F. Enl. 7/4/61 at Charleston, Kanawha Co., WV. Det. as a mounted guard at White Sulphur Springs, WV 8/13-14/61. Capt. and paroled near Charleston, WV 5/10/65. Age: 27. 5'5". Dark complexion, dark hair, gray eyes. Postwar res. Craig Co., Va. D. 1914 near Union, WV.

AMICK, A. J.: Pvt. Co. C. Enl. 5/1/62 at White Sulphur Springs, Greenbrier Co., WV. Det. as a scout to go with Capt. Thurmond on Sept.-Oct. 1862 roll. Det. as a "support miner" 10/2/62. To work upon Govt. work at Narrows, New River, Va. Nov.-Dec. 1862 roll. "Work on fortifications in Dept. of Western Virginia. Nature of service: mechnic."

AMICK, F. M.: Pvt./Corp. Co. C. Enl. 5/1/62 at White Sulphur Springs, Greenbrier Co., WV. Absent on a scout on Nov.-Dec. 1862 roll. Present on rolls from Jan. 1863 to Apr. 1863. NFR. PWR says enl. 1861.

AMICK, JAMES A.: Co. C. Enl. 1861. Postwar roster.

AMICK, JOHN A.: Pvt. Co. C. Res. of Nicholas Co., WV. Date and place of enlistment not stated. Capt. in Nicholas Co., WV 3/7/64. POW Camp Chase, OH 4/4/64. Sent to City Point, Va. for exchange 2/25/65. Age: 18. Farmer: 5'11". Dark complexion, dark hair, blue eyes. See also: Co. A. Hounshell's Battn. Va. Cav. An original member of Jeb Stuart UCV Camp, Fayette Co., WV. Gave Raven's Eye, WV as res. Alive in 1909.

AMICK, MADISON: Pvt. Co. C. Enl. 9/20/62 at Charleston, Kanawha Co., WV. On a scout on Nov.-Dec. 1862 roll. AWOL on Nov.-Dec. 1863 roll. Farm laborer, Fayette Co., WV 1860. Age: 18.

AMOS, MARTIN V.: Pvt. Co. A. Enl. 5/22/61 at Poca, Putnam Co., WV. KIA at Lewisburg, WV 5/23/62.

ANDERSON, ACEL: Pvt. Co. I. Enl. 8/14/61 at Boone C.H., Boone Co., WV. Enlisted previously in Capt. M. C. Sh(illegible) Co. Deserted 11 or 12/15/61.

ANDERSON, HIRAM HINES: Pvt. Co. H. B. 12/9/39 in Botetourt Co. (now Craig), Va. son of John and Matilda Max Anderson. Married Mary Frances Beard 9/16/58 at Salem, Va. Enl. 4/28/62 at White Sulphur Springs, Greenbrier Co., WV. In hosp. at Charlottesville, Va., for acute diarrhea 6/8/64. Furlo. for 60 days 10/26/64. In hosp. at Charlottesville, Va. for debilitas 12/27/64 to 2/27/65. UDC report says dischd. 1865. UDC report lists him in Co. K, 54th Va. Inf. for four years. Living in Craig Co., Va., in 1898 at age of 59. D. 8/19/1917. Age: 79. Bur. Brickey Cem., Rt. 311 near Mountain View Church, Craig Co., Va.

ANDERSON, JAMES W.: Pvt./Corp. Co. A. Farm laborer, Charleston, WV 1860. Age: 18. Enl. 6/7 or 9/61 at Camp Tompkins, Kanawha Co., WV. Sick at White Sulphur Springs, WV 8/20/61. Present on Sept.-Oct. 1861 roll. Apptd. 3rd corp. 3/1/63. Left sick at Beverly, WV on Mar.-Apr. 1863 roll. Present on Aug.-Dec. 1863 roll. Capt. at Winchester, Va. 9/19/64. POW Pt. Lookout, MD 9/24/64. Exchd. 3/15/65. Took oath at Charleston, WV and sent north 4/17/65. Age: 23. Farmer. 5'11'. Dark complexion, dark hair, hazel eyes.

ANDERSON, JOHN M.: Pvt. Co. B. Enl. 7/16/62 at Union, Monroe Co., WV. Capt. at Winchester, Va. 9/19/64. POW Pt. Lookout, MD 9/24/64. Exchd. 11/1/64. Took oath 5/12/65. Age: 24. 5'8½". Light complexion, dark hair, blue eyes, dark whiskers.

ANKRUM, ROBERT I.: Co. G. Was buried with flag of Co. G draped over his casket after the war.

ANKRUM, RICHARD: Pvt./Corp. Co. B. Enl. 6/1/61 at Ripley, Jackson Co., WV. Appears as a 4th Corp. on Sept.-Oct. 1862 roll. On det. service on Nov.-Dec. 1863 roll. WIA (head) New Market, Va. 5/15/64. Capt. at Winchester, Va. 9/19/64. POW Pt. Lookout, MD. 9/24/64. Exchd. 3/15/65. Took oath and sent north 5/18/65. Age: 30. Farmer. 5'4". Dark complexion, black hair, gray eyes. Marital status: single.

ANKRUM, WILLIAM R.: Pvt. Co. B. Enl. 7/1/61 at Ripley, Jackson Co., WV. Capt. at Lewisburg, WV 5/23/62. POW Camp Chase, OH 5/30/62. Exchd. 8/25/62. Another card says capt. in Highland Co. 5/29/62 or 64.

ARBAUGH, JAMES O.: Pvt. Co. D. Date and place of enlistment not stated. Took oath at Charleston, WV 5/11/65. Age: 21. 5'5". Light complexion, light hair, gray eyes, light whiskers. Also served as a Pvt. in Thurmond's Co.

ARBOGAST, MICHAEL: Pvt. Co. D. Enl. 5/30/61 at Summersville, Nicholas Co., WV. AWOL 7/28/61.

ARCHER, L.: Pvt. Co. C. Date and place of enlistment not stated. Capt. at Winchester, Va. 9/19/64. POW Pt. Lookout, MD 9/24/64. Exchd. 1/17/65.

ARMISTEAD, JAMES A.: Pvt./Corp. Co. B. Enl. 6/1/61 at Ripley, Jackson Co., WV. In hosp. at Charlottesville, Va. 1/2/62 to 5/19/62. Capt. at Winchester, Va. 9/19/64. POW Pt. Lookout, MD 9/24/64. Joined U.S. service 10/15/64.

ARMISTEAD, J. H.: Pvt. Co. E. Enl. 6/8/61 at Charleston, Kanawha Co., WV. Left sick at Gauley Bridge, WV 7/27/61. NFR.

ARMISTEAD, JAMES M.: Pvt. Co. E. Res. Roane Co., WV. Enl. 7/10/61 at Porter's Shoals, WV. Left sick at White Sulphur Springs, WV 8/20/61. Capt. in Roane Co., WV 11/20/62. Age: 49. POW Wheeling, WV 11/30/62. POW Camp Chase, OH 12/2/62. Transf. to Cairo, Ill. for exchange 12/2/62. Farmer. 5'6½". Fair complexion, dark hair, blue eyes, dark gray whiskers.

ARMISTEAD, JOSIAH: Pvt. Co. E. Enl. 5/21/63 at Camp Stonewall. Capt. in Roane Co., WV 1/12/65. Age: 24. POW Wheeling, WV. POW Cumberland, MD. POW Camp Chase, OH 2/13/65. D. at Camp Chase, OH of pleuratis (pleurisy?) 3/6/65. Bur. grave #1580 ½ mile so. of Camp Chase, OH. Farmer. Fair complexion, brown hair, gray eyes.

ARMISTEAD, J. W.: Pvt. Co. E. Enl. 7/10/61 at Porter's Shoals, WV. Left sick at White Sulphur Springs, WV 8/20/61. NFR.

ARMISTEAD, M. S.: Pvt. Co. E. Enl. 7/8/61 at Charleston, Kanawha Co., WV. Left sick at White Sulphur Springs, WV 8/20/61. NFR.

ARMISTEAD, THOMAS: Pvt. Co. E. Enl. 7/10/61 at Porter's Shoals, WV. Left sick at White Sulphur Springs, WV 8/20/61. Believed capt. at Lewisburg, WV 5/23/62. Present on rolls from Dec. 1863 to Nov. 1864. Paroled at Lewisburg, WV 4/26/65. Age: 19. Farmer. 5'8". Dark complexion, dark hair, dark eyes.

ARMISTEAD, WILLIAM: Chief Cook. Colored. Co. H. Mulatto. Barber. Charleston, WV. Age: 44, in 1860. Enl. 5/8/61 at Charleston, Kanawha Co., WV. Absent on furlo. on Nov.-Dec. 1863 roll. Present on Apr. 1864 roll. Took oath at Charleston, WV 8/11/65. Gave age as 40. Farmer. 5'6". Black complexion, black hair, dark eyes. "At battle of New Market, Va. he got caught up in excitement and picked up a rifle and joined the 22nd regt. in the fight." Apparently misfiled in CSR under Armistead Paul.

ARMSTEAD, ALFRED A.: Co. B. Age: 21. Res. Ripley, WV. 1860.

ARMSTRONG, GEORGE W.: Co. B. Age: 22. (16 in 1860 census) Res. Jackson Co., WV.

ARMSTRONG, ISAAC H.: Pvt. Co. A. Enl. 5/22/61 at Poca, Putnam Co., WV. Sick at White Sulphur Springs, WV 8/20/61. Det. as a teamster on Sept.-Oct. 1862 roll. Served as regt. teamster. Deserted at Charleston, WV Jan. 1865. Took oath at Charleston, WV 1/12/65. Age: 22. Farmer. 5'6". Sandy complexion, light hair, blue eyes.

ARMSTRONG, MATHIAS BENSON: 1st Lt. Co. B. Age: 28 in 1850.

ARMSTRONG, VIRGIL S.: 1st Lt. Co. B. Lawyer, Ripley, WV 1860, age 23. Capt. of Co. G, 17th Va. Cav. 1/28/63 until close of war. D. 10/28/1917.

ARMSTRONG, WILLIAM T.: Pvt. Co. A. Enl. 10/1/62 at Charleston, Kanawha Co., WV. Absent sick on Apr.-Oct. 1862 roll. Deserted and took oath at Charleston, WV 1/12/65. Age: 24. Farmer. 5'8½". Dark complexion, black hair, gray eyes.

ARNOLD, ALANSON: Corp./2nd Lt. Co. H. B. in MA. Clerk, Charleston, WV 1860. Age: 33. Enl. 5/1/61 at Charleston, Kanawha Co., WV. Served as a Pvt. until 8/8/61. Then to Corp. Elected 2nd Lt. 9/5/63. MWIA at Cold Harbor, Va. 6/3/64. Bahlmann said he believed him to be bur. at Hollywood Cem., Richmond, Va.

ARNOLD, ENOS S.: Ord. Sgt. Co. H. B. 5/10/20 Berkshire Co., MASS. Educated there. Clerk in dry-goods store, New York, NY 1838. Retail dry goods, Charleston, WV 1842. Salt business 1848-51. Sheriff, Kanawha Co., WV 1858. Enl. 1861 at Charleston, Kanawha Co., WV. Apptd. Q.M. Dept. at White Sulphur Springs, WV, then Richmond, Va., then Salem, Va. until 1865. Retail dry goods, Charleston, WV 1865. Wholesale business 1877. Retired from business 1887. Still alive 1891.

ARRINGTON, A. J.: Pvt. Co. B. Enl. 5/?/62 at White Sulphur Springs, Greenbrier Co., WV. Joined from desertion at Camp McCausland, WV 7/5/62. Took oath at Charleston, WV 12/14/64. Age: 25. Farmer. 6'. Dark complexion, dark hair, blue eyes. Res. Craig Co., Va. 1898 at age 64.

ARRITT, JOHN M.: Pvt. Co. G. Enl. 7/1/62 at Union, Monroe Co., WV. Deserted 10/22/62. Also spelled Arritte. PWR says enl. Mar. 1862 at Lewisburg, WV. Age: 23.

ARRITT, MICHAEL JEFFERSON: Pvt. Co. G. B. Alleghany Co., Va. 12/25/42. Enl. 5/1/62 at White Sulphur Springs, Greenbrier Co., WV. Sick at Bonsacks, Va. July 1862. Listed as a deserter on Aug. 1862 broadside. Left 5/24/62. Capt. at White Sulphur Springs, WV 8/26/63. Age: 20. POW Camp Chase, OH Sept. 1863. Transf. to Rock Island, Ill. 1/24/64. Exchd. 3/2/65. Farmer. 5'9¾". Dark complexion, black hair, brown eyes. D. Alleghany Co., Va. 7/2/1906.

ARTHUR, ALEXANDER JR.: Pvt., Co. K. Day laborer, Fayette Co., WV 1860. Age: 17. Enl. 1861. D. in winter 1861-62.

ARTHUR, JAMES CALVIN: Pvt. Co. K. B. 3/14/28. Enl. 6/6/61 at Charleston, Kanawha Co., WV. In jail for treason at Fayetteville, WV 8/30/61. Postwar res. Craig Co., Va. D. at Charleston, WV of tuberculosis 10/12/1900. Bur. Spring Hill Cem., Charleston, WV.

ARTHUR, MARSHALL: Pvt. Co. K. Enl. 6/6/61 at Charleston, Kanawha Co., WV. Mustered 8/31/61. NFR. Day laborer, Fayette Co., WV 1860. Age: 15.

ARTHUR, WILLIAM: Pvt. Co. K. Res. Fayette Co., WV 1860. Age: 17. Enl. 1861. D. in winter of 1861-62.

ASBERRY, LORENZO D.: Pvt. Co. D/Co. G. Res. Greenbrier Co., WV. Enl. 7/4/61 at Charleston, Kanawha Co., WV. Capt. at Lewisburg, WV 5/23/62. POW Camp Chase, OH 5/30/62. Exchd. 8/25/62. 5'8¾". Fair complexion, light hair, gray eyes. Postwar roster says deserted to the enemy.

ASBURY, JOSEPH: Pvt. Co. A. Enl. 5/27/61 at Poca, Putnam Co., WV. Present on July-Aug. 1861 roll. NFR.

ASHUEL, DANIEL P.: Pvt. Co. A/Co. C. B. circa 1835. Enl. 5/1/62 at White Sulphur Springs, Greenbrier Co., WV. Capt. at Winchester, Va. 9/19/64. POW Pt. Lookout, MD 9/27/64. Exchd. 3/15/65. Postwar res. Craig Co., Va. Bur. Ross Cem., Rt. 42, near Maywood on Fred Keffer farm. May also spell name Ashwell.

ATKINS, GEORGE: 2nd Sgt. Co. I. Enl. 6/29/61 at Charleston, Kanawha, Co., WV. Present on 8/31/61 roll. NFR.

ATKINSON, ANDERSON: Pvt. Co. C. Enl. 1861. PWR.

ATWELL, EUPHRONEUS M.: Co. B. Entered Richmond Old Soldiers Home 11/11/1902 at age 78. D. 5/15/1904. Bur. Hollywood Cem., Richmond, Va.

AUSTIN, CHARLES NEWTON: 2nd Lt. Co. B. B. 1832 Augusta Co., Va. Res. Harrison Co., WV 1846. Grad. Cleveland, OH Medical College 1856. Physician, Ripley, WV, 1860, age 28. Enl. 6/6/61 at Ripley, Jackson Co., WV. Principal organizer of Co. B, 22nd Va. Inf. Elected 2nd Lt. 6/1/61. Det. for special duty in hosp. at White Sulphur Springs, WV 8/15/61. Ordered to establish a hospital there for the sick and wounded. Comm. Assistant Surg. Promo. Surg. 1862. Assigned Surgeon, Edgar's Battn. May 1862. Served with that unit until retired from service for physical disability spring 1864. CSR says resgnd. at Camp Wharton 1/9/65. Another card says resgnd. Aug. 1861. PWR says resgnd. 10/10/61. Postwar: Doctor, Lewisburg, WV. Still alive in 1899.

AUSTIN, ISAAC S.: Pvt. Co. I. Enl. 5/1/62 at White Sulphur Springs, Greenbrier Co., WV. Absent sick on Sept.-Oct. 1864 roll. NFR.

AUSTIN, JAMES: Pvt. Swann's Co. Enl. 5/8/61 at Charleston, Kanawha Co., WV. Age: 38. Co. transf. to Co. K., 59th Va. Inf. 7/25/61.

AUSTIN, SAMUEL HUNTER: Lt. Co. B. B. near Tinkling Springs, Augusta Co., Va. Son of Dr. A. M. Austin. Res. West Milford, Harrison Co., WV 1843. Attended V.M.I. 1856-60. Studied medicine under his father at Winchester Medical College 1860-61. Attended medical lectures at the College of Va. Enl. 6/1/61 at Ripley, Jackson Co., WV. First assgnd. to 36th Va. Inf. Elected 2nd Lt. 6/10/61. Transfd. 22nd Va. Inf. July 1861. Promo. 1st Lt. 5/23/62. Tendered resignation 1/5/63. Resignd. 7/22/63. Re-enl. in 20th (19th?) Va. Cav. as acting Assistant Surg. at White Sulphur Springs, WV hosp., spring 1863. Served in 20th Va. Cav. spring of 1864. Grad. College of Va. Mar. 1866. Post grad. course at Philadelphia, Pa. Grad. Va. Medical College 1866. Postwar: Doctor, Lewisburg, WV. U.S. Examining Surg., Lewisburg, WV since 1881. Served 2 three years terms as Commissioner of Schools at Lewisburg, WV. Four years Chairman, Democratic Executive Committee. Surg. Maj. in 2nd WV Inf. National Guard. D. 11/16/1926 Charleston, WV. Bur. Old Stone Church Cem., Lewisburg, WV.

AUSTIN, WILLIAM: Pvt. Co. I. Enl. 5/1/62 at White Sulphur Springs, Greenbrier Co., WV. Capt. at Winchester, Va. 9/19/64. POW Pt. Lookout, MD 9/24/64. Exchd. 3/15/65.

AYERS, CHARLES: Co. F. Listed as deserter on Aug. 1862 broadside. Bolens Bend.

AYERS, CHRISTOPHER J.: Pvt. Co. F. Enl. 7/4/61 at Charleston, Kanawha Co., WV. AWOL 8/21/61. Res. Monroe Co., WV.

AYERS, JAMES MADISON: Pvt./1st Sgt. Co. F. B. 1843 Dickson's Springs, near Pickaway, WV. Res. Greenbrier Co., WV 1861. Enl. 7/4/61 at Charleston, Kanawha Co., WV. Promo. Corp. Winter 1861. At Hoke's Mill in Greenbrier Co., WV for Gen. H. A. Wise. Working Nichols Mill on July 1862 roll. Promo. Ord. Sgt. Summer 1862. On furlo. by reason of recruitment on Sept.-Oct. 1862 roll. In Charlottesville, Va. hosp. for acute bronchitis 8/17/64. Transf. to Lynchburg, Va. hosp. 8/18/64. Capt. and paroled at Charleston, WV 5/15/65. Miller. 5'4". Fair complexion, dark hair, blue eyes. PWR says enl. 4/11/61 (9 days before turning 18). Promo. Sgt. Maj. on New Market, Va. battlefield 5/15/64 and afterwards promo. Adj. Helped bury 214 Federal soldiers in one grave at New Market, Va. Capt. at Cedar Creek, Va. but escaped from captors following night, slipping away after dark. Went to Massanutten (?) Mountain where spent 3 or 4 days and nights, finally getting to Page Valley at Luray, Va. and joined McCausland's Brig. and fought a battle. Brought in with McCausland's men and rejoined 22nd Va. Inf. 11 days after capture. WIA (slight) at Monocacy, MD. Sgt. Maj. and Acting Adj. until close of war. Postwar res. Greenbrier Co., WV, farming, measuring lumber, etc. Res. Green Sulphur District, Summers Co., WV 1882 or 1883. Store clerk and carpenter. Res. Hinton, WV 1890. Apptd. deputy clerk co. court. Later mercantile business. City recorder 1893-94. Deputy clerk Summers Co. Circuit Court 1896. Still alive 1899. D. prior to 1914. Also spelled Ayres.

AYERS, JOHN L.: Pvt. Co. F. Res. Kanawha Co., WV. Enl. 7/4/61 at Charleston, Kanawha Co., WV. At home sick on surg. cert. 7/27/61. Took oath and sent north. 5'10". Light complexion, light hair, gray eyes.

AYERS, LEWIS A.: Co. F. Postwar roster. Res. Monroe Co., WV.

AYERS, NEWTON M.: Pvt. Co. F. Enl. 7/4/61 at Charleston, Kanawha Co., WV. At home sick on surg. cert. 7/27 or 29/61. NFR.

AYERS, D. G.: NFR.

AYLIFFE, SKELTON: Pvt. Co. K. Enl. 2/9/63 or 64 at Lewisburg, Greenbrier Co., WV. Confusing record which indicates desertion at Camp Gauley, Greenbrier Co., WV either 4/15/62 or 1864 or April 28 although listed as present on 5/13/64 roll.

BABER, ISAAC K.: Pvt. Co. A. Res. Kanawha Co., WV. Enl. 9/18/62 at Charleston, Kanawha Co., WV. Capt. at Winchester, Va. 9/19/64. POW Pt. Lookout, MD 9/24/64. Took oath at Pt. Lookout, MD 6/23/65. 5'6". Florid complexion, auburn hair, blue eyes. J. B. Strange UCV.

BABER, WILLIAM G.: Pvt. Co. A. Enl. 6/5/61 at Camp Tompkins, Kanawha Co., WV. Sick at White Sulphur Springs, WV 8/20/61. Still present 4/30/63. NFR.

BAHLMANN, WILLIAM FREDERICK: 1st Sgt./Capt./2nd Lt. Co. K. B. 8/24/36 New York City, NY. Attended school in New York. Taught school in Va. 1855-58. Attended Universities of Munster and Munich in Germany for two years. Tutored in Louisiana until 1861. Enl. 6/6/61 at Charleston, Kanawha Co., WV. Studying law at enlistment. Adj. 27th Brig. Va. Militia Sept. 1861. A.Q.M. at hosp. at Princeton, WV. Returned to 22nd Va. Inf. Dec. 1861. WIA (shot through right arm and hand) and capt. at Lewisburg, WV 5/23/62. Age: 25. In U.S. hosp. at Gallipolis, OH for vulvirus sclopet 6/11/62. POW Camp Chase, OH 7/8/62. Exchd. 8/25/62. Married Lydia Abbott 12/9/62. WIA (minie ball in left knee) and capt. at Droop Mountain, WV 11/6/63. Permanently disabled at Droop Mountain and recommended for retirement by commdg. officer. Retired 12/31/64 due to his wounds. Spent remainder of war as Enrollment Officer at Lewisburg, WV. 5'10". Fair complexion, brown hair, hazel eyes. Postwar: Principal, Lexington Missouri High School. In 1875 began teaching German and Latin at Central Missouri State Teacher's College. Superintendent, Independence, MO schools 9 years. Returned to Central MO State College 13 yrs. Attd. Vet. reunion Charleston, WV in 1924 and gave address as Charleston, WV. D. 3/17/1930 at the age of 93. Late in his years he wrote his war memoirs "Down In The Ranks", which were not published until Oct. 1970, in the *Journal of the Greenbrier Historical Society.* PWR says Brvt. 2nd Lt. 12/?/61. Elected 1st Lt. 4/?/63. Elected Capt. 11/11/63. Left arm paralyzed by pressure of crutch June 1864. Prov. Marshall Bristol, TN Fall 1864.

BAILEY, ALEXANDER P.: Pvt. Co. G. Enl. 7/3/61 at Charleston, Kanawha Co., WV. Capt. at Winchester, Va. 9/19/64. POW Pt. Lookout, MD 9/23/64. Exchd. 3/15/65. Age: 29. 5'6". Dark complexion, light hair, gray eyes. Res. Wyoming Co., WV.

BAILEY, D.: Pvt. Co. H. Res. Wyoming Co., WV. Enl. 7/11/63 at Lewisburg, Greenbrier Co., WV. Joined by transfer. WIA (severe hip) at White Sulphur Springs, WV 8/26-27/63. Took oath at Charleston, WV and sent north 4/27/65. Age: 22. Farmer. 6'. Fair complexion, black hair, blue eyes.

BAILEY, FIELDEN: Pvt./Sgt. Co. A. Farm laborer, Tyler Mountain, WV 1860. Age: 22. Enl. 5/27/61 at Poca, Putnam Co., WV. Elected 4th Sgt. 7/30/61. WIA (severe arm) at White Sulphur Springs, WV 8/26-27/63. Capt. at Winchester, Va. 9/19/64. POW Pt. Lookout, MD 9/24/64. Exchd. 3/15/65.

BAILEY, HENRY W.: 2nd Lt. Co. A. PWR.

BAILEY, ISHAM T.: Pvt. Co. A. Enl. 6/14/61 at Camp Tompkins, Kanawha Co., WV. Unaccounted for as of 7/26/61. Present 11/1/62. Deserted in "face of the enemy" 12/13/62. Age: 23. Farmer.

BAILEY, JAMES LAWRENCE: Pvt./Ord. Sgt. Co. K. B. 4/1/43 Lewisburg, WV. Moved to Palmyra, MO. Attended school there and ran away at age 18 to enlist. Enl. 2/10/63 at Lewisburg, Greenbrier Co., WV. Det. in Commissary Dept. at Giles C.H., Va. 2/28/63. Present 10/31/64. Postwar druggist, Selma, AL. Then druggist at Demopolis, AL. Married 1871. D. 4/28/1913 at U.S. Army and Navy Hosp., Hot Springs, ARK. Bur. in his Confed. uniform. Photo in *Confederate Veteran* #21.

BAILEY, JAMES O.: Pvt./Sgt. Co. G. Enl. 7/7/61 at Charleston, Kanawha Co., WV. Deserted 11/9/63. Wded. and on parole 12/30/63. (Was probably WIA at Droop Mountain, WV). Res. Wyoming Co., WV

BAILEY, JOHN P.: Pvt. Co. G. Enl. 9/20/62 at Charleston, Kanawha Co., WV. Sent to hosp. at Union, WV 12/27/62. Present 12/31/62. Had a furlo. expire 1/24/64.

BAILEY, JOSEPH BROOKS: Pvt. Co. A. Res. Kanawha Co., WV. Laborer, Tyler Mountain, WV 1860. Age: 17. Enl. 11/1/62 at Raleigh C.H. (Beckley), WV. Sick at Marlin's Bottom (Marlinton, WV 4/30/63. In Chimborazo Hosp. for rubeola 5/31/64. Trans. to Farmville, Va. hosp. 6/2/64. Ret. to duty 6/14/64. Capt. at Poolesville, MD 7/15/64. POW Elmira, NY. 5'11½". Fair complexion, red hair, hazel eyes.

BAILEY, LUCIAN: Pvt. Co. G. Postwar roster. Enl. Mar. 1862 at Lewisburg, WV. Res. Wyoming Co., WV. Surrendered Apr. 1865.

BAILEY, MILES H.: Pvt./Sgt. Co. G. Enl. 10/1/62 at Charleston, Kanawha Co., WV. Absent wded. since 10/19/64. Listed as a POW. (Probably WIA and capt. at Cedar Creek, Va. 10/19/64). Age: 28. 5'6". Fair complexion, light hair, blue eyes. Brother of Theodore F. Bailey. Also spelled Myles. Res. Wyoming Co., WV.

BAILEY, MORRIS: Pvt. Co. A. Enl. 6/30/61 at Camp Tompkins, Kanawha Co., WV. Unaccounted for as of 7/22/61.

BAILEY, PETER: Pvt. Co. A. WIA (right elbow) in 1863.

BAILEY, PLEASANT: Pvt. Co. A. B. 9/6/35 Kanawha Co., WV. Res. Putnam Co., WV. Farm laborer, Tyler Mountain, WV 1860. Enl. 5/27/61 at Poca, Putnam Co., WV. Capt. at Winchester, Va. 9/19/64. POW Elmira, NY. Relsd. 5/15/65. 5'9". Florid complexion, dark hair, blue eyes. Active in vet. organizations. Res. Putnam Co., WV 1910 at age 75. In a photo of R. E. Lee UCV Camp #887, Charleston, WV taken Aug. 1911. Entered Richmond, Va. Old Soldiers' Home 5/9/16 at age 81. D. 1/30/1920 Richmond, Va. Old Soldiers Home. Body taken by relatives. Another card says P. Bailey - capt. near Washington, D.C. 7/13/64. POW Elmira, NY 7/23/64. Rlsd. 5/15/65 and took oath.

BAILEY, ROBERT AUGUSTUS ('GUS'): Capt./Maj. Co. K. Son of Circuit Judge Edward P. Bailey of Lewisburg, WV. Capt. 142nd Regt., 27th Brig. Va. Militia, 5th Div. 6/6 or 7/61. Enl. 6/6/61 at Charleston, Kanawha Co., WV. Promo. to Maj. 11/23/61. (Records show him as Lt. Col./Col. 14th Va. Cav. 9/5/62.) Maj. in command of Dept. at Lewisburg, WV 3/1/63. MWIA and capt. at Droop Mountain, WV 11/6/63. D. 11/11/63 at Beard house at Hillsboro, WV. Bur. Old Stone Church Cem., Lewisburg, WV. Attorney, Fayette Co., WV 1860.

BAILEY, T.: Pvt. Co. H. Enl. 7/11/63 at Lewisburg, Greenbrier Co., WV. Joined by transfer. Deserted 8/26/63 at White Sulphur Springs, WV. Took oath at Charleston, WV 4/24/65. Age: 38. Farmer. 6'2". Dark complexion, brown hair, grey eyes.

BAILEY, THEODORE F.: Sgt./Jr. 2nd Lt. Co. G. B. 8/16/39 at Baileysville, WV. Enl. 7/3/61 at Charleston, Kanawha Co., WV. Served as a Pvt. until apptd. 2nd Sgt. 8/20/61. Elected 2nd Lt. 5/1/62. Resignd. due to ill health 4/1/64. Another source says transf. to Thurmond's Co. in 1863. Ancestors say after he regained health enlisted in 8th Va. Cav. Postwar: Farmer. Married in 1866. Taught first free school, Baileysville 1866. Wyoming Co. Superintendent of Free Schools 1870-71. Established first Negro school in Wyoming Co., WV 1870. Mercantile business, 1870. Merchant and postmaster 1872. Justice of the Peace and Deputy Sheriff, Baileysville District 1872. Lived in Baileysville, Wyoming Co., WV after war. Ran hotel 1876. Two terms Wyoming Co. Board of Health 1883. Apptd. Deputy Sheriff 1885. Ran Bailey gristmill. Lumbering, early 1900's. Ran Bailey Hotel 1908-20. D. 4/24/1926 of pneumonia. Bur. Baileysville, WV. Brother of Miles Bailey.

BAILEY, WILLIAM HENRY: Pvt. /2nd Lt. Co. A. Enl. 6/6/61 at Camp Tompkins, Kanawha Co., WV. Elected 3rd Sgt. 6/6/61. WIA (leg) in 1863. Prom. to 2nd Lt. 11/24/63. Capt. at Winchester, Va. 9/19/64. POW Ft. Delaware. D. at Ft. Delaware of chronic diarrhea 5/2/65. Bur. Finn's Point Nat. Cem., N.J.

BAKER, CHARLES O.: Pvt. Co. K. Enl. 2/10/63 at Lewisburg, Greenbrier Co., WV. Entitled to bounty being underage when enlisted. Deserted at Staunton, Va. 5/12/64. Twin brother of George Baker. Middle name is Otis.

BAKER, "CUSSY": Pvt. Co. H. PWR.

BAKER, GEORGE: Pvt. Co. K. Enl. 2/10/63 at Lewisburg, Greenbrier Co., WV. Entitled to bounty being underage when enlisted. Deserted at Staunton, Va. 5/12/64. Bahlmann says enl. 1862. Twin brother of Charles Otis Baker. Says middle initial is A.

BAKER, JOHN R. S.: Pvt./Sgt. Co. D. Enl. 5/30/61 at Summersville, Nicholas Co., WV. Left sick at Beverly, WV 4/25/63. Det. service scouting in Craig Co., Va. Present on 10/31/64 roll. POW and took oath at Charleston, WV 5/10/65. Age: 25. 5'4". Dark complexion, black hair, dark whiskers.

BAKER, LEWIS L.: Pvt. Co. D. Res. Summersville, Nicholas Co., WV. Enl. 5/30/61 at Summersville, Nicholas Co., WV. Capt. at Lewisburg, WV 5/23/62. Age: 20. Exchd. 8/25/62. Det. service scouting in Craig Co., Va. 4/12/64. Deserted at Goshen Depot, Va. 10/31/64. 5'11". Dark complexion, black hair, hazel eyes.

BAKER, MICHAEL M.: Pvt. Co. D. Res. Summersville, Nicholas Co., WV. Enl. 8/1/62 at White Sulphur Springs, Greenbrier Co., WV. Left sick at Beverly, WV 4/25/63. Capt. at Strasburg, Va. 10/19/64. Listed as MIA at Cedar Creek, Va. 10/31/64. POW Pt. Lookout, MD. Relsd. and took oath 5/12/65. Farmer.

BAKER, NAPOLEON ('NIP') ALEXANDER: Pvt./Sgt. Co. K. Enl. 4/24/62 at White Sulphur Springs, Greenbrier Co., WV. To 2nd Corp. 12/1/62. Present on 10/31/64 roll. Listed as POW. Took oath at Charleston, WV 5/11/65. Age: 22. 5'9¾". Fair complexion, brown hair, blue eyes. Postwar res. Gallatin, MO 1912. PWR says also served in Edgar's Battn.

BAKER, ROBERT S.: Pvt. Co. D. Enl. 5/30/61 at Summersville, Nicholas Co., WV. Left sick at Lewisburg, WV 4/13/63. Capt. Jan. 1864.

BAKER, WILLIAM: Pvt. Co. A. PWR.

BAKER, WILLIAM A.: Pvt. Co. K. Enl. 4/24/62 at White Sulphur Springs, Greenbrier Co., WV. Left sick at Rock (Red?) Sulphur hosp. 11/1/62. Deserted near Staunton, Va. 5/12/64. Brother of George and Charles Otis Baker.

BAKER, WILLIAM M.: Pvt. Co. F. Enl. 6/17/62 at Camp McCausland, Monroe Co., WV. Assd. wagoner July 1862. Driver of regt. ambulance 12/31/62. Paroled at Appomattox, Va. 4/9/65. Took oath at Charleston, WV 6/12/65. Age: 24. Teamster. 6'2". Dark complexion, dark hair, black eyes, black whiskers.

BALDWIN, DAVID: Pvt. Co. E. Enl. 6/8/61 at Charleston, Kanawha Co., WV. AWOL 6/25/61.

BALDWIN, D. S.: Pvt. Co. H. Enl. 7/23/61 at Pitman's Ferry, ARK in Co. D, 15th Arkansas. Transf. to 22nd Va. 1/23/62. Capt. at Cold Harbor, Va. 6/3/64. POW Elmira, NY 7/12/64. Exchd. 3/10/65.

BALDWIN, JOEL: Pvt. Co. I. Enl. 6/29/61 at Charleston, Kanawha Co., WV. Present on 8/31/61 roll. NFR.

BALDWIN, JOHN: Pvt. Co. I. Enl. 6/29/61 at Charleston, Kanawha Co., WV. Present on 8/31/61 roll. NFR.

BALDWIN, SHEFFY: Co. H. Res. Staunton, Va. PWR.

BALDWIN, THOMAS MACON: Co. B. NFR.

BALES, JOHN: Pvt. Co. D. Enl. 11/9/62 at Red Sulphur Springs, Monroe Co., WV Transf. to Co. F, 36th Va. Inf. 4/6 or 7/3. Records of 36th Va. Inf. say enl. in 22nd Va. Inf. 10/23/62 in Nicholas Co., WV. Age: 21 in 1865. 5'5". Gray eyes and dark hair. Also spelled Bail.

BALES, LEFTRIDGE: Pvt. Co. D. Enl. 5/30/61 at Summersville, Nicholas Co., WV. Capt. Jan. 1864.

BALLARD, B. S.: Sgt. Co. I. Enl. 6/29/61 at Charleston, Kanawha Co., WV. Present on 8/31/64 roll. NFR.

BALLARD, JOHN F.: Pvt. 2nd Lt. Co. I. Res. Boone Co., WV. Enl. 6/29/61 at Charleston, Kanawha Co., WV. Entitlted to $50 bounty. Elected 3rd Lt. 4/11/63. WIA (serious breast) at White Sulphur Springs, WV 8/26-27/63. WIA (face/flesh) and capt. at Winchester, Va. 9/19/64. In Baltimore, MD hosp. 10/26/64. Transf. to Washington, D.C. hosp. 11/2/64. POW Ft. Delaware 12/16/64. Relsd. 6/17/65. 5'9". Light complexion, light hair, blue eyes. In a photo of R. E. Lee UCV Camp #887, Charleston, WV taken Aug. 1911.

BALLARD, WILLIAM A.: Pvt. Co. A. Res. Monroe Co., WV. Enl. 8/8/62 at White Sulphur Springs, Greenbrier Co. WV. Deserted at Packs Ferry, WV 10/17/62. Surrendered to Siber's cavalry. Age: 22. 5'11". Light hair and gray eyes.

BALLINGER, J. B.: Co. F. Deserted on Aug. 1862 broadside. Bolens Bend.

BALLINGER, ELI: Rank Unknown. Co. I. Date and place of enl. not stated. Capt. at Raleigh, WV 6/26/62. Arrived at Gallipolis, OH 9/18/62. Age: 48. 6'1". Light complexion, light hair, blue eyes.

BARBEE, ANDREW RUSSELL: Capt./Lt. Col. Co. A. B. 12/9/27 at Hawsburg, Rappahanock Co., Va. Son of Andrew Russell Barbee. Tanner by trade. Educated at Petersburg, Va. studying medicine under Dr. John J. Thompson of Luray, Va. Univ. of Pennsylvania (medicine) 1848-49. Richmond Medical College 1849-50. Grad. of Univ. of Pennsylvania M.D. Apr. 1851. Located first at Flint Hill, Rappahannock Co., Va. then to Criglersville, Madison Co., Va. 1852. Marr. Margaret A. G. Thompson, daughter of Dr. Thompson May 1852. Moved to Poca Bottom, Putnam Co., WV to plant, farm, and practice medicine and surgery. Opposed secession. Enl. 5/12/61 at Poca, Putnam Co., WV as Capt. WIA (serious injury through elbow joint causing bone to break and severing ulnar nerve) at Scary Creek, WV 7/17/61. Commission burnt on the *Julia Maffitt* 7/24/61. To Lt. Col. 5/1/62. WIA (severe left arm and hit in right hip by gun barrel, being driven into the groin causing paralysis of the great muscles of the hip and thigh and affecting the leg) at White Sulphur Springs, WV 8/26-27/63. Ordered to report to Gen. Kemper 5/16/64, having a thorough knowledge of surgery and medicine. Assgnd. to duty in Confederate Medical Dept. until close of war. In a postwar interview he claimed he succeeded to command of the 22nd Va. following Col. Patton's death 9/19/64. Retired from service Apr. 1865. Paroled at Charleston, WV 5/9/65. Age: 36. 6'10". Florid complexion, dark hair, gray eyes, dark whiskers. Postwar: Doctor. Moved to Pt. Pleasant, WV from Buffalo, WV 1868. Elected WV State Senator 1880-84 although ill with pyaemic blood poisoning caused by an abrasion of his index finger while engaged in a life saving operation on a patient, often rendering Barbee unconscious and ill during the election. Apptd. WV Board of Health 1881. Twice nominated for U.S. Congress but defeated. Apptd. Pres. of U.S. Examining Board for his area 1885. Presiding surgeon U.S. Examining Board. Member of Mason Co., WV Medical Assoc., Galen Co., OH Medical Assoc., WV State Medical Asso., Ohio Valley Medical Asso. D at Pruntytown, WV of heart disease 1903. Bur. Lone Oak Cem., Pt. Pleasant, WV.

BARBER, KEMP: Pvt. Co. A. PWR.

BARGER, PETER A.: Pvt. Co. I. Enl. 5/1/64 at Camp Gauley, Greenbrier Co., WV. Returned from desertion Apr. 1864. Sick at Montgomery White Sulphur Springs, Va., hosp. 10/31/64.

BARGER, WILLIAM H.: Pvt. Date and place of enl. not stated. Paroled at Charleston, WV 6/5/65. Age: 39. 6'2". Dark complexion, dark hair, blue eyes, dark whiskers.

BARKER, DANIEL: Pvt. Co. I. Enl. 11/1/62 at Charleston, Kanawha Co., WV. Present on 3/4/63 roll. NFR.

BARKER, NOAH B.: Pvt./Corp. Co. I. Enl. 6/24/61 at Charleston, Kanawha Co., WV. Entitled to bounty. Sick at Montgomery White Sulphur Springs, Va. July 1862. Sick at Lewisburg, WV 4/30/63. Sick at home in Boone Co., WV 1/30/64. Absent wded. 10/31/64.

BARKER, RUSSELL T.: Pvt./Sgt. Co. I. Enl. 6/29/61 at Charleston, Kanawha Co., WV. Entitled to bounty. Capt. at Winchester, Va. 9/19/64. POW Pt. Lookout, MD 9/24/64. Exchd. 3/15/65. Transf. to Wayside Hosp., Farmville, Va. 3/22/65. Took oath at Charleston, WV 5/1/65. Age: 33. 6'¾". Dark complexion, black hair, hazel eyes.

BARKSDALE, WILLIAM LEIGH: Surgeon. B. 11/11/36 Halifax Co., Va. Attended Samuel Davis Institute. University of Va., medicine. Grad. Jefferson Medical College 1858. Practiced medicine, Lewisburg, WV 1858. Enl. 1861 as Surgeon, W. L. Jackson's Cav. Signed documents as Brig. Surgeon of the 22nd Va. Inf. 1864. Postwar res. Lewisburg, WV. Moved to Va. to practice medicine for 5 years. Later moved to WV. Res. Alderson, WV Jan. 1874, medicine. Res. Hinton, WV 1892. Spent 2 years in lumbering and helped develop Brownstone industry. Still alive in 1908. D. 12/17/1913 at Hinton, WV after a few hours ilness.

BARLEY, ALEXANDER: Pvt. Co. G. Postwar roster. Also spelled Barly.

BARLOW, WILLIAM H.: Pvt. Swann's Co. Enl. 5/22/61 at Charleston, Kanawha Co., WV. Age: 20. Co. transf. to Co. K, 59th Va. Inf. 7/25/61.

BARNES, E. W.: Pvt. Co. B. Res. Roane Co., WV. Enl. 6/?/63 at Lewisburg, Greenbrier Co., WV. Capt. in Roane Co., WV 1/15/64. Age: 30. Arrested by state troops. Sent to Cumberland, MD 4/27/64. Farmer. 5'10". Dark complexion, brown hair, gray eyes.

BARNHART, BENJAMIN: Pvt. Co. F. Enl. 5/1/62 at Charleston, Kanawha Co., WV. Capt. at Cold Harbor (Totopotomoy), Va. 5/31/64. Gave up voluntarily, POW Pt. Lookout, MD. Transf. to Elmira, NY 7/8/64. Sick in Elmira hosp. 5/29/65. D. at Elmira of variola 5/31/65. Bur. Elmira, Woodlawn Cem., grave #2906.

BARR, (REV. J. C.?): Chaplain. Co. H. Res. Charleston, WV. Clergyman. Chaplain to Kanawha Riflemen 1861. Postwar res. Charleston, WV. 1st Presbyterian Church (South).

BARR, JOHN W.: Pvt. Co. B. Enl. 9/22/62 at Charleston, Kanawha Co., WV. Present on 6/10/64 roll. NFR.

BARRETT, ANDREW J.: 1st Lt. Co. I. Enl. 6/21/61 at Boone C.H., Boone Co., WV. Elected Lt. 4/5/64. In hosp. at Lynchburg, Va. 9/30/64. Absent on surg. cert. 10/19/64. Absent sick 10/31/64. Took oath at Charleston, WV and sent north 1/10/65. Age: 20. Clerk. 5'. Dark complexion, dark hair, gray eyes.

BARRETT, SAMUEL: Pvt. Co. K. Enl. 6/6/61 at Charleston, Kanawha Co., WV. Det. as a teamster 8/2/61. On re-enlistment furlo. 12/30/63. Det. brigade wagoner 4/1/64. WIA at Winchester, Va. 9/19/64. Res. Fayette Co., WV 1860. Age: 19.

BARTON, NORMAN P.: Pvt. Co. H. Enl. 5/8/61 at Charleston, Kanawha Co., WV. Transf. to Co. I, 8th Va. Cav. 8/15/61. Entered Richmond, Va. Old Soldiers Home 1/22/20 at age 81. D. 12/22/23 at Richmond, Va. Old Soldiers Home. Registered as Herman P. Barton. Bur. Hollywood Cem., Richmond, Va.

BASHAM, GREEN B.: Pvt. Co. I. Enl. 6/29/61 at Charleston, Kanawha Co., WV. AWOL 7/26/61. Returned from desertion 9/15/62. Deserted near Staunton, Va. 6/28/64.

BASHAM, JAMES H.: Pvt. Co. I. Enl. 6/29/61 at Charleston, Kanawha Co., WV. Present 8/31/61. NFR.

BASS, SIDNEY B.: Co. B. D. 6/16/1918 at age 78. Bur. Maury Cem., Richmond, Va.

BAWYER, DANIEL: Pvt. Co. B. Enl. 5/?/62 at White Sulphur Springs, Greenbrier Co., WV. Deserted 11/16/62.

BAYRS, CHARLES: Pvt. Co. F. Date and place of enl. not stated. Listed as having deserted on a roll dated 7/14/62 at Camp McCausland, WV.

BAYS, JOHN: Pvt. Co. I. Enl. 6/29/61 at Charleston, Kanawha Co., WV. Absent from the co. 7/26/61. NFR.

BEAN, WILLIAM M.: Pvt. Co. F. Enl. 7/4/61 at Charleston, Kanawha Co., WV. Listed as a prisoner charged with presenting a pistol at a sentinel. Had not had a trial 8/14/61. Sick on surg. cert. 8/22/61. Det. to ferry at Greenbrier Bridge, WV 10/28/63. Deserted about 5/11/64. Capt. at Hagerstown, MD 7/13/64. POW Elmira, NY 7/28/64. Killed at Gap Mills, WV 1890.

BEANE, THOMAS N.: Pvt. Co. K. B. circa 1821 Craig Co., Va. Enl. 10/15/63 at Lewisburg, Greenbrier Co., WV. Transf. to Co. F 4/?/64. Deserted about 5/11/64. Postwar res. Craig Co., Va.

BEARD, DAVID M.: Pvt. Co. K. B. circa 1841 Craig Co., Va. Res. Craig Co., Va. Enl. 4/24/62 at White Sulphur Springs, Greenbrier Co., WV. Capt. at Droop Mountain, WV 11/6/63. Age: 22. POW Camp Chase, OH 11/18/62. Transf. to Pt. Lookout, MD 2/29/64. Relsd. 6/19/65. Farmer. 5'10½". Fair complexion, dark hair, brown eyes. Res. Craig Co., Va. 1898 at 58. Bur. Ross Cem., Rt. 42, near Maywood on Fred Keffer farm, Craig Co., Va.

BEARD, JOHN: Pvt. Co. K. B. circa 1849 Craig Co., Va. Enl. 5/27/62 in Greenbrier Co., WV. Entitled to bounty being underage when enlisted. Deserted at Staunton, Va. 5/12/64. Postwar res. Craig Co., Va.

BEARD, JOSIAH C.: 4th Corp. Taylor's Co. Enl. 6/19/61 at Frankford, Greenbrier Co., WV. Age: 20. Co. transf. to Wise Brigade 7/10/61.

BEARD, SAMUEL W.: Co. K. Res. Pocahontas Co., WV. PWR. Pvt. Enl. 1862.

BEARD, WILLIAM: Pvt. Co. K. Enl. 1862.

BECKETT, JAMES DANIEL: Co. G./Co. F. B. 8/28/47 at Peterstown, WV. D. 2/13/1933 at Pickaway, WV.

BECKETT, WILBUR F.: Pvt. Co. E./Co. F. B. 5/30/44 in Monroe Co., WV. Enl. 10/31/63 at Dublin, Pulaski Co., Va. Paroled at Charleston, WV 6/21/65. Age: 20. 5'10". Fair complexion, auburn hair, blue eyes. PWR says Co. F. Res. Mercer Co., WV. Still alive 1894.

BECKETT, WILLIAM W.: Pvt. Co. F. Enl. 4/15/64 at Camp Gauley, Greenbrier Co., WV. Joined by transfer or order 4/1/64. Present 10/31/64. NFR.

BECKLEY, ISAAC C.: Corp. Co. K. Enl. 6/6/61 at Charleston, Kanawha Co., WV. Det. to militia service by order of Gen. Beckley 8/8/61. Bahlmann transf. to Capt. Neville Beckley's Co., 23rd Battn. Va. Inf.

BECKWITH, OSKER J.: Pvt. Co. B. Res. Jackson Co., WV. Enl. 9/22/62 at Charleston, Kanawha Co., WV. On det. service on Nov.-Dec. 1863 roll. Capt. in Jackson Co., WV by Capt. King 1/2/64. POW Camp Chase, OH 2/6/64. Escaped enroute from Camp Chase, OH to Ft. Delaware 1 or 2/22/64. Listed as a prisoner 4/1/64. Returned from desertion 4/?/64. Capt. at Winchester, Va. 9/19/64. POW Pt. Lookout, MD 9/22/64. Exchd. 3/15/65. Age: 21. Farmer. 5'11". Light complexion, light hair, blue eyes.

BEIRNE, ANDREW: Pvt. Co. H. Enl. 3/24/62 at Lewisburg, Greenbrier Co., WV. Capt. at Winchester, Va. 9/19/64. POW Pt. Lookout, MD 9/24/64. Paroled at Charleston, WV 5/10/65. Age: 34. Merchant. 5'10". Fair complexion, dark hair, blue eyes.

BELCHER, LEWIS: Pvt. Co. K. Res. Logan Co., WV. Enl. 6/29/61 at Charleston, Kanawha Co., WV. Capt. in Logan Co., WV. 7/11/62. Arrested by 4th (West) Va. Inf. Charged as a bushwhacker and horse thief. POW Camp Chase, OH 7/31/62. Exchd. 10/13/62. Age: 38. Farmer. 5'10". Dark complexion, dark hair, black eyes.

BELCHER, RICHARD: Pvt. Co. G. Enl. 7/3/61 at Charleston, Kanawha Co., WV. Absent on furlo. 7/24/61. Listed as unfit for duty on surg. cert.

BELCHER, WILLIAM H.: Pvt. Co. I. Enl. 6/29/61 at Charleston, Kanawha Co., WV. AWOL 7/26/61.

BELL, ADDISON: Pvt. Co. D. Enl. 5/30/61 at Summersville, Nicholas Co., WV. Absent with leave 7/20/61. "Supposed to be deranged."

BELL, JOSEPH D.: Pvt. Taylor's Co. Enl. 6/19/61 at Frankford, Greenbrier Co., WV. Age: 38. Co. transf. to Wise Brigade 7/10/61. Postwar res. Craig Co., Va.

BELL, R. FOLLIS: Pvt. Co. C. Enl. 5/1/62 at White Sulphur Springs, Greenbrier Co., WV. Deserted 4/?/63. Listed as deserted on Aug. 1862 broadside, Craig Co. Va. Postwar res. Craig Co., Va. Broadside says first initial is B.

BENNETT, C.: Pvt. Co. H. Res. Christiansburg, Va. Enl. 4/28/62 at White Sulphur Springs, Greenbrier Co., WV. Sick in Greenbrier Co., WV on Nov.-Dec. 1863 roll. Capt. at Cold Harbor (Gaines Mill), Va. 6/5/64. POW Elmira, NY 7/12/64. Resid. and took oath 7/3/65. 5'8". Florid complexion, dark hair, hazel eyes.

BENNETT, CHARLES: Pvt. Co. B. Enl. 6/1/61 at Ripley, Jackson Co., WV. POW on roll dated 4/1/64.

BENNETT, JESSE: Pvt. Co. K. Enl. 4/14/62 at White Sulphur Springs, Greenbrier Co., WV. Deserted 5/26/62, Potts Creek, Va. on Aug. 1862 broadside. Sick in hosp. at Union, WV on Nov.-Dec. 1862 roll. Left sick at Greenbrier River, WV on Mar.-Apr. 1863 roll. Capt. at Winchester, Va. 9/19/64. POW Pt. Lookout, MD 9/23/64. Exchd. 3/15/65. Postwar res. Craig Co., Va.

BENNETT, JOHN: Pvt. Co. B. B. 1845. Res. Jackson Co., WV. Enl. 9/22/62 at Charleston, Kanawha Co., WV. Capt. and WIA at Droop Mountain, WV 11/6/63. Minie ball entered front of thigh, left 4 inches above knee joint, came out behind one inch higher, passing to inside of bone. In hosp. at Grafton, WV for wound 11/20/63. POW Camp Chase, OH 1/13/64. Transf. to Ft. Delaware 3/14/64. D. at Ft. Delaware of measles 4/15/64. Bur.: Ft. Delaware on Jersey shore opposite post. Finn's Point Nat. Cem. N.J. Age: 19. Farmer. 5'11". Fair complexion, light hair, gray eyes.

BENNETT, THOMAS L.: Co. D. Entered Richmond, Va. Old Soldiers Home 3/6/26 at age 80. D. 1/1/1928 Richmond, Va. Old Soldiers Home. Body taken by relatives.

BENNETT, W. A.: Pvt. Co. B. Enl. 5/?/62 at White Sulphur Springs, Greenbrier Co., WV. Deserted on roll dated 11/16/62. Aug. 1862 broadside says assigned 5/2/62 and never reported to co.

BENSON, WILLIAM E.: Pvt. Co. G. Res. Greenbrier Co., WV. Enl. 3/28/64 at Dublin, Pulaski Co., Va. Deserted at Martinsburg, WV in 1864. Capt. at Hagerstown, MD 7/30/64. POW Ft. Delaware. Took oath 1/23/65. 5'11". Dark complexion, brown hair, gray eyes.

BENTZ, ANDREW: Pvt. Co. H. PWR.

BENTZ, HENRY: Pvt. Co. H. Enl. 5/8/61 at Charleston, Kanawha Co., WV. Absent reason unknown on 8/31/61 roll. Bur. Spring Hill Cem., Charleston, WV.

BESS, J. C.: Co. D. On Allegheny Co. pension list.

BESS, JESSE V.: Pvt. Co. A. Enl. 7/24/62 at Camp McCausland, Salt Sulphur Springs, Monroe Co., WV. Sick from 12/31/62 to 2/?/63. Capt. and WIA at Winchester, Va. 9/19/64. Very serious fracture of the right humorous, lower third of left arm. Splint and simple dressing applied. At Chimborazo Hosp. for wound 2/15/65.

BESS, JOHN L.: Pvt. Co. G. B. 1837 Alleghany Co., Va. Enl. 5/1/62 at White Sulphur Springs, Greenbrier, WV. WIA (left arm) at Lewisburg, WV 5/23/62. Listed as AWOL from hosp. on Aug. 1862 broadside. Sick at Montgomery White Sulphur Springs, Va. 11/11/62. Sick in hosp. on 2/20/63 roll. Left sick at Lewisburg, WV 4/3/63. Dischd. 7/8/63 due to disability of left arm due to his wound in the battle of Lewisburg, WV. Age: 25. Farmer. 5'7". Fair complexion, dark hair, blue eyes. Postwar: Schoolteacher. Married 1870. D. 1913 at age 76. PWR says his shoulder wound consisted of a bullet resting against an artery and the surgeon feared a rupture would result, with possible death from bleeding, if removal was attempted. Therefore, the bullet never was removed. His wound ended his infantry service but he continued to serve the military throughout the war by making and checking wagon train manifests. Teacher at Rich Patch Community School, Alleghany Co., Va. 1860. PWR says dischd. at Bungers Mill, WV 1863 then Alleghany Co. Home Guard supply corps. Bur. in family cem. Allegheny Co., Va.

BIAS, RILENT: Pvt. Co. I. Enl. 6/29/61 at Charleston, Kanawha Co., WV. Present 8/31/61. NFR.

BIBBEE, JOSEPH: Pvt. Co. B. Enl. 9/22/62 at Charleston, Kanawha Co., WV. Present 1/10/64. NFR.

BIBBY, ROBERT: 2nd Lt. Swann's Co. Enl. 5/8/61 at Charleston, Kanawha Co., WV. Age: 23. Co. transf. to Co. K, 59th Va. Inf. 7/25/61.

BICKLEY, J. BENTON: Pvt. Co. A. D. 12/22/64. Bur. Camp Chase, OH Cem., grave #675.

BIDDLE, J. M.: Co. F. In Winchester, Va. hosp. for wded. arm 1864.

BINS, JOHN B. C.: Pvt. Co. I. Enl. 6/29/61 at Charleston, Kanawha Co., WV. AWOL 7/26/61.

BLACK, JAMES: Pvt. Co. B. Enl. 1/?/64 at Camp Gauley, Greenbrier Co., WV. Listed as having deserted on 11/30/64 roll.

BLACK, JOHN: Pvt. Co. K. Enl. 9/18/64 at Winchester, Va. Capt. at Fishers Hill, Va. 9/22/64. POW Pt. Lookout 10/1/64. Joined U.S. service 10/12/64.

BLACKWELL, JAMES: Pvt. Co. B. Enl. not stated. Deserted in Apr. 1865. Took oath at Charleston, WV and sent north 4/19/65. Age: 38. Shoemaker. 5'8". Dark complexion, dark hair, hazel eyes.

BLACKWELL, WILLIAM B.: Pvt. Co. B. Enl. 5/?/62 at White Sulphur Springs, Greenbrier Co., WV. Aug. 1862 broadside says assigned 5/2/62 and never reported to co. Listed as having deserted on 11/16/62 roll. Listed as sick at home on 11/30/64 roll.

BLAINE, WILLIAM CHARLES: Pvt. Co. H. B. Mar. 1823 at Charleston, WV. In 1853 sold gold in Calif. for 2 years and had moderate success. Returned to Charleston, WV to care for invalid sister. Asst. Marshall, Charleston, WV 1860. Enl. 5/8/61 at Charleston, Kanawha Co., WV. Present 8/31/61. PWR: Albemarle Co., Va. UDC. Claimed served entire war. Returned to farm near Charleston, WV. Repaired houses. Marr. 1866 to Elizabeth Jane Champe. D. 2/15/1905 at Charleston, WV.

BLAINE, W. J.: Co. C. Enl. 1861.

BLAKE, CHARLES A.: Pvt. Co. K. Enl. 6/6/61 at Charleston, Kanawha Co., WV. MWIA at Scary Creek, WV 7/17/61. D. 7/18/61.

BLAKE, DICKINSON: Pvt. Co. K. Enl. 6/6/61 at Charleston, Kanawha Co., WV. Present 9/4/61. NFR. Day laborer, Fayette Co., WV 1860. Age: 18. Bahlmann says "left."

BLAKE, HAMILTON: Pvt. Co. K. Enl. 6/6/61 at Charleston, Kanawha Co., WV. Absent cause unknown 9/4/61. NFR. Farm laborer, Fayette Co., WV 1860. Age: 16.

BLAKE, MAJOR CLAUDIUS: Pvt./1st Sgt. Co. K. B. 1842. Res. Loup Creek area, Fayette Co., WV. Enl. 6/1/61 at Charleston, V. WIA at Lewisburg, WV 5/23/62. At Montgomery White Sulphur Springs, Va. hosp. 1/?/63 to 3/11/63. Left sick at ewisburg, WV on Mar.-Apr. 1863 roll. Capt. at Droop Mountain, WV 11/6/63. Age: 21. POW Camp Chase, OH 11/18/63. ansf. to Ft. Delaware 3/4/64. Relsd. 6/19/65. Farmer. 5'5". Fresh complexion, sandy hair, gray eyes. D. 1899. Bur. Bud lake Cem., Beckwith, Fayette Co., WV.

BLAKER, G. W.: Pvt. Co. K. Enl. 10/15/63 at Lewisburg, Greenbrier Co., WV. Deserted 12/25/63.

BLAKER, WILLIAM F.: Pvt. Co. G. Enl. 8/10/62 at Union, Monroe Co., WV. Absent on furlo. on 2/28/63 roll. Absent sick on 4/20/63 roll. NFR.

BLANCIT, THOMAS J.: Pvt. Co. K. Enl. 6/6/61 at Charleston, Kanawha Co., WV. On det. service 9/1/62. NFR. Bahlmann says "left in 1861."

BLANKENSHIP, CHARLES W.: Sgt. Co. H. Enl. 7/3/61 at Charleston, Kanawha Co., WV. Left sick at White Sulphur Springs, WV 8/30/61. NFR.

BLANKENSHIP, J. H.: Pvt. Co. E. Enl. 7/1/61 at Sandy, Kanawha Co., WV. WIA (slight) in 1862 Kanawha Valley, WV campaign. Accidentally wded. and on furlo. on 10/31/64 roll. Surrendered at Charleston, WV and took oath there and sent north 3/4/65. Age: 43. Farmer. 5'11". Red complexion, dark hair, blue eyes.

BLANKENSHIP, LEWIS W.: Pvt. Co. G. Enl. 7/3/61 at Charleston, Kanawha Co., WV. Left sick at White Sulphur Springs, WV 8/20/61. Re-enl. 1/28/64. Present on arrest 4/30/64. Deserted 5/4/64.

BOADEN, SAMUEL A.: Pvt. Co. F. Res. Monroe Co., WV. Enl. 7/4/61 at Charleston, WV. WIA (slight arm) at White Sulphur Springs, WV 8/26-27/63. On sick furlo. on rolls from 12/31/63 to 4/30/64. Surrendered at Charleston, WV 3/15/65. Took oath and sent north at Charleston, WV 3/20/65. Age: 24. Farmer: 5'5". Red complexion, light hair, blue eyes. Still alive 1894. Also spelled Bowden.

BOARD, JAMES A.: Pvt. Co. B. B. 1841. Res. Jackson Co., WV. Enrolled 1864 Botetourt Co., Va. Det. duty at White Sulphur Springs, WV hosp. at surrender.

BOBBITT, JAMES H.: Pvt. Co. A. Enl. 8/11/62 at Salt Sulphur Springs, Monroe Co., WV. Sick on 12/31/62 roll. Sick on 2/28/63 roll. Transf. to 19th Va. Cav. and given up as a deserter from that regt.

BOGGESS, JOHN G.: B. 5/3/36 Monroe Co., WV. Sheriff, Fayette Co., WV 1856-58. Deputy Sheriff, Boone Co., WV 1861. WIA (minie ball passing through right lung) at Scary Creek, WV 7/17/61. WIA (pistol shot in neck causing loss of right eye) at Lewisburg, WV 5/23/62. Dischd. due to wd. 1862. Res. Kanawha Co., WV 1862. Elected city councilman, Charleston, WV 1864 and served two years. Re-elected 1871-72. Dentist, Charleston, WV. Bur. Spring Hill Cem., Charleston, WV.

BOGGESS, WILLIAM: Pvt. Co. B. Enl. 9/22/62 at Charleston, Kanawha Co., WV. In Chimborazo Hosp. for rubeola 6/2/64 to 7/18/64. Capt. at Winchester, Va. 9/19/64. POW Pt. Lookout, MD 9/23/64. Exchd. 3/15/65. Took oath at Charleston, WV and sent north 4/22/65. Age: 27. Farmer. 5'3". Florid complexion, brown hair, blue eyes.

BOGGS, JAMES E.: Pvt./Sgt. Co. E. Enl. 6/8/61 at Charleston, Kanawha Co., WV. Served as Pvt. until 7/3/61 then apptd. 3rd Sgt. On det. service on 3/4/63 roll. Det. service at Lewisburg, WV 4/1/64. Det. Provost Guard at Lewisburg, WV 5/1/64. One source says deserted near Lewisburg, WV to Gen. Crook's pickets. Capt. in Kanawha Co., WV by Home Guard 5/27/64. POW Camp Chase, OH. Paroled at Charleston, WV 5/29/65. Age: 36. Farmer. 5'11¾". Brown hair, gray eyes, sandy whiskers.

BOGGS, JOHN R.: Pvt. Co. B. Enl. 12/11/62 at Union, Monroe Co., WV. Shown present on 11/30/64 roll. NFR.

BOGGS, SAMPSON: Pvt. Co. E. Enl. 6/8/61 at Charleston, Kanawha Co., WV. Sick at Montgomery White Sulphur Springs, Va. July 1862. Capt. by Union Home Guards 9/20/62 and paroled same day. Capt. at Droop Mountain, WV 11/6/63.

BOGGES, ENOCH: Pvt. Co. I. Enl. 2/?/63 at Lewisburg, Greenbrier Co., WV. Present 10/31/64. NFR.

BOITNOTT, DAVID WILLIAM: Co. B. Postwar res. Roanoke Co., Va. Entered Richmond, Va. Old Soldiers Home 9/20/29 at age 83. D. 12/20/1929 Richmond, Va. Old Soldiers Home. Bur. Hollywood Cem., Richmond, Va.

BOLES, CHARLES T.: Pvt. Co. C. Enl. 10/26/64 at Richmond, Va. Paroled at Charleston, WV 5/11/65. Age: 36. Light complexion, gray hair, blue eyes, gray whiskers.

BOLEY, D.: Pvt. Co. I. Date and place of enl. not stated. At Giles C.H. (Pearisburg), Giles Co., Va. July 1862. D. at Giles C.H. Va. Claim filed 11/3/63.

BOLEY, JOHN W.: Pvt. Co. K. Enl. 5/1/62 at White Sulphur Springs, Greenbrier Co., WV. Sick in hosp. on Nov-Dec. 1862 roll. Sick in hosp. at Red Sulphur Springs, WV on Jan.-Apr. 1863 roll. Capt. at Winchester, Va. 9/19/64. POW Pt. Lookout, MD 9/23/64. D. at Point Lookout, MD of gastro intestinitis 2/4/65.

BOLEY, WILLIAM AUGUSTUS: Pvt. Co. K. B. 4/13/46. Enl. 2/24/64 at Camp Gauley, Greenbrier Co., WV. Capt. at Winchester, Va. 9/19/64. POW Pt. Lookout, MD 9/23/64. Exchd. 1/17/65. Also served in 54th Va. Postwar: C & O Railroad employee. D. 4/13/1903. Bur. Palestine Presbyterian Church, Alleghany Co., Va.

BOLTON, HENRY: Pvt. Co. B. Enl. 10/?/63 at Lewisburg, Greenbrier Co., WV. Capt. at Winchester, Va. 9/19/64. POW Pt. Lookout, MD 9/24/64. Took oath 1/26/65.

BONES, EDWARD T.: Pvt./Sgt. Co. K. Enl. 4/15/62 at Lewisburg, Greenbrier Co., WV. Capt. at Lewisburg, WV 5/23/62. POW Camp Chase, OH. Exchd. 8/25/62. 2nd Sgt. at Camp Bailey 11/23/63. Capt. at Winchester, Va. 9/19/64. POW Pt. Lookout, MD 9/23/64. Exchd. 3/15/65.

BOOKER, A. J.: Pvt. Co. I. Enl. 5/1/62 at White Sulphur Springs, Greenbrier Co., WV. D. 8/21/63 of typhoid fever in hosp. at Pearisburg, Giles Co., Va.

BOOKER, WILLIAM: Pvt. Co. I. Enl. 5/1/62 at White Sulphur Springs, Greenbrier Co., WV. Present 3/4/63. NFR.

BOOTH, HIRAM A.: Pvt. Co. E. Enl. 5/1/62 at White Sulphur Springs, Greenbrier Co., WV. On wded. sick leave on rolls dated 12/31/63 and 10/31/64. Paroled at Charleston, WV 6/3/65. Age: 21. 5'9". Dark complexion, black hair, black eyes.

BOOTH, JESSE: Pvt. Co. B. Res. Jackson Co., WV. PWR.

BOOTH, JOHN: Pvt. Swann's Co. Date and place of enl. not stated. Appears on a report of a mounted guard at White Sulphur Springs, WV 8/13-14/61. Co. transf. to Co. K, 59th Va. Inf. 7/25/61.

BOOTHE, J. H.: Pvt. Co. E. Enl. 6/8/61 at Charleston, Kanawha Co., WV. Left sick at White Sulphur Springs, WV 8/20/61. NFR.

BOOTHE, JIMMISON EDWARD: Pvt. Co. E. Enl. 7/18/61 at Porter's Shoals, WV. Left sick at White Sulphur Springs, WV 8/20/61. WIA (slight) in 1862 Kanawha Valley, WV campaign. Det. to hosp. at Union, WV on June-Mar. 1863 roll. WIA (severe shoulder) at White Sulphur Springs, WV 8/26-27/63. Deserted 2/?/64. Also called James.

BOOTHE, JOHN T.: Pvt./Corp. Co. E. Enl. 5/1/62 at White Sulphur Springs, Greenbrier Co., WV. On wded. sick leave 12/31/63. Present on Mar.-Apr. 1864 roll. Absent wded. on 10/31/64 roll. Paroled at Charleston, WV 6/30/65. Age: 25. 5'10". Dark complexion, dark hair, dark eyes, dark whiskers.

BOOTHE, STEPHEN: Pvt. Co. E. Enl. 7/18/61 at Porter's Shoals, WV. Left sick at White Sulphur Springs, WV 8/20/61. Paroled at Lewisburg, WV 4/27/65. Age: 24. Farmer, 5'11¾". Fair complexion, brown hair, hazel eyes.

BOSTIC, B. R.: Pvt. Co. B. Enl. 5/?/62 at White Sulphur Springs, Greenbrier Co., WV. Deserted at Union, WV 5/23/62. Listed deserted 5/23/62 near White Sulphur Springs, WV on Aug. 1862 broadside. Joined from desertion 2/28/63. Present 4/30/63. NFR.

BOSTICK, CAPERTON: Pvt. Co. B. Enl. 5/?/62 at White Sulphur Springs, Greenbrier Co., WV. Deserted at Union, WV 5/23/62. Listed as desrted 5/23/62 near White Sulphur Springs, WV on Aug. 1862 broadside. Joined from desertion 2/28/63. Present 4/30/63. NFR.

BOSWELL, MARTIN W.: Pvt. Co. H. Asst. hotel keeper, Charleston, WV 1860. Age: 23. Enl. 5/8/61 at Charleston, Kanawha Co., WV. Acting as company commissary since 8/22/61. NFR.

BOTTIMORE, WILLIAM G.: Pvt. Co. K. Enl. 6/6/61 at Charleston, Kanawha Co., WV. Present on 9/4/61 roll. Transf. to 45th Va. Inf. according to Bahlmann. Day laborer, Fayette Co., WV 1860. Age: 20.

BOUSE, ISAAC W.: Corp. Co. E. Enl. 6/8/61 at Charleston, Kanawha Co., WV. Asst. surg. of regt. Left as a nurse with the sick at Locust Lane, WV 8/28/61. Capt. at Locust Lane, Fayette Co., WV 1861. POW Camp Chase, OH.

BOWDEN, CAPERTON: Pvt. Co. F. Enl. 7/4/61 at Charleston, Kanawha Co., WV. AWOL 8/8/61. D. prior to 1894. PWR says name is Cape Bowden.

BOWDEN, J.: Pvt. Co. F. Date and place of enl. not stated. Det. as a scout for Capt. Thurmond on July 1862 roll.

BOWEN, DUDLEY: Pvt. Co. G. Date and place of enl. not stated. Surrendered at Guyandotte, WV. Mar. 1865. Took oath and sent north 4/29/65. Age: 27. Farmer. 5'6½". Dark complexion, fair hair, brown eyes.

BOWERS, J. J.: Sgt. Co. C. Enl. 5/1/62 at White Sulphur Springs, Greenbrier Co., WV. On scout on 12/31/62 roll. KIA at White Sulphur Springs, WV 8/26/63. PWR says enl. 1861.

BOWERS, WILLIAM L.: Pvt./Sgt. Co. I. Enl. 5/1/62 at White Sulphur Springs, WV, Greenbrier Co., WV. Present 10/31/64. NFR.

BOWLING, THOMAS A.: Pvt. Taylor's Co. Enl. 6/19/61 at Frankford, Greenbrier Co., WV. Age: 32. Co. transf. to Wise Brigade 7/10/61.

BOWYER, BENTON: Pvt. Co. G. Res. Covington, Alleghany Co., Va. Enl. 5/1/62 at White Sulphur Springs, Greenbrier Co., WV. Deserted 10/20/62. Present 12/31/62. Capt. at White Sulphur Springs, WV 8/26/63. Age: 22. POW Camp Chase, OH. Transf. to Rock Island, IL. Relsd. 6/16/65. Farmer. 5'9¾". Dark complexion, auburn hair, dark eyes. PWR says d. of disease at home.

BOWYER, DAVID: Pvt. Co. G. Enl. 5/1/62 at White Sulphur Springs, Greenbrier Co., WV. Present 4/30/63. According to *History of Alleghany Co.* was in Camp Chase, OH for nine months and died of disease. Res. Alleghany Co., Va.

BOWYER, JOHN: Pvt. Co. H. Enl. 7/22/62 in Monroe Co., WV. Det. to shuck corn on 1/1/64 roll. Took oath at Charleston, WV 4/11/65. Age: 21. Farmer. 6'. Fair complexion, fair hair, gray eyes.

BOWYER, LEWIS C.: Pvt. Co. F. Enl. 7/6/62 at Camp McCausland, Monroe Co., WV. Deserted about 5/1/64. Surrendered at Charleston, WV Mar. 1865. Took oath at Charleston, WV 3/11/65. Age: 33. Farmer. 5'11". Red complexion, dark hair, gray eyes.

BOWYER, SAMUEL: Pvt. Co. A. Enl. 4/10/62 at Brushy Ridge, Greenbrier Co., WV. Deserted at White Sulphur Springs, WV 5/1/62. Left sick at White Sulphur Springs, WV 5/3/62.

BOYD, A. MAT: Pvt. Co. F. Enl. 3/6/64 at Camp Gauley, Greenbrier Co., WV. WIA (slight face) at Cold Harbor, Va. 6/4/64. At Chimborazo Hosp. for very serious left temple wound 6/3-28/64. On sick furlo. on 10/31/64 roll. D. prior to 1894.

BOYD, CHARLES: Co. C. Enl. 1861.

BOYD, ROBERT: Pvt. Co. F. Enl. 7/4/61 at Charleston, Kanawha Co., WV. Sick on surg. cert. 8/12/61. WIA (severe thigh) at White Sulphur Springs, WV 8/26-27/63. Later at home due to wound. PWR says d. from White Sulphur Springs wd.

BOYD, WILLIAM: Pvt. Co. F. Enl. 5/1/62 at White Sulphur Springs, Greenbrier Co., WV. Det. to attend to his wounded brother 12/31/63. Wded. and at Charlottesville, Va. hosp. 10/31/64. Postwar res. Monroe Co., WV. Still alive 1894.

BOYER, D.: Co. B. Assigned 5/2/62 and never reported to co. according to Aug. 1862 broadside.

BOYLES, CALEB: Pvt. Taylor's Co. Enl. 6/19/61 at Frankford, Greenbrier Co., WV. Age: 29. Co. transf. to Wise Brigade 7/10/61.

BRACKMAN, D. M.: Pvt. Co. B. Enl. 6/?/63 at Lewisburg, Greenbrier Co., WV. Present 11/30/64. Listed as a rebel deserter surrendering at Charleston, WV Mar. 1865. Took oath and sent north 3/28/65. Age: 19. Farmer: 6'2". Fair complexion, dark hair, gray eyes.

BRACKMAN, JAMES W.: Pvt. Co. B. Enl. 5/?/62 at White Sulphur Springs, Greenbrier Co., WV. Conscript. At home on 11/30/64 roll. Listed as a rebel deserter surrendering at Charleston, WV Mar. 1865. Took oath and sent north 3/28/55. Age: 23. Farmer. 6'2½". Fair complexion, dark hair, hazel eyes. Bur: Spring Hill Cem., Charleston, WV.

BRACKMAN, MICHAEL: Pvt. Co. B. Enl. 6/?/63 at Lewisburg, Greenbrier Co., WV. At home wded. on 11/30/63 roll. WIA (face) at New Market, Va. 5/15/64. Listed as a rebel deserter surrendering at Charleston, WV Mar. 1865. Took oath at Charleston, WV and sent north 3/28/65. Age: 38. Farmer. 5'11". Fair complexion, fair hair, hazel eyes.

BRADFORD, HENRY A.: Sgt. Co. H. Clerk, Charleston, WV 1860. Age: 22. Enl. 5/8/61 at Charleston, Kanawha Co., WV. Served as a Pvt. until elected 5th Sgt. 7/27/61. Accidentally shot by Tom Grant of Co. H on Cotton Hill 11/2/61. D. 11/4/61 at the Warner house, Beckwith, WV. Bur. Dickinson Cem. near Fayetteville, WV.

BRADLEY, CORNELIUS: Pvt./Corp. Co. F. Enl. 7/5/62 at Camp McCausland, Monroe Co., WV. KIA at Cedar Creek, Va. 10/19/64.

BRADLEY, JOHN A.: Pvt. Co. A. Laborer, Charleston, WV 1860. Age: 25. Enl. 5/27/61 at Poca, Putnam Co., WV. Present 8/31/61. NFR.

BRADLEY, LINDSEY: Pvt./Musician. Co. F. B. circa 1835 Craig Co., Va. Cabinetmaker. Enl. 5/1/62 at White Sulphur Springs, Greenbrier Co., WV. Present 10/31/64. Postwar res. Craig Co., Va. Bur. Fairview Cem., Rt. 42 on F. B. Given farm near Level Green Church.

BRAFFORD, GEORGE W.: Pvt./Sgt. Co. K. Res. Fayette Co., WV. Enl. 6/6/61 at Charleston, Kanawha Co., WV. Capt. at Droop Mountain, WV 11/6/63. Age: 31. POW Camp Chase, OH 11/18/63. Transf. to Ft. Delaware 3/14/64. Relsd. 6/19/65. Farmer. 5'8½". Florid complexion, auburn hair, gray eyes. Farmer, Fayette Co., WV 1860.

BRAGG, JACOB: Pvt. Co. E. Res. Dublin, Va. Enl. 6/8/61 at Charleston, Kanawha Co., WV. Left sick at White Sulphur Springs, WV 8/20/61. Present on Sept. 1862 roll. Capt. at Droop Mountain, WV 11/6/63. Age: 21. POW Camp Chase, OH 11/18/63. Transf. to Ft. Delaware 2/19/64. At Chimborazo Hosp. as a furlo. prisoner with chronic diarrhea 3/3/65. Furlo. from Chimborazo Hosp. for ten days 3/12/65. Farmer. 6'. Light complexion, light hair, gray eyes.

BRAGG, JAMES: Co. E. B. 11/14/39. D. at Webster, WV 7/23/1917. Bur. Bragg Cem. near Cowen, WV.

BRAGG, JAMES H.: Co. E. B. in Va. 2/22/47. D. of disease 7/7/64. Bur. Spring Hill Cem., Huntington, WV.

BRAGG, JESSE: Pvt. Co. E. Enl. 6/8/61 at Charleston, Kanawha Co., WV. Left sick at White Sulphur Springs, WV 8/20/61. NFR.

BRAGG, JESSE JR.: Pvt. Co. E. Enl. 6/10/61 at Charleston, Kanawha Co., WV. Left sick at White Sulphur Springs, WV 8/20/61. NFR.

BRAGG, JOSEPH: Pvt. Co. E. Enl. 5/1/62 at White Sulphur Springs, Greenbrier Co., WV. Sick on rolls from 1/2/63 to 4/30/63. Present 12/31/63. Capt. at Winchester, Va. 9/19/64. POW Pt. Lookout, MD 9/23/64. D. at Pt. Lookout, MD of pneumonia 2/4/65. Bur. POW graveyard at Pt. Lookout, MD.

BRAGG, N. G.: Pvt. Co. H. Date and place of enlistment not stated. Capt. at Droop Mountain, WV 11/6/63. POW Ft. Delaware 3/4/64. Exchd. 7/27/65.

BRAGG, WILLIAM: Pvt. Co. A. Enl. 9/15/62 at Charleston, Kanawha Co., WV. Deserted in Raleigh Co., WV 10/16/62. Returned from desertion 4/?/64. Capt. at Winchester, Va. 9/19/64. POW Pt. Lookout, MD 9/27/64. D. at Pt. Lookout, MD of chronic diarrhea. Bur. POW graveyard at Pt. Lookout, MD.

BRANDSTELLER, XAVIER: Musician. Co. H. Enl. 1861. Also spelled Brodstretter. PWR.

BRANSFORD, SILAS W.: Pvt. Taylor's Co. Enl. 6/19/61 at Frankford, Greenbrier Co., WV. Age: 26. Co. transf. to Wise Brigade 7/10/61.

BRANT, WILLIAM M.: Pvt. Co. F. Enl. 5/1/62 at White Sulphur Springs, Greenbrier Co., WV. Sick at Giles C.H., Va. on July 1862 roll. Capt. at Winchester, Va. 9/19/64. POW Pt. Lookout, MD 9/23/64. Exchd. 3/15/65.

BRATTON, TILFORD: Pvt. Co. D. Enl. 7/21/62 at Camp McCausland, Monroe Co., WV. Det. to work in Salt Petre Cave on July 1862 roll. Det. to work in nitre cave 12/31/62 to 2/28/63.

BRAUGHMAN, JOHN L.: Pvt. Co. D. Enl. 5/1/62 at White Sulphur Springs, Greenbrier Co., WV. Was allowed to drive a team for two days 10/13/62 and had not reported back as of 12/31/62 roll. Present 4/30/63. Recvd. pay 9/10/63.

BRAUGHMAN, WILLIAM: Pvt. Co. I. Enl. 5/1/62 at White Sulphur Springs, Greenbrier Co., WV. Sick at Giles C.H., Va. on July 1862 roll. Present 12/31/62. Wded. and in hosp. 8/26/63. (Probably wded. at White Sulphur Springs, WV 8/26/63). Wded. and sick at home in Botetourt Co., Va. on rolls from 4/1/64 to 10/31/64.

BREEDEN, FRANCIS M.: Pvt. Co. A. Farm laborer, Charleston, WV 1860. Age: 23. Res. Poca, Putnam Co., WV. Enl. 5/27/61 at Poca, Putnam Co., WV. Surrendered ("gave up") at Charleston, WV 9/7/63. POW Camp Chase, OH 9/17/63. Transf. to Rock Island, IL 1/22/64. Took oath 5/9/64. Age: 29. Carpenter. 5'7¾". Florid complexion, dark hair, dark eyes.

BREEDEN, HAMILTON V.: Pvt. Co. A. B. Page Co., Va. Res. Kanawha Co., WV. Farm laborer, Charleston, WV 1860. Age: 22. Enl. 5/27/61 at Charleston, Kanawha Co., WV. Capt. at Lewisburg, WV 5/23/62. POW Camp Chase, OH. Exchd. 8/25/62. At home sick on 11/1/62 roll. Arrested at home by Capt. Baggs' Snakehunters 11/7/62. Age: 25. 5'8". Fair complexion, light hair, blue eyes.

BREEDEN, JAMES E.: Pvt. Co. E. Enl. 6/8/61 at Charleston, Kanawha Co., WV. Capt. at Winchester, Va. 9/19/64. POW Pt. Lookout, MD 9/23/64. Exchd. 3/15/65. Took oath at Charleston, WV and sent north 4/14/65. Age: 23. Farmer. 5'7". Fair complexion, fair hair, blue eyes.

BREEDLOVE, LINDSEY: Pvt. Swann's Co. Enl. 5/15/61 at Charleston, Kanawha Co., WV. Age: 21. Co. transf. to Co. K, 59th Va. Inf. 7/25/61. PWR says "arm broken in fight at Cheat Mountain, WV."

BREEDLOVE, ROBERT: Pvt. Swann's Co. B. 9/13/45 Pittsylvania Co., Va. Enl. 6/29/61 at Charleston, Kanawha Co., WV. Age: 21. Co. transf. to Co. K, 59th Va. Inf. 7/25/61. PWR says twice captured. First time he and other POW's captured their own captors the second night. Captured second time 12/9/63. POW until 6/10/65. Postwar farmer, Big Horse Creek, Lincoln Co., WV. Surveyor. School Trustee.

BREW, BENJAMIN: Pvt. Co. E. Enl. 3/24/64 at Dublin, Pulaski Co., Va. On wded. furlo. on 10/31/64 roll.

BRICKEY, JOHN BRADSHAW: Pvt. Co. E. Res. Roanoke Co., Va. Enl. 5/1/62 at White Sulphur Springs, Greenbrier Co., WV. Conscript. Left sick in Maryland in July 1864. Capt. at Poolesville, MD 7/16/64. "Claims to have been a Union man and wished to renew his allegiance and return to his family in Roanoke Co., Va." POW Elmira, NY 7/23/64. In Elmira Hosp. for chronic diarrhea 8/18/65 to 9/12/65. Took oath at Elmira, NY 9/11/65. Transportation provided to Bonsack Station, Va. 5'11". Dark complexion, black hair, gray eyes. Postwar res. Craig and/or Roanoke Co., Va.

BRIDGETT, JACOB H.: Enl. 10/?/64 at New Market, Va. Postwar roster. Co. G. Res. Alleghany Co., Va. Surrendered Apr. 1865.

BRIGHT, SPENCER: Musician. Co. D. Enl. 10/1/62 at Charleston, Kanawha Co., WV. AWOL 2/28/63.

BRILES, ANDREW: Pvt. Co. C. Enl. 10/26/64 at Richmond, Va. Took oath at Charleston, WV 6/19/65. Age: 43. 5'8". Dark complexion, dark hair, gray eyes, dark whiskers.

BRILLHEART, ADAM: Pvt. Co. E. Enl. 3/3/64 at Dublin, Pulaski Co., Va. Conscript. (Two different death dates). D. at Chimborazo Hosp. of disease 8/3/64 or D. of disease in hosp. at Staunton, Va. 9/?/64.

BRINEGER, DAVID: Pvt. Co. G. Enl. 7/3/61 at Charleston, Kanawha Co., WV. Left his co. on the road from the turnpike to Miller's Ferry, WV 7/21/61. Listed as AWOL on 9/31/61 roll.

BRITE, JAMES H.: Pvt. Co. H. B. circa 1831 Craig Co., Va. Enl. 4/28/62 at White Sulphur Springs, WV. Capt. at Winchester, Va. 9/19/64. POW Pt. Lookout, MD 9/24/64. Exchd. 3/15/65. Postwar res. Craig Co., Va. (This is probably the same man as James Henry Britts who applied for pensions in Giles and Bland Co., Virginia and D. 1/6/1910).

BRODT, JOHN TAYLOR: Pvt. Co. H. B. in NY. Schoolteacher, Charleston, WV 1860. Age: 40. Enl. 5/8/61 at Charleston, Kanawha Co., WV. Recvd. rations in Jan. and June 1862. On det. service 4/2-12/62. D. 1890. Bur. Spring Hill Cem., Charleston, WV. Also spelled Brott.

BROGEN, CREED: Co. B. Attended a vet. reunion in Charleston, WV in 1924 and gave address as Crow, Va.

BROOKS, JAMES M.: Pvt. Co. A. Enl. 4/10/62 at Brushy Ridge, Greenbrier Co., WV. Det. service on 11/1/62 roll. Det. service as bridge builder on 12/31/62 roll. Det. service in Engineer Corps at Dublin Depot, Va. and T.R.R. on 4/30/63 roll.

BROOKS, NAPOLEON B.: Pvt./1st Sgt. Co. H. Laborer, Coalsmouth, WV 1860. Age: 23. Enl. 7/10/61 at Coalsmouth (St. Albans), Kanawha Co., WV. Sick at White Sulphur Springs, WV 8/20/61. Present on July-Aug. 1862 roll. Capt. at Winchester, Va. 9/19/64. POW Pt. Lookout, Md. 9/24/64. Relsd. by order of the President and took oath 2/25/65.

BROOKS, W. B.: Pvt. Co. H. PWR.

BROUN, THOMAS LEE: Not in CSR. Pvt. Co. H. B. 12/26/23 at Loudoun Co., Va. Res. of WV 9/15/50. Grad. Univ. of Va. Studied law at Charleston, WV under George W. Summers and Albert Gallatin Jenkins 1850-51. Admitted to Kanawha Co. Bar Jan. 1852. Practiced law in Boone Co., WV and U.S. Supreme Court of Appeals and U.S. Court of Charleston, WV. Law partnership with George S. Patton ("Broun & Patton"). Apptd. attorney for Pres. of Coal River Navigation Co. 1857 succeeding (Gen.) William S. Rosecrans. Democrat. One of the editors of the *Kanawha Valley Star*. Enl. 4/?/61 in Co. H at Charleston, Kanawha Co., WV. Det. service recruiting in Boone Co., WV June-July 1861. Acting Maj. in 3rd Regt., Wise Brigade Aug. 1861. Typhoid pneumonia at Sewell Mountain, WV Oct. 1861 which rendered him unfit for service. Later served as commandant and Maj. Quartermaster at Dublin Depot, Va. Volunteer aide to Col. Jones of the 60th Va. Inf. Seriously wded. at Cloyd's Mountain, Va. in May of 1864. Returned to Charleston, WV June 1865. Marr. Mary M. Fontaine 6/8/66. Moved to New York City, NY 1866. Returned to Charleston, WV and resumed law practice 1870. Member of Camp Patton #1, CV to which he delivered a patriotic address at Charleston, WV on Memorial Day 6/6/88. Res. of Putnam Co., WV 1910 at age 63. Camp Thomas L. Broun, U.S.C.V., #193, Charleston, WV was named after him.

BROWN, A. J.: Pvt. Co. B. Res. Jackson Co., WV. Enl. 3/1/62 at Lewisburg, Greenbrier Co., WV. Substitute. In Farmville, Va. hosp. with rubeola 3/16/65 to 4/3/65. PWR says mustered 1863 for 2 years.

BROWN, ALFRED D.: Pvt. Co. A. PWR.

BROWN, ANDREW D.: Pvt. Swann's Co./Co.H. Enl. 6/30/61 at Charleston, Kanawha Co., WV. Age: 20. Co. transf. to Co. K, 59th Va. Inf. 7/25/61. Transf. to Co. H, 22nd Va. 3/19/62. Returned to Swann's Co., Edgar's Battalion 7/8/62.

BROWN, CHARLES: Corp. Co. K. Bur. in a national cem. somewhere in MO.

BROWN, GEORGE: Pvt. Co. I. Res. Boone Co., WV. Enl. 8/22/61 at Meadow Bluff, Greenbrier Co., WV. Capt. at Droop Mountain, WV 11/6/63. Age: 21. POW Camp Chase, OH 11/18/63. Transf. to Ft. Delaware 2/29/64. Relsd. 6/19/65. Farmer. 6'½". Florid complexion, dark hair, blue eyes.

BROWN, GEORGE L.: Pvt. Co. D. Enl. 5/30/61 at Summersville, Nicholas Co., WV. AWOL 7/28/64.

BROWN, GEORGE N.: Pvt. Co. A. Enl. 5/27/62 at Salt Sulphur Springs, Monroe Co., WV. Dischd. for disability at Lewisburg, WV 3/28/63. Took oath at Charleston, WV 4/10/65. Age: 29. Farmer. 6'2". Fair complexion, dark hair, blue eyes.

BROWN, H.: Pvt. Co. K. Date and place of enl. not stated. Capt. at Winchester, Va. 9/19/64. POW Pt. Lookout, MD 9/23/64.

BROWN, JAMES A.: Co. H, 22nd Va. Inf. B. 4/7/40. D. 5/2/1916. Bur: Blacks Chapel Cem., Camden-On-Gauley, WV. Not in CSR.

BROWN, JAMES M.: Co. G. Enl. 1863 for 2 years. Promo. to Sgt. Cav. Postwar roster.

BROWN, JOHN A.: Pvt. Co. A. Enl. 9/1/62 at Salt Sulphur Springs, Monroe Co., WV. Deserted at Packs Ferry, WV 10/17/62.

BROWN, JOHN C.: Pvt. Taylor's Co. Enl. 6/19/61 at Frankford, Greenbrier Co., WV. Age: 29. Co. transf. to Wise Brigade 7/10/61.

BROWN, JOHN F.: Pvt. Taylor's Co. Enl. 6/19/61 at Frankford, Greenbrier Co., WV. Age: 23. Co. transf. to Wise Brigade 7/10/61.

BROWN, JOHN H.: Pvt. Taylor's Co. Enl. 6/19/61 at Frankford, Greenbrier Co., WV. Age: 19. Co. transf. to Wise Brigade 7/10/61.

BROWN, JOHN STEWARD: Pvt. Co. B. B. Noble Co., OH 5/12 or 16/40. Moved to Jackson Co., WV. Res. Jackson Co., WV. Enl. 6/1/61 at Ripley, Jackson Co., WV. WIA and capt. at Lewisburg, WV 5/23/62. Arrested in Jackson Co., WV by 11th Va. 1/2/63. Age: 22. POW Camp Chase, OH 1/13/63. Paroled. Farmer. 5'11". Fair complexion, light hair, blue eyes, sandy whiskers. Horse trader and wholesaler 1915. Purchased horses for U.S. government.

BROWN, JONATHAN M.: Pvt. Co. F. Res. Mountain Home, Craig Co., Va. Enl. 5/1/62 at White Sulphur Springs, Greenbrier Co., WV. Sick in Lewisburg, WV hosp. on Mar.-Apr. 1863 roll. Capt. at Droop Mountain, WV 11/6/63. POW Camp Chase, OH 11/18/63. Relsd. 5/8/65. Took oath 5/13/65. Age: 27. Shoemaker. 5'7¾". Dark complexion, black hair, gray eyes.

BROWN, JOSEPH ALLINE: Pvt./2nd Lt. Co. H. B. Huntersville, WV 7/12/41. Res. Amelia Co., WV. Apptd. Corp. 2/2/63. Sick in Lewisburg, WV on 4/30/63 roll. Present 6/30/63. WIA at Droop Mountain, WV 11/6/63. Absent 1/1/64. Present 4/30/64. In Charlottesville, Va. hosp. 6/9-16/64. Elected 2nd Lt. 7/?/64. Capt. at Winchester, Va. 9/19/64. POW Ft. Delaware. Relsd. 6/17/65. 5'8". Light complexion, light hair, blue eyes. After 1865 music teacher Baylor Female College, Independence, Kansas. Life insurance agent Austin, TX 1890-1912. Clerk for Congressman, J. P. Buchanan of TX. Attended Veterans Reunion at Culpepper, Va. 1920. Middle name also spelled Alleine. D. 1/27/1929 at Washington, D.C. Bur. Arlington Nat. Cem.

BROWN, JOSEPH M.: Co. H. Listed as Assistant to Q.M. on 4/26/61 orders for Kanawha Riflemen.

BROWN, THOMAS: Pvt. Co. I. Res. Boone Co., WV. Enl. 9/15/62 at Charleston, Kanawha Co., WV. Capt. at Buckhannon, WV 5/10 or 12/63. POW Camp Chase, OH 5/21/63. Transf. to Johnson's Island, OH 6/14/63. Transf. to Pt. Lookout, MD 11/30/63. Took oath at Charleston, WV 4/21/65. Age: 18. Farmer. 5'8¼". Fair complexion, light hair, blue eyes.

BROWN, T. R.: Pvt. Co. C. Date and place of enl. not stated. Capt. at Summersville, WV 10/28/64. Age: 18. Arrested by Capt. Ramsey. POW Camp Chase, OH 12/27/64. Transf. to New Orleans, LA for exchange 5/2/65. In POW hosp. at Vicksburg for chronic diarrhea 5/21-22/65. 5'5". Fair complexion, light hair, gray eyes.

BROWN, WILLIAM M.: Pvt. Co. B. Enl. 10/15/63 at Lewisburg, Greenbrier Co., WV. Det. on Provost Guard in Alleghany Co., Va. 1/10/64. Present on 4/1/64 roll. At home sick on 11/30/64 roll.

BROWN, WILLIAM W.: Pvt./Sgt. Co. A. Enl. 10/12/62 at Salt Sulphur Springs, Monroe Co., WV. Absent sick on rolls dated 12/31/63 and 4/1/64. WIA and capt. at Winchester, Va. 9/19/64. Fracture to left leg caused by shell — lower one-third of thigh amputated. In Pt. Lookout, MD hosp. 1/8/65. Transf. to U.S. Hosp. at Pt. Lookout, MD 1/28/65. Recvd. from Hammond Gen. Hosp. 3/31/65. Relsd. 6/3/65.

BRUFFMAN, WILLIAM: Pvt. Co. I. Date and place of enl. not stated. WIA (serious arm) at White Sulphur Springs, WV 8/26-27/63.

BRUMBAUGH, DAVID: Pvt. Co. B. Enl. 10/?/63 at Lewisburg, Greenbrier Co., WV. Det. in Signal Corps on rolls dated 4/15/64 to 10/30/64. Det. as a gunsmith on 10/31/64 roll. Det. as an armorer at Wytheville, Va. on 11/30/64 roll.

BRYAN, THOMAS A.: 2nd Lt. Co. A. Laborer, Coalsmouth, WV 1860. Age: 24. Enl. 5/22/61 at Poca, Putnam Co., WV. Served as a Pvt. until elected 2nd Lt. 7/12/61. Apptd. 8/19/61. Resgnd. from the regt. 4/1/62. Elected Capt. of Artillery (Bryan's Battery) in Monroe Co., WV 12/?/61 and 3/26/62.

BRYANT, HARRISON: Pvt. Co. D. Enl. 5/30/61 at Summersville, Nicholas Co., WV. KIA at Lewisburg, WV 5/23/62.

BRYANT, THOMAS H.: Pvt. Co. K. Enl. 4/1/64 at Camp Gauley, Greenbrier Co., WV. Deserted near Lexington, Va. 6/25/64. Paroled at Charleston, WV 5/11/65. Age: 19. 6'. Light complexion, dark hair, gray eyes.

BUCKLAND, L.: Co. F. Postwar roster.

BURCH, JOSEPH: Pvt. Company Unknown. Date and place of enl. not stated. On a list of rebel deserters sent from City Point, , Va. dated 4/17/65. Transportation ordered to Baltimore, MD and Philadelphia, PA.

BURCHETT, J.: Pvt. Co. D. Date and place of enl. not stated. Listed as POW and admitted to Farmville, Va. hosp. for debilitas 4/7 or 10/65 to 6/15/65. Paroled from hosp. 1865.

BURDETT, ABNER G.: Pvt. Co. A. Enl. 4/10/62 at Brushy Ridge, Greenbrier Co. WV. Paroled at Charleston, WV 5/10/65. Age: 21. 5'8". Light complexion, light hair, gray eyes.

BURDETT, ANDREW H.: Pvt. Co. F. Enl. 4/4/64 at Camp Gauley, Greenbrier Co., WV. Deserted about 5/8/64.

BURDETT, GEORGE W.: Pvt. Co. E/Co. A. Enl. 7/4/61 at Charleston, Kanawha Co., WV. On mounted guard at White Sulphur Springs, WV 8/13-14/61. Capt. at Lewisburg, WV 5/23/62. POW Camp Chase, OH 5/30/62. Exchd. 8/25/62. Paroled at Lewisburg, WV 4/25/65. Age: 26. Farmer. 6'1". Fair complexion, brown hair, hazel eyes. Postwar res. Fayette Co., WV. Still alive 1894.

BURDETT, JAMES M.: Pvt. Co. F. Enl. 8/6/61 at White Sulphur Springs, Greenbrier Co., WV. D. Mercer Co., WV sometime prior to 9/11/63. PWR says died at Princeton, WV 1861. Also says middle initial is L.

BURDETT, JOHN J.: Pvt. Co. A. Enl. 4/1/64 at Monroe Draft, Greenbrier Co., WV. Paroled at Lewisburg, WV 4/25/65. Age: 19. Farmer. 6'. Brown hair and blue eyes.

BURDETTE, LOUIS: Pvt. Co. A. PWR.

BURDETTE, ROBERT A.: Pvt. Co. A. Enl. 4/10/62 at Brushy Ridge, Greenbrier Co., WV. Capt. at Winchester, Va. 9/19/64. POW Pt. Lookout, MD 9/23/64. Exchd. 3/15/65.

BURDETTE, WILLIAM: Pvt. Co. A. Enl. 4/10/62 at Brushy Ridge, Greenbrier Co., WV. Transf. to Herndon's Battalion 9/25/62.

BURFORD, ROBERT K.: Pvt. Co. A. Enl. 9/15/62 at Charleston, Kanawha Co., WV. Deserted at Shady Spring, Raleigh Co., WV 11/2/62.

BURFORD, STEPHEN N.: Pvt. Co. A. B. Monroe Co., WV. Farmer, Tyler Mountain, WV 1860. Age: 24. Res. Kanawha Co., WV. Enl. 5/?/61 at Charleston, Kanawha Co., WV. Sick at White Sulphur Springs, WV 8/20/61. Present on 11/1/62 roll. Absent wd-ed. (leg) on 12/31/63 roll. Present on 4/30/64 roll. Deserted at Beverly, WV about 8/28/64. Took oath and relsd. 8/31/64. Age: 29. Farmer. 5'11". Fair complexion, light hair, gray eyes.

BURGER, WILLIAM G.: Pvt. Co. E. Enl. 5/1/62 at White Sulphur Springs, Greenbrier Co., WV. Capt. at Winchester, Va. 9/19/64. POW Pt. Lookout, MD 9/23/64. Exchd. 3/15/65.

BURGESS, HENRY C.: Pvt. Co. K. Enl. 6/6/61 at Charleston, Kanawha Co., WV. Absent cause unknown on 9/4/61 roll. Pres-ent on 11/1/62 roll. Deserted while on furlo. 2/1/64. Bahmann says joined U.S. service. Res. and farmer, Fayette Co., WV 1860. Age: 27. Middle name is Clay.

BURGESS, WASHINGTON: Pvt. Co. K. Enl. 6/6/61 at Charleston, Kanawha Co., WV. Sick on 9/1/61 roll. Present on 11/1/61 roll. WIA at Winchester, Va. 8/19/64. Bahlmann says MIA and believed MWIA in Valley of Va. 1864. Farmer, Fayette Co., WV 1860. Age: 28.

BURKE, W. H.: Co. G. Postwar roster. On Alleghany Co. pension list.

BURKETT, ALBERT: Pvt. Co. A. PWR.

BURKHOLDER, JAMES F.: Pvt. Co. I. Enl. 5/1/62 at White Sulphur Springs, Greenbrier Co., WV. Det. as regt. wagoner on 4/30/63 roll. Present on 1/31/64 roll. Teamster.

BURKS, M. M.: Pvt. Co. G. Enl. 7/7/61 at Charleston, Kanawha Co., WV. Transf. from Capt. Sarver's Co. Left sick at White Sulphur Springs, WV 8/20/61. NFR.

BURNS, D. H. M.: Pvt. Company unknown. B. 7/6/41. Date and place of enl. not stated. Appears on a list of prisoners surrend-ered by Gen. R. E. Lee at Lynchburg, Va. 4/13/65 and paroled. D. 7/25/1908. Bur. Mt. Pleasant Methodist Church, Alleghany Co., Va.

BURNS, HARVEY: Pvt. Co. A. Enl. 5/25/61 at Camp Tompkins, Kanawha Co., WV. Deserted 7/17/61.

BURNSIDE, ELIJAH: Pvt. Co. I. Res. Putnam Co., WV. Enl. 8/22/61 at Meadow Bluff, Greenbrier Co., WV. Capt. at Trace Fork of Mud River, Putnam Co., WV 2/26/62. POW Camp Chase, OH. Exchd. 8/25/62. Considered a "very bad man" when cap-tured. Rejoined regt. Capt. at Winchester, Va. 9/19/64. POW Pt. Lookout, MD 9/24/64. Exchd. 3/15/65. Farmer.

BURNSIDES, GEORGE: Pvt./Quartermaster. Co. I. Enl. 6/22/61 at Charleston, Kanawha Co., WV. Deserted 9/18/62.

BURWELL, ANDREW J.: Pvt. Co. F. Enl. 7/4/61 at Charleston, Kanawha Co., WV. AWOL 8/21/61. Present on 10/31/62 roll. Corp. 12/15/63. Capt. at Winchester, Va. 9/19/64. Took oath at Charleston, WV 4/4/65. Age: 21. Farmer. 5'6". Fair complex-ion, fair hair, blue eyes. D. 1894.

BUSH, ADAM: Pvt. Co. B. Res. Alleghany Co., Va. Enl. 2/?/63 at Lewisburg, Greenbrier Co., WV. Substitute for J. J. Stull. Deserted 12/?/63. Took oath and sent north. Absent sick on 1/10/64 roll. Age: 53. Farmer. 5'9". Light complexion, gray hair, blue eyes.

BUSH, ALCHANEY A.: Pvt. Co. G. Enl. 5/1/62 at White Sulphur Spring, Greenbrier Co., WV. Det. to make shoes for brigade on 2/28/63 roll. Working in govt. shoe shop on 12/31/63 roll. Capt. at Winchester, Va. 9/19/64. POW Pt. Lookout, MD 9/23/64. Exchd. 3/15/65. Res. Alleghany Co., Va. PWR says middle initial is W and may be same man as A. W. Bush.

BUSH, A. W.: Co. A. Res. Oklahoma 1911.

BUSH, JAMES S.: Pvt. Co. G. Res. Alleghany Co., Va. Enl. 9/20.62 at Charleston, Kanawha Co., WV. Deserted 10/29/62. Capt. in Kanawha Co., WV by Home Guards 5/28/63. Age: 30. POW Camp Chase, OH 6/15/63. Transf. to Ft. Delaware 7/14/63. Farmer. 5'9¾". Dark complexion, light hair, blue eyes.

BUSHONG, DANIEL H.: Pvt. Co. B. Enl. 5/?/62 at White Sulphur Springs, Greenbrier Co., WV. KIA at New Market, Va. 5/15/64.

BUTCHER, JAMES A.: Pvt./Sgt. Co. E. Res. Clay Co., WV. Enl. 6/8/61 at Charleston, Kanawha Co., WV. Capt. in Clay Co., WV 2/15/64. POW Camp Chase, OH 2/24/64. Transf. to Ft. Delaware 3/14/64. Took oath 6/7/65. Age: 23. Farmer. 5'8". Dark complexion, black hair, black eyes. Middle name probably Addison.

BUTCHER, NATHAN D.: Pvt. Co. A. Enl. 5/13/61 at Camp Tompkins, Kanawha Co., WV. Left sick at Marlin's Bottom (Marlinton), WV on 4/30/63 roll. Present on 12/31/63 roll. Teamster. Paroled at Charleston, WV 5/10/65. Age: 35. 5'10". Dark complexion, dark hair, dark eyes, dark whiskers.

BUTLER, S. A.: Pvt. Company unknown. Date and place of enl. not stated. Listed as a deserter sent to Washington, D.C. by Army of the Potomac dated 3/22/65. Took oath 3/29/65. Transportation provided to Philadelphia, Pa.

BUTTERWORTH, W. J.: Pvt. Co. E. Enl. 6/8/61 at Charleston, Kanawha Co., WV. Left in Kanawha Co., WV 7/24/61. NFR.

BUTTLER, W. F.: Pvt. Co. E. Enl. 6/8/61 at Charleston, Kanawha Co., WV. Left sick at White Sulphur Springs, WV 8/20/61. NFR.

BYER, WILLIAM E.: Pvt. Co. B. Res. Jackson Co., WV. Enl. 5/?/62 at White Sulphur Springs, Greenbrier Co., WV. Joined from desertion at Camp McCausland, WV 7/12/62. Also spelled Byers.

CABELL, HENRY CLAY: Pvt. Co. H. Enl. 6/5/61 at Charleston, Kanawha Co., WV. Had two week leave of absence and had not returned on 8/31/61 roll. Present on July-Aug. 1862 roll. Transf. to 8th Va. Cav. 7/11/63. Served in Co. G, 8th Va. Cav. until POW paroled at Lewisburg, WV 4/27/65. Age: 28.

CABELL, ROBERT: Pvt. Co. H. Enl. 5/8/61 at Charleston, Kanawha Co., WV. Absent cause unknown on 8/31/61 roll. NFR. PWR says "didn't go."

CALDWELL, ALEXANDER L.: Pvt. Co. C. Enl. 5/1/62 at White Sulphur Springs, Greenbrier Co., WV. Under arrest for desertion on 3/4/63 roll. Deserted at Red Sulphur Springs, WV 9/1/63. Absent sick on 4/18/64 roll. Deserted 6/10/64. Postwar res. Craig Co., Va. Bur. Caldwell Cem., Hutton farm, Rt. 624, Craig Co., Va.

CALDWELL, ARCHIBALD McHENRY: 1st Corp. Co. C. B. 10/14/43. Enl. 5/1/62 at White Sulphur Springs, Greenbrier Co., WV. Present on 10/31/64 roll. UDC report says enl. Mar. 1862 at Camp Buckner, WV. Soon afterward elected 1st Corp., short time later made Sgt. and held that position until close of war. Was in 13 battles. WIA in the Valley of Virginia. Capt. at Winchester, Va. 9/19/64. Dischd. from service at Montgomery Co., Va. at close of war. Claimed he served 4 years with 22nd Va. Inf. and was a member of Pickett's (?) Div. Postwar res. Simmondsville, Va. Res. Craig Co., Va. 1898 at age 54. D. 2/10/1927 at age 83 years, 3 months, 26 days. Bur. Ross Cem., near Maywood, on Fred Keffer farm, Craig Co., Va.

CALDWELL, ARTER PRESTON: Pvt. Co. C. B. 1/18/44 Craig Co., Va. Enl. 5/1/62 at White Sulphur Springs, Greenbrier Co., WV. Sick at White Sulphur Springs, WV hosp. on July 1862 roll. In Chimborazo Hosp. for very serious left cheek 6/5/64. Returned to duty 6/11/64. Res. Craig Co., Va. 1898 at age 54. D. 10/7/1928. Bur. Caldwell Cem., Rt. 42, near Gravel Hill Church Rd., Craig Co., Va.

CALDWELL, CEPHAS: Listed in UDC booklet *Craig's Share In The War Between The States*. Res. Craig Co., Va.

CALDWELL, ENOS WASHINGTON B.: Pvt. Co. C. B. 2/22 or 26/46. Enl. 3/1/64 at Camp Gauley, Greenbrier Co., WV. In Farmville, Va. hosp. for rubeola 3/16-27/65. Res. Craig Co., Va. 1898 at age 51. D. 10/27/1921 at Craig Hollow Springs, Va. Interment at Level Green, Va. Bur. Caldwell Cem., Rt. 42, near Gravel Hill Church Rd., Craig Co., Va. Middle name is either Brooks or Banks.

CALDWELL, F. B.: Pvt. Co. C. Enl. 10/?/64 for ½ years. Not in CSR. Res. Craig Co., Va. 1898 at age 49.

CALDWELL, FLOYD: Listed in UDC booklet *Craig's Share In The War Between The States*. Res. Craig Co., Va. (This may be F. B. Caldwell).

CALDWELL, GEORGE: Pvt. Co. C. B. ca. 1832 Craig Co., Va. Enl. 5/1/62 at White Sulphur Springs, Greenbrier Co., WV. Capt. at Winchester, Va. 9/19/64. POW Pt. Lookout, MD 9/24/64. Exchd. 3/15/65. Res. Craig Co., Va. 1898 at age 65. Letter written by George Caldwell at Salt Sulphur Springs, WV 6/6/62 appears in *Hard Times Vol. I.*

CALDWELL, GEORGE C.: Sgt. Co. C. B. ca. 1833 Craig Co., Va. Enl. 5/1/62(?) at White Sulphur Springs, Greenbrier Co., WV. Deserted at Camp McCausland, WV 7/10/62. Det. arresting deserters 1/10-16/63.

CALDWELL, HAMILTOIN B.: Pvt. Co. K. B. ca. 1839 Craig Co., Va. Enl. 4/24/62(?) at White Sulphur Springs, Greenbrier Co., WV. Sick at Beverly, WV on Mar.-Apr. 1863 roll. KIA at Droop Mountain, WV 11/6/63. Was married to Adeline Elmore who later married William "Billy" Smith.

CALDWELL, JACKSON: Listed in UDC booklet *Craig's Share In The War Between The States*. Res. Craig Co., Va.

CALDWELL, JACOB S.: Pvt. Co. C. B. 7/23/30 Craig Co., Va. Enl. 6/22/62(?) at Camp McCausland, Monroe Co., WV. D. at Lewisburg, WV of typhoid fever 3/9 or 10/63. Latter date appears on grave. William Reynolds, Sr. administrator. Estate settled by Sarah Jane, widow, 3/23/63 Craig Co. Court Order Book. Bur. Ross Cem., Rt. 42, near Maywood, on Fred Keffer farm, Craig Co., Va.

CALDWELL, LLOYD: Pvt. Co. C. B. ca. 1826 Craig Co., Va. Enl. 10/26/64 at Richmond, Va. PWR says enl. Apr. 1862 for 3 years and also served in 30th Battn. Va. Sharpshooters. Also with Craig Co. police, 2nd District 12/22/62(?). Res. Craig Co., Va. 1898 at age 72.

CALDWELL, MADISON: Pvt. Co. K. B. ca. 1842 Craig Co., Va. Enl. 4/24/62 at White Sulphur Springs, Greenbrier Co., WV. Deserted at Staunton, Va. 5/12/64.

CALDWELL, N. T.: Rank unknown. Co. H. Appears as deserted 5/28/62 Craig Co., Va. on 8/20/62 broadside.

CALDWELL, ROBSON: Pvt. Co. C. B. ca. 1820 Craig Co., Va. Enl. 10/11/64 near New Market, Va. A PWR says enl. Apr. 1863 for 2 years. Res. Craig Co., Va. 1898 at age 78. (First name also listed as Robinson and Robison).

CALDWELL, WILLIAM A.: Rank unknown. Co. H. Enl. 9/13/62 at Kanawha Co., WV. Capt. at Winchester, Va. 9/19/64. POW Pt. Lookout, MD. Took oath at Pt. Lookout, MD and relsd. 2/25/65. Postwar res. Craig Co., Va.

CALDWELL, WILLIAM H.: Pvt. Co. H. Enl. 6/8/61 at Coalsmouth (St. Albans), Kanawha Co., WV. Absent cause unknown on 8/26/61 roll. Present on Sept.-Oct. 1862 roll. Capt. at Winchester, Va. 9/19/64. Relsd. by order of the President 2/22/65. Postwar res. Craig Co., Va. (This could be the same man as William A. Caldwell as record is very similiar).

CALE, JAMES T.: Pvt. Co. C. B. ca. 1845 Craig Co., Va. Enl. 2/20/63 at Lewisburg, Greenbrier Co., WV. Present on 10/31/64 roll. Postwar res. Craig Co., Va.

CALLAGHAN, D.: Pvt. Co. D. Enl. 7/21/62 at Camp McCausland, Monroe Co., WV. NFR.

CALWELL, H. C.: Pvt. Co. E. Enl. 3/19/64 at Camp Gauley, Greenbrier Co., WV. Det. to Signal Corps 4/1/64 to 10/31/64. Postwar res. Craig Co., Va.

CAMBELL, VALENTINE: Pvt. Co. A. Enl. 8/7/62 at (Salt?) Sulphur, WV. Deserted at Charleston, WV 9/20/62.

CAMPBELL, JAMES: Rank unknown. Co. F. Res. Monroe Co., WV pre-war and postwar. D. sometime prior to 1914.

CAMPBELL, JOHN: 1st Sgt./Capt. Co. F. Enl. 7/4/61 at Charleston, Kanawha Co., WV. Elected 1st Lt. 8/22/61. Promo. to Capt. 2/4/63. WIA at Cedar Creek, Va. 10/19/64. Gunshot wd. of left shoulder and left hip. In Lynchburg, Va. hosp. 10/21/64. In Charlottesville, Va. hosp. 11/30/64. Transf. to Petersburg (Pearisburg?), Giles Co., Va. 11/26/64. Paroled at Charleston, WV 6/12/65. Age: 32. 5'5". Florid complexion, brown hair, blue eyes. Postwar res. Bloomington, Illinois.

CAMPBELL, JOHN F.: 1st Sgt./2nd Lt. Co. D. Enl. 5/30/61 at Summersville, Nicholas Co., WV. Elected 2nd Lt. 5/1/62. Capt. at Lewisburg, WV 5/23/62. POW Camp Chase, OH. Transf. to Johnson's Island, OH 7/3/62. Exchd. 9/1/62. Capt. at Winchester, Va. 9/19/64. POW Ft. Delaware. Relsd. 6/17/65. 5'9". Dark complexion, dark hair, dark eyes.

CAMPBELL, JOHN H.: Pvt. Co. F. Res. Monroe Co., WV. Enl. 8/6/61 at White Sulphur Springs, Greenbrier Co., WV. Capt. at Lewisburg, WV 5/21 or 23/62. Age: 23. POW Camp Chase, OH. Exchd. 8/25/62. At home by reason of having a substitute on Mar.-Apr. 1863 roll. 5'7½". Light complexion, sandy/dark hair, blue eyes, sandy whiskers. Postwar res. McL_____? Co., Missouri. Still alive 1914.

CAMPBELL, L. M.: Pvt. Co. H. B. Amherst Co., Va. Enl. 7/18/62 at Camp McCausland, Monroe Co., WV. Transf. from Jackson's Cav. Co. died at Lewisburg, WV of pneumonia 3/16/63. Initials may be reversed and should possibly read M. L. Campbell.

CAMPBELL, NATHAN R.: Pvt. Co. D. Enl. 5/30/61 at Summersville, Nicholas Co., WV. Left sick at White Sulphur Springs, WV 8/20/61. Bur. Newbern, Va.

CAMPBELL, PALEY L.: Pvt. Co. D. Enl. 5/1/62 at White Sulphur Springs, Greenbrier Co., WV. Sick in Greenbrier Co., WV on 5/31/62 roll. In Chimborazo Hosp. for acute diarrhea 6/8/64. Capt. at Winchester, Va. 9/19/64. POW Camp Chase, OH 9/27/64. Exchd. 3/15/65. Paroled at Charleston, WV 4/24/65. Age 28. 5'10". Fair complexion, dark hair, light eyes, dark whiskers.

CAMPBELL, ROBERT S.: 2nd Lt./1st Lt. Co. D. Enl. 5/30/61 at Summersville, Nicholas Co., WV. Left sick at White Sulphur Springs, WV 8/20/61. Elected 1st Lt. 5/1/62. In hosp. at Montgomery White Sulphur Springs, Va. on July 1862 roll. WIA at Fayetteville, WV 9/10/62. Absent due to wd. 10/31/62 to 12/31/62 rolls. Present on 2/28/63 roll. KIA and capt. at Winchester, Va. 9/19/64.

CAMPBELL, WILLIAM: Pvt./Hosp. Steward. Co. F. Res. Monroe Co., WV. Enl. 7/4/61 at Charleston, Kanawha Co., WV. On special duty in Commissary Dept. 8/14/61. Regt. steward. Letter dated 11/9/64 request leave of absence for William Campbell, M.D., hosp. steward of the 22nd Va., to accept adj. surg. of the local and reserve forces of West and Southwest Va. Letter written by W. Leigh Barksdale, 22nd Va. Inf. Brig. Surg. On furlo. on 10/31/64 roll. Postwar res. Ronceverte, WV. D. sometime prior to 1914.

CAMPBELL, WILLIAM C.: Pvt. Co. F. Date and place of enlistment not stated. Rcvd. pay for period 4/10/64 to 6/8/64.

CAMPBELL, WILLIAM H.: Pvt./Sgt. Co. E. Enl. 7/25/61 at Porter's Shoals. Relieved of guard duty 8/14/61 by A. H. Collision at request of Capt. Swann, he being sick. Left sick at White Sulphur Springs, WV 8/20/61. Present on Sept.-Oct. 1862 roll. Absent sick but since returned on 10/31/64 roll.

CAMPS, JAMES T.: Pvt. Co. I. Enl. not stated. Left sick at White Sulphur Springs on July 1862 roll. NFR.

CANTERBURY, CLAYTON: Co. I.

CANTERBURY, SAMUEL: Pvt. Co. I. Enl. 6/29/61 at Charleston, Kanawha Co. WV. Present on 8/31/61 roll. Entered Richmond, Va. Old Soldiers Home 8/18/1911 at age 82. Dischd. 6/26/1913. Registered as having served in Co. K.

CANTERBURY, WILLIAM: Co. I.

CANTLEY, AUGUSTUS: Pvt. Co. I. Enl. 6/29/61 at Charleston, Kanawha Co., WV. Capt. in Fayette Co., WV 9/13/61. POW Camp Chase, OH. Exchd. 8/25/62.

CANTLEY, W. E.: Pvt. Co. H. Enl. 11/8/61 at Cotton Hill, Fayette Co., WV. Sick in hosp. at Giles C.H. (Pearisburg), Va. on rolls July-Aug. 1862 to 12/31/62. Present on 2/28/63 roll. Paroled at Charleston, WV 6/5/65. Age: 19. 5'9". Fair complexion, light hair, blue eyes.

CARINGLER, JOHN: Exchd. from Capt. Amick's Co. 3/10/64 for William H. Halstead.

CARNEAL, WILTON POLK: Co. G. Enl. 1862 for 3 years. Wounded. PWR.

CARNIFEX, WILLIAM ALEXANDER: Pvt. Co. D. Enl. 3/17/64 at Camp Gauley Greenbrier Co., WV. Dischd. by order of Maj. Gen. Breckinridge 4/18/64. PWR says "Defected and came home" after his father, Benjamin Franklin Carnifex of 60th Va. Inf. was KIA at Cedar Creek, Va. 10/19/64. Res. Fayette Co., WV 1860. Age: 14.

CARPER, ALLEN M.: Pvt. Co. K. B. 6/9/23. Enl. 2/1/63 in Monroe Co., WV. Entitled to a bounty being underage when volunteered. Deserted at Staunton, Va. 5/12/64. Res. Craig Co., Va. D. 5/25/1902. Bur. Carper Cem. at Camp Mitchell Rd. on hill above Rt. 615, Craig Co., Va. Bahlmann says enl. 1862.

CARPER, A. M.: Pvt. Co. C. Enl. 4/?/61 for 4 years. Res. Craig Co., Va. 1898 at age 74. PWR.

CARPER, CHARLES M.: Pvt. Co. K. B. 2/2/49. Enl. 4/24/62 at White Sulphur Springs, Greenbrier Co., WV. Deserted 5/1/62 on Aug. 1862 broadside, John's Creek, Craig Co., Va. Deserted at Staunton, Va. 5/12/64. Res. Craig Co., Va. D. 5/14/1913. Bur. Carper Cem., at Camp Mitchell Rd. on hill abo e Rt. 615, Craig Co., Va.

CARPER, D. C.: Pvt., Co. K. Enl. 1862.

CARPER, FLEMING BOWYER: Listed in UDC booklet *Craig's Share In the War Between the States.* D. 9/23/61. Bur. Carper-McClung Cem., Craig City, near Va. Highways shop, Craig Co., Va.

CARPER, JAMES ALLEN: Listed in UDC booklet *Craig's Share In the War Between the States.* Res. Craig Co., Va.

CARR, CLAUDIUS: Pvt. Taylor's Co. Enl. 6/19/61 at Frankford, Greenbrier Co., WV. Age: 19. Co. transf. to Wise Brigade 7/10/61.

CARR, FLOYD G.: Pvt. Co. C. B. ca. 1845 Craig Co., Va. Enl. 5/1/62 at White Sulphur Springs, Greenbrier Co., WV. In hosp. at Union, WV on 12/31/62 roll. Continued to report sick up to and including 10/31/64 roll. D. 3/31/1928 at age 82. Bur. Carr Cem., Rt. 658, John's Creek, Robert Dunbar farm, Craig Co., Va.

CARR, JAMES S.: Corp. Taylor's Co. Enl. 6/19/61 at Frankford, Greenbrier Co., WV. Age: 23. Co. transf. to Wise Brigade 7/10/61.

CARR, JOHN GAY: Pvt./3rd Lt./2nd Lt. Co. H. B. Albemarle Co., Va. Farmer, Charleston, WV 1860, age 27. Enl. 5/8/61 at Charleston, Kanawha Co., WV. Served as Pvt., until elected 5th Sgt. 6/6/61. Elected 2nd Lt. 5/1/62. Comm. 2nd Lt. 7/27/61 (?). Apptd. Provost Marshall on July-Aug. 1862 roll. KIA (bullet through brain) at White Sulphur Springs, WV 8/26-27/63. Age: 31. Bur. Old Stone Church Cem., Lewisburg, WV.

CARR, JOHN O.: Pvt. Co. H. Enl. 1861. "Didn't go."

CARR, JOHN W.: Pvt. Co. I. B. ca. 1838 Craig Co., Va. Enl. 4/9/64 at Dublin, Pulaski Co., Va. Deserted at Rockfish Gap, Va. 6/16/64.

CARR, SAMUEL BILL: Pvt. Co. C. B. ca. 1840 Craig Co., Va. Enl. 5/1/62 at White Sulphur Springs, Greenbrier Co., WV. Dischd. at Camp Royal Oak 8/29/62. Bur. Cedar Hill Cem., Covington, Va.

CARRON, GEORGE: Pvt. Taylor's Co. Enl. 6/19/61 at Frankford, Greenbrier Co., WV. Age: 26. Co. transf. to Wise Brigade 7/10/61.

CARSON, JOHN A.: Pvt. Co. G. Res. Alleghany Co., Va. Enl. 5/1/62 at White Sulphur Springs, Greenbrier Co., WV. Sick at Covington, Va. on July 1862 roll. AWOL from hosp. on Aug. 1862 broadside. Capt. in Nicholas Co., WV 2/25/63. Bushwhacker. Relsd. 3/27/63. Capt. at Fayetteville, WV 3/?/63. Deserter. POW Camp Chase, OH 3/24/63. Age 32. Farmer. 5'11". Dark complexion, dark hair, blue eyes.

CARSON, MARTIN V. B.: Pvt. Co. K. B. Monroe Co., WV. Enl. 4/5/62 at Lewisburg, WV. Capt. at Lewisburg, WV 5/23/62. Age: 24. POW Camp Chase, OH 5/30/62. Exchd. 8/25/62. At home on sick furlo. from 10/20/62 to 4/?/63. Dischd. at Camp Stonewall 5/25/63. Another card says dischd. 7/2/63 for hypertrophy of the heart. Farmer. 6'4". Light complexion, black hair, dark eyes, dark whiskers.

CARSON, WILLIAM A.: Pvt. Co. G. Enl. 5/1/62 at White Sulphur Springs, Greenbrier, Co., WV. Brigade wagoner on July 1862 roll. D. at Montgomery Ferry (Kanawha Falls), WV 10/10/62.

CART, ANDREW J.: Pvt. Co. A. Enl: 8/26/61 at Camp Gauley (Carnifex Ferry), Nicholas Co., WV. Present on 8/31/61 roll. NFR.

CART, JAMES A.: Pvt. Co. A. Enl. 5/22/61 at Poca, Putnam Co., WV. Capt. at Winchester, Va. 9/19/64. POW Pt. Lookout, MD.

CART, JOHN A. C.: Pvt./Sgt. Co. F. Res. Monroe Co., WV. Enl. 7/4/61 at Charleston, Kanawha Co., WV. At home sick 8/7/61. Present on 10/31/62 roll. Capt. at Winchester, Va. 9/19/64. POW Pt. Lookout, MD 9/23/64. Exchd. 3/15/65. Paroled at Charleston, WV 6/12/65. Age: 30. 5'8½". Dark complexion, dark hair, blue eyes. D. sometime prior to 1914.

CART, JOHN W.: Pvt./Sgt. Co. A. Enl. 5/22/61 at Poca, Putnam Co., WV. Paroled at Lewisburg, WV 4/25/65. Age: 22. Farmer. 5'8". Brown hair and blue eyes. B. in Va. Res. Fry, WV 1924 and attended a vet. reunion in Charleston, WV that year.

CART, URIAH: Pvt. Co. E. Res. Clay Co., WV. Enl. 5/1/62 at Richmond, Va. Exchd. to Co. E 12/12/62. Capt. in Clay Co., WV 2/15/64. Age: 21. POW Camp Chase, OH 2/24/64. Transf. to Ft. Delaware 3/14/64. Relsd. and took oath 6/19/65. Farmer. 5'1". Dark complexion, dark hair, dark eyes.

CARTER, ARCHIBALD: Pvt. Co. F. Enl. 9/12/?? at Red Sulphur Springs, Monroe Co., WV. Deserted at Camp McCausland, WV 7/20/62. Listed as deserter on Aug. 1862 broadside, Monroe Co., WV. Sick at home on 10/31/62 roll. Present on 12/31/62 roll. Sick at home near Union, WV on 2/25/63 roll. Present on Mar.-Apr. 1863 roll. Capt. at Winchester, Va. 9/19/64. POW Pt. Lookout, MD 9/23/64. Exchd. 3/15/65. Paroled at Charleston, WV 6/12/65. Age: 29. 5'9". Light complexion, dark hair, black eyes.

CARTER, CHARLES: Pvt. Co. B. Enl. 6/1/61 at Ripley, Jackson Co., WV. Absent in hosp. on 9/1/61 roll. NFR.

CARTER, FREDERICK K.: Pvt. Co. A. Enl. 6/24/61 at Poca, Putnam Co., WV. Regt. blacksmith on 11/1/62 roll. Absent sick on 10/31/64 roll. Age: 30. Applied for a pension in 1870 at the age of 60. Res. of Denmark. Farmer. Res. Kerrs Creek, WV in 1870 census. Applied for pension 4/20/1900 at age 60. Res. Denmark. Farmer. Res. Kerr's Creek District 1910 at age 70. D. 12/14/1919 at Kerrs Creek at the age of 81. Bur. near his home on Kerrs Creek, WV.

CARTER, HENRY: Pvt. Co. B. Enl. 6/1/61 at Ripley, Jackson Co., WV. Absent in hosp. on 9/1/61 roll. NFR.

CARTER, JAMES: Pvt./Sgt. Co. B. Enl. 6/1/61 at Ripley, Jackson Co., WV. Absent in hosp. on 9/1/61 roll. Present on 11/16/62 roll. Sick in hosp. on rolls dated 2/28/63 and 4/30/63. Present on 1/10/64 roll. Paroled at Charleston, WV 5/12/65. Age: 24. 6'1". Dark complexion, dark hair, black eyes, dark whiskers.

CARTER, JAMES PETTUS: Pvt. Co. G. Enl. 12/28/62 at Union, Monroe Co., W.V. Deserted 11/15/63. Took oath and sent north of the Ohio River 1863. Age: 26. Carpenter. 6'. Light complexion, brown hair, hazel eyes. D. 11/77.

CARTER, JOHN: Pvt. Co. B. Res. Ripley, WV. Enl. 6/1/61 at Ripley, Jackson Co., WV. In hosp. at White Sulphur Springs, WV 9/1/61. Capt. at Ripley, WV 9/28/62. POW Camp Chase, OH 10/12/62. Exchd. 10/27/62. Capt. at Ripley, WV by state troops 11/22/64. Age: 30. POW Ft. Delaware 3/14/64. Relsd. 6/19/65. Farmer. 6'3". Fresh and sandy complexion, dark hair, blue eyes, sandy whiskers.

CARTER, JOHN D.: Pvt. Swann's Co. Enl. 5/8/61 at Charleston, Kanawha Co., WV. Age: 34. Co. transf. to Co. K, 59th Va. Inf. 7/25/61.

CARTER, JOHN M.: Pvt. Co. G. Enl. 5/1/62 at White Sulphur Springs, Greenbrier Co., WV. Deserted 5/1/62.

CARTER, LEWIS V.: Pvt. Co. A. Enl. 6/3/61 at Poca, Putnam Co., WV. Unaccounted for as of 7/24/61.

CARTER, RUPERT K.: Pvt. Co. A. Enl. 5/27/62 at Salt Sulphur Springs, Monroe Co., WV. Deserted at Lewisburg, WV 4/12/63. AWOL on 12/31/63 roll. Sick on rolls dated 4/30/64 and 10/31/64. Took oath at Charleston, WV 4/10/65. Age: 27. Farmer 5'11". Fair complexion, dark hair, gray eyes.

CARTER, THOMAS H.: Pvt./Sgt. Co. A. Enl. 5/22/61 at Poca, Putnam Co., WV. Sick at White Sulphur Springs, WV 8/20/61. Present on 11/1/62 roll. Took oath at Charleston, WV 4/20/65. Age: 31. Farmer. Fair complexion, fair hair, gray eyes.

CARTMILL, JAMES H.: Pvt. Co. A. Enl. 6/3/61 at Poca, Putnam Co., WV. WIA (severe thigh) at White Sulphur Springs, WV 8/26-27/63. Absent due to wd. on rolls dated 12/31/63 and 4/30/64. Took oath at Charleston, WV 4/24/65. Age: 26. Farmer. Fair complexion, fair hair, hazel eyes.

CARTMILL, JOHN G.: Pvt. Co. A. Res. Putnam Co., WV. Enl. 10/26/62 at Charleston, Kanawha Co., WV. Left sick at Marlin's Bottom (Marlinton), WV on 12/31/63 roll. Capt. In Boone Co., WV 2/9/64. Age: 20. POW Camp Chase, OH 3/7/64. Transf. to Ft. Delaware 3/14/64. Relsd. 6/19/65. Farmer. 6'. Light complexion, dark hair, hazel eyes.

CARTWELL, PETE: Co. A. Bahlmann says WIA and capt. at Droop Mountain, WV 11/6/63. Barbee says wded. in foot and "seems to get wded. in every fight."

CARY, JOHN M.: Pvt. Co. F. Date and place of enlistment not stated. Rcvd. clothing for 1864 period. Paroled at Charleston, WV 5/11/65. Age: 36. 5'8". Dark complexion, black hair, black eyes, black whiskers.

CASTO, ISAAC J.: Pvt. Co. B. Res. Jackson Co., WV. Cooper, Jackson Co., WV. 1860 age: 38. Enl. 7/1/61 at Ripley, Jackson Co., WV. WIA at Lewisburg, WV 5/23/62. At home sick on rolls dated 4/1/64 and 5/1/64. PWR says mustered out after being WIA. Bur. Crane Cem., Fairplain, WV.

CAUGHMAN, J. A.: Pvt. Co. H. Res. Greenbrier Co., WV. Enl. 4/28/62 at White Sulphur Springs, Greenbrier Co., WV. Det. shoemaker at Lewisburg, WV 12/3/63. Det. shoemaker for First Brigade 1/1/64 to 4/30/64. Capt. at Cold Harbor (Gaines Mill), Va. 6/3/64. Exchd. 9/18/64. In Chimborazo Hosp. for debility 9/22-25/64.

CAUSLING, SAMUEL: Co. G. Postwar roster.

CAVENDISH, GEORGE W.: Pvt. Co. C. Res. Fayette Co., WV. Date and place of enlistment not stated. Capt. in Greenbrier Co., WV 11/10/63. Age: 18. POW Camp Chase, OH 11/18/63. Transf. to Ft. Delaware 2/29/64. Took oath and relsd. 6/19/65. Farmer. 5'8½". Florid complexion, dark hair, gray eyes.

CAVENDISH, JAMES W.: Pvt. Co. C. B. 3/?/45 Lookout, WV. Enl. 7/1/63 at Packs Ferry, Monroe Co., WV. Capt. at Droop Mountain, WV 11/6/63. Age: 19. POW Camp Chase, OH 11/18/63. Transf. to Ft. Delaware 2/29/64. Farmer. 5'10". Fresh complexion, sandy hair, blue eyes. Resided: Fayette Co., WV. D. at Ft. Delaware of measles 4/15/64. bur. Jersey shore opposite Ft. Delaware, Finn's Point Nat. Cem., N.J.

CAVENDISH, JOHN M.: Sgt. Co. C. Res. Fayette Co., WV. Date and place of enlistment not stated. Listed as a POW in U.S. camp at Cross Lanes, WV 9/20/61. POW Camp Chase, OH 9/28/61. Relsd. and took oath 6/6/62. Age: 36. Dark hair and red whiskers. Farmer. PWR says enl. 1861 at age 36.

CAVENDISH, JOSEPH: 2nd Sgt. Co. C. B. in Va. Res. Fayette Co., WV 1860. Age: 38. Enl. 1861.

CAVES, LEWIS: Pvt. Co. A. Enl. 5/22/61 at Poca, Putnam Co., WV. Present on 8/31/61 roll. NFR.

CHAMBERS, ANDREW: Pvt. Co. A. Enl. 7/22/61 at Salt Sulphur Springs, Monroe Co., WV. Deserted at Lewisburg, WV 4/12/63.

CHAMBERS, JOHN: Pvt. Co. H. Clerk, Charleston, WV 1860. Age: 18. Enl. 5/8/61 at Charleston, Kanawha Co., WV. Det. to A.C.S. Dept. at Lewisburg, WV on 12/5/63 roll. Det. to A.C.S. Dept., First Brigade, Lewisburg, WV on 1/1/64 roll. Capt. at Cold Harbor (Gaines Mill), Va. 6/3/64. Transf. to Elmira, NY 7/9/64. Relsd. 6/16/65.

CHAMBERS, ROBERT: Pvt. Co. I. Enl. 6/29/61 at Charleston, Kanawha Co., WV. D. sometime prior to 1/8/63. Death claim presented by Robert Chambers, father.

CHAMBERS, WILLIAM F.: Pvt./Sgt. Co. I. Enl. 6/29/61 at Charleston, Kanawha Co., WV. Served as 2nd Sgt. until promo. to O. Sgt. 4/11/63. WIA (slight head) at Cold Harbor, Va. 6/?/64. Took oath at Charleston, WV 5/1/65. Age: 26. Farmer. 5'9", dark complexion, fair hair, hazel eyes.

CHAMP, ERASTUS F.: Pvt. Co. C. B. ca. 1834 Craig Co., Va. Date and place of enlistment not stated. Dischd. at Camp McCausland, WV 7/10/62. Middle initial may be "C".

CHANDLER, MORRIS CHAPMAN: Pvt. Co. E. Enl. 6/8/61 at Charleston, Kanawha Co., WV. Left sick at White Sulphur Springs, WV 8/20/61. Present on 5/1/62 roll. KIA at Droop Mountain, WV 11/6/63. Tombstone says Co. G. Bur. Teays Hill Cem., St. Albans, WV.

CHANDLER, WILLIAM S.: Capt. Co. I. Enl. 6/29/61 at Charleston, Kanawha Co., WV. Elected Capt. 6/29/61. Resgnd. 10/1/61. Listed as a prisoner as of 12/27/61.

CHAPMAN, AND. J.: Pvt. Co. D. Enl. 5/30/61 at Summersville, Nicholas Co., WV. AWOL 7/28/61.

CHAPMAN, GEORGE: Pvt. Co. D. Res. Nicholas Co., WV. Enl. 5/30/61 at Summersville, Nicholas Co., WV. Capt. in Nicholas Co., WV while on furlo. 11/1/62. Capt. in Clay Co., WV 1/9/64. Age: 21. Listed as a deserter by Capt. McNeil. POW Camp Chase, OH 2/2/64. Left prison and joined U.S. Navy 7/20/64. Farmer. 5'10". Florid complexion, brown hair, gray eyes.

CHAPMAN, J. W.: Co. G. Postwar roster.

CHAPMAN, JAMES W.: Pvt. Co. D. Enl. 5/30/61 at Summersville, Nicholas Co., WV. AWOL 7/28/61.

CHASE, BENJAMIN: 1st Sgt./Capt. Co. B. Son of John H. Chase. Brother of Henry, Jonathan, and Lawrence. Res. Ripley, WV. Revenue Officer, 1860. Age: 25. Enl. 6/1/61. Place of enl. not stated. Colorbearer. In hosp. at White Sulphur Springs, WV 9/1/61. Elected Capt. 5/1/62. KIA (shot through heart) at Lewisburg, WV 5/23/62.

CHASE, HENRY F.: Pvt. Co. B. B. 4/15/36 Ripley, WV. Son of John H. Chase. Brother of Benjamin, Jonathan, and Lawrence. Res. Mill Creek, WV, 1860. Age: 24. Enl. 6/?/63 at Lewisburg, Greenbrier Co., WV. Det. service on 1/10/64 roll. Capt. at Winchester, Va. 9/19/64. POW Pt. Lookout, MD 9/24/64. Exchd. 3/15/65. Paroled at Lewisburg, WV 4/26/65. Age: 28. Millwright. 6'. Fair complexion, dark hair, blue eyes. PWR says mustered in 1862. Postwar flour mill business, Jackson Co., WV. Res. Mill Creek, WV, 1883. D. 1905, Ripley, WV from illness brought on by exposure during the war. Photo in *Confederate Veteran* #23.

CHASE, JONATHAN: Pvt./Corp. Co. B. Son of John H. Chase. Brother of Benjamin, Henry, and Lawrence. Enl. 10/15/62 at Charleston, Kanawha Co., WV. Capt. and WIA (very serious left hand fracture) by minie ball) at Winchester, Va. 9/19/64. POW Pt. Lookout, MD. In Baltimore, MD hosp. 10/13/64. Transf. to Gen. Hosp. 10/25/64. Exchd. 10/30/64. Paroled at Charleston, WV 5/10/65. Age: 27. 5'8". Sandy complexion, sandy hair, blue eyes.

CHASE, LAWRENCE: Pvt. Co. B. Son of John H. Chase. Brother of Benjamin, Henry, and Jonathan. Enl. 6/?/63 at Lewisburg, Greenbrier Co., WV. Capt. at Winchester, Va. 9/19/64. POW Pt. Lookout, MD 9/23/64. Exchd. 3/15/65. Paroled at Charleston, WV 5/16/65. Age: 32. 5'9". Sandy complexion, brown hair, blue eyes, sandy whiskers.

CHEWNING, CHARLES: Pvt. Co. H. Farm manager, Charleston, WV 1860 at age 18. Enl. 5/13/61 at Coalsmouth (St. Albans), Kanawha Co., WV. KIA at Lewisburg, WV 5/23/62. William Bahlmann claims he was killed by a mis-directed Confederate artillery shell.

CHILDERS, JOHN WALKER: Pvt. Co. A. Enl. 5/22/61 at Poca, Putnam Co., WV. Unaccounted for as of 7/24/61. Present on 11/1/62 roll. Left sick at Marlin's Bottom (Marlinton), WV on 4/30/63 roll. Absent wded. on 12/31/63 roll. Present on 4/30/64 roll. Capt. at Winchester, Va. 9/19/64. POW Pt. Lookout, MD 9/27/64. Exchd. 3/15/65. Took oath at Charleston, WV 4/20/65. Age: 22. Farmer. 5'8". Fair complexion, fair hair, blue eyes. Bur. Childers Cem. near Bills Creek, Putnam Co., WV.

CHILDRESS, WILLIAM: Co. B. Drew pension in Cross Co., Ark. D. 10/22/1904.

CHILTON, GEORGE STEPTOE: Lt./Capt. Co. E. B. 7/13/33 in Charleston, WV. Clerk, Coalsmouth (St. Albans), WV 1860. Age 27. Res. Kanawha Co., WV. Enl. 6/8/61 at Charleston, Kanawha Co., WV. Served as 2nd Sgt. until elected 3rd Lt. 7/3/61. Elected Capt. 5/1/62. WIA (shot through right jaw) at Droop Mountain, WV 11/6/63. Det. on G.C.M. on 4/1/64 roll. Capt. at Winchester, Va. 9/19/64. POW Pt. Lookout, MD 9/25/64. Relsd. 6/17/65. "Was seriously wded. in the war." 5'10". Florid complexion, dark hair, blue eyes. D. 12/24/1917 at Charleston, WV. Bur. Teays Hill Cem., St. Albans, WV. Appears in a photo of R. E. Lee Camp #887, Charleston, WV Aug. 1911.

CHURCH, ANDREW: Pvt. Co. K. B. in Va. Enl. 10/15/63 at Lewisburg, Greenbrier Co., WV. Det. brig. blacksmith on 12/30/63 roll. Govt. blacksmith on 4/1/64 roll. Det. brig. blacksmith on 5/1/64 roll. Paroled at Lewisburg, WV 4/25/65. Age: 25. Blacksmith. 5'9". Dark complexion, dark hair, black eyes.

CLARK, H. F.: Pvt. Co. B. Mustered 1862 for 3 years. Res. Jackson Co., WV.

CLARK, OBEDIAH: Pvt. Co. A. Enl. 6/25/61 at Poca, Putnam Co., WV. Unaccounted for as of 7/24/61. Deserted at Charleston, WV 10/8/62.

CLARK, TIMOTHY: Pvt. Co. I. Enl. 6/29/61 at Charleston, Kanawha Co., WV. Present on 8/31/61 roll. NFR.

CLARK, V. F.: Co. G. Enl. 1861 at Lunenburg, Va. for 4 years. Postwar roster.

CLARKSON, ALEXANDER Q.: Pvt. Co. H. Res. Charleston, WV 1860. Age: 15. Enl. 6/24/61 at Coalsmouth (St. Albans), Kanawha Co., WV. Joined by transfer. Absent in Montgomery Co., Va. on 1/1/64 roll. Present on 4/1/64 roll. Capt. at Winchester, Va. 9/19/64. POW Pt. Lookout, MD 9/24/64. Exchd. 2/13/65. Paroled at Charleston, WV 6/8/65. Age: 20. 5'9". Fair complexion, dark hair, black eyes.

CLARKSON, ERASMUS: Pvt. Co. K. Enl. 4/24/62 at White Sulphur Springs, Greenbrier Co., WV. Sick in hosp. on 12/30/63 roll. D. at Union, WV of typhoid pneumonia 1/14/64. Bahlmann claims KIA at Winchester, Va. 9/19/64.

CLAY, HIRAM (or HYRAM): Pvt. Co. G. Enl. 7/3/61 at Charleston, Kanawha Co., WV. Left his co. on the road from the turnpike to Millers Ferry, WV 7/24/61. Listed as AWOL on 9/31/61 roll. Dischd. 8/13/62.

CLEMENS, AHART F.: Pvt. Co. D. Res. Nicholas Co., WV. Enl. 5/30/61 at Summersville, Nicholas Co., WV. At home sick 7/28/61. Capt. in Nicholas Co., WV 10/11/61. POW Camp Chase, OH. Exchd. 8/25/62. Capt. in Nicholas Co., WV by the 11th O.V.I. 1/5/63. Age: 29. Present on 12/31/63 roll. Capt. at Winchester, Va. 9/19/64. POW Pt. Lookout, MD. Exchd. 3/15/65. Farmer. 5'7". Fair complexion, auburn hair, gray eyes.

CLEMENS, FRANCIS H.: Pvt. Co. D. Enl. 5/30/61 at Summersville, Nicholas Co., WV. Sick at home 7/28/61. Present on 10/31/62 roll. MWIA (severe arm) at White Sulphur Springs, WV 8/26-27/63. D. 10/1/63.

CLEMENS, JOHN W.: Pvt. Co. D. Enl. 10/8/62 at Charleston, Kanawha Co., WV. Present on 10/31/64 roll. NFR.

CLENENDON, ARGLEN H.: Pvt. Co. G. Enl. 9/20/62 at Charleston, Kanawha Co., WV. Capt. at Winchester, Va. 9/19/64. POW Pt. Lookout, MD 9/20/64. D. at Pt. Lookout, MD of acute bronchitis 2/24/65. Bur. POW graveyard at Pt. Lookout, MD.

CLENDENON, JAMES: Co. G. Postwar roster. Pvt. Enl. 1861 at Charleston, WV. Age: 30. Res. Wyoming Co., WV. Capt. at New Market, Va. 5/15/64.

CLENDENON, M. H.: Pvt. Co. G. Enl. 9/22/64 at Dublin, Pulaski Co., Va. Capt. and and WIA (serious flesh wd. to left axilla - shell - affecting major muscles) at Cedar Creek, Va. 10/19/64. POW Pt. Lookout, MD. Sent to Baltimore, MD hosp. 11/23/64. Transf. to U.S. Gen. Hosp. 1/28/65. Exchd. 2/10/65. A postwar roster says enl. 1862 for 3 years and was WIA at Strasburg, Va. This may be Milton Lee Clendenen. B. 1833. D. 1915.

CLINGHAM, DAVID: Pvt. Taylor's Co. Enl. 6/19/61 at Frankford, Greenbrier Co., WV. Age: 37. Co. transf. to Wise Brigade 7/10/61.

COCKE, WILLIAM: Co. H. Claimed served 3 years and 8 mos. Was in 20 battles and had two ribs broken, but only lost 10 days out of entire service. Claimed to be one of "Stonewall" Jackson's boys and furnished the litter bearers to bring him off field. Postwar res. Huntington, WV. Still alive 11/?/1928.

COE, ALFRED S.: Pvt. Co. B. Enl. 6/1/61 at Ripley, Jackson Co., WV. AWOL 6/22/61. Capt. in Jackson Co., WV by 11th Va. Regt. 1/2/63. Age: 24. Bushwhacker. POW Camp Chase, OH 1/7/63. Exchd. 3/28/63. 5'8¾". Fair complexion, dark hair, gray eyes.

COFER, GEORGE N.: Pvt./2nd Lt. Co. D. B. Wood Co., WV. Enl. 5/30/61 at Summersville, Nicholas Co., WV. Elected 2nd Lt. 5/1/62. WIA (very serious right shoulder & hand) at Cedar Creek, Va. 10/19/64. Sent to Lynchburg, Va. hosp. At Lynchburg, Va. hosp. wded. 10/31/64. Entered Richmond, Va. Old Soldiers Home 8/16/1918 at age 81. D. 7/2/1919. Bur. Hollywood Cem., Richmond, Va.

COFER, WILLIAM R.: Sgt. Co. D. Enl. 5/1/62 at White Sulphur Springs, Greenbrier Co., WV. Capt. at Winchester, Va. 9/19/64. POW Pt. Lookout, MD 9/23/64. Exchd. 3/15/65. Drew pension in White Co., Ark. 1901.

COFFMAN, ABRAHAM: Pvt. Co. D. Enl. 8/6/61 at White Sulphur Springs, Greenbrier Co., WV. Joined by transfer. Transf. from Capt. Riggs' Co. by order of Col. Tompkins 8/6/61. NFR.

COFFMAN, JAMES M.: Pvt. Co. B. Enl. 5/?/62 at White Sulphur Springs, Greenbrier Co., WV. Conscript. At home sick on rolls from 1/10/64 to 11/30/64. NFR.

COLE, CHARLES M.: Pvt. Co. C. Res. Fayette Co., WV. Enl. 5/1/62 at White Sulphur Springs, Greenbrier Co., WV. Capt. in Fayette Co., WV by the 47th O.V.I. 4/24/62. Age: 36. Exchd. 8/25/62. Det. brig. wagoner 5/1/62 (?). Capt. at Winchester, Va. 9/19/64. POW Pt. Lookout, MD 9/24/64. 6'2½". Dark complexion, hazel eyes, dark whiskers. Farmer. PWR says enl. 1861.

COLE, G. W.: Pvt. Co. C. Enl. 5/1/62 at White Sulphur Springs, Greenbrier Co., WV. Deserted at Camp Dickinson 11/1/62. Postwar res. Victor, WV. An original member of Jeb Stuart UCV Camp, Fayette Co., 1909. Farmer, Fayette Co., WV 1860. Age: 30. PWR says enl. 1861 and name is George.

COLE, WILLIAM H.: Pvt. Co. C. Res. Fayette Co., WV. Enl. 5/1/61 at White Sulphur Springs, Greenbrier Co., WV. Capt. in Greenbrier Co., WV by the 47th O.V.I. 4/28/62. Exchd. 8/25/62. WIA (very serious through middle of right-leg) and capt. at Winchester, Va. 9/19/64 or 9/25/64. D. 10/1/64. Bur. C.S. Cem., Winchester, Va., grave #44. Age: 26. 6'4". Dark complexion, dark hair, gray eyes, dark whiskers.

COLEMAN, JACOB V.: Pvt. Co. C. Res. Fayette Co. WV. 1860. Age: 21, Enl. 1861.

COLEMAN, JAMES W.: Pvt. Co. C. Farmer, Fayette Co., WV 1860. Age: 34. Enl. 1861.

COLEMAN, S. C.: Co. C. Capt. at Carnifex Ferry, WV 9/11/61. Attended a vet. reunion at Beckley, WV 9/12/1918 and gave res. as Beckley, WV.

COLEMAN, SETON L.: Pvt. Co. C. Day laborer, Fayette Co., WV 1860. Age: 25. Enl. 1861. Also served as a Pvt. in Thurmond's Co.

COLLINS, GEORGE: Pvt. Co. C. B. ca. 1845 Craig Co., Va. Enl. 2/18/63 at Lewisburg, Greenbrier Co., WV. KIA at White Sulphur Springs, WV 8/26-27/63.

COLLINS, M. V.: Co. K. Deserted 5/25/62 on Aug. 1862 broadside. Potts Creek, Craig Co., Va.

COLLINS, SAMUEL: Pvt. Co. K. Enl. 3/8/64 at Camp Gauley, Greenbrier Co., WV. Dischd. for being underage 4/8/64.

COLLISON, A. H.: Pvt./Corp. Co. E. Enl. 6/18/61 at Charleston, Kanawha Co., WV. On mounted guard at White Sulphur Springs, WV 8/13-14/61, relieving Pvt. Campbell at request of Capt. Swann. Capt. by Union Home Guard (Capt. King) and paroled same day 9/20/62.

COMBS, COMTE T.: Pvt. Co. B. Enl. 6/1/61 at Ripley, Jackson Co., WV. In hosp. at White Sulphur Springs, WV on 9/1/61 roll. NFR.

COMBS, EDGAR: Pvt. Co. B Enl. 9/?/62 at Charleston, Kanawha Co., WV. Det. brigade teamster 4/1/64. Capt. at Winchester, Va. 9/19/64. POW Pt. Lookout, MD 9/27/64. Exchd. 3/15/65. Paroled at Charleston, WV 6/9/65. Age: 20. 5'10". Fair complexion, brown hair, gray eyes.

COMBS, WILLIAM H.: Pvt. Co. B. Enl. 6/1/61 at Ripley, Jackson Co., WV. Worked as a nurse at hosp. at White Sulphur Springs, WV 10/1/61 to 11/1/61. Paroled at Charleston, WV 6/9/65. Age: 54. 5'10". Dark complexion, gray hair, blue eyes. Bur. Combs Cem. near Nat, Mason Co., WV.

CONNELL, JOHN: 1st Lt. Co. E. Enl. 6/8/61. Place of enlistment not stated. Elected 1st Lt. 6/8/61. Resgnd. 7/1/61.

COOK, HARRY: Co. G. Postwar roster. Pvt. Enl. 1861 at Charleston, WV. Age: 18. Res. Wyoming Co., WV. Capt. at New Market, Va. 5/15/64.

COOK, IRVINE: 1st Sgt. Co. G. Enl. 5/?/61 in Kanawha Co., WV. Capt. at New Market, Va. 5/15/64. Postwar roster. Res. Wyoming Co., WV. Age: 21 in 1861.

COOK, JAMES A.: 2nd Lt./Capt. Co. G. Res. Wyoming Co., WV 1860, store clerk, age 22. Recruiting volunteers at Oceana, WV May 1861 along with William T. Sarver and Erastus A. Duncan. Enl. 7/3/61 at Charleston, Kanawha Co., WV. Elected 2nd Lt. 7/8/61. Elected Capt. 1/24/62. In Charlottesville, Va. hosp. for diarrhea 6/12-17/64. In hosp. at Danville, Va. 7/15/64. Resgnd. 11/22/64 at Staunton, Va. Listed as commdg. regt. 11/30/64. Farmer. Known as "Captain Jim" postwar when he managed "The Mountain House" resort in Oceana, WV. Still so engaged in 1889.

COOK, JAMES H.: Pvt. Co. G. Res. Wyoming Co., WV. Enl. 5/1/62 at White Sulphur Springs, Greenbrier Co., WV. In hosp. on 11/13/62 roll. Deserted. Took oath at Charleston, WV and sent north 5/31/65. Age: 30. Farmer. 6'. Fair complexion, black hair, black eyes.

COOK, JAMES R.: Pvt. Co. G. Res. Wyoming Co., WV. Enl. 7/3/61 at Charleston, Kanawha Co., WV. Relsd. from military prison at Lynchburg, Va. in July 1864. Deserted from guard at Lynchburg, Va. 10/31/64. Took oath at Charleston, WV 5/5/65. Age: 24. Farmer. 6'. Fair complexion, fair hair, blue eyes.

COOK, JAMES T.: Pvt. Taylor's Co. Enl. 6/19/61 at Frankford, Greenbrier Co., WV. Age: 21. Co. transf. to Wise Brigade 7/10/61.

COOK, JOHN H.: Pvt. Co. G. Enl. 7/7/61 at Charleston, Kanawha Co., WV. Left sick at White Sulphur Springs, WV 7/20/61. Transf. from Capt. Sarver's Co. 7/25/61. NFR.

COOK, JOHN K.: Corp. Co. G. Enl. 5/1/62 at White Sulphur Springs, Greenbrier Co., WV. Had a furlo. which twice expired 1/25/64 on 10/31/64 roll. Middle name is probably Kemper. PWR says enl. 1861 at Charleston, WV. Age: 22. Res. Wyoming Co., WV. Capt. at New Market, Va. 5/15/64.

COOK, MATTHEW E.: Pvt. Co. G. Enl. 10/20/62 at Fayetteville, Fayette Co., WV. On scout on 4/1/64 roll. Capt. at Winchester, Va. 9/19/64. POW Pt. Lookout, MD 9/23/64. Joined U.S. service 10/12/64.

COOK, R. E.: Co. G. Enl. 5/?/61 in Kanawha Co., WV. Postwar roster. Res. Wyoming Co., WV. Age: 22 in 1861. Capt. at New Market, Va. 5/15/64.

COOK, SAMUEL H.: Pvt. Co. D. Enl. 5/30/61 at Summersville, Nicholas Co., WV. WIA at Lewisburg, WV 5/23/62. Did not return to service due to wd. D. in Nicholas Co. (or Greenbrier Co.) of fever while at John Cook's 9/20/63.

COOK, WALTON: Pvt. Co. H. Enl. 1861. "Didn't go."

COOK, WILLIAM H.: Pvt./1st Sgt. Co. G. Enl. 7/3/61 at Charleston, Kanawha Co., WV. AWOL having left his co. at Clifton (Pratt), WV 7/24/61. Present on 11/13/62 roll. Furlo. expired 1/18/63 and did not return. Listed as a deserter.

COOK, WILLIAM J.: Pvt./1st Sgt. Co. G. Enl. 7/7/61 at Charleston, Kanawha Co., WV. Left sick at White Sulphur Springs, WV 7/24/61 and transf. from Capt. Sarver's Co. 7/25/61. Deserted 3/4/64.

COOK, WILLIAM N.: Pvt. Co. H. Enl. 5/31/61 at Coalsmouth (St. Albans), Kanawha Co., WV. In hosp. at Montgomery White Sulphur Springs, Va. from July to Aug. 1862. Present on 12/31/62 roll. Sick at Montgomery White Sulphur Springs, Va. on rolls from 1/1/64 to 4/30/64. Present on 12/28/64 roll. In hosp. at Charlottesville, Va. 6/9/64 to 7/16/64.

COOK, WILLIAM R.: Corp. Co. G. Enl. 7/3/61 at Charleston, Kanawha Co., WV. Left sick at White Sulphur Springs, WV on a fifteen day furlo. 8/11/61. NFR.

COOK, WILLIAM S.: Pvt. Co. G. Enl. 7/3/61 at Charleston, Kanawha Co., WV. Left sick at White Sulphur Springs, WV 7/20/61. NFR.

COONEY, GEORGE: Pvt. Co. A. Enl. 5/22/61 at Poca, Putnam Co., WV. Present on 8/31/61 roll. NFR.

COOPER, ANDREW: Pvt. Co. F. Enl. 8/20/62 at Camp Lynchmill. On furlo. on 8/20/63 roll. D. in service prior to 8/13/63.

COOPER, CREED: Pvt. Co. F. Res. Roanoke, Va. Enl. 5/1/62 at White Sulphur Springs, Greenbrier Co., WV. Capt. at Droop Mountain, WV 11/6/63. POW Camp Chase, OH 11/15/63. Transf. to Ft. Delaware 2/29/64. D. at Ft. Delaware of inflammation of lungs 6/1/65. Bur. Jersey shore of Ft. Delaware, Finn's Point Nat. Cem., N.J. Age: 34. Farmer. 5'9". Florid complexion, black hair, black eyes.

COOPER, GEORGE W.: Pvt. Co. F. Enl. 7/5/62 at Camp McCausland, Monroe Co., WV. Capt. at Droop Mountain, WV 11/6/63. Age: 19. POW Camp Chase, OH 11/18/63. Transf. to Ft. Delaware 2/29/64. Relsd. 6/19/65. Farmer. 5'11". Florid complexion, dark hair, brown eyes. Res. Roanoke, Va.

COOPER, JAMES: Pvt. Swann's Co. Enl. 5/8/61 at Charleston, Kanawha Co., WV. Age: 21. Co. transf. to Co. K, 59th Va. Inf. 7/25/61.

COOPER, JOEL: Pvt. Co. F. Enl. 5/1/62 at White Sulphur Springs, Greenbrier Co., WV. Capt. at Droop Mountain, WV 11/6/63. Age: 32. POW Camp Chase, OH 11/18/63. Transf. to Ft. Delaware 2/29/64. Relsd. 6/19/65. Farmer. 5'8¼". Fresh complexion, sandy hair, dark eyes. Res. Roanoke, Va. Member of Hupp-Deyerle CV Camp, Salem, Va. 1909. Postwar res. Roanoke Co., Va.

COOPER, J. W.: Pvt. Co. H. Enl. 5/8/61 at Charleston, Kanawha Co., WV. Paroled at Charleston, WV 6/5/65. Age: 24. 6'. Light complexion, light hair, gray eyes, light whiskers.

COOPER, THOMAS R.: Pvt./Sgt. Co. G. Enl. 7/3/61 at Charleston, Kanawha Co., WV. At home wded. on 12/31/63 roll. Present on 4/1/64 roll. Paroled at Charleston, WV 5/10/65. Age: 24. 5'11". Fair complexion, light hair, blue eyes. Postwar res. Ansted, WV. An original member of Jeb Stuart UCV camp, Fayette Co., WV. 1909.

COOPER, WILLIAM H.: Corp. Co. I. Enl. 6/29/61 at Charleston, Kanawha Co., WV. AWOL 7/26/61.

CORDER, WILLIAM: Pvt. Co. G. Enl. 7/7/61 at Charleston, Kanawha Co., WV. Transf. from Capt. Sarver's Co. 7/25/61. Present on 9/31/61 roll. NFR.

CORLEY. WILLIAM M.: Pvt. Co. E. Cooper, Charleston, WV 1860. Age: 20. Enl. 6/8/61 at Charleston, Kanawha Co., WV. Capt. at Jenkins River, Va. 1/23/62. POW Camp Chase, OH Exchd. 8/25/62. KIA at Belle Grove (Cedar Creek), Va. 10/19/64.

CORRELL, ROBERT F.: Pvt. Taylor's Co. Enl. 6/19/61 at Frankford, Greenbrier Co., WV. Age: 17. Co. transf. to Wise Brigade 7/10/61.

CORWATT, DAVID: Pvt. Co. G. Enl. 7/3/61 at Charleston, Kanawha Co., WV. Left his co. at Charleston, WV about 7/15/61. AWOL on 9/3/61 roll.

COTTRELL, THOMAS: Pvt. Co. E. Enl. 7/1/61 at West Fork and/or 5/1/62 at White Sulphur Springs, WV. Capt. on Sandy Creek, Calhoun Co., WV by 11th Va. 1/3/63. Age: 22. POW Camp Chase, OH 1/12/63. Relsd. 3/4/63. 5'7". Dark complexion, black hair, black eyes. Res. Calhoun Co., WV.

COTTRILL, ELIJAH W.: Pvt. Co. B. Enl. 6/1/61 at Ripley, Jackson Co., WV. At home sick 7/12/61. On list of a mounted guard at White Sulphur Springs, WV 8/13-14/61 and relieved of guard by W. H. Justice at request of Capt. Swann, he being sick. At home sick since 7/12/61 on 9/1/61 roll.

COUNTS, SAMUEL FRANKLIN: Pvt./2nd Lt. Co. A. B. 6/15/42 Flatwoods, WV. Res. Kanawha Co., WV. Laborer, Tyler Mountain, WV 1860. Enl. 5/22/61 at Poca, Putnam Co., WV. Sick at White Sulphur Springs, WV 8/20/61. Present on 11/1/62 roll. Elected 2nd Lt. 12/6/63. Capt. in Boone Co., WV 2/8/64. Age: 21. POW Camp Chase, OH 3/7/64. Transf. to Ft. Delaware 3/14/64. WIA (severe head) at Cold Harbor, Va. 6/3/64. Capt. at Winchester, Va. 9/19/64. While POW jumped off train car about 20 miles below Pittsburg, PA. Got home 6/20/65. Farmer. 6'5½". Fair complexion, light hair, dark eyes. Postwar merchant, Scott's Depot, WV. Res. Putnam Co., WV. 1910 at age 67. Still alive 3/16/1928. Bur. Teays Hill Cem., St. Albans, WV.

COUNTS, WILLIAM H.: Pvt. Co. A. Farm laborer, Tyler Mountain, WV 1860. Age: 21. Enl. 5/22/61 at Poca, Putnam Co., WV. Left sick at White Sulphur Springs, WV 8/20/61. NFR.

COX, E.: Co. H. In Winchester. Va. Gen. Hosp. 1864.

COX, FRANCIS M. (FRANK): Pvt. Co. H. Student, Charleston, WV 1860. Age: 19. Enl. 5/8/61 at Charleston, Kanawha Co., WV. Ex-corp. reduced in rank 2/4/64. WIA (very serious thigh)? at Kernstown, Va. 7/24/64. In hosp. at Winchester, Va. wded. 7/25/64 to 8/4/64. Paroled at Charleston, WV 6/7/65. Age: 24. 5'11". Fair complexion, dark hair, gray eyes. D. 3/8/1911.

COX, GEORGE G.: Co. A. Entered Richmond, Va. Old Soldiers Home 11/17/1902 at age 68. D. 11/19/1902. Bur. Maury Cem., Richmond, Va.

COZART, WILLIAM H.: Pvt. Co. G. Enl. 7/3/61 at Charleston, Kanawha Co., WV. Scouted for Thurmond July 1862. Absent wded. on 12/30/63 roll. Present on 4/1/64 roll. Ambulance driver on 10/21/64 roll. Took oath at Charleston, WV 5/9/65. Age: 36. Farmer. Sandy complexion, brown hair, blue eyes. Res. Wyoming Co., WV.

CRAFT, BENJAMIN NEWTON: Pvt. Co. K. B. ca. 1847 Craig Co., Va. Enl. Apr. 1863 for 2 years. Res. Craig Co., Va. 1898 at age 51. D. 2/1/1924 at Clifton Forge, Va. at age 78. Bur. Drummond Cem., Rt. 617, at foot of Craig Co. Mountain on Barbour's Creek, Craig Co., Va.

CRAFT, FREDERICK: Listed in UDC booklet *Craig's Share in the War Between The States.* Res: Craig Co., Va. (This may be James F. Craft who is bur. Caldwell Cem., Rt. 615, on Fred Looney farm, Craig Co., Va.).

CRAFT, GEORGE: Photo in *Hard Times, Vol. III.* Res. Craig Co., Va. NFR.

CRAFT, HENRY: Co. I. Listed as deserter on Aug. 1862 broadside. Allegheny Co., Va.

CRAFT, HENRY: Co. K. Bur. Waugh Cem., Rt. 616, near the Lee Bush home, Cuba Rd., New Castle, Va.

CRAFT, JACOB: Pvt. Co. H. B. ca. 1835 Craig Co., Va. Enl. 9/5/62 in Monroe Co., WV. Transf. to Co. K 3/5/63. Det. brigade teamster on 12/30/63 roll. Brigade wagoner on 4/1/64 roll. Took oath at Charleston, WV 12/14/64. Age: 33. Surrendered at Augusta, Ga. 5/25/65. Farmer. 6'1". Dark complexion, dark hair, blue eyes. Bur. Waugh Cem., Rt. 616, Cuba Rd., New Castle, Va.

CRAFT, JAMES H.: Pvt. Co. K. B. 1845 Craig Co., Va. Enl. 3/9/62 at Lewisburg, Greenbrier Co., WV. WIA at Gardner's Farm, Va. 5/30-31/64. In Chimborazo Hosp. for very serious flesh wd. of right arm 6/1/64. Res: Craig Co., Va. 1898 at age 53. Bur. Caldwell Cem., Rt. 615, on Fred Looney farm, Craig Co., Va.

CRAFT, WILLIAM H.: Pvt. Co. A. Res. Craig Co., Va. Enl. 4/10/62 at Brushy Ridge, Greenbrier Co., WV. Deserted at Packs Ferry, WV 10/17/62. Present on rolls for 12/31/62 and Sept. 1863.

CRAGO, GEORGE R.: Pvt. Co. A. Enl. 5/22/61 at Poca, Putnam Co., WV. Deserted at Buffalo, WV 5/27/61.

CRAGO, HARRISON: Pvt. Co. A. Enl. 5/22/61 at Poca, Putnam Co., WV. Away with leave on 8/31/61 roll in order to nurse his twin brother, Madison, who was wded. at Scary Creek, WV 7/17/61.

CRAGO, MADISON: Pvt. Co. A. Enl. 5/22/61 at Poca, Putnam Co., WV. WIA (severe hip & thigh by musketballs) at Scary Creek, WV 7/17/61.

CRAWFORD, J. D.: Co. C. Listed as deserter on Aug. 1862 broadside. Craig Co., Va.

CRAWFORD, JOHN LEFFEL, JR.: Pvt. Co. C. B. Craig Co., Va. Enl. 3/17/63 at Lewisburg, Greenbrier Co., WV. Absent sick & ordered to the regt. 3/16/63. In Charlottesville, Va. hosp. 6/8/64. D. in hosp. at Charlottesville, Va. of acute dysentery (rubeola) 6/21/64.

CRAWFORD, JOHN LEFFEL, SR.: Pvt. Co. C. B. 3/24/32 Craig Co., Va. Enl. 5/1/62 at White Sulphur Springs, Greenbrier Co., WV. PWR says enl. 3/27/62. Sick at Montgomery White Sulphur Springs, Va. July 1862. On arrest and confined for desertion 4/6/63. Absent sick on 10/31/64 roll. Res. Craig Co., Va. 1898 at age 67. D. 6/24/1911.

CRAWFORD, PATTERSON: Pvt. Co. G. Enl. 3/15/64 at Dublin, Pulaski Co., Va. Deserted at Salem, Va. on 10/31/64 roll.

CREMEANS, LUTHER: Pvt. Co. G. Enl. 2/29/64 at Staunton, Va. Had a leave of absence and had not returned on 10/31/64 roll.

CRIGLER, ZACHARIAH F.: Pvt. Co. F. Res. Monroe Co., WV. Enl. 7/4/61 at Charleston, Kanawha Co., WV. On special duty to Commdg. Dept. to 8/13/61. At home sick on surg. cert. on 8/14/61 roll. Postwar roster says transf. D. sometime prior to 1914.

CRINER, GEORGE: Pvt. Co. C. Res. Craig Co., Va. Enl. 5/28/62 at Red Sulphur Springs, Monroe Co., WV. Formerly a member of the 27th Va. Ordered to return to the 27th Va. 7/17/63. KIA at Droop Mountain, WV 11/6/63.

CRISAMORE, FRANKLIN: Pvt. Taylor's Co. Enl. 6/19/61 at Frankford, Greenbrier Co., WV. Age: 19. Co. transf. to Wise Brigade 7/10/61.

CROCKETT, WILLIAM: Pvt. Swann's Co. Enl. 5/8/61 at Charleston, Kanawha Co., WV. Age: 18. Co. transf. to Co. K., 59th Va. Inf. 7/25/61.

CROOKSHANKS, JOHN: Pvt. Co. D. B. in Greenbrier Co., WV. Res. Nicholas Co., WV. Enl. 5/30/61 at Summersville, Nicholas Co., WV. Dischd. 5/15/62 at Camp Success being overage required by law for military duty. Age: 41. Took oath at Charleston, WV 7/15/65. Farmer. Age: 41. 5'11". Dark complexion auburn/brown hair, blue/gray eyes, light whiskers.

CROSS, SAMUEL S.: Pvt. Co. I. Enl. 5/2/62 at White Sulphur Springs, Greenbrier Co., WV. Brigade wagoner July 1862. Det. as Provost Guard at Lewisburg, WV on 4/30/63. Returned from desertion on 11/11/63 roll.

CROSS, THOMAS: Pvt. Co. I. Enl. 5/2/62 at White Sulphur Springs, Greenbrier Co., WV. At foundry at Pattensburg, Va. July 1862. Present on 3/4/63 roll.

CROUCH, CHRISTOPHER: Pvt. Swann's Co. Enl. 5/15/61 at Charleston, Kanawha Co., WV. Age: 24. Co. transf. to Co. K, 59th Va. Inf. 7/25/61.

CROUCH, (DAVIS) D.: Co. H. D. 2/20/1916.

CROW, WILLIAM: Sgt./1st Lt. Co. B. Res. Ripley, WV Enl. 6/1/61 at Ripley, Jackson Co., WV. Elected Sgt. 5/1/62. Promo. to 2nd Lt. 5/23/62. Promo. to 1st Lt. 7/22/63. Capt. at Winchester, Va. 9/19/64. POW Ft. Delaware 9/25/64. Relsd. 6/17/65. 5'11". Dark complexion, dark hair, gray eyes.

CRUM, HENRY: Pvt. Co. B. Enl. 9/22/62 at Charleston, Kanawha Co., WV. Det. service on 1/10/64 roll. Deserted. Capt. in Roane Co., WV 3/19/64. POW Camp Chase, OH. Relsd. 12/29/64 and desired to take oath. Age: 26. Farmer. 5'9". Fair complexion, light hair, blue eyes.

CRUM, MARTIN: Pvt. Co. B. Enl. 10/15/62 at Charleston, Kanawha Co., WV. Present 1/10/64 roll.

CRYMES, GEORGE A.: Co. B. Entered Richmond, Va. Old Soldiers Home 5/23/1927 at age 83. D. 6/25/1926 (?). Bur. Hollywood Cem., Richmond, Va.

CUMB, WM.: Pvt. Co. H. B. ca. 1823 Craig Co., Va. Enl. 10/9/64 for six months. Living in Craig Co., Va. in 1898 at age 78.

CUNNINGHAM, JOHN V.: Pvt. Co. C. Enl. 3/25/64 at Camp Gauley, Greenbrier Co., WV. Paroled at Lewisburg, WV 4/26/65. Age: 18. Merchant. Light complexion, light hair, hazel eyes. Res. Blue Grove, TX 1925.

CUNNINGHAM, WILLIAM D.: Pvt. Co. C. Enl. 5/1/62 at White Sulphur Springs, Greenbrier Co., WV. On scout by order of Gen. Williams on 12/31/62 roll. Present on 3/4/62 roll. WIA at Droop Mountain, WV 11/6/63. Absent due to wd. 12/31/63 roll. Deserted 1/11/64.

CURD, JAMES A.: Pvt. Co. D. Enl. 3/1/64 at Camp Gauley (Greenville), Monroe Co., WV. In Chimborazo Hosp. for very serious right hand 6/5-9/64. Capt. at Winchester, Va. 9/19/64.

CURIGAN, JAMES: Co. C. D. 6/22/1915 at Vanderpool, Highland Co., Va.

CURRY, WILLIAM N.: Pvt. Taylor's Co. Enl. 6/19/61 at Frankford, Greenbrier Co., WV. Age: 23. Co. transf. to Wise Brigade 7/10/61.

CUSHMAN, WILLIAM H.: Pvt. Co. H. Dentist, Charleston, WV 1860. Age: 24. Enl. 5/8/61 at Charleston, Kanawha Co., WV. Wagon master at hosp. at White Sulphur Springs, WV 10/7-15/61. Steward at hosp. at White Sulphur Springs, WV 10/15/61 to 11/1/61. Sick at Lewisburg, WV on 6/30/63 roll. Det. in Signal Corps 12/30/63 to 9/30/64. Det. in Signal Corps and duty as courier on 11/8/64 roll. Paroled at Charleston, WV 6/9/65. Age: 30. 5'7". Dark complexion, dark hair, dark eyes, dark whiskers. Bur. Spring Hill Cem., Charleston, WV. Postwar roster says middle initial is A.

CUSTER, GEORGE W.: Pvt. Co. B. Enl. 8/?/62 at Lewisburg, Greenbrier Co., WV. Paroled at Charleston, WV 5/12/65. Age: 20. 6'2". Dark complexion, dark hair, dark eyes, light whiskers.

DACON, JOHN O.: Pvt. Co. F. Res. Roanoke, Va. Enl. 5/1/62 at White Sulphur Springs, Greenbrier Co., WV. Capt. at Droop Mountain, WV 11/6/63. Age: 23. POW Camp Chase, OH 11/18/63. Transf. to Ft. Delaware 2/29/64. Took oath at Ft. Delaware 6/19/65. Farmer. 6'1⅓". Florid complexion, dark hair, dark eyes.

DAILY, WILLIAM: Pvt. Taylor's Co. Enl. 6/19/61 at Frankford, Greenbrier Co., WV. Age: 46. Co. transf. to Wise Brigade 7/10/61.

DAIR, GOODALL: Pvt. Co. G. Enl. 7/3/61 at Charleston, Kanawha Co., WV. Dischd. 7/3/62.

DALLAS, JAMES: Pvt. Co. E. Enl. 5/1/62 at White Sulphur Springs, Greenbrier Co., WV. Present on 4/30/63 roll. NFR.

DALLAS, W. R.: Pvt. Co. E. Enl. 5/1/62 at White Sulphur Springs, Greenbrier Co., WV. Det. at hosp. on Sept.-Oct. 1862 roll. NFR.

DAME, ROBERT C.: Pvt. Co. H. Res. Craig Co., Va. Enl. 3/4/63 at Lewisburg, Greenbrier Co., WV. WIA (serious foot) at White Sulphur Springs, WV 8/26-27/63. In hosp. with wd. on 12/3/63 roll. Present on 4/1/64 roll. On a July 1864 list of Co. B, Ward's Battn., C.S.A. prisoners recently released from military prison at Lynchburg, Va. Deserted from battn. Took oath at Charleston, WV 12/14/64. Age: 23. Farmer. 5'7". Dark complexion, light hair, blue eyes.

DAMRON, HENRY C.: Pvt. Co. G. Enl. 8/15/62 at Union, Monroe Co., WV. Det. service for surgeon on 4/30/63 roll. Present on 12/30/63 roll. Det. as clerk for regt. 4/30/64 roll. Paroled at Charleston, WV 5/10/65. Age: 35. 6'. Light complexion, light hair, blue eyes, light whiskers. Postwar roster says died of disease. Res. Alleghany Co., Va.

DAMRON, J. K.: Pvt. Co. F. Enl. 10/7/?? at Camp Beauregard. In Chimborazo Hosp. for debility 6/6/64. D. at Chimborazo Hosp. of measles (rubeola) 6/12/64.

DAMRON, SAMUEL A.: Pvt. Co. G. B. 3/12/38. Enl. 5/1/62 at White Sulphur Springs, Greenbrier Co., WV. At Chimborazo Hosp. for gunshot wd. of left forearm 5/27/64 to 6/6/64. Absent due to wd. on 10/31/64 roll. Paroled at Charleston, WV 5/10/65. Age: 21 or 27. 5'8". Light complexion, dark hair, blue eyes, dark whiskers. D. 7/4/1919. Bur. Palestine Presbyterian Church, Allegheny Co., Va. Res. Alleghany Co., Va.

DAMRON, STANNARD L.: Pvt. Co. G. Enl. 5/1/62 at White Sulphur Springs, Greenbrier Co., WV. Sick at Union, WV 8/29/62. Dischd. 12/22/62. Res. Alleghany Co., Va. Age: 25 at enl. PWR says D. of disease at home.

DAMRON, WILLIAM L.: Pvt. Co. G. Enl. 5/1/62 at White Sulphur Springs, Greenbrier Co., WV. Hosp. steward at Salt Sulphur Springs, WV July 1862. Det. as battn. steward on 11/13/62 roll. Res. Alleghany Co., Va. Age: 30. PWR says D. of disease at home 1863.

DARRING, JAMES A.: Pvt. Co. F. Enl. 5/1/62 at White Sulphur Springs, Greenbrier Co., WV. Absent on furlo. on 2/2863 roll. NFR.

DAUGHERTY, E. A.: Pvt. Co. G. Enl. 5/1/61 at White Sulphur Springs, Greenbrier Co., WV. Wded. and sick in hosp. since 5/24/62 on 11/13/62 roll. (probably WIA at Lewisburg, WV 5/23/62). Sick at Montgomery White Sulphur Springs, Va. July 1862. Dischd. 3/30/63.

DAUGHERTY, J. H.: Pvt. Co. G. Enl. 5/1/62 at White Sulphur Springs, Greenbrier Co., WV. Sick leave from regt. surgeon on 11/13/62 roll. Dischd. 3/23/63.

DAVENPORT, E. R.: Pvt. Co. H. Enl. 9/3/62 in Monroe Co., WV. Capt. at Cold Harbor (Gaines Mill), Va. 6/3/64. POW Pt. Lookout, MD 6/11/64. Transf. to Elmira, NY 7/12/64. Transf. for exchange 2/25/65. Paroled at Charleston, WV 4/28/65. Age: 22. In Richmond, Va. hosp. 5/2/65. 5'10". Fair complexion, light hair, blue eyes.

DAVIDSON, C. H.: Co. K. Res. Rockbridge Co., Va. PWR.

DAVIDSON, STEPHEN: Pvt. Co. A. Farm manager, Tyler Mountain, WV 1860. Age: 26. Enl. 5/22/61 at Pcoa, Putnam Co., WV. Colorbearer of regt. Paroled at Charleston, WV 5/29/65. Age: 35. 5'8". Fair complexion, brown hair, blue eyes. Postwar res. Forest Hill District, Summers Co., WV. Also listed as Stephenson.

DAVIS, HIRAM G.: Pvt. Co. C. Shoemaker, Fayette Co., WV 1860. Age: 25. Enl. 5/10/61.

DAVIS, J. H.: Pvt. Co. G. Enl. 7/3/61 at Charleston, Kanawha Co., WV. Left his co. at Charleston, WV about 7/15/61. Served as a teamster 1863-64.

DAVIS, J. P.: Pvt. Co. G. Enl. 7/3/61 at Charleston, Kanawha Co., WV. Left his co. at Clifton (Pratt), WV 7/24/61. NFR.

DAVIS, JEFF: Colored Cook. Co. H. Worked with colored cook William Armistead. PWR.

DAVIS, JOHN W.: Co. G. Entered Richmond, Va. Old Soldiers Home 7/14/99 at age 68. Dischd. 7/10/1901.

DAVIS, WILLIAM H.: Pvt. Co. B. Enl. 5/?/62 at White Sulphur Springs, Greenbrier Co., WV. Deserted 5/2/62 near White Sulphur Springs, WV. Listed as deserter on 11/10/62 roll.

DAWEWOOD, A. M.: Co. H. Deserted 5/28/62 on Aug. 1862 broadside. Craig Co., Va.

DEAL, C. THOMAS: Sgt./1st Lt. Co. C. Res. Ravenseye, WV. Enl. 5/1/62 at White Sulphur Springs, Greenbrier Co., WV. On scout on 12/31/62 roll. Present on 3/4/63 roll. Capt. at Winchester, Va. 9/19/64. Age: 28. POW Pt. Lookout, MD 9/24/64. Exchd. 3/15/65. Farmer. 5'7". Dark complexion, dark hair, dark eyes. Postwar res., Sewell, WV then Charleston, WV. D. 1923. Brother of Joe, Peter, Lafayette, and Thad Deal.

DEAL, JOSEPH: Pvt. Co. C. Res. Ravenseye, WV. Enl. 4/24/64 at Camp Gauley, Greenbrier Co., WV. In Charlottesville, Va. hosp. for acute diarrhea 6/10-17/64. Capt. at Winchester, Va. 9/19/64. POW Pt. Lookout, MD 9/24/64. Exchd. 3/13/65. Paroled at Lewisburg, WV 4/26/65. Age: 22. Farmer. 5'7". Dark complexion, dark hair, dark eyes. Res. Santa Barbara, CA ca. 1880. Brother of Tom, Peter, Lafayette, and Thad Deal. Res. Hondo, CA 1926. Middle initial is H.

DEAL, PETER W.: Pvt. Co. C. Res. Ravenseye, WV. Exchanged from Co. C, 60th Va. Inf. to 22nd Va. Inf. for Simon Thomas 4/24/64. Enl. 4/24/64 at Camp Gauley, Greenbrier Co., WV. Capt. at Rockville, MD 7/11/64. Gave himself up to a PA cav. regt. and desired to take oath and return to Fayette Co., WV. POW Elmira, NY 7/23/64. D. of typhoid fever at Elmira, NY 2/15/65. Bur. Woodland Cem., Elmira, NY. Brother of Tom, Joe, Lafayette, and Thad Deal. Farm laborer, Fayette Co., WV 1860. Age: 21.

DEAL, RICHARD LAFAYETTE: Pvt./Corp. Co. C. Res. Ravenseye, WV. Enl. 5/1/62 at White Sulphur Springs, Greenbrier Co., WV. Deserted 7/10/64. Capt. at Hagerstown, MD 7/13/64. POW Old Capitol Prison, Washington, D.C. 7/17/64. Sent to Washington, D.C. 7/20/64. POW Elmira, NY 7/23/64. Relsd. 11/18/64. Postwar res. and carpenter, Rainelle, WV. Married 1889. Attended the 70th anniversary of the Carnifex Ferry, WV battle in 1931 and claimed he had fought in the battle, indicating he may have served in 1861. At the anniversary he was said to be "hale and active and rode 150 miles in one day in an automobile to attend the celebration. He delighted thousands with a few remarks and the old rebel yell. He was the hero of the day." At that time (9/10/1931) was one of the last 3 surviving vet. of the battle. Brother of Tom, Joe, Peter, and Thad Deal. Farm laborer, Fayette Co., WV 1860. Age: 21.

DEAL, THAIDUS H.: Pvt. Co. C. Res. Ravenseye, WV. Enl. 9/26/62 at Camp Charleston, Kanawha Co., WV. Conscript. On scout by order of Gen. Williams on 12/31/62 roll. Capt. at Rockville, MD 7/15/64. Gave up voluntarily to a PA cav. regt. and desired to take oath and return to Fayette Co., WV. POW Elmira, NY 7/23/64. Relsd. 5/19/65. Took oath at Elmira, NY 5/29/65. 5'10". Fair complexion, auburn hair, hazel eyes. Postwar farmer near Corless, WV. Married 1868 and had 13 children. Entered Richmond, Va. Old Soldiers Home 9/14/1924 at age 80. D. 6/25/1925. Remains taken by relatives. Brother of Joe, Peter, Lafayette, and Tom Deal.

DEAL, THOMAS: Pvt. Co. E. Enl. 10/12/62 at Cotton Hill, Fayette Co., WV. Paroled at Charleston, WV 4/27/65. Age: 20. 5'3". air complexion, fair hair, blue eyes. Farm laborer, Fayette Co., WV 1860.

DEAN, BARTLEY: Pvt. Co. H. Enl. 5/8/61 at Charleston, Kanawha Co., WV. Absent cause unknown on 8/31/61 roll. NFR.

DEAN, WILLIAM H.: Pvt. Co. C. Res. Fayette Co., WV. Enl. 5/?/61 or 3/30/64 at Camp Gauley, Greenbrier Co., WV. Conscript. Capt. at Hagerstown, MD 7/13/64. Gave up voluntarily to Ft. Winchester Provost Marshal at Hagerstown, MD. POW Washington, D.C. 7/20/64. POW Elmira, NY 7/25/64. Relsd. 11/18/64. 5'10". Light complexion, auburn hair, hazel eyes.

DEAN, WILLIAM W.: Exchd. from Capt. W. D. Thurmond's Co. 10/1/64 for T. P. Hendrick.

DEARRING, WILLIAM D.: Pvt. Taylor's Co. Enl. 6/19/61 at Frankford, Greenbrier Co., WV. Age: 26. Co. transf. to Wise Brigade 7/10/61.

DEEDS, JOHN: Pvt. Co. E. Enl. 5/1/62 (?) at White Sulphur Springs, Greenbrier Co., WV. Deserted 5/22/62. Listed as deserter on Aug. 1862 broadside.

DEGRUYTER, MARTIN F.: Pvt. Co. D. B. 7/5/31 Veulo-on-the-Meuse, Holland. Educated Germany. Res. Clifton, WV 1854. Merchant, Charleston, WV 1860. Enl. 5/30/61 at Summersville, Nicholas Co., WV. Acting Commissary for Wise Legion since 6/13/61 on 8/31/61 roll. Transf. to 36th Va. Inf. as AQM 8/10/61. AQM says appt. Capt. Commissary 50th Va. Inf. 9/?/61. Division Commissary 10/?/61. Capt. QM 50th Va. Inf. Apptd. AQM, Dublin, Va. and remained there until seriously ill near close of war. Postwar res. Charleston, WV. Res. Covington, KY for 5 years. Commission business Cinn., OH. Res. Charleston, WV 1872. Wholesale malt liquor business. Still alive 1899. Bur. Spring Hill Cem., Charleston, WV. Name misspelled in CSR as DeGinyter and DeGryter.

DEISHER, ABSALOM P.: Pvt. Co. I. Enl. 5/1/62 at White Sulphur Springs, Greenbrier Co., WV. Sick at Giles C.H. (Pearisburg), Va. July 1862. Employed as a nurse at Giles C.H., Va. on 8/15/62 roll. Det. to tender the sick at Giles C.H., Va. on rolls dated 11/11/62 and 12/31/62. Det. as ward master at Giles C.H. Hosp. on 3/4/63 roll. Det. to hosp. at Union, WV 8/13/63 to 12/31/63. Present on 4/1/64 roll. Capt. at Winchester, Va. 9/19/64. POW Pt. Lookout, MD 9/24/64. Exchd. 3/15/65.

DEISHER, EVANS: Co. I. Listed as deserter on Aug. 1862 broadside. Botetourt Co., Va.

DEISHER, GEORGE: Co. I. Listed as deserter on Aug. 1862 broadside. Botetourt Co., Va.

DEMPSEY, JOHN E.: Pvt. Co. K. Enl. 6/6/61 at Charleston, Kanawha Co., WV. Det. as scout with Capt. Thurmond July 1862. AWOL on 11/1/62 roll. NFR. Res. Fayette Co., WV 1860. Age: 18. Bahlmann says transf. as a Pvt. in Capt. Thurmond's Co. Middle name is Edward.

DENT, GARLAND: Pvt. Co. I. Enl. 6/22/61 at Charleston, Kanawha Co., WV. AWOL 7/26/61.

DEWS, SAMUEL STONE: 1st Lt. Co. C. B. 3/25/35 Halifax Co., Va. Enl. 6/6/61 at Charleston, Kanawha Co., WV. Elected 1st Lt. 6/6/61. Resgnd. 8/5/61. Successor: E. C. Peters. Became a Capt. in 60th Va. Inf. PWR says WIA twice and capt. during war. Postwar res. Victor, WV. Res. Ansted, WV 1880. Member Jeb Stuart UCV Camp, Fayette Co., 1909. Farmer, New Hope, WV 1915. Farmer, Fayette Co., WV 1860.

DIAL, MOSES B.: Pvt. Co. G. Enl. 7/3/61 at Charleston, Kanawha Co., WV. Present on 9/3/61 roll. NFR.

DIAMOND, WILLIAM H. H.: Pvt. Co. K. Enl. 6/6/61 at Charleston, Kanawha Co., WV. Sick in hosp. at White Sulphur Springs, WV 8/21/61 on 9/4/61 roll. NFR. Day laborer, Fayette Co., WV 1860. Age: 19. Bahlmann says "left."

DICKERSON, JOHN WESLEY MICHAEL: Rank unknown. Co. C. B. Appomattox Co., Va. 7/1/47. Enl. Spring 1861. Sent home with typhoid fever. Re-enl. in 25th Va. Battn. Mustered out 4/9/65 at Appomattox, Va. One of the three surviving C.S.A. vet. at the Carnifex Ferry battle celebration in 1931. D. Rainelle, WV 8/3/1932 at the age of 89. Bur. McClung Cem., Hico, Fayette Co., WV. Not in CSR. Day laborer, Fayette Co., WV 1860. Postwar res. Clifty, WV.

DICKERSON, PAUL: Pvt. Co. A. Enl. 7/14/61 at Camp Tompkins, Kanawha Co., WV. Sick at Montgomery White Sulphur Springs, Va. July 1862. WIA (severe head) at White Sulphur Springs, WV 8/26-27/63. WIA (gunshot wd. to left side and pelvis) at Droop Mountain, WV 11/6/63. Absent due to wd. on rolls dated 12/31/63 and 4/30/64. In Farmville, Va. hosp. 3/21/64 to 8/20/64. Desired to join Invalid Corps. Paroled at Charleston, WV 5/6/65. Age: 30. 5'10". Dark complexion, dark hair, blue eyes, dark whiskers.

DICKERSON, R.: Pvt. Co. B. Date and place of enlistment not stated. Rcvd. pay 5/1/62 to 8/31/62.

DICKINSON, HENRY B.: Capt. Co. C. B. 1835. Enl. 7/15/61 at Charleston, Kanawha Co., WV. Elected Capt. 5/1/62. WIA at Cedar Creek, Va. 10/19/64. Absent due to wd. on 10/31/64 roll. Paroled at Charleston, WV 5/5/65. Age: 30. 5'6". Fair complexion, light hair, gray eyes, light whiskers. D. 1912. Bur. Dickinson Cem. near Fayetteville, WV. Physician, Fayette Co., WV 1860.

DIDDLE, JOHN M.: Sgt. Co. F. Enl. 7/4/61 at Charleston, Kanawha Co., WV. Sick at home on surg. cert. 8/13/61. Working at Nichols factory July 1862. Det. to work in woolen factory 10/31/62 to 2/25/63. Present on Mar.-Apr. 1863 roll. Wded. (very serious right leg) and admitted to Charlottesville, Va. hosp. 8/2/64. Still in hosp. on 10/31/64 roll. Took oath at Lewisburg, WV 4/27/65. Age: 26. Farmer. 5'10". Dark complexion, dark hair, dark eyes. Postwar res. Putnam Co., WV.

DILLEY, JACOB: Pvt. Co. D. Enl. 3/17/64 at Camp Gauley, Greenbrier Co., WV. Capt. at Winchester, Va. 9/19/64. POW Pt. Lookout, MD 9/23/64. Exchd. 3/15/65. Paroled at Charleston, WV 5/1/65. Age: 19. 6'1". Fair complexion, dark hair, hazel eyes.

DILWORTH, ANTHONY: Pvt. Co. B. Age: 37 in 1850. Res. Jacksion Co., WV.

DILWORTH, CHARLES "CES": Pvt. Co. B. Res. Jackson Co., WV. KIA.

DILWORTH, JOSEPH W.: Pvt./Sgt. Co. B. Res. Jackson Co., WV. Enl. 6/1/61 at Ripley, Jackson Co., WV. Capt. at Lewisburg, WV 5/23/62. POW Camp Chase, OH. Exchd. 8/25/62. Capt. at Ripley, WV 1/22/64. Age: 22. POW Camp Chase, OH 2/7/64. Transf. to Ft. Delaware 3/17/64. Relsd. 6/7/65. Printer. 5'11". Fresh complexion, brown hair, blue eyes.

DILWORTH, JOSIAH S.: Pvt. Co. B. Res. Ripley, WV 1860. Age: 20. Enl. 6/1/61 at Ripley, Jackson Co., WV. Capt. at Lewisburg, WV 5/23/62. POW Camp Chase, OH. Exchd. 8/25/62. In Montgomery White Sulphur Springs, Va. hosp. for wd. 1/10/64 to 2/10/64. On a list of deserters who surrendered at Charleston, WV in Feb. 1865. Took oath at Charleston, WV 2/17/65. Age: 25. Farmer. 6'. Fair complexion, light hair, blue eyes.

DIXON, W. T.: Pvt. Co. B. Res. Jackson Co., WV. Mustered 1861 for 4 years.

DODD, FLOYD W.: Pvt. Co. A. B. ca. 1841 Craig Co., Va. Enl. 7/9/62 at Salt Sulphur Springs, (Camp McCausland), Monroe Co., WV. Transf. to 27th Va. Inf. 2/18/63.

DODD, LORENZO R.: 1st Lt. Co. I. B. ca. 1833 Craig Co., Va. Enl. 5/1/62 at White Sulphur Springs, Greenbrier Co., WV. Elected 1st Lt. 5/1/62. Capt. at Lewisburg, WV 5/23/62. Age: 29. POW Camp Chase, OH 5/30/62. Exchd. 8/25/62. WIA (slight hip) at White Sulphur Springs, WV 8/26-27/63. Resgnd. at Camp Gauley due to disability 2/25/64. 5'11". Fair complexion, black hair, dark eyes, black whiskers. Res. Craig Co., Va. and/or Bland Co., Va.

DODDRIDGE, J. E. JR.: Pvt. Co. H. Enl. 1861.

DODDRIGE, JOHN M.: Pvt. Co. H. Enl. 5/1/61 at Charleston, Kanawha Co., WV. Acting commissary sgt. for brigade July 1862. Det. in Commissary Dept. July-Aug. 1862. Det. in A.C.S. Dept. on Sept.-Oct. and Nov.-Dec. 1862 rolls. Det. in A.C.S. Dept. at Saltville, Va. on 4/30/63 roll. Capt. at Winchester, Va. 9/19/64. POW Pt. Lookout, MD 9/24/64. Took oath at Pt. Lookout, MD and relsd. 6/12/65. 5'9-7/8". Light complexion, brown hair, gray eyes.

DODDRIDGE, PHILIP: Pvt. Co. H. B. 1835 Charleston, WV. Boat builder and pilot on Kanawha River 1853. Boatman, Charleston, WV 1860. Age: 25. Enl. 5/8/61 at Charleston, Kanawha Co., WV. Absent cause unknown on 8/31/61 roll. Det. in P.O. Dept. on Sept.-Oct. 1862 roll. Det. in telegram dept. on 12/31/62 roll and in telegram office on 2/28/63 roll. Det. in telegram office in Richmond, Va. on rolls from 4/30/63 to 12/25/64. Postwar res. Charleston, WV. D. 1894. Bur. Spring Hill Cem., Charleston, WV.

DODDS, THOMAS J.: Pvt. Co. D. enl. 5/25/63 at Camp Stonewall. Wded. in hosp. on 10/31/64 roll. Paroled at Charleston, WV 5/10/65. Age: 20. 5'9". Fair complexion, dark hair, blue eyes.

DONAHOE, ISAAC A.: Pvt. Co. C. Enl. 3/13/64 at Camp Gauley, Greenbrier Co., WV. In Chimborazo Hosp. for chronic diarrhea 6/1-26/64. Deserted 7/20/64. Took oath at Charleston, WV 12/13/64. Age: 19. Farmer. 5'. Fair complexion, light hair, gray eyes.

DONALDSON, JOHN P.: 2nd Lt./Capt. Co. H. Res. Philadelphia, PA. Res. Charleston, WV ca. 1858. Salesman in wholesale and retail hardware business. Enl. 5/8/61 at Charleston, Kanawha Co., WV Served as 2nd Sgt. until comm. 2nd Lt. 7/27/61. Elected 2nd Lt. 8/1/61. Elected Capt. 5/1/62. Sick in Lewisburg, WV 6/30/63. WIA (side and shoulder) at Droop Mountain, WV 11/6/63. Capt. at Cold Harbor (Gaines Mill), Va. 6/3/64. POW Pt. Lookout, MD. Transf. to Ft. Delaware 6/23/64. Relsd. 6/10/65. Paroled at Ft. Delaware 6/10/65. 5'11½". Fair complexion, brown hair, gray eyes. Postwar businessman Cinn., OH. William Bahlmann and Albert Singleton met him there postwar. Member Dry Goods Firm. Businessman, Philadelphia, PA 1901. Stricken with illness and went to sanitarium at Atlantic City, NJ. D. 7/22/1901.

DONNALLY, ANDREW MAT: Pvt. Co. H. Enl. 5/8/61 at Charleston, Kanawha Co., WV. Capt. Jan. 1864 head of Big Sycamore Creek, Clay Co., WV. Escaped and returned to regt. Capt. at Cold Harbor (Gaines Mill), Va. 6/3/64. POW Pt. Lookout, MD. Transf. to Elmira, NY 7/12/64. Transf. for exchange 3/10/65. Paroled at Elmira, NY 3/10/65. Still alive 1901.

DONNALLY, ANDREW VAN: Pvt. Co. H. Enl. 6/1/61 at Charleston, Kanawha Co., WV. Transf. from Capt. Hale's Artillery Co. at White Sulphur Springs, WV 8/17/61. Capt. Jan. 1864 head of Big Sycamore Creek, Clay Co., WV. Escaped and returned to regt. Det. by Gen. Echols First Brigade on 4/30/64 roll. Capt. at Cold Harbor (Gaines Mill), Va. 6/2-3/64. POW Pt. Lookout, MD 6/11/64. POW Elmira, NY 7/12/64. Paroled at Elmira, NY 3/10/65 and transf. for exchange same day. Still alive 1901.

DOOLEY, JOHN L.: Pvt. Co. I. Enl. 5/1/62 at White Sulphur Springs, Greenbrier Co., WV. Sick in Lewisburg, WV on 4/30/63 roll. Capt. at Winchester, Va. 9/19/64. POW Pt. Lookout, MD 9/24/64. Exchd. 3/15/65.

DOOSING, JOHN W.: Pvt. Co. E. B. 1846. Enl. 3/4/64. at Dublin, Pulaski Co., Va. Conscript. $60 deducted from wages for lost Enfield rifle on 10/31/64 roll. D. 1917.

DORSEY, BENJAMIN B.: Pvt./Corp. Co. D. Enl. 5/30/61 at Summersville, Nicholas Co., WV. WIA (severe breast) at White Sulphur Springs, WV 8/26-27/63. Paroled at Lewisburg, WV 4/25/65. Age: 25. Farmer. 5'8". Fair complexion, brown hair, hazel eyes.

DORSON, HYATT: Pvt. Co. E. Enl. 10/12/62 at Cotton Hill, Fayette Co., WV. Present on 10/31/64 roll. NFR.

DOSS, GEORGE: Co. I.

DOSS, HARVEY: Pvt. Co. I. Res. Boone Co., WV. Enl. 6/22/61 at Charleston, Kanawha Co., WV. Absent sick 7/26/61. Not able for service. Took oath at Charleston, WV 5/1/65. Age: 27. Farmer. 5'4". Fair complexion, fair hair, gray eyes.

DOSS, JAMES A.: Sgt. Co. I. Enl. 6/29/61 at Charleston, Kanawha Co., WV. Deserted at Charleston, WV 10/15/62.

DOSS, WILLIAM: Co. I.

DOTSON, DANIEL: Pvt. Co. I. Enl. 6/22/61 at Charleston, Kanawha Co., WV. AWOL 7/28/61.

DOTSON, ELIJAH D.: Pvt./Corp./Sgt. Co. E. B. 1830. Enl. 6/8/61 at Charleston, Kanawha Co., WV. Served as corp. at organization of co. Guarding prisoners sent to Richmond, Va. on 8/31/61 roll. Promo. to ensign 4/30/64. At Chimborazo Hosp. for very serious arm fracture 6/5-12/64. On wded. furlo. on 10/31/64 roll. D. 1917. Bur. Teays Hill Cem., St. Albans, WV.

DOTSON, ISAAC F.: Pvt. Co. D. Enl. 5/30/61 at Summersville, Nicholas Co., WV. On det. service feeding brig. stock since Nov. 1863 on rolls dated 4/12/64 and 4/30/64. Present 10/31/64 roll. Paroled at Charleston, WV 5/10/65. Age: 25. 5'8". Light complexion, brown hair, blue eyes, dark whiskers.

DOTSON, JAMES A.: Pvt. Co. D. Enl. 8/29/61 at Camp Gauley (Carnifex Ferry), Nicholas Co., WV. Transf. from Capt. Jones' co. by order of Gen. Floyd 8/29/61. NFR.

DOTSON, THOMAS C.: Pvt. Co. D. Enl. 8/29/61 at Camp Gauley (Carnifex Ferry), Nicholas Co., WV. Transf. from Capt. Jones' co. by order of Gen. Floyd 8/29/61. NFR.

DOTSON, WILLIAM: Pvt. Co. I. Enl. 6/22/61 at Charleston, Kanawha Co., WV. AWOL 7/26/61.

DOUGLAS, JOSHUA: Pvt. Co. E. Enl. 7/2/61 on Sandy. AWOL 7/24/61.

DOWDY, WILLIAM H.: Pvt. Co. I. Enl. 5/2/62 at White Sulphur Springs, Greenbrier Co., WV. Conscript. Enl. 6/1/62 in 36th Va. Inf. at Giles Co., Va. Transf. to 22nd Va. Inf. 8/1/62. Under arrest on rolls dated 8/31/63 and 4/10/64. On a list of Co. B, Ward's Battn. of prisoners at Lynchburg, Va. Deserted 9/1/64 at Leetown, Va. although listed as capt. at Smithfield, Va. 8/20/64. POW Camp Chase, OH 9/2/64. Transf. to Chicago, IL 3/20/65. Desired to take oath. Enlisted in U.S. army.

DRAKE, WILLIAM: Co. H. Deserted 5/4/62 on Aug. 1862 broadside. Greenbrier Co., WV.

DRANSFIELD, GEORGE: Pvt. Co. A. Date and place of enl. not stated. Forage master 8/1/62 to 9/10/62. Paroled at Charleston, WV 6/12/65. Age: 19. 5'8". Light complexion, light hair, blue eyes.

DRESSLER, CHARLES H.: Pvt. Co. K. B. 1/2/27. Enl. 5/1/62 (?) at White Sulphur Springs, Greenbrier Co., WV. D. at Montgomery White Sulphur Springs, Va. 7/12/62. According to a postwar source died 11/26/84. Bur. Mt. Pleasant U. Methodist Church, Covington, Va.

DRISCALL, JOSEPH W.: Pvt. Co. I. Date and place of enlistment not stated. Left sick at White Sulphur Springs, WV July 1862. NFR.

DRUMMOND, CHARLES T.: Pvt. Co. H. B. 4/10/22. Enl. 3/?/62 for 2 years. WIA (right hip). Res. Craig Co., Va. 1898 at age 76. D. 6/29/22 at age 100.

DRYDEN, JOHN: Pvt. Co. H. B. 12/27/17. Deputy Clerk in circuit and county court, Charleston, WV 1860. Age: 41. Secretary of Kanawha Rifleman 4/19/61. QM Sgt. Kanawha Rifleman 4/26/61. Enl. 36th Va. Inf. as Acting Commissary 8/19/61. D. 7/15/82. Bur. Spring Hill Cem., Charleston, WV. Lot 49 Section 5.

DUDDING, CHARLES A.: Pvt. Co. A. Res. Putnam Co., WV. Enl. 5/22/61 at Poca, Putnam Co., WV. Det. service on 2/28/63 roll. Present on 4/30/63 roll. Det. as ferryman at Greenbrier River on 4/30/64 roll. Took oath at Charleston, WV 3/25/65. Age: 27. Farmer. 6'. Dark complexion, black hair, gray eyes.

DUDDING, JOHN T.: 1st Sgt. Co. A. B. 1834. Laborer, Coalsmouth (St. Albans), WV 1860, Age: 26. Enl. 5/22/61 at Poca, Putnam Co., WV. WIA (cannonball) at Scary Creek, WV 7/17/61. Taken to a Charleston, WV hosp. Res. Rocky Fork, Kanawha Co., WV. Farmer. Two years Justice Kanawha and Putnam counties. 3 years Kanawha Co. Board of Education.

DUDDING, WILLIAM S.: Pvt. Co. A. Res. Coalsmouth (St. Albans), WV 1860. Age: 17. Enl. 7/17/61 at Scary Creek, Putnam Co., WV. Unaccounted for since 7/24/61 on 8/31/61 roll. Present on 11/1/62 roll. Left sick at Marlin's Bottom (Marlinton), WV on 4/30/63 roll. On wded. furlo. on 12/31/63 roll. Present on 4/30/64 roll.

DUKE, RICHARD L.: Sgt. Co. A. Date and place of enlistment not stated. Rcvd. pay 1/1/62 to 6/30/62.

DULING, JOHN R.: Pvt. Co. H. Farm laborer, Charleston, WV 1860. Age: 18. Enl. 5/8/61 at Charleston, Kanawha Co., WV. Capt. Jan. 1864 head of Big Sycamore Creek, Clay Co., WV. Escaped and returned to regt. Listed as capt. by the enemy on 4/30/64 roll. D. sometime prior to 1901.

DUMAS, JOHN GERRARD: Pvt. Swann's Co. Enl. 5/17/61 at Charleston, Kanawha Co., WV. Age: 44. Co. transf. to Co. K, 59th Va. Inf. 7/25/61.

DUNBAR, ISAAC E.: Pvt. Co. D. Enl. 5/1/62(?) at White Sulphur Springs, Greenbrier Co., WV. Deserted (probably at Lewisburg, WV) 5/23/62(?) on 10/31/62 roll.

DUNBAR, JOHN M.: Pvt. Co. D. Enl. 5/1/62 (?) at White Sulphur Springs, Greenbrier Co., WV. Deserted (probably at Lewisburg, WV) 5/23/62(?).

DUNBAR, SAMUEL N.: Pvt. Co. D. Enl. 5/1/62(?) at White Sulphur Springs, Greenbrier Co., WV. Sick at Dublin Depot, Va. July 1862. Deserted from hosp. at Dublin, Va. on 10/31/62 roll.

DUNBAR, WILLIAM T.: Pvt. Co. F. Date and place of enlistment not stated. Rcvd. clothing 4 qr. of 1864. Paroled at Charleston, WV 6/12/65. Age: 43. 5'8". Dark complexion, gray hair, hazel eyes.

DUNCAN, ERASTUS A.: Capt. Co. G. Res. Oceana, WV 1860 sheriff, age 30. Enl. 7/3/61 at Charleston, Kanawha Co., WV. Elected Capt. 7/8/61. D. 12/?/61. Another source says died 3/20/1917. Recruiting volunteers at Oceana, WV May 1861, with William T. Sarver and James A. Cook.

DUNCAN, JAMES A.: Pvt. Co. C. B. 1839 Green Sulphur Springs, WV. Blacksmith. Enl. 5/1/62(?) at White Sulphur Springs, Greenbrier Co., WV. Capt. at Winchester, Va. 9/19/64. POW Pt. Lookout, MD 9/24/64. Exchd. 3/15/65. Paroled Charleston, WV 6/12/65. Age: 41. 5'10". Dark complexion, black hair, hazel eyes. PWR says born 1827 Craig Co., Va. D. 4/21/1916 of old age at Green Sulphur Springs, WV. Bur. Halidan Cem., near Green Sulphur Springs, WV.

DUNCAN, JAMES B.: Pvt. Taylor's Co. Enl. 6/19/61 at Frankford, Greenbrier Co., WV. Age: 23. Co. transf. to Wise Brigade 7/10/61.

DUNCAN, JOHN LAWSON: Pvt. Co. C. B. ca. 1838 Craig Co., Va. Res. Level Green, Craig Co., Va. Enl. 5/1/62 at White Sulphur Springs, Greenbrier Co., WV. Capt. at Droop Mountain, WV 11/6/63. POW Camp Chase, OH 11/18/63. In Camp Chase, OH hosp. as "never vaccinated" 2/29/64 to 4/22/64. Transf. to Ft. Delaware 3/4/64. Transf. to Johnson's Island, OH 9/13/64. D. at Johnson's Island, OH 12/21/64 and buried there in grave #146. Age: 34. Farmer. 5'9½". Fair complexion, dark hair, brown eyes. Court records indicate he was 36 when he died.

DUNN, JOHN H.: Pvt. Co. F. Enl. 10/6/63 or 11/1/63 at Camp Beauregard. Res. Monroe Co., WV. Appears on a list of rebel deserters. Took oath and sent North 3/21/65. Age: 26. Merchant. 5'8". Fair complexion, dark hair, black eyes.

DUNN, LEWIS A.: Pvt. Co. F. B. in Giles Co., Va. 11/7/34. Res. Monroe Co., WV. Brother of William A. Dunn. Enl. 5/1/62(?) at White Sulphur Springs, Greenbrier Co., WV. Another source says enlisted in 1861. Surrendered at Charleston, WV March 1865. Took oath at Charleston, WV 3/20/65. Age: 30. Farmer. 6'. Fair complexion, gray hair, dark eyes. Postwar res. Mercer Co., WV. D. sometime prior to 1914.

DUNN, RUFUS K.: Pvt. Co. F. Res. Monroe Co., WV. Enl. 7/4/61 at Charleston, Kanawha Co., WV. Det. to take prisoners to Richmond, Va. 8/28/61. Present on 10/31/62 roll. Exchd. to Bryan's Battery on Mar.-Apr. 1863 roll. Went West after the war.

DUNN, WILLIAM A.: Pvt. Co. F. Res. Monroe Co., WV. Brother of Lewis A. Dunn. Enl. 7/4/61 at Charleston, Kanawha Co., WV. Capt. at Winchester, Va. 9/19/64. POW Pt. Lookout, MD 9/23/64. D. at Pt. Lookout, MD of pneumonia 1/23/65. Bur. grave #849 in Pt. Lookout, MD POW graveyard.

DUNN, W. L.: Ord. Sgt. Co. F. Res. Monroe Co., WV. D. sometime prior to 1914.

DUNNAVANT, JOHN: Pvt./Corp. Co. I. Enl. 5/2/61 at Charleston, Kanawha Co., WV. Present on 7/8/63 roll. NFR.

DUNSMORE, JOHN: Pvt. Co. F. Res. Monroe Co., WV. Enl. 5/1/62 at White Sulphur Springs, Greenbrier Co., WV. Driving brigade wagon on 4/1/64 roll. Det. service on 4/30/64 roll. Paroled at Lewisburg, WV 5/10/65. Age: 45. Farmer. 5'10". Dark complexion, light hair, dark eyes.

DYCHE, ELIJAH: Pvt. Co. A. B. in Va. Enl. 4/?/62 at Lewisburg, Greenbrier Co., WV. Another source says enlisted in 1861 at Harpers Ferry, WV (although that is unlikely). Det. service as brigade butcher on 12/31/63 roll. Present on 4/30/64 roll. In Charlottesville, Va. hosp. for chronic diarrhea 6/12-17/64. Paroled at Lewisburg, WV 4/25/65 and/or paroled at Charleston, WV 5/10/65. Age: 39. Butcher. 5'8" or 5'7½". Dark complexion, brown hair, black eyes, brown whiskers.

DYE, CORNELIUS H.: Pvt. Co. B. Enl. 6/1/61 at Ripley. Jackson Co., WV. On det. service on 6/10/64 roll. Capt. at Winchester, Va. 9/19/64. POW Pt. Lookout, MD 9/24/64. Exchd. 3/15/65. Took oath and sent North 4/25/65. Age: 27. Farmer. 5'10½". Fair complexion, black hair, blue eyes.

EADES, JOSHUA: Pvt. Co. F. Enl. 6/4/61 at Charleston, Kanawha Co., WV. Sick at home on surg. cert. 8/22/61. Killed by a freight train at Alderson, WV about 1890.

EADS, AUGUSTUS C.: Pvt. Co. C. Res. Monroe Co., WV, 1857. Enl. 8/3/61 at Union, WV in 108th Regt. Va. Militia. Enl. 3/24/62 at Union, WV. In Capt. A. B. McNeers Co. 108th Regt. Va. Militia. Enl. 6/15/62 at Salt Sulphur Springs, Monroe Co., WV. Accidently wded. and sent home on furlo. on July 1862 roll. Capt. at Cedar Creek, Va. 10/19/64. POW Pt. Lookout, MD. Has a court-martial record while under Gen. Early. Paroled at Charleston, WV 6/12/65. Farmer, Fayette Co., WV 1860. Age: 40.

EADS, ESOM C.: Pvt. Co. H/Co. F. B. ca. 1832 Craig Co., Va. Wheelwright. Res. Craig Co., Va. Enl. 4/28/62 at White Sulphur Springs, Greenbrier Co., WV. Transf. to Co. F 3/13/63. MWIA (severe head) at White Sulphur Springs, WV 8/26-27/63. D. at White Sulphur Springs, WV 8/29/63.

EADS, W. J.: Pvt. Co. C. Enl. 6/29/62(?) at Salt Sulphur Springs, Monroe Co., WV. Wded. at Salt Sulphur Springs, WV and at home on 3/4/63 roll.

EAGLE, CHARLES: Pvt./Corp./Capt. Co. E. Enl. 6/10/61(?) or 7/27/61 at Gauley Bridge, Fayette Co., WV. Another card says enl. 5/1/62 at White Sulphur Springs, Greenbrier Co., WV. Left sick at White Sulphur Springs, WV 8/20/61. Present on Sept.-Oct. 1862 roll. Promo. to Capt. 4/1/64. At Stuart Hosp., Richmond, Va. for vulvius sclopet 6/4/64 to 11/22/64. Reported as wded. possibly at or near Staunton, Va. on 10/31/64 roll. In Charlottesville, Va. Hosp. for very serious leg wd. 11/23/64 to 1/21/65. At Pettigrew Gen. Hosp., Raleigh, NC for debilitas 2/22/65 to 3/6/65.

EAGLE, DAVID: Pvt. Co. E. Enl. 6/8/61 at Charleston, Kanawha Co., WV. Left sick at White Sulphur Springs, WV 8/20/61. Present on 3/4/63 roll. Det. brigade shoemaker on 12/31/63 roll. Capt. at Winchester, Va. 9/19/64. POW Pt. Lookout, MD 9/23/64. Exchd. 10/30/64. Listed as a paroled prisoner 4th qr. 1864. Took oath at Union, WV 7/15/65. Age: 31. 5'9½". Dark complexion, black hair, black eyes, black whiskers.

EAGLE, JOHN: Pvt. Co. E. Res. Clay Co., WV. Enl. 6/8/61 at Charleston, Kanawha Co., WV. Left sick at White Sulphur Springs, WV 8/20/61. Present on Sept.-Oct. 1862 roll. In Charlottesville Va. hosp. for debilitas 1/20-21/65. Capt. in Clay Co., WV 2/12/65. POW Camp Chase, OH 2/24/65. Age: 28. Farmer. 5'9". Fair complexion, dark hair, gray eyes.

EAGLE, STUART: Pvt. Co. E. Enl. 5/1/62 at White Sulphur Springs, Greenbrier Co., WV. WIA (serious both legs) at White Sulphur Springs, WV 8/26-27/63. Present on rolls dated 12/30/63 and 10/31/64.

EAKIN, JOHN E.: Pvt. Co. E. Enl. 5/1/62 at White Sulphur Springs, Greenbrier Co., WV. Sick at Giles C.H. (Pearisburg), Va. since May 1862 on July 1862 roll. Sick in hosp. on Sept.-Oct. 1862 roll. Det. in nitre service on rolls dated 12/31/63 and 10/31/64.

EASLEY, ISAAC HOWSON: 2nd Lt. Co. K. Enl. 6/6/61 at Charleston, Kanawha Co., WV. Elected 2nd Lt. 6/6/61. Resgnd. 11/20/61. Bahlmann says went into salt petre and groundhog dept. and returned to farm.

EDDY, GEORGE W.: Pvt. Co. F. Enl. 5/1/62(?) at White Sulphur Springs, Greenbrier Co., WV. Det. in hosp. at White Sulphur Springs, WV on 12/31/63 roll. Present on 4/1/64 roll. Paroled at Charleston, WV 6/12/65. Age: 42. 5'11". Dark complexion, dark hair, blue eyes, dark whiskers.

EDMONDS, SAMUEL D.: Pvt. Co. B. Res. Cow Run, Mason Co., WV. Enl. 6/?/63 at Lewisburg, Greenbrier Co., WV. Originally enl. in Edgar's Battn. in Nov. 1862 at Lewisburg, WV. Transf. to Co. B, 22nd Va. Inf. ("traded with Ramsey"). Rcvd. clothing 5/13/64. Participated in New Market, Va. battle 5/15/64. Wrote an unpublished memoir of his service. Entered Richmond, Va. Old Soldiers Home 4/20/1916 at age 72. Dischd. 7/28/1916.

ELKIN, RILANT: Pvt. Co. I. Enl. 6/22/61 at Charleston, Kanawha Co., WV. AWOL 7/26/61.

ELKINS, DANIEL: 2nd Lt./1st Lt. Co. I. Enl. 6/29/61 at Charleston, Kanawha Co., WV. Elected 2nd Lt. 6/29/61. Elected 1st Lt. 3/1/62. Dropped 5/2/62.

ELLIOTT, JAMES: Pvt. Co. B. Res. Jackson Co., WV. Enl. 6/1/61 at Ripley, Jackson Co., WV. Det. service on 1/10/64 roll. Capt. in Roane Co., WV 1/15/64. POW Camp Chase, OH 2/2/64. Transf. to Ft. Delaware 3/17/64. Relsd. 6/20/65. Age: 23. Farmer. 5'9½". Fair complexion, dark hair, hazel eyes.

ELLIOTT, THOMAS B.: Pvt. Co. D. Enl. 5/1/62 at White Sulphur Springs, Greenbrier Co., WV. Sick at Beverly, WV 4/25/63. WIA (serious hand and leg) at White Sulphur Springs, WV 8/26-27/63. Present on 12/31/63 roll. In Winchester, Va. hosp. for debilitas 8/1-3/64. Capt. at Winchester, Va. 9/19/64. POW Pt. Lookout, MD 9/24/64. Exchd. 2/13/65. D. at Oak Hill, WV. 10/20/1920 at the age of 80.

ELLIS, MARION F.: Enl. 6/10/61 Nicholas Co., WV in Co. F, 36th Va. Inf. Served with 22nd Va. Inf. Apr. 1862. Elected Sgt., Co. F, 36th Va. Inf. 5/13/62 and remained with that regt. until he died at Pt. Lookout, MD of heart trouble. Bur. Pt. Lookout, MD on Jersey shore. Age: 32. 5'9½". Gray eyes and light hair.

ELMORE, CLIFTON WALKER: Pvt. Co. D. Res. New Castle, Craig Co., Va. Enl. 6/18/63 at Camp Jackson. PWR says enl. June 1862. Reportedly fought at New Market, Va. Hated war. In Chimborazo Hosp. 5/25/64 to 6/2/64. Transf. to Farmville, Va. hosp. for serious right foot wd. Furlo. from hosp. for 40 days on 6/7/64 roll. At home on sick furlo. on 10/31/64 roll. Res. Craig Co., Va. in 1898 at age 53. Postmaster, New Castle, Va. D. 4/21/1940 at age 95, 10 days. Bur: West View Cem., old Rt. 42, New Castle, Va. PWR says WIA at the battle of the Wilderness (?). Considered himself Craig Co.'s last C.S.A. veteran.

ELMORE, DEXTER O. (or S.): Pvt. Co. D. Enl. 5/30/61 at Summersville, Nicholas Co., WV. Sick at home since 7/28/61 on 8/31/61 roll. Sick since Sept. 1861 on 12/31/62 roll. NFR. Possible res. Craig Co., Va.

ELMORE, ELIAS: Listed in UDC booklet *Craig's Share in the War Between the States*. Res. Craig Co., Va.

ELMORE, JACOB: Pvt. Co. G. Res. Craig Co., Va. B. ca. 1825 Craig Co., Va. Enl. 11/28/63 at Dublin, Pulaski Co., Va. Deserted at Salem, Va. on 10/31/64 roll. Capt. in Randolph Co., WV 9/28/64. Age: 39. Took oath at Charlesburg, WV and sent North 10/5/64. 5'8". Fair complexion, dark hair, blue eyes.

ELMORE, JOEL BROWNLEW: Co. K. PWR.

ELSWICK, JOHN: Pvt. Co. G. Enl. 7/3/61 at Charleston, Kanawha Co., WV. On a list of mounted guard at White Sulphur Springs, WV 8/13-14/61. Left sick at White Sulphur Springs, WV 8/20/61. NFR.

ENGLES, JOSEPH: Pvt. Co. E. Enl. 3/30/64 at Dublin, Pulaski Co., Va. Conscript. Capt. at Winchester, Va. 9/19/64. POW Pt. Lookout, MD 9/23/64. Relsd. by order of the President 1/26/65. Postwar res. Roanoke Co., Va.

ERSKIN, A. M.: Pvt. Co. F. Enl. 6/9/62(?) at Camp McCausland, Monroe Co., WV. At home sick near Rock Point, WV on Mar.-Apr. 1863 roll. Under arrest on 4/1/64 roll. Deserted about 5/11/64.

ERSKINE, CALVIN: Pvt. Co. G. Res. Monroe Co., WV. Enl. 5/1/62 at White Sulphur Springs, Greenbrier Co., WV. Left sick at Greenbrier Bridge, Pocahontas Co., WV on 4/30/63 roll. Deserted while on furlo. on 12/31/63 roll. Surrendered, took oath, and sent North of the Ohio River 11/15/63. Age: 32. Blacksmith. 5'8". Light complexion, brown hair, gray eyes.

ERSKINE, GEORGE WASHINGTON: Pvt. Co. G. Res. Monroe Co., WV. Enl. 5/1/62 at White Sulphur Springs, Greenbrier Co., WV. Sick in hosp. on 11/13/62 roll. Capt. at White Sulphur Springs, WV. 8/26-27/63. Age: 30. POW Camp Chase, OH 9/7/63. Transf. to Rock Island, IL 1/22/64. Took oath at Rock Island, IL 6/16/65. Farmer. 5'9¾". Dark complexion, dark hair, black eyes. Res. Greenbrier and Monroe counties of WV until circa. 1920. Applied for pension 1920.

ERSKINS, M.: Pvt. Co. K. Date and place of enlistment not stated. Dischd. at Camp McCausland, WV 7/15/62.

ESTEP, ANDREW JACKSON: Pvt. Co. I. Res. Boone Co., WV. B. ca. 1833 Boone Co., WV. Married 1861. Enl. 6/22/61 at Charleston, Kanawha Co., WV. Absent 7/26/61. Not able for service being dispeptic. Present on Sept.-Oct. 1862 roll. Capt. at Droop Mountain, WV 11/6/63. Age: 31. POW Camp Chase, OH 11/18/63. Transf. to Ft. Delaware 3/14/64. Relsd. and took oath at Ft. Delaware 6/20/65. 6'2". Florid complexion, dark hair, gray eyes.

ESTEP, MARTIN: Pvt. Co. I. Res. Boone Co., WV. Enl. 6/22/61 at Charleston, Kanawha Co., WV. Capt. at Strasburg, Va. 10/19/64. POW Pt. Lookout, MD 10/23/64. Exchd. 3/28/65. Married postwar in Boone Co., WV. Brother of William Estep.

ESTEP, MASTIN: Pvt. Co. I. Res. Boone Co., WV 1860. Enl. 6/1/63 at Lewisburg, Greenbrier Co., WV. WIA (slight hand) at White Sulphur Springs, WV 8/26-27/63. Took oath at Charleston, WV 5/2/65. Age: 16. Farmer. 5'5". Dark complexion, dark hair, black eyes. Also spelled name Maston. (Also listed under this name is M. Estep, Pvt. Co. I, 22nd Va. Inf., age: 25. Farmer. 6'. dark complexion, hazel eyes, dark hair. Also spelled Estees and Esteses). Entered Richmond, Va. Old Soldiers Home 8/17/1916 at age 69. Dropped 9/14/1917. Registered as serving in Co. C.

ESTEP, MILES: Pvt. Co. I. B. circa 1831. Res. Kanawha Co., WV 1850. Enl. 6/22/61 at Charleston, Kanawha Co., WV. Sick at Lewisburg, WV on 4/30/63 roll. Deserted near Staunton, Va. 6/28/64. Surrendered May 1865. Took oath and sent North 5/27/65. Age: 23. Farmer. 6'. Dark complexion, dark hair, hazel eyes.

ESTEP, SAMUEL: Pvt. Co. I. B. ca. 1835. Married 1855 in Boone Co., WV. Enl. 3/1/64 at Camp Gauley, Greenbrier Co., WV. Courtmartial record dated 8/15/64. Surrendered May ;1865. Took oath at Charleston, WV 5/2 or 22/65. Age: 17. Farmer. Fair complexion, fair hair, hazel eyes. Res. Boone Co., WV.

ESTEP, WILLIAM: Pvt. Co. I. B. ca. 1833. Enl. 6/22/61 at Charleston, Kanawha Co., WV. WIA (slight hand) at White Sulphur Springs, WV 8/26-27/63. Deserted in Alleghany Co., Va. 5/4/64. Brother of Martin Estep.

ESTILL, DAVIS H.: Pvt. Co. H. Res. Charleston, WV in 1860. Age: 16. Enl. 5/20/63 at Lewisburg, Greenbrier Co., WV. Capt. at Cold Harbor, Va. 6/3/64. In Charlottesville, Va. hosp. for acute dysentery 6/9-17/64. Capt. at Winchester, Va. 9/19/64. POW Pt. Lookout, MD 9/24/64. D. at Pt. Lookout, MD of chronic diarrhea 1/5/65. Bur. grave #763 in POW graveyard at Pt. Lookout, MD.

ESTILL, WILLIAM H.: Pvt. Co. H. B. 1838. Res. Charleston, WV 1860. Enl. 5/8/61 at Charleston, Kanawha Co., WV. Absent cause unknown on 8/31/61 roll. Det. in Commissary Dept. at Lewisburg, WV on 4/30/63 roll. Det. in A.C.S. Dept. at Lewisburg, WV on rolls dated 6/30/63 and 12/3/63. Det. in A.C.S. Dept., First Brigade on rolls dated 1/1/64 and 12/28/64. Paroled at Lewisburg, WV 4/27/65. Age: 27. Steamboat. 5'11". Dark complexion, dark hair, brown eyes. D. 1872.

EVANS, A.: Co. C. D. 8/27/64 at Ft. Delaware. Bur. Finn's Point Nat. Cem., NJ. Served as a Pvt.

EVANS, BARNABUS: Pvt. Co. G. Enl. 7/3/61 at Charleston, Kanawha Co., WV. Dischd. by G. M. McDonald, surg. of the 22nd Va., at White Sulphur Springs, WV for disability 7/11/61.

EVANS, EDWARD R.: Co. G. Alleghany Co. pension rolls.

EVANS, LEWIS: Pvt. Co. B. Enl. 5/?/62 at White Sulphur Springs, Greenbrier Co., WV. Conscript. Paid as a teamster 7/1/63 to 9/30/63 and 12/1/63 to 1/31/64. Det. Brig. teamster on 1/10/64 roll. Present on 11/30/64 roll. D. 8/10/1912 at Rich Hollow. Age: about 90. Res. Fayette Co., WV for many years. Bur. Holley Cem.

EVANS, OLIVER PERRY: Pvt. Co. B. Son of Ephraim S. and Ruama (Wright) Evans. Res. Jackson Co., WV 1860. Age: 18. Enl. 6/1/61 at Ripley, Jackson Co., WV. In hosp. at White Sulphur Springs, WV on 9/1/61 roll. Apptd. V.M.I. cadet 9/9/61. Dischd. to attend V.M.I. 1/3/62. Promo. for gallantry on the battlefield. Comm. Capt. Flag bearer for the V.M.I. cadets at the battle of New Market, Va. 5/15/64. Nickname is "Big."

EVANS, TIMOTHY PATTERSON: Pvt. Co. G. Enl. 5/1/62 at White Sulphur Springs, Greenbrier Co., WV. Listed as wded. on 10/31/64 roll. This could be the man postwar rosters refer to as Patrick Evans of Co. G. Patrick Evans was res. Botetourt Co., Va. Age: 22 at enl. Surrendered Apr. 1865.

FALL, B. F.: Pvt. Co. B. Enl. 5/?/62 at White Sulphur Springs, Greenbrier Co., WV. Conscript. WIA and capt. at Lewisburg, WV. 5/23/62.

FARLEY, NOAH D.: Pvt. Co. I. Res. Giles Co., Va. Enl. 6/22/61 at Charleston, Kanawha Co., WV. Capt. at Jenkins River, Va. 1/23/62. POW Camp Chase, OH 1/22(?)/62. Returned to regt. 10/30/62. Dischd. at Camp Sam Miller 2/1/63. Capt. in Boone Co., WV 11/1/63. Age: 47. Bushwhacker. POW Camp Chase, OH 11/13/63. Transf. to Ft. Delaware 2/29/64. Took oath and relsd. at Ft. Delaware 6/20/65. Farmer. 6'3". Dark complexion, dark hair, blue eyes.

FARLEY, PERRY: Pvt. Co. I. Enl. 6/22/61 at Charleston, Kanawha Co., WV. Paid as a teamster 5/31/63 to 9/30/63. Regt. wagoner 8/31/63 to 12/31/63. Paid as a teamster 2/1/64. Brig. wagoner 4/1/64. Paroled at Charleston, WV 5/10/65. Age: 24. 6'2". Fair complexion, light hair, blue eyes.

FARMER, CORNELIUS H.: Pvt. Co. A. Enl. 6/5/61 at Camp Tompkins, Kanawha Co., WV. Present on 8/31/61 roll. NFR.

FARRALL, JOHN A.: Pvt. Co. K. Res. Greenbrier Co., WV. B. ca. 1828 Craig Co., Va. Enl. 4/15/62 at Lewisburg, Greenbrier Co., WV or 5/1/62 at White Sulphur Springs, Greenbrier Co., WV. Capt. at Droop Mountain, WV 11/6/63. Age: 40. POW Camp Chase, OH 11/18/63. Transf. to Ft. Delaware 2/29/64. Took oath and relsd. at Ft. Delaware 6/20/65. Laborer. 5'6". Fair complexion, dark hair, brown eyes. D. 1902. Bur. Sizer Cem., Pott's Creek, Va. Also spelled Ferrel.

FARRIER, W. L.: Pvt. Co. C. Enl. Feb. 1862 for 3 years. Res. Craig Co., Va. 1898 at age 68.

FAUDRE, JOHN L.: Pvt. Co. A. Res. Kanawha Co., WV. Enl. 6/3/61 at Camp Tompkins, Kanawha Co., WV. On special duty as a teamster en route from White Sulphur Springs, WV on 8/31/61 roll. Regt. wagon master on 11/1/62 roll. Left sick at Marlin's Bottom (Marlinton), WV on 4/30/63 roll. Deserted in the face of the enemy 12/13/63. Took oath and sent North 12/31/63. Age: 31. Farmer. 5'5". Light complexion, dark hair, blue eyes. (Also spelled Fondreu and Fondeu). William Balhmann refers to this man as Fandry.

FAUGHT, JOHN N. P.: Pvt. Co. I. Enl. either 4/28/62 or 5/1/62 at White Sulphur Springs, Greenbrier Co., WV. Sick at Monterey, Va. on 4/30/63 roll. WIA (slight arm) at White Sulphur Springs, WV 8/26-27/63. In Staunton, Va. hosp. for rheumatism on 1/16/64 roll. Sick at home in Botetourt Co., Va. on 4/1/64 roll. Present on 4/30/64 roll. Capt. at Winchester, Va. 9/19/64. POW Pt. Lookout, MD 9/24/64. Exchd. 3/15/65.

FAUGHT, PHILLIP A.: Pvt./Sgt. Co. I. Enl. 5/1/62 at White Sulphur Springs, Greenbrier Co., WV. Rcvd. clothing on list dated 5/13/64. NFR.

FEINE, CARL M.: Musician. Co. H. Enl. 5/8/61 at Charleston, Kanawha Co., WV. Absent cause unknown on 8/31/61 roll. NFR. PWR says "didn't go."

FELLERS, JAMES D.: Corp. Co. K. Enl. 6/6/61 at Charleston, Kanawha Co., WV. Absent cause unknown on 9/4/61 roll. NFR. Res. Fayette Co., WV 1860. Age: 26. Bahlmann says "left."

FELLERS, JESSE K.: Pvt. Co. K. Enl. 6/6/61 at Charleston, Kanawha Co., WV. Absent cause unknown on 9/4/61 roll. NFR. Res. Fayette Co., WV 1860. Age: 22. Bahlmann says "left."

FELLERS, WILLIAM C.: Pvt. Co. K. Son of Otey Fellers. Enl. 6/6/61 at Charleston, Kanawha Co., WV. MWIA at Scary Creek, WV 7/17/61. Farm laborer, Fayette Co., WV 1860. Age: 19.

FERGUSON, ROLEN J.: Pvt. Co. G. Enl. 3/16/64 at Dublin, Pulaski Co., Va. In Charlottesville, Va. hosp. for debilitas 6/25-28/64. Deserted from hosp. on 10/31/64 roll.

FERMON, JAS. P.: Co. H. Deserted 5/3/62 on Aug. 1862 broadside. Craig Co., Va.

FERRELL, ELI: Pvt. Co. I. Possible res. Craig Co., Va. Enl. 6/22/61 at Charleston, Kanawha Co., WV. Deserted near Staunton, Va. 6/28/64.

FERRELL, LEVI: Pvt. Co. I. Res. Boone Co., WV. Possible res. Craig Co., Va. Enl. 6/22/61 at Charleston, Kanawha Co., WV. Sick at Lewisburg, WV on 4/30/63 roll. Present on rolls dated 8/31/63 to 12/31/63. Appears on a list of deserters surrendered at Charleston, WV. Mar. 1865. Took oath at Charleston, WV and sent North 3/13/65. Age: 25. Farmer. 5'7". Dark complexion, dark hair, blue eyes.

FERRELL, OLIVER: Pvt. Co. I. Possible res. Craig Co., Va. Enl. 6/22/61 at Charleston, Kanawha Co., WV. Sick at home in Boone Co., WV on 4/1/64 roll. Sick on 10/31/64 roll.

FINK, JAMES H.: Pvt. Co. F. Res. Monroe Co., WV. Enl. 7/13/61 at Camp Two-Mile, Kanawha Co., WV. Sick at home since 8/7/61 on 8/31/61 roll. Postwar roster says died at Salt Sulphur Springs, WV 1862 and gives middle initial as "A".

FINK, JOHN (or JONATHAN) L.: Pvt. Co. F. Enl. 7/13/61 at Camp Two-Mile, Kanawha Co., WV. At home sick since 8/22/61 on 8/31/61 roll. D. at Salt Sulphur Springs, WV. Claim filed 11/4/64.

FINK, REUBEN R.: Pvt. Co. G. Enl. 7/3/61 at Charleston, Kanawha Co., WV. Present on 11/13/62 roll. NFR.

FINK, RICHARD P.: Pvt. Co. B. B. 4/2/23 Mercer Co., WV. Another source says 4/2/63 Monroe Co., WV. Enl. 2/?/64 at Lewisburg, (Camp Gauley), Greenbrier Co., WV. WIA (right arm) at New Market, Va. 5/15/64. Paroled at Charleston, WV 5/10/65. Age: 44. 5'4½". Light complexion, dark hair, blue eyes, sandy whiskers. Postwar Miller at Blaker's Mill, Greenbrier Co., WV. PWR says fought at Dry Creek and Droop Mountain.

FINK, WILLIAM: Co. F. Res. Monroe Co., WV. Postwar roster.

FIRESTONE, WILLIAM L.: Pvt. Co. I. Res. Botetourt Co., Va. Enl. 5/1/62 at White Sulphur Springs, Greenbrier Co., WV. Capt. at Lewisburg, WV 5/23/62. POW Camp Chase, OH. Exchd. 8/25/62. Deserted in Botetourt Co., Va. 6/20/64. Capt. at Beverly, WV 8/15/64. Gave himself up. Relsd. and took oath 8/18/64. Age: 21. 5'8". Fair complexion, dark hair, gray eyes.

FISHER, CHRISTOPHER: Pvt. Co. F. Enl. 5/1/62 at White Sulphur Springs, Greenbrier Co., WV. Regt. butcher July 1862. At home sick on surg. cert. on 10/31/62 roll. Present on 2/28/63 roll. Paid as a teamster 7/31/63 to 9/30/63. Regt. butcher on 4/1/64 roll. Paroled at Charleston, WV 5/10/65. Age: 29. 6'. Dark complexion, brown hair, black eyes, brown whiskers.

FISHER, HENRY: Pvt. Co. I. Enl. either 6/?/61 at Charleston, Kanawha Co., WV or 5/2/62 at White Sulphur Springs, Greenbrier Co., WV or 9/15/62(?) at Charleston, Kanawha Co., WV. Serving sentence for G.C.M. at Dublin, Va. on 8/31/63 to 12/31/63 rolls. Present on 4/1/64 roll. Found guilty of desertion 8/9/64. Executed by sentence of military court on 10/31/64 roll.

FISHER, ROBERT S.: Pvt. Co. A. Enl. 5/22/61 at Poca, Putnam Co., WV. Present on 8/31/61 roll. NFR.

FISHER, WILLIAM: Chaplain. Field and Staff. Res. Craig Co., Va. Date and place of enlistment not stated. Apptd. June 1862. Resgnd. either 11/1/63 or 12/7/63.

FITZHUGH, NICHOLAS: 2nd Lt./3rd Lt. Co. H. B. 1/4/23. Attd. Mt. Ovis Academy, Charleston, WV. Lawyer, Charleston, WV. 1860, age 37. Enl. 5/22/61 at Charleston, Kanawha Co., WV. Mustered into C.S. service July 1861. Elected 2nd Lt. 5/22/61. Resgnd. 1861. Listed as a Capt. on a note for payment of a horse dated 10/10/63. PWR says served as a Maj. and Adj., Jenkin's Cav. Brig. Mustered out at Appomattox, Va. 1865. This information confirmed by Gen. John McCausland in 1922. In an 1865 photo taken in Charleston, WV of C.S.A. veterans. Bur. Spring Hill Cem., Charleston, WV.

FITZWATER, DAVID F.: Pvt. Co. A. Enl. 6/5/61 at Camp Tompkins, Kanawha Co., WV. In Charlottesville, Va. hosp. for icterosis 1/2/62. Returned to duty 2/2/62. WIA (severe thigh) at White Sulphur Springs, WV 8/26-27/63. AWOL on 12/31/63 roll.

FITZWATER, HENRY: Pvt. Co. E. Res. Clay Co., WV. Enl. 4/1/63 at Lewisburg, Greenbrier Co., WV. Paroled at Lewisburg, WV 5/10/65. Age: 30. Farmer. 6'. Fair complexion, black hair, blue eyes.

FITZWATER, JEFFERSON: Pvt. Co. E. Res. Clay Co., WV. Enl. 10/12/62 at Cotton Hill, Fayette Co., WV. In Chimborazo Hosp. for rubeola 6/6-24/64. Capt. at Winchester, Va. 9/19/64. POW Pt. Lookout, MD 9/23/64. Exchd. 3/15/65. In Farmville, Va. hosp. until 3/22/65. Took oath at Charleston, WV and sent North 4/15/65. Age: 27. Farmer. 5'7". Dark complexion, dark hair, gray eyes.

FITZWATER, LEMUEL: Pvt. Co. E. Enl. 4/1/62 at Lewisburg, Greenbrier Co., WV. Present on 5/1/64 roll. NFR.

FITZWATER, LEVI: Pvt. Co. E. B. in Va. Enl. 4/12/62 at Cotton Hill, Fayette Co., WV. Also listed as enl. at Gauley Bridge, Fayette Co., WV. Paroled at Lewisburg, WV 4/25/65. Age: 27. Farmer. 5'9". Fair complexion, brown hair, hazel eyes.

FITZWATER, THOMAS C.: Pvt. Co. A. Res. Kanawha Co., WV. Enl. 6/5/61 at Camp Tompkins, Kanawha Co., WV. Left sick at Beverly, WV on 4/30/62 roll. Deserted in face of the enemy 12/13/63. Took oath at Charleston, WV and sent north 12/31/63. Age: 20. Farmer. 5'8". Light complexion, dark hair, hazel eyes.

FITZWATER, T. L.: Pvt. Co. E. Res. Clay Co., WV. Enl. 3/6/62 at Lewisburg, Greenbrier Co., WV. Sick at hosp. but since returned on 10/31/64 roll. Paroled at Lewisburg, WV 5/10/65. Age: 26. Farmer. 6'. Fair complexion, light hair, dark eyes.

FLACK, ALEX: Pvt. Co. B. Enl. 2/20/64 at Lewisburg (Camp Gauley), Greenbrier Co., WV. D. from accidental wd. at Winchester, Va. 8/16/64.

FLEMING, JAMES: Pvt. Co. F. Enl. 5/1/62(?) at White Sulphur Springs, Greenbrier Co., WV. Serving as a clerk, Col. McCausland commanding brig. on July 1862 roll. Clerking for Gen. A. G. Jenkins on 10/28/63 roll. Last date paid 10/17/64.

FLESHER, HENRY: Pvt. Co. B. Res. Greenbrier Co., WV. Enl. either 7/12/62 at Camp McCausland, Monroe Co., WV or 7/16/62 at Union, Monroe Co., WV. Sick at home on 2/28/63 roll. Present on 4/30/63 roll. WIA (slight arm) at New Market, Va. 5/15/64. Capt. at Piedmont, Va. 6/4/64. Relsd. on oath 4/4/65. 5'8½". Dark complexion, black hair, gray eyes.

FLESHMAN, CHARLES H.: Pvt. Taylor's Co. Enl. 6/19/61 at Frankford, Greenbrier Co., WV. Age: 23. Co. transf. to Wise Brigade 7/10/61.

FLESHMAN, THOMAS J.: Fife Major. Co. K. B. 10/1/40 Greenbrier Co., Enl. either 4/15/62 at Brushy Ridge, Greenbrier Co., WV or 5/1/62(?) at White Sulphur Springs, Greenbrier Co., WV. Deserted 9/23/64. Postwar sources refer to this man as Thompson H. and Thomas N. Fleshman.

FLINT, EZEKIEL: Pvt. Co. B. Enl. 5/?/62 at White Sulphur Springs, Greenbrier Co., WV. Conscript. WIA (severe leg) at White Sulphur Springs, WV 8/26-27/63. Absent wded. on 1/10/64 roll. At home wded. on rolls from 4/1/64 to 11/30/64.

FLOWERS, AUGUSTUS: Pvt. Co. B. Res. Jackson Co., WV. Enl. 9/22/62 at Charleston, Kanawha Co., WV. At Wayside Hosp., Richmond, Va. 5/26-28/64. In Chimborazo Hosp. for measles 5/28/64 to 6/24/64. Capt. at Winchester, Va. 9/19/64. POW Pt. Lookout, MD 9/23/64. Exchd. 3/15/65. Listed as deserter in Apr. 1865. Took oath at Charleston, WV and sent North 4/17/65. Age: 20. Farmer. 5'6". Dark complexion, light hair, gray eyes.

FLOWERS, JAMES K.: Pvt. Co. K. Enl. either 8/1/63 or 8/15/63 at Narrows, Va. Entitled to a bounty being underage when enlisted. In Chimborazo Hosp. for amputation of thumb 6/4 or 5/64. Furlo. for 30 days 6/17/64. Issued shoes 8/31/64.

FLOWERS, J. R.: Pvt. Co. B. Res. Jackson Co., WV.

FLOYD, J.: Pvt. Co. H. Enl. 6/1/63 at Lewisburg, Greenbrier Co., WV. In Charlottesville, Va. hosp. for acute dysentery 6/9-18/64. Transf. to convalescent camp 6/18/64.

FOGUS, HENRY: Pvt. Taylor's Co. Enl. 6/19/61 at Frankford, Greenbrier Co., WV. Age: 32. Co. transf. to Wise Brigade 7/10/61.

FOGUS, JOHN: Pvt. Taylor's Co. Enl. 6/19/61 at Frankford, Greenbrier Co., WV. Age: 32. Co. transf. to Wise Brigade 7/10/61.

FORBER, GEORGE W.: Pvt. Co. G. Res. Boone Co., WV. Enl. 7/1/61 at Charleston, Kanawha Co., WV. Absent on surg. cert. from 7/28/61 until fit for duty. Capt. in Kanawha Co., WV by 28th O.V.I. 1/10/63. Age: 52. Farmer. 5'7". Dark complexion, gray hair, gray eyes.

FORD, RENNICK C.: Pvt. Taylor's Co. Enl. 6/19/61 at Frankford, Greenbrier Co., WV. Age: 24. Co. transf. to Wise Brigade 7/10/61.

FORREN, JOHN: Pvt. Co. F. Res. Monroe Co., WV. Enl. 2/15/63(?) at Lewisburg, Greenbrier Co., WV. Capt. at Winchester, Va. 9/19/64. POW Pt. Lookout, MD 9/23/64. Exchd. 3/15/65. Postwar res. Greenbrier Co., WV. Still alive 1914.

FORSYTH, ABRAM: Pvt. Co. C. Enl. 5/1/62(?) at White Sulphur Springs, Greenbrier Co., WV. Det. as a scout on rolls for July 1862 and Sept.-Oct. 1862. Also served with Capt. Thurmond's Co. Farmer, Fayette Co., WV 1860. Age: 30. PWR says enl. 5/10/61.

FORSYTH, SAMUEL: Pvt. Co. C. Day laborer, Fayette Co., WV 1860. Age: 34. Enl. 5/10/61.

FOSTER, JAMES: Pvt. Co. A. Enl. 4/10/62(?) at Brushy Ridge, Greenbrier Co., WV. Dischd. at Packs Ferry, Monroe Co., WV 11/3/62.

FOSTER, JAMES E.: Pvt. Co. G. Enl. 7/20/62 on New River. Dropped from the rolls by order on 10/31/64 roll.

FOSTER, JOSEPH: Pvt. Co. I. Enl. 6/22/61 at Charleston, Kanawha Co., WV. AWOL 7/26/61.

FOUTS, W. M.: Pvt. Co. H. Enl. 4/28/62 at White Sulphur Springs, Greenbrier Co., WV. Joined from desertion at Camp McCausland, WV. Sick at home in Roanoke, Va. July 1862. Sick in Craig Co., Va. on rolls from 4/30/63 to 1/1/64. Nurse at Montgomery White Sulphur Springs, Va. on rolls from 4/1/64 to 4/30/64. Sick in Craig Co., Va. on 12/25/64 roll.

FOX, ISAAC L.: Pvt. Co. F. Enl. 3/25/63 at Lewisburg, Greenbrier Co., WV. MIA White Sulphur Springs, WV 8/26-27/63. On sick leave on 12/31/63 roll. Present on 4/1/64 roll. Deserted about 5/8/64. Postwar res., Monroe Co., WV.

FOX, WILLIAM H.: Co. G. Entered Richmond, Va. Old Soldiers Home 3/28/1914 at age 76. At own request 4/13/1914.

FRAME, JAMES A.: Pvt. Co. E. Res. Kanawha Co., WV Enl. 6/8/61 at Charleston, Kanawha Co., WV. Left sick at White Sulphur Springs, WV 8/20/61. Hosp. nurse since May 1862 on July 1862 roll. WIA (slight) in 1862 Kanawha Valley campaign. Capt. in Pocahontas Co., WV 4/2/63. Listed as AWOL on 4/30/63 roll. POW Camp Chase, OH 4/13/63. Exchd. 5/13/63. In Petersburg, Va. hosp. for debilitas, as a paroled prisoner, 5/17/63 to 6/6/63. Age: 24. Lumber merchant. 5'4½". Fair complexion, light hair, blue eyes, sandy whiskers. Pension application rejected in Johnson Co., Ark. 1897.

FRAME, M. W.: Pvt. Co. E. Enl. 6/8/61 at Charleston, Kanawha Co., WV. Left sick at White Sulphur Springs, WV 8/20/61. Present on Sept.-Oct. 1862 roll. WIA (severe leg) at White Sulphur Springs, WV 8/26-27/63 roll. Rcvd. clothing 5/13/64. (Also listed under this name in the CSR is a Mortimer W. Frame, Co. C. 22nd Va. who appears as a prisoner capt. by Capt. King's Union Home Guard and paroled same day 9/20/62).

FRANKLIN, JOHN W.: Pvt. Co. A. Res. Flatwoods, WV. Enl. 6/5/61 at Camp Tompkins, Kanawha Co., WV. Det. on special guard duty by order of Gen. Wise 7/20/61. Deserted July 1861. Rejoined regt. as wagon master Sept. 1862. Dischd. 11/4/62. Was willing to take the oath.

FRANKLIN, WILLIAM H. ("Billy"): Pvt. Co. K. Enl. 6/6/61 at Charleston, Kanawha Co., WV. Left sick at Lewisburg, WV on Mar.-Apr. 1863 roll. Present on 12/30/63 roll. WIA (shot through lung) at New Market, Va. 5/15/64. In Farmville, Va. hosp. for vul. sclop. chest. D. at Wayside Hosp., Farmville, Va. 10/14/64. Effects turned over to his brother. Farm laborer, Fayette Co., WV 1860. Age: 22.

FRIDLEY, SAMPSON: On Alleghany Co., Va. pension rolls.

FRIDLEY, WILLIAM H.: Pvt. Co. G. Enl. 6/26/61 at Lewisburg, Greenbrier Co., WV or 12/14/62 (or 63) near Union, Monroe Co., WV. Deserted 5/8/64. Res. Allegheny Co., WV. Age: 21 in 1862.

FRIDLY, JOHN: Pvt. Co. A. Res. Alleghany Co., Va. Enl. 4/10/62 at Brushy Ridge, Greenbrier Co., WV. Left sick at Camp Bartow, WV on 4/30/63 roll. Capt. at Beverly, WV 11/30/63. Age: 30. Listed as deserter. POW Camp Chase, OH 12/7/63. Desired to take oath 6/10/64. In Camp Chase, OH hosp. for variola - never vaccinated 10/20/64. D. at Camp Chase, OH of pneumonia 4/3/65. Bur. grave #1813 in Camp Chase, OH POW cem. Farmer. 6'3". Dark complexion, dark hair, blue eyes.

FRY, JAMES H., JR.: Pvt. Co. H. Enl. 5/8/61 at Charleston, Kanawha Co., WV. Present on 8/31/61 roll. Bur. Spring Hill Cem., Charleston, WV.

FRY, PHILIP F.: Co. I. Listed as deserter on Aug. 1862 broadside, Greenbrier Co., WV.

FRY, SAMUEL CLARK: Pvt. Co. K. B. 1/25/36. Enl. either 4/15/62 at Lewisburg, Greenbrier Co., WV or 5/1/62 at White Sulphur Springs, Greenbrier Co., WV. Sick at Montgomery White Sulphur Springs, Va. July 1862. Present on 12/30/62 roll. Capt. at Winchester, Va. 9/19/64. POW Pt. Lookout, MD 9/23/64. Exchd. 3/15/65. D. 8/13/1904. Bur. Mt. Pleasant, Va. Methodist Church, Covington, Va.

FRY, W. A.: Sgt. Co. A. D. at Ft. Delaware 7/16/64. Bur. Finn's Point Nat. Cem., NJ.

FUGATE, WILLIAM M.: Pvt. Co. E. Enl. 6/8/61 at Charleston, Kanawha Co., WV. Sick in Roane Co., WV., 7/1/61. NFR.

FUQUA, WILLIAM M.: Pvt. Co. B. Enl. 5/?/62 at White Sulphur Springs Greenbrier Co., WV. Conscript. Capt. at Winchester, Va. 9/19/64. POW Pt. Lookout, MD 9/23/64. Relsd. by order of the President 1/26/65. Took oath 1/26/65 and transportation provided to Springfield, OH.

GADD, ANDREW P.: Pvt./Corp. Co. G. B. 1829. Res. Mercer Co., WV. Farmer. Enl. 9/8/62 in Mercer Co., WV. Capt. at Winchester, Va. 9/19/64. POW Pt. Lookout, MD 9/23/64. Exchd. 3/15/65. Paroled at Charleston, WV 6/27/65. Age: 37. 5'1" or 5'9". Fair complexion, dark hair, blue eyes. D. 1909. PWR says enl. May 1861 at Charleston, WV. Age: 20. Res. Wyoming Co., WV.

GADD, SANFORD P.: Pvt./Corp. Co. G. Enl. 7/7/61 at Charleston, Kanawha Co., WV and/or 5/1/62 at White Sulphur Spring, Greenbrier Co., WV. Transf. from Capt. Sarver's Co. 7/25/61. Left sick at White Sulphur Springs, WV 8/20/61. Present on 3/13/62 roll. KIA at New Market, Va. 5/15/64. PWR says res. Wyoming Co., WV and age 21 at enl.

GALLISPIE, WILLIAM C.: Pvt. Swann's Co. Enl. 6/7/61 at Charleston, Kanawha Co., WV. Age: 23. Co. transf. to Co. K, 59th Va. Inf. 7/25/61.

GARMAN, ADAM M.: Pvt. Co. E. B. 1/25/32 Craig Co., Va. Enl. 5/1/62 at White Sulphur Springs, Greenbrier Co., WV. Sick in hosp. on Sept.-Oct. 1862 roll. Det. service on 4/30/63 roll. D. 4/18/1908. Bur. Hypes Cem., Rt. 42, Ray Givens Farm near Bethel Churches, Craig Co., Va.

GARMAN, JOHN: Pvt. Co. E. Res. Craig Co., Va. Enl. 5/1/62 at White Sulphur Springs, Greenbrier Co., WV. Sick in Giles C.H. (Pearisburg), Va. hosp. since May 1862 on July 1862 roll. In hosp. sick on rolls from Sept.-Oct. 1862 to 3/4/63. Det. service on 4/30/63 roll. Sick at Giles C.H. (Pearisburg), Va. hosp. on 12/31/63 roll. Sick on 4/1/64 roll. Listed as a deserter from First Brig., Army of Southwestern Va., Gen. Echols, dated Apr. 1864. Deserted from the hosp. Oct. 1864. Went to the enemy.

GARMON, JOSEPH: Pvt. Co. E. Enl. 5/1/62 at White Sulphur Springs, Greenbrier Co., WV. Sick in Giles C.H. (Pearisburg), Va. hosp. since May 1862 on July 1862 roll. Det. in hosp. on Sept.-Oct. 1862 roll. D. 2/21/63. Claim filed (presented) to Edward M. Clark, atty.

GAYLOR, JEREMIAH: Pvt. Co. A. Res. Kanawha Co., WV. Laborer, Tyler Mountain, WV 1860. Age: 18. Enl. 7/2/61 at Camp Tompkins, Kanawha Co., WV or 10/1/62 at Charleston, Kanawha Co., WV. Unaccounted for since 7/4/61. Present on 11/1/62 roll. Capt. at Winchester, Va. 9/19/64. POW Pt. Lookout, MD Relsd. from Pt. Lookout, MD. 3/12 or 14/65. Exchd. 3/15/65. Transportation provided to Romney, WV. Farmer.

GAYLOR, JOHN: Pvt. Co. A. Enl. 5/22/61 at Charleston, Kanawha Co., WV. Away with sick leave at White Sulphur Springs, WV 8/20/61. NFR.

GEARHART, DANIEL: Pvt. Co. A. Enl. 5/22/61 at Charleston, Kanawha Co., WV. Away with sick leave at White Sulphur Springs, WV 8/20/61. Present on 11/1/62 roll. Det. brig. shoemaker on 12/31/63 roll. Det. service on 4/30/64 roll. WIA (severe finger) at Cold Harbor, Va. 6/2-3/64. Paroled at Lewisburg, WV 4/25/65 and/or at Charleston, WV 5/10/65. Age: 36. 5'5" or 5'7". Shoemaker. Dark complexion, dark hair, black eyes.

GEE, JOSEPHUS W.: Pvt. Taylor's Co. Enl. 6/19/61 at Frankford, Greenbrier Co., WV. Age: 22. Co. transf. to Wise Brigade 7/10/61.

GENANY, JOHN: Pvt. Co. E. Enl. 5/1/62 at White Sulphur Springs, Greenbrier Co., WV. Det. wagoner on 1/2/63 roll. Det. brig. wagoner on 3/4/63 roll. Sick on 4/30/63 roll. Det. brig. wagoner on rolls dated 12/31/63 to 4/1/64. Deserted while on sick furlo. 5/12/64. In Charlottesville, Va. hosp. for rheumatism 6/12/64. Transf. to convalescent camp 6/13/64.

GEORGE, JOHN P.: Pvt. Co. K. Enl. 4/15/62 at Lewisburg, Greenbrier Co., WV. Sick in hosp. on 11/1/62 roll. At home on sick furlo. on Nov.-Dec. 1862 roll. Present on 2/28/63 roll. Deserted near Lexington, Va. 6/23/64. In Lee Camp Soldiers Home, Richmond, Va. 1/18/1917.

GEORGE, THOMAS A.: Res. Green Sulphur Springs, WV. D. 1920 at Crimora, Augusta Co., Va. Also served with Capt. Thurmond's Co.

GIBBONY, ALBERT HOLLER: Pvt. Co. H. B. 4/16/45. Enl. 4/23/62 at Lewisburg, Greenbrier Co., WV. Det. as clerk on A. A. Gen. office, District of Gauley, on July-Aug. 1862 roll. Clerk in A. A. Gen., Dept. of Heth's Div., on rolls from 1/1/64 to 6/30/64. (Also spelled Gibboney). D. 6/1/1917. Bur. Wytheville, Va.

GIBSON, BOYD: Pvt. Co. I. Enl. 6/22/61 at Charleston, Kanawha Co., WV. AWOL from 7/26/61 to 8/21/61. Deserted about 11/15/61. Must have returned because also listed as deserted 3/28/64.

GIBSON, ELIJAH: Pvt. Co. I. Enl. 6/22/61 at Charleston, Kanawha Co., WV. Absent from 7/26/61 to 8/21/61. Deserted 11/15/61. Listed as a deserter on list dated Apr. 1864 for First Brig., Army of Southwest Va., Gen. Echols.

GIBSON, GEORGE: Pvt. Co. A. Enl. 5/22/61 at Charleston, Kanawha Co., WV. and/or 9/15/62 at Charleston, Kanawha Co., WV. Unaccounted for as of 7/24/61. Present on 11/1/62 roll. Left sick at Marlin's Bottom (Marlinton), WV on 4/30/63 roll. Present on 12/31/63 roll. Capt. at Winchester, Va. 9/19/64. POW Pt. Lookout, MD 9/24/64. Exchd. 3/15/65.

GIBSON, GILBERT: Pvt. Co. G. Enl. 7/3/61 at Charleston, Kanawha Co., WV. Left sick at White Sulphur Springs, WV 8/20/61. Sick at Giles C.H. (Pearisburg), Va. July 1862. Present on 11/13/62 roll. At home sick 12/20/62. Capt. in Wyoming Co., WV by Home Guard 1/25/63. Age: 21. POW Camp Chase, OH 2/23/63. Exchd. 3/25/63. With a disability at Camp Sam Jones for general disability of long standing 7/3/63. Dishd. at Packs Ferry, Monroe Co., WV for disability 7/19/63. 5'6¾". Dark complexion, black hair, gray eyes.

GILLASPIE, GEORGE: Pvt. Co. B. Res. Jackson Co., WV. Enl. 6/1/61 at Ripley, Jackson Co., WV. Capt. at Lewisburg, WV 5/23/62. Age: 23. POW Camp Chase, OH. Transf. to Johnson's Island, OH 9/6/62. Exchd. 11/22/62. 5' (1 or 7)½". Dark complexion dark hair, blue eyes. Also spelled Gillispie. Another card says capt. in Jackson Co., WV.

GILLIAM, CHARLES W.: 3rd Corp. Taylor's Co. Enl. 6/19/61 at Frankford, Greenbrier Co., WV. Age: 18. Co. transf. to Wise Brigade 7/10/61. Also spelled Gilliam.

GISH, EDWARD G.: Pvt. Co. B. Enl. 5/?/62 at White Sulphur Springs, Greenbrier Co., WV. Conscript. Capt. at Winchester, Va 9/19/64. POW Pt. Lookout, MD 9/23/64. Took oath and relsd. by order of the President 1/26/65.

GISH, GEORGE R.: Co. I. Enl. 1864. Served 6 months.

GISH, WILLIAM: Pvt. Co. B. Enl. 5/?/64 at Richmond, Va. Capt. at Winchester, Va. 9/19/64. POW Pt. Lookout, MD 9/24/64. Took oath and relsd. by order of the President 1/26/65.

GIVEN, JOHN: Pvt. Co. B. B. 3/13/24. Killed by Federal partisans while home on leave. D. 5/7/62. Bur. Triplett Cem., near Cowen, WV. Another source says bur. Given Cem., Upper Glade, Webster Co., WV.

GIVENS, JOHN F.: Co. C. Res. Craig Co., Va. Bur. Caldwell Cem., Rt. 624, on Holland Caldwell farm, Craig Co., Va.

GIVENS, SAMUEL: Pvt. Co. K. Res. Craig Co., Va. Enl. 4/15/62 at Lewisburg, Greenbrier Co., WV. Capt. at Lewisburg, WV 5/23/62. Age: 21. POW Camp Chase, OH 5/30/62. D. at Camp Chase, OH 7/26/62, grave #2087. 5'11". Dark complexion, black hair, light blue eyes, dark whiskers. Also listed res. Alleghany Co., Va.

GLEASON, PLEASANT: Pvt. Co. E. Res. Roanoke Co., Va. Enl. either 3/28/64 at Dublin, Va. or 3/29/64 in Monroe Co., WV. Conscript. In Chimborazo Hosp. for paratitis 5/28/64 to 6/22/64. Absent sick on 10/21/64 roll. Postwar res. Roanoke Co., Va. Also spelled Gleeson.

GLEESON, PAUL F.: Pvt. Co. G. B. 1832. Enl. 5/1/62 at White Sulphur Springs, Greenbrier Co., WV. Deserted 10/20/62. Present on 12/31/62 roll. Working on Cow Pasture Bridge on 12/31/63 roll. Det. to work on railroad on 4/1/64 roll. Present on 4/30/64 roll. Det. as police of Alleghany Co., Va. on 10/31/64 roll. Also spelled Gleason. D. 1913. Bur. Oakland Presbyterian Church, near Selma, Va. Res. Alleghany Co., Va.

GLOVER, JOHN H.: Co. F. Res. Monroe Co., WV. D. sometime prior to 1914. Postwar roster.

GOAD, J. H.: Pvt. Co. E. Enl. 6/8/61 at Charleston, Kanawha Co., WV AWOL 7/1/61.

GOFF, CHRISTOPHER: Pvt. Co. K. Enl. 4/15/62 at Lewisburg, Greenbrier Co., WV. Capt. and WIA (very serious compound fracture of tibia & fibula) at Winchester, Va. 9/19/64. In U.S. field hosp. at Winchester, Va. for amputation of lower third of thigh. D. of pycemia 10/1/64.

GOODE, ALEXANDER C.: Pvt. Co. H. B. ca. 1841 Craig Co., Va. Enl. 4/28/62 at White Sulphur Springs, Greenbrier Co., WV. Was assigned to and joined this co. 4/28/62. Capt. at Cold Harbor (Gaines Mill), Va. 6/3/64. POW Pt. Lookout, MD. Transf. to Elmira, NY 7/12/64. Exchd. 3/10/65.

GOODE, JOHN HENRY: Pvt. Co. H. B. ca. 1846 Craig Co., Va. Enl. 4/6/64 in Greenbrier Co., WV. Capt. at Winchester, Va. 9/19/64 and/or capt. at Fisher's Hill, Va. 9/22-23/64. POW Pt. Lookout, MD 10/1/64. Exchd. 10/30/64.

GORE, FLOYD S.: Pvt./2nd Lt. Co. G. Res. Logan Co., WV. Enl. 7/3/61 at Charleston, Kanawha Co., WV. Transf. from Capt. Stone's Co., Co. C, 36th Va. Inf., 8/10/61. Left sick at White Sulphur Springs, WV 8/20/61. Was elected 5/1/62. WIA at Droop Mountain, WV 11/6/63 (arm and leg broken). Promo. 11/22/64. Dischd. 1864 at Staunton, Va. Took oath at Charleston, WV 4/29/65. Age: 26. Farmer. 5'8". Dark complexion, dark hair, blue eyes.

GRAHAM, LANTY J.: Pvt. Co. A. Enl. 4/10/62 at Brushy Ridge, Greenbrier Co., WV. Capt. at Lewisburg, WV 5/23/62. Age: 20. Exchd. and left sick in Miss. Not heard from after Nov. 1862. 5'10". Fair complexion, mouse color hair, dark gray eyes, light whiskers.

GRANT, JAMES O.: Co. H. Postwar roster. B. in Ky. Gentleman, Charleston, WV 1860. Age: 20. D. sometime prior to 6/26/91.

GRANT, THOMAS T.: Pvt./Sgt. Co. H. Enl. 5/31/61 at Coalsmouth (St. Albans), Kanawha Co., WV. Det. as guard at Lewisburg, WV on 4/1/64 roll. Det. as Provost Guard at Lewisburg, WV on 4/30/64 roll. Capt at Winchester, Va. 9/19/64. POW Pt. Lookout, MD 9/24/64. In Pt. Lookout, MD hosp. for diarrhea 10/5-7/64. D. (probably at Pt. Lookout, MD) 10/7 or 8/64. Was a first cousin of U.S. Grant.

GRAY, GABRIEL: Res. Culpepper, Va. 1853 grad. V.M.I. Postwar res. Fincastle, Va. Mentioned as Col. by W. B. Myers. A. G. 9/11/63. G. T. Gray apptd. Chaplain Sept. 1862.

GRAY, JAMES: Co. H. Res. Tornado, WV 1871.

GRAY, SAUNDERS: Pvt. Co. K. Enl. 8/15/63 at Giles C.H. (Pearisburg), Va. Wded. at Potts Creek and in hosp. on 12/30/63 roll. Rcvd. clothing at Montgomery White Sulphur Springs, Va. 9/27/64. Wded. at Sweet Springs, WV 12/20/63 but since returned on 10/31/64 roll.

GREENWAY, JOSEPH: Listed in UDC booklet *Craigs Share in the War Between the States.* Res. Craig Co., Va. Believed bur. Spraker's Farm, top of Sinking Creek Mountain, Rt. 42, on the old Os Cooper place, Craig Co., Va.

GREENWAY, RICHARD: Pvt. Co. E. Res. Craig Co., Va. Enl. 5/1/62 at White Sulphur Springs, Greenbrier Co., WV. Listed as a deserter Aug. 1862 broadside. KIA at Droop Mountain, WV 11/6/63.

GREGORY, F. J. (DR.): Pvt. Co. E. Date and place of enlistment not stated. Found unable to perform duties of a soldier in the field. Suffering from phithisis pulmonaris 2/19/65. Asked for apptd. as hosp. steward, Medical Director's Office, Richmond, Va. 2/23/65. Paroled 4/21 (?)/65. Numerous medical papers on file in CSR.

GREGORY, HENRY: Pvt. Co. A. Grocer, Coalsmouth, WV 1860. Age: 21. Enl. 5/22/61 at Charleston, Kanawha Co., WV. Sick at White Sulphur Springs, WV 8/20/61. Present on 11/1/62 roll. KIA at White Sulphur Springs, WV 8/26-27/63. Bur. Rust Cem., Nitro, WV.

GREGORY, JOHN: Pvt. Taylor's Co. Enl. 6/19/61 at Frankford, Greenbrier Co., WV. Age: 16. Co. transf. to Wise Brigade 7/10/61.

GREYBILL, A.: Pvt. Co. B. Enl. 5/?/62 at White Sulphur Springs, Greenbrier Co., WV. Conscript. Sick at Montgomery White Sulphur Springs, Va. July 1862. Sick in hosp. on 11/16/62 roll. NFR.

GREYBILL, GEORGE: Pvt. Co. B. Enl. 5/?/62 at White Sulphur Springs, Greenbrier Co., WV. Sick at Montgomery White Sulphur Springs, Va. July 1862. Sick in hosp. on 11/16/62 roll. NFR.

GRIFFIN, GEORGE C.: Pvt. Co. B. Enl. 7/1/61 at Ripley, Jackson Co., WV. Det. as a cook in hosp. at White Sulphur Springs, WV on 9/1/61 roll. Listed as a POW on card dated 6/27/62.

GRIFFIN, MICHAEL: Pvt. Co. F. Res. Monroe Co., WV. Enl. 7/4/61 at Charleston, Kanawha Co., WV. Present on 8/31/61 roll. Believed still alive 1914.

GRIFFIN, SANDIFORTH M.: Musician. Co. H. B. in England. Civil engineer, Malden WV in 1860. Age: 41. Enl. 5/8/61 at Charleston, Kanawha Co., WV. Absent cause unknown on 8/31/61 roll. NFR.

GRIFFIN, WILLIAM R.: Pvt. Co. B. Enl. 7/1/61 at Ripley, Jacksoin Co., WV. In hosp. at White Sulphur Springs, WV on 9/1/61 roll. NFR.

GRIFFITH, EDWIN: Pvt. Co. I. Enl. 6/22/61 at Charleston, Kanawha Co., WV. Capt. in Mercer Co., WV 6/23/62. POW Camp Chase, OH. Exchd. 8/25/62. Listed as died on 4/30/63 roll.

GRIFFITH, N. G.: Pvt. Co. H. Enl. 1861. PWR says "didn't go."

GRIFFITS, JACOB VAN: Pvt. Co. A. Laborer, Charleston, WV 1860. Age: 24. Enl. 5/22/61 at Charleston, Kanawha Co., WV. AWOL 7/31/61. Name probably Griffith.

GRIM, JOHN J.: Pvt. Co. B. Enl. 10/1/61 at Sewell Mountain, Fayette Co., WV. Was authorized to raise a co. by Gen. Williams and had not reported back on 11/16/62 roll. Present on 12/31/62 roll. In arrest at Dublin Depot, Va. on 4/30/63 roll. Det. service on 1/10/64 roll. Supposed to be a prisoner on 4/1/64 roll.

GRIM, SIMON: Pvt. Co. B. Enl. 7/1/61 at Ripley, Jackson Co., WV. Dischd. at Camp McCausland, WV 7/16/62.

GROVES, DAVID: Pvt. Co. G. Enl. 3/5/64 at Staunton, Va. Deserted 4/12/64.

GROVES, GEORGE W.: Pvt. Swann's Co. Apprentice, Coalsmouth, WV 1860. Enl. 5/8/61 at Charleston, Kanawha Co., WV. Age: 21. Co. transf. to Co. K, 59th Va. Inf. 7/25/61.

GROVES, JAMES R.: Pvt. Swann's Co., Apprentice, Coalsmouth, WV 1860. Enl. 5/8/61 at Charleston, Kanawha Co., WV. Age: 19. Co. transf. to Co. K, 59th Va. Inf. 7/25/61.

GRUBB, GEORGE W.: Pvt. Co. F. Enl. 3/16/64 at Camp Gauley, Greenbrier Co., WV. Two months pay taken by sentence of courtmartial. Rcvd. clothing for 4 qr. 1864.

GRUBB, JOHN W.: Pvt. Co. F. Enl. 5/1/62 at White Sulphur Springs, Greenbrier Co., WV. WIA (severe foot) at White Sulphur Springs, WV 8/26-27/63. AWOL on 4/30/64 roll. Two months wages taken by sentence of court-martial on 10/31/64 roll. Order placed for a coffin on 11/30/64 roll. PWR says served 3 years.

GUILL, RICHARD: Corp. Swann's Co. B. Farmville, Va. 6/15/25. Res. WV ca. 1850. Lumber and dock business. Married 12/10/51. Carpenter, Charleston, WV 1860, age: 34. Enl. Charleston, Kanawha Co., WV. Date of enlistment not stated. Age: 36. Co. transf. to Co. K, 59th Va. Inf. PWR says WIA (slight) in first battle. Postwar: Ran for mayor of Charleston, WV. Nine years president of Charleston, WV school board. D. 11/22/1904. Bur. Spring Hill Cem., Charleston, WV.

GUNN, WILLIAM R.: Pvt. Co. B. Res. Ripley, WV 1860, lawyer.

GUNNOE, ISAAC: Pvt. Co. I. Res. Kanawha Co., WV. Enl. 6/22/61 at Charleston, Kanawha Co., WV. Det. brig. wagoner on rolls from 8/31/63 to 4/1/64. Deserted near Staunton, Va. 6/28/64. Capt. in Cabell Co., WV by 7th WV Cav. 2/13/65. Another card says capt. in Boone Co., WV. POW Camp Chase, OH 2/20/65. Bushwhacker. Took oath at Camp Chase, OH 6/13/65. Age: 23 or 24. 5'10". Dark complexion, dark hair, black eyes.

GWINN, ANDREW L.: Pvt. Co. F. Enl. 9/12/62(?) or 10/31/62 at Red Sulphur Springs, Monroe Co., WV. Dischd. on undated surg. cert. of disability.

GWINN, JOSEPH H.: Pvt. Co. F. Enl. 7/23/63(?) or 12/31/63 at Camp Smith. KIA at Winchester, Va. 9/19/64.

GWINN, OLIVER W.: Pvt./Corp. Co. F. Res. Monroe Co., WV. Enl. 7/4/61 at Charleston, Kanawha Co., WV. Sick in hosp. at White Sulphur Springs, WV 8/20/61. Present on 10/31/62 roll. Apptd. Corp. 2/14/63. Present on 10/31/64 roll. Listed as a deserter. Took oath at Charleston, WV 1/30/65. Age: 30. Farmer. 6'½". Fair complexion, dark hair, blue eyes. D. near Lowell, NY 1/1/1906.

GWINN, WILLIAM A.: Sgt./Corp. Co. F. Res. Monroe Co., WV. Enl. 7/4/61 at Charleston, Kanawha Co., WV. Sick in hosp. at White Sulphur Springs, WV 8/20/61. Det. as a scout for Thurmond July 1862. Present on 10/31/62 roll. Capt. at Lewisburg, WV 10/22/63 and/or capt. at Droop Mountain, WV 11/6/63. Age: 25. POW Camp Chase, OH 11/18/63. Transf. to Ft. Delaware 2/29/64. Took oath and relsd. at Ft. Delaware 6/20/65. D. at Ft. Delaware of typhoid fever 7/2/64. Bur. Finn's Point Nat. Cem., NJ. Farmer. 6'2½". Dark complexion, dark hair, dark eyes.

HACKELTHORN, CHARLES: Pvt. Co. B. Enl. 6/1/61 at Ripley, Jackson Co., WV. Left June 25 on leave of absence for a one day leave and was taken prisoner while on leave. Capt. in Jackson Co., WV in 1861 according to Sept.-Oct. 1862 roll.

HAGERMAN, JAMES C.: Pvt. Co. H. Clerk, Charleston, WV 1860. Age: 23. Enl. 5/8/61 at Charleston, Kanawha Co., WV. Absent cause unknown 8/31/61. NFR. PWR says "didn't go."

HALE, JOHN PETER: Capt. Hale's Co./Kanawha Artillery/Co. H. B. 5/1/24 Ingle's Ferry, Va. Res. Kanawha Valley, WV 1840. Attd. Mercer Academy, Charleston, WV 1841-42. Studied medicine under Dr. Spicer Patrick. Grad. of Univ. of PA Medical School 1845. Medical practice with Dr. Spicer Patrick, Charleston, WV. Gave up medicine and entered salt business 1847. Created largest salt business in nation. Enl. 7/8/61 at Charleston, Kanawha Co., WV. Fought at Scary Creek, WV 7/17/61. Resgnd. 8/21/61. Later voluntarily acted as an asst. surg. Organized Bank of WV, Charleston, WV, 1863. President of Kanawha Salt Co., 1864. Built first steam packet on upper Kanawha River, 1864. Introduced first brick making machinery in Kanawha Valley and financed laying of first brick street in Charleston, WV 1870. Organized first gas company in Charleston, WV 1870. Constructed steamers, 1878. Once owned all the ferries in Charleston, WV. Chiefly responsible for building first State Capitol building in Charleston, WV. Mayor of Charleston, WV 1871. Built Hale House, 1872. President and part owner Charleston, WV's first daily newspaper. Engaged in coal and timber, 1880's. Wrote many books in 1880's on Kanawha Co. history. Founded WV Historical and Antiquarian Society, 1890. D. 7/11/1902. Bur. Spring Hill Cem., Section A, Charleston, WV.

HALFPENNY, DAVID C. C.: Pvt. Co. F. En. 7/4/61 at Charleston, Kanawha Co., WV. Sick at Lewisburg, WV hosp. on Jan.-Feb. 1863 roll. Appears on a list of rebel deserters Dec. 1863. Probably was capt. 12/31/63. Took oath and sent North. Age: 21. Blacksmith. 6'1". Dark complexion, dark hair, gray eyes.

HALL, ALEXANDER: Pvt. Co. B./Co. C. Enl. 7/1/61 at Ripley, Jackson Co., WV. In Charlottesville, Va. hosp. for paratitis (mumps) 1/2-20/62. Listed as a prisoner on Apr.-Oct. 1864 roll. Appears on a list of deserters Apr. 1865. Took oath 4/10/65. Age: 28. Farmer. 6'. Dark complexion, dark hair, blue eyes.

HALL, HENRY L.: Pvt. Co. G. Age: 48. Enl. 5/1/62 at White Sulphur Springs, Greenbrier Co., WV. Conscript. Left sick at Pearisburg, Va. May 1862. Capt. in Alleghany Co., Va. 5/17/62. AWOL from hosp. on Aug. 1862 broadside. POW Camp Chase, OH 5/30/62. Exchd. 8/25/62. Left sick at Beverly, WV on Mar.-Apr. 1863 roll. Present on Nov.-Dec. 1863 roll. On scout 4/1/64. Present on May-Oct. 1864 roll. (Also listed under this name is H. L. Hone who was capt. at Winchester, Va. 9/19/64. POW Pt. Lookout, MD 9/23/64. Paroled at Pt. Lookout, Md. and transferred for exchange 3/15/65. 5'10". Fair complexion, gray hair, brown eyes. Res. Nicholas Co., WV. This man's record also appears to be mixed with a Henry G. Hall. Physical description is his as is the capture record in Alleghany Co., Va.

HALL, LEDFORD ALLEN: Pvt. Co. G. B. 6/27/47 Potts Creek, Va. Enl. 7/30/62 at New River. At Charlottesville, Va. hosp. for rubeola 6/10/64 to 7/20/64. D. 11/11/1911 at Clifton Forge, Va.

HALLERBRIGHT, MOSES: Pvt. Co. I. Enl. 5/2/62 at White Sulphur Springs, Greenbrier Co., WV. Sick at Montgomery White Sulphur Springs, Va. Jan.-Feb. 1863. (This is probably an incorrect spelling and should be Moses Albright).

HALSTEAD, AMOS: Pvt. Co. I. Res. Boone Co., WV. Enl. 5/2/62(?) at White Sulphur Springs, Greenbrier Co., WV. Court-martial 8/15/64. Present under sentence of court-martial on Sept.-Oct. 1864 roll. Took oath at Charleston, WV 7/14/65. Age: 40. 5'4". Dark complexion, black hair, blue eyes. PWR says enl. 5/10/61 in Co. C.

HALSTEAD, BASIL: Pvt. Co. I. Enl. 6/22/61 at Charleston, Kanawha Co., WV. Sick in Fayette Co., WV with measles on 8/31/61 roll. Present on Sept.-Oct. 1862 roll. POW on Sept.-Oct. 1864 roll. Paroled at Charleston, WV 5/26/65. Age: 26. 5'7". Dark complexion, brown hair, gray eyes, sandy whiskers. Also spelled Bazzle.

HALSTEAD, BENJAMIN: Pvt. Co. I. Enl. 6/22/61 at Charleston, Kanawha Co., WV. Sick at White Sulphur Springs, WV on June-Aug. 1861 roll with measles. NFR.

HALSTEAD, DAVIS: Corp. Co. I. Res. Boone Co., WV. Enl. 6/22/61 at Charleston, Kanawha Co., WV. Capt. at Droop Mountain, WV 11/6/63. Age: 24. POW Camp Chase, OH 11/18/63. Transf. to Ft. Delaware 2/29/64. Took oath at Ft. Delaware and relsd. 6/20/65. Farmer. 5'10½". Florid complexion, dark hair, gray eyes.

HALSTED, IVERSON: Pvt. Co. I. Enl. 9/15/62 at Charleston, Kanawha Co., WV. Entitled to a $50 bounty. Deserted near Staunton, Va. 6/28/64.

HALSTEAD, JOHN JAMES: Pvt. Co. C. B. 10/27/37. Enl. 5/1/62(?) at White Sulphur Springs, Greenbrier Co., WV. Evidence suggests this man was capt. at Carnifex Ferry, WV 9/11/61 which would indicate an earlier enl. date. Returned from captivity 3/10/63. PWR says also served as Capt. of cav. Postwar res. Cross Lanes, WV. Sheriff, merchant, real estate dealer, Nicholas Co., WV. D. 1931. Another PWR says enl. 5/10/61. WIA at Meadow Bluff, WV. Capt. at Carnifex Ferry, WV 9/10/61. POW Camp Chase, OH for 5 mos. Later Capt., Co. A, Hounshell's Battn. D. 4/12 or 23/1931.

HALSTEAD, JOSEPH: Pvt. Co. I. Enl. 5/1/62 at White Sulphur Springs, Greenbrier Co., WV. Capt. at Winchester, Va. 9/19/64. POW Pt. Lookout, MD. Paroled at Pt. Lookout, MD and exchd. 2/18/65. In Chimborazo Hosp. for rheumatism 3/5-12/65. Appears on a list of rebel deserters Apr. 1865. Took oath at Charleston, WV and sent North 4/18/65. Age: 20. Farmer. 5'4". Dark complexion, dark hair, gray eyes. Res. Newport, Giles Co., Va. (on oath) and Boone Co., WV (when capt.).

HALSTEAD, MORDICA E.: Pvt. Co. C. B. 1808. Date and place of enl. not stated. MWIA (serious - both legs and side) at White Sulphur Springs, WV 8/26/63. D. 9/17/63. Death claim filed 1/14/64 by M. Halstead, mother. (First name is probably spelled Mordecai). Bur. C.S.A. Cem., Lewisburg, WV. Monument to him at Halsted-Crest Cem., on Saturday Rd., Fayette Co., WV. PWR says enl. 5/10/61. Farmer, Fayette Co., WV 1860. Age: 50.

HALSTEAD, PEMBROKE: Pvt. Co. I. Enl. 6/22/61 at Charleston, Kanawha Co., WV. Present on 8/31/61 roll. NFR.

HALSTEAD, WILLIAM A.: Pvt. Co. C. Res. Fayette Co., WV. Enl. 5/1/62 at White Sulphur Springs, Greenbrier Co., WV. Capt. in Greenbrier Co., WV by 47th O.V.I. 4/15/62. Age: 21. POW Camp Chase, OH. Exchd. 8/25/62. Rejoined regt. and was exchd. for John Caringler of Capt. Amick's Co. 3/10/64. 5'10". Fair complexion, dark hair, hazel eyes. PWR says enl. 5/10/61. Farm laborer, Fayette Co., WV 1860.

HAMBRICK, ANDERSON R.: Pvt. Co. I. Enl. 5/1/62 at White Sulphur Springs, Greenbrier Co., WV. Sick at Giles C.H. (Pearisburg), Va. July 1862. In jail in Botetourt Co., Va. 1863. Sick at Beverly, WV on Mar.-Apr. roll. Present on Aug.-Dec. 1863 roll. Capt. at Winchester, Va. 9/19/64. POW Pt. Lookout, MD 9/24/64. Exchd. 3/15/65. Entered Richmond, Va. Old Soldiers Home 10/1/1909 at age 75. D. 3/26/1916. Sent to Botetourt Co., Va. for burial.

HAMBRICK, GILSON: Pvt./Musician/Fifer. Co. E. Enl. 6/18/61 at Charleston, Kanawha Co., WV. Left sick at White Sulphur Springs, WV 8/20/61. Present on Sept.-Oct. 1862 roll. Paroled at Lewisburg, WV 4/26/65. Age: 20. Farmer. 5'10". Fair complexion, dark hair, dark eyes. (Also spelled Jilson).

HAMILTON, ISAAC: Pvt. Co. D. Enl. 5/1/62 at White Sulphur Springs, Greenbrier Co., WV. Deserted in the Kanawha Valley, WV 10/28/62.

HAMILTON, J. P.: Co. G. Died in service. Postwar roster.

HAMILTON, JAMES T.: Pvt. Co. A. Laborer, Tyler Mountain, WV 1860. Age: 20. Res. Kanawha Co., WV. Enl. 9/15/62 at Charleston, Kanawha Co., WV. Deserted in the face of the enemy 12/13/63. Took oath and sent North 12/31/63. Age: 23. Farmer. 5'8". Dark complexion, dark hair, blue eyes.

HAMILTON, JAMES W.: Pvt. Co. I. Enl. 5/1/62 at White Sulphur Springs, Greenbrier Co., WV. Deserted in Botetourt Co., Va. 6/26/64.

HAMILTON, JOHN A.: Pvt. Co. I. Enl. 5/1/62 at White Sulphur Springs, Greenbrier Co., WV. Sick at home in Botetourt Co., Va. Apr. 1864. In Charlottesville, Va. hosp. for acute diarrhea 6/10/64 to 8/27/64.

HAMILTON, JOHN P.: Pvt. Co. A. Res. Tyler Mountain, WV 1860. Age: 18. Enl. 6/5/61 at Camp Tompkins, Kanawha Co., WV. Sick at White Sulphur Springs, WV 8/20/61. Present on Sept.-Oct. 1862 roll. Wded. on Aug.-Dec. 1863 roll. Furlo. due to wds. on Feb.-Mar. 1864 roll. In Charlottesville, Va. hosp. for gunshot wd. of right hand 7/26/64. Transf. to Lynchburg, Va. hosp. 7/26/64. Paroled at Amherst Co., Va. 5/29/65. On Bedford Co. pension list. D. 1891.

HAMILTON, SAMUEL D.: Pvt. Co. A. B. Allegheny Co., Va. Res. Allegheny Co., Va. Enl. 4/10/62 at Brushy Ridge, Greenbrier Co., WV. Capt. at Lewisburg, WV 5/23/62. Age: 18. POW Camp Chase, OH. Exchd. 8/25/62. Sick at Alleghany Co., Va. on Mar.-Apr. 1863 roll. Dischd. on surg. cert. for general debility 8/18/63 at Camp Price. (Must have re-enlisted in Co. A) Present on Feb.-Mar. 1864 roll. On furlo. 3/16/64 to 4/1/64. In Winchester, Va. hosp. as "G. D. Hamilton", for febrile billious 7/30/64. Appears on a final disposition of conscripts at Camp Lee, Va. as a volunteer in Chapman's Battery. Farmer. 5'10" or 5'11". Dark/fair complexion, dark hair, hazel/brown eyes.

HAMILTON, WILLIAM B.: Pvt. Co. D. Enl. 5/30/61 at Summersville, Nicholas Co., WV. AWOL 7/28/61.

HAMLETT, JAMES A.: Pvt. Co. D. Enl. 9/29/63 at Camp Beauregard. In Chimborazo Hosp. for rubeola 6/1-11/64.

HAMLETT, WILLIAM H.: Pvt. Co. D. Enl. 5/1/62 at White Sulphur Springs, Greenbrier Co., WV. Det. as a nurse at Sweet Springs, WV 12/20/63. Present on 4/30/64 roll.

HANAHAN, LAWRENCE ("LARRY"): Pvt. Co. F/Co. G. Enl. 7/4/61 at Charleston, Kanawha Co., WV. Postwar res: Charleston, S.C.

HANCOCK, ADAM: Listed in UDC booklet *Craig's Share in the War Between the States*. Res. Craig Co., Va.

HANDCOCK, STEPHEN: Pvt. Co. K. B. 2/4/24 Halifax Co., Va. PWR says B. 1827 Craig Co., Va. Res. New Castle, Va. Miller. Enl. 3/30/64 at Camp Gauley, Greenbrier Co., WV. WIA at Cold Harbor, Va. 6/6/64. Vulvus. sclopet. - severe gunshot wd. of leg. In Stuart Hosp., Richmond, Va. 6/4-14/64. UDC report says enl. 1863 at Craig Co., Va. for two years. Fought at New Market and Cold Harbor. WIA in both legs. Served with Echols and Breckenridge. D. 8/10/1900 New Castle, Va. Bur. Waugh Cem., Rt. 616, Cuba Rd., New Castle, Va. Also spelled Hancock.

HANDLEY, ALEXANDER G.: 2nd Lt./1st Lt. Co. A. B. 2/21/36. Res. Boone Co., WV. Enl. 5/24/61 at Buffalo, Putnam Co., WV. Elected 1st Lt. 5/1/62. Promo. 1st Lt. 5/23/62. Capt. in Roane Co., WV 2/29/64. Age: 28. Another card says capt. in the Kanawha Valley, WV 2/29/64. POW Camp Chase, OH 3/7/64. Transf. to Ft. Delaware on which card says capt. in the Kanawha Valley, WV 2/29/64. POW Camp Chase, OH 3/7/64. Transf. to Ft. Delaware on which card says capt. in Boone Co., WV. Farmer. 5'9". Florid complexion, dark hair, dark eyes. D. Greenbrier Co., WV. Bur. Handley Cem., Sleepy Hollow Golf Course, Putnam Co., WV.

HANDLEY, BENNIAH F.: Pvt. Co. A. B. 9/8/43. Enl. 9/15/62 at Charleston, WV. D. in hosp. at Union, WV 2/25/63. Bur. 2/25/63. A PWR says died at Camp Chase, OH 2/25/63 and bur. Handley Cem., Sleepy Hollow Golf Course, Putham Co., WV.

HANNA, DAVID: Pvt. Co. D. B. 9/23/24. Res. Craig Co., Va. Enl. 5/30/61 at Summersville, Nicholas Co., WV. Sick at home 7/28/61. Res. Lower Craig Creek, Craig Co., Va. 1870. D. 3/21/94. Bur. Hannah Cem., Craig Co., Va.

HANNA, WILLIAM: Pvt. Co. D. Enl. 5/30/61 at Summersville, Nicholas Co., WV. Sick at home 7/28/61. NFR.

HANNAH, JAMES: Pvt. Co. D. Enl. 10/8/62(?) at Charleston, Kanawha Co., WV. Present on Mar.-Apr. 1863 roll. NFR.

HANNAH, JOHN M.: Pvt. Swann's Co. Enl. 4/25/61 at Charleston, Kanawha Co., WV. Age: 24. Co. transf. to Co. K, 59th Va. Inf. 7/25/61.

HANNAH, JOHN P.: Pvt. Swann's Co./Co. H. Enl. 4/25/61 at Buffalo, Putnam Co., WV. Age: 22. Transf. from Co. K, 59th Va. Inf. to Co. H. 22nd Va. Inf. 3/18/62. Returned to Swann's Co., Edgar's Battn. 7/6/62. Det. to commissary dept. reported to co. 3/7/63. Since transf. to Co. A. Edgar's Battn. 3/19/63.

HANNAH, JOSEPH: B. 1823 (or 5/12/17). Res. Lower Craig's Creek, Craig Co., Va. 1870. Bur. Hannah Cem., Craig Co., Va.

HANNAH, SAMUEL: Pvt. Co. H. Enl. 7/22/62 at Camp McCausland, Monroe Co., WV. Deserted at Camp Royal Oak 7/27/62. Deserted 7/27/62 on Aug. 1862 broadside. Greenbrier Co., WV. Also spelled Hanna.

HANNAHAN, LARRY: Pvt. Co. K. Enl. 6/17/62 at Union, Monroe Co., WV. Left sick at Hightown on Mar.-Apr. 1863 roll.

HANSFORD, CARROL MORRIS: Corp. Co. H. B. 1836 Kanawha Co., WV. Enl. 5/8/61 at Charleston, Kanawha Co., WV. On clothing receipt 3/13/64. Last entry on 4/28/64 roll. PWR says transf. to Co. I, 26th Va. Battn. 8/27/63 with rank of Corp. Promo. Ord. Sgt. Capt. at Cold Harbor, Va. 6/?/64. POW Pt. Lookout, MD. Transf. Elmira, N.Y. Relsd. 3/2/65. Postwar res. St. Albans, WV. Still alive in 1899. Bur. Teays Hill Cem., St. Albans, WV.

HANSFORD, FELIX G.: Pvt. Co. H. B. in Kanawha Co., WV. Merchant, Cannelton, WV 1860. Age: 30. Enl. 3/10/62 at Lewisburg, Greenbrier Co., WV. Conscript. Det. as clerk in A.Q.M. Dept. at Lynchburg, Va. on Jan.-Feb. 1863 roll. Medical discharge 7/5/63 for pulmonary phthisis (wasting of body - possibly tuberculosis). Dischd. at Camp Sam Jones 7/8/63. Age: 35. Clerk/merchant. 5'9". Dark complexion, dark hair, gray eyes.

HANSHAW, JAMES: Pvt. Co. E. Enl. 9/25/62 at Charleston, Kanawha Co., WV and/or Cotton Hill, Fayette Co., WV. Deserted on the road to Belle Grove (Cedar Creek), Va. 10/8/64.

HANSHAW, SILAS S.: Pvt. Co. E. Res. Clay Co., WV. Enl. 5/1/62 at White Sulphur Springs, Greenbrier Co., WV. Left with the wded. at Woodstock, Va. 10/20/64 by order of a board of officers chosen for that purpose. $60 deducted from wages for one Enfield rifle lost. On deserted list Mar. 1865. Took oath 5/5/65. Age: 25. Farmer. 5'11". Dark complexion, fair hair, blue eyes.

HANSHAW, WILLIAM: Pvt. Co. E. Enl. 10/12/62 at Cotton Hill. Fayette Co., WV. MWIA at Belle Grove (Cedar Creek), Va. 10/19/64. D. 10/30/64.

HANSON, ANTHONY: Pvt. Co. E. Enl. 9/18/62 at Charleston, Kanawha Co., WV. Present on Sept.-Oct. 1862 roll. NFR.

HANSON, HIRAM: Co. H. Deserted 5/23/62 on Aug. 1862 broadside. Greenbrier Co., WV.

HAPLENSTATE, S.: Pvt. Co. B. Res. Jackson Co., WV. In hosp. at Winchester, Va. July, 1864 for febris.

HARDBARGER, GEORGE W.: Pvt. Co. I. Enl. 5/1/62 at White Sulphur Springs, Greenbrier Co., WV. Present on Jan.-Feb. 1863 roll. NFR.

HARDSOCK, GEORGE H.: Pvt. Co. D. Enl. 3/7/64 at Camp Gauley, Greenbrier Co., WV. Capt. at Cold Harbor (Gaines Mill), Va. 5/31/64. POW Pt. Lookout, MD. Transf. to Elmira, NY 7/25/64. D. at Elmira, NY of chronic diarrhea 1/18/65. Bur. grave #1576 at Elmira, NY.

HARE, ROBERT P.: Pvt./Corp. Co. H. B. in New York. Carpenter, Charleston, WV 1860. Age: 25. Enl. 5/8/61 at Charleston, Kanawha Co., WV. Elected Corp. 7/27/61. Det. to go to Summersville, WV to repair steam mill 8/30/61. Det. in Ordnance Dept. at Newbern, Va. (undated). Det. in salt petre works at Saltville, Va. on Jan.-Feb. 1863 roll. Det. as engineer for Tenn. & Va. Railroad Apr. 31-Oct. 31, 1864 roll.

HARFORD, HARVY: Pvt. Co. F. Enl. 1/23/64(?) at Camp Gauley, Greenbrier Co., WV. NFR.

HARKINS, DANIEL: Pvt. Co. H. Enl. 4/25/61 at Charleston, Kanawha Co., WV. Transf. from Co. K 59th Va. Inf. 3/19/62. WIA at Gardner's Farm, Va. 5/30-31/64. Severe side and shoulder. Also on wded. list 5/27/64. WIA at Cold Harbor, Va. 6/4/64. In Chimborazo Hosp. for very serious side wd. (bullet entered near 22nd vertabrae making exit) 6/1/64 to 7/1/64. Age: 21. (at enlistment). Age given as 22 when wded., which does not correspond with enlistment age. Cooper.

HARLOW, HENRY JACKSON: Co. D. B. Louis Co., Va. 1838. D. 3/5/1903. Bur. Zion, Va.

HARMAN, THOMAS H.: Pvt. Co. A. Enl. 5/22/62 at White Sulphur Springs, Greenbrier Co., WV. WIA (severe thigh and arm) sick at White Sulphur Springs, WV 8/26-27/63. Capt. at Winchester, Va. 9/19/64. POW Pt. Lookout, MD 9/24/64. Wrote a letter from Pt. Lookout, MD 11/21/64 requesting to take oath and to go to Jackson County as his parents were dead and all his brothers and sisters lived in Ohio. Exchd. 3/15/65. Paroled at Charleston, WV 5/11/65. Age: 22. 5'10". Dark complexion, dark hair, blue eyes. D. at Winfield, WV. Bur. Woods Cem., ½ mile east of Grapevine School, 100 feet west of Rt. 34 in WV.

HARRAH, JOHN A.: Pvt. Co. D. B. in Va. Laborer. Enl. 5/1/62 at White Sulphur Springs, Greenbrier Co., WV. Sick in Greenbrier Co., WV on Nov. 1863 roll. Surrendered at Charleston, WV Mar. 1865. Took oath at Charleston, WV 3/20/65. Age: 21. Farmer. 5'7". Dark complexion, dark hair, blue eyes. Res. Lexington, Va. 1870, age 26. Filed for pension 1914.

HARRAH, WASHINGTON: Pvt. Co. C. Enl. 5/1/62 at White Sulphur Springs, Greenbrier Co., WV. Absent sick (undated). NFR.

HARRIS, A. M.: Pvt. Co. K. Date and place of enlistment not stated. Made flour barrels in Christiansburg, Va. and Salem, Va. A letter requested that he be det. as a cooper in Christiansburg, Va. while on active duty. It also stated he had sufficient timber to make 200 barrels which were needed to be made while the troops were inactive, on an approximate 90 day det. Cooper.

HARRIS, ROBERT ("BOB"): Pvt. Co. K. Enl. 6/6/61 at White Sulphur Springs, Greenbrier Co., WV. Sick on surg. cert. 9/12/61. Present on Sept.-Oct. 1862 roll. KIA (shot through heart) at New Market, Va. 5/15/64. Res. Fayette Co., WV 1860. Age: 16.

HARRIS, WILLIAM: Pvt. Co. A. B. in Greenbrier Co., WV. Enl. 4/10/62 at Brushy Ridge, Greenbrier Co., WV. Capt. at Lewisburg, WV 5/23/62. Age: 28. POW Camp Chase, OH. Apparently exchd. Left sick at Lewisburg, WV on Mar.-Apr. 1863 roll. Medical dischd. at Camp Sam Jones 7/2/63, having one ankle crippled & disabled & atrophy of corresponding leg. Farmer. 5'9". Fair complexion, light hair, brown eyes.

HARRIS, WILLIAM: Pvt. Co. K. Res. Fayette Co., WV 1860. Age: 14. Enl. 1862.

HARRIS, WILLIAM A.: Pvt. Co. F. B. 10/15/44 Monroe Co., WV. Enl. 7/4/61 at Charleston, Kanawha Co., WV. Dischd. 8/17/61. On surg. cert. as crippled condition of his feet and ankle which lames him when required to walk. Age: 26. 5'9". Light complexion, light hair, blue eyes. D. 10/23/1914. Bur. Cedar Hill Cem., Covington, Va. Tombstone says Co. G. Postwar roster says middle initial is "H".

HARRISON, HENRY: Pvt. Swann's Co./Co. E. Enl. 7/17/61 at Charleston, Kanawha Co., WV. Det. as a mounted guard at White Sulphur Springs, WV 8/13-14/61. Det. as a nurse in hosp. at White Sulphur Springs, WV 11/18/61 to 12/30/61.

HARRISON, JAMES M.: Pvt. Co. A. Farmer, Coalsmouth, WV 1860. Age: 33. Enl. 6/4/61 at Camp Tompkins, Kanawha Co., WV. NFR.

HARVEY, WILLIAM: 5th Sgt. Co. G. Enl. 7/3/61 at Charleston, Kanawha Co., WV. Det. as nurse for the sick at Locust Lane, WV 8/20/61. Capt. at White Sulphur Springs, WV 8/26-27/63. Age: 22. POW Camp Chase, OH 9/7/63. Transf. to Rock Island, IL 1/22/64. Remark while at Rock Island, IL: "Tired of C.S.A." Took oath at Rock Island, IL 6/16/65. Farmer. 6" or 5'11". Florid/light complexion, dark brown hair, gray eyes. Res. Sunhill, Wyoming Co., WV. Postwar roster says captured at Winchester, Va. 9/19/64.

HARWOOD, M. L.: Pvt. Company, date and place of enlistment not stated. Clothing receipt roll says "Witness - H. C. Dawson".

HASH, THOMAS: Pvt. Co. E. Enl. 2/26/64 at Grayson C.H., Va. Dischd. on surg. cert. of disability. Paper not on file. Probably from Galax, Grayson Co., Va.

HASSLER, FREDERICK W. B.: Sgt. Co. B. Grandson of F. R. Hassler, organizer and first supt. of U.S. Coast Survey. Res. Ripley, WV, 1860, age: 20. Enl. 6/1/61 at Ripley, Jackson Co., WV. On list of deserters at Charleston, WV Feb. 1865. Took oath at Charleston, WV and sent North 2/16/65. Age: 25. Merchant. 5'5¼". Fair complexion, dark hair, blue eyes. Claimed he was watching over family properties when war broke out and was forced to join C.S. army. Postwar partnership with his brother, Brvt. Col. F. R. Hassler, who was one of the first to invest money and establish a factory in the new state of WV. Res. Ripley, WV, 1869. Wrote and published memoir of his service in *Historical Magazine* 1869.

HASSLER, FREDERICK R.: Pvt. Co. H. Grandson of F. R. Hassler, organizer and first supt. of U.S. Coast Survey. Enl. 8/8/61 at White Sulphur Springs, Greenbrier Co., WV. Was enrolled 6/16/61 and served in Capt. Sweeney's Co. until disbanded and then re-enlisted in this co. 8/8/61. Postwar partnership with brother, Frederick W. B. Hassler, one of the first to invest money and establish a factory in the new state of WV.

HATCHER, A. J.: Corp. Co. G. Enl. 7/3/61 at Charleston, Kanawha Co., WV. Left sick at White Sulphur Springs, WV 8/25/61. Present on Nov.-Dec. 1862 roll. Transf. to Derrick's Battn. Feb. 1863.

HATCHER, COLUMBUS W.: Pvt. Swann's Co. Tinner, Charleston, WV 1860. Enl. 4/25/61 at Charleston, Kanawha Co., WV. Age: 24. Co. transf. to Co. K, 59th Va. Inf. 7/25/61.

HATCHER, JOSEPH T.: Pvt. Co. D. Res. Greenbrier Co., WV. Enl. 5/1/62 at White Sulphur Springs, Greenbrier Co.. WV. Left sick at Lewisburg, WV 4/13/63. Capt. at Droop Mountain, WV 11/6 or 8/63. Age: 19. POW Camp Chase, OH 11/18/63. Transf. to Ft. Delaware 2/29/64. Exchd. 2/27/65. In Richmond, Va. hosp. 3/4-11/65. Farmer. 5'8". Florid complexion, light hair, dark eyes.

HAWKINS, CHARLES M.: Pvt. Co. H. B. ca. 1845 Craig Co., Va. Enl. 3/19/63 at Lewisburg, Greenbrier Co., WV. Capt. at Winchester, Va. 9/19/64. POW Pt. Lookout, MD 9/24/64. Exchd. 3/15/65. Middle name is probably Nelson.

HAWKINS, HENRY C.: Pvt. Co. K. Enl. 6/6/61 at Charleston, Kanawha Co., WV. Absent cause unknown on June-Sept. 1861 roll. NFR. Res. Fayette Co., WV 1860. Age: 19.

HAWKINS, LEWIS PHILLIP: Pvt. Co. H. B. ca. 1846 Craig Co., Va. Enl. 12/11/63 in Monroe Co., WV. Capt. at Fishers Hill, Va. 9/22/64. POW Pt. Lookout, MD 10/1/64. D. at Pt. Lookout, MD of typhoid fever 11/6/64. Age: 18. Bur. grave #628 at Pt. Lookout, MD POW Cem. Was a conscript at Camp Lee, Richmond, Va. Court records list him as Phillip Lewis.

HAWVER, SAMUEL J.: Corp. Co. C. Enl. 5/1/62 at White Sulphur Springs, Greenbrier Co., WV. Sick during summer of 1862. Sick in hosp. at Montgomery White Sulphur Springs, Va. July, 1862. Present on Nov.-Dec. 1862 roll. Capt. at Winchester, Va. 9/19/64.

HAWVER, SAMUEL: PWR says enl. 5/10/61. Day laborer, Fayette Co., WV 1860. Age: 18.

HAYES, BENJAMIN: Pvt. Taylor's Co. Enl. 6/19/61 at Frankford, Greenbrier Co., WV. Age: 27. Co. transf. to Wise Brigade 7/10/61.

HAYES, JOHN B.: Pvt. Taylor's Co. Enl. 6/19/61 at Frankford, Greenbrier Co., WV. Age: 21. Co. transf. to Wise Brigade 7/10/61.

HAYES, SAMUEL C.: Pvt. Taylor's Co. Enl. 6/19/61 at Frankford, Greenbrier Co., WV. Age: 16. Co. transf. to Wise Brigade 7/10/61.

HAYNES, WILLIAM: Pvt. Co. E. Student, Sissonville, WV 1860. Age: 17. Enl. 6/18/61 at Charleston, Kanawha Co., WV. Left sick at White Sulphur Springs, WV 8/20/61. NFR.

HAYS, R. G.: Pvt. Co. E. Date and place of enlistment not stated. Det. as regt. wagoner July 1862. NFR.

HAZELWOOD, JOHN JAMES: Co. B. Chapter 59 UDC.

HAZELWOOD, JOHN W.: Pvt. Co. D. Enl. 4/7/64 at Camp Gauley, Greenbrier Co., WV. Capt. at Cold Harbor (Gaines Mill), Va. 6/3/64. POW Pt. Lookout, MD 6/11/64. Transf. to Elmira, NY 6/12/64. D. at Elmira, NY of chronic diarrhea 10/7/64. Bur. grave #585, Section G, Woodlawn Cem., Elmira, NY.

HEDRICK, GEORGE P.: Pvt./Orderly Sgt. Co. K. Enl. 6/6/61 at Charleston, Kanawha Co., WV. Run over by train on Va. Central Railroad. Arm amputated at Charlottesville, Va. hosp. due to wd. in elbow and ankle 1/3/62. D. 1/10/62. Caught by telegraph wires and thrown from top of a box car. William Bahlmann says an arm and a leg amputated and he died from shock. Left a widow and one child.

HEDRICK, JOHN F.: Pvt. Co. B. Enl. 5/1/62 at White Sulphur Springs, Greenbrier Co., WV. Conscript. Deserted 5/2/62 on Aug. 1862 broadside, Alleghany Co., Va. Deserted on Sept.-Oct. 1862 roll. Returned on Mar.-Apr. 1863 roll. Capt. at Winchester, Va. 9/19/64. POW Pt. Lookout, MD 9/24/64. Exchd. 3/15/65. Paroled at Lewisburg, WV 4/26/65. Age: 21. Farmer. 6'. Fair complexion, black hair, dark eyes.

HELMES, THOMAS: Pvt. Co. K. Enl. 6/11/63 at Meadow Bluff, Greenbrier Co., WV. Deserted 12/25/63.

HELMINTOLLER, JORDAN: Pvt. Co. K. B. Alleghany Co., Va. 1842 Enl. 8/7/61 at Union, Monroe Co., WV. Sick at hosp. May 1862. Unaccounted for as of 11/1/62. Listed as deserted on Jan.-Feb. 1863 roll. PWR says disability discharge for measles 1864. Farmer and stonemason, Alleghany Co., Va. D. at age 72 ca. 1914.

HELMINTOLLER, WILLIAM: Pvt. Co. A. B. 3/12/29 Alleghany Co., Va. Enl. 8/10/62 at Salt Sulphur Springs, Monroe Co., WV. On Provost Guard in Alleghany Co., Va. Deserted at Packs Ferry, WV 10/17/62. Returned to regt. Left sick at Lewisburg, WV on Mar.-Apr. 1863 roll. On det. service on Apr.-Oct. 1864 roll. D. 8/18/09 Alleghany Co. Va. Bur. in a cem. at Rts. 614 & 18, Alleghany Co., Va.

HELMINTOLLER, WILLIAM: Pvt. Co. A. Date and place of enlistment not stated. Deserted at Jacksons River, Va. on Apr.-Oct. 1864 roll. Also spelled Helmantoller. (This is probably the same William Helmintoller as listed previously).

HELMS, GEORGE W.: Co. K. Deserted 5/3/62 on Aug. 1862 broadside. Paint Back Valley. Also spelled Holms.

HELMS, JOHN L.: Pvt. Co. C. Res. Craig Co., Va. Enl. 6/7/63 at Camp Jackson. Det. as brig. wagoner. Deserted 7/31/64. Took oath at Charleston, WV 12/13/64. Age: 36. Farmer. 5'11". Light complexion, light hair, blue eyes. Also spelled Helloms.

HELMS, MASON V. ("MASE"): Pvt. Co. K. B. ca. 1835 Craig Co., Va. Enl. 4/24/62 at White Sulphur Springs, Greenbrier Co., Va. Deserted 5/3/62 on Aug. 1862 broadside, Craig's Creek, Va. Joined from desertion 10/12/62. KIA at Droop Mountain, WV 11/6/63. Shot through head and body and finger shot off. D. 11/7/62. Left a widow and one or two children. William Bahlmann called him "a most efficient soldier." Also spelled Holms.

HENDERSON, HARVEY: Pvt. Co. E. Res. Roanoke Co., Va. Date and place of enlistment not stated. Sick at Giles C.H. (Pearisburg), Va. May 1862. Capt. at Droop Mountain, WV 11/6/63. Age: 20. POW Camp Chase, OH 11/18/63. Transf. to Ft. Delaware 3/4/64. Took oath at Ft. Delaware and relsd. 6/20/65. Farmer. 5'10". Florid complexion, sandy hair, blue eyes.

HENDERSON, RODNEY SHELTON: PWR.

HENDRICK, OBEDIAH: Pvt. Co. C. Enl. 9/25/62 at Camp Charleston, Kanawha Co., WV. Sick in hosp. on Sept.-Oct. 1862 roll. Present on Nov.-Dec. 1862 roll. WIA (severe neck and arm) at White Sulphur Springs, WV 8/26-27/63. Capt. at Droop Mountain, WV 11/6/63. Farmer, Fayette Co., WV 1860. Age: 41. PWR says enl. 5/10/61. Name also spelled Hedrick.

HENDRICK, T. P.: Pvt. Co. C. Enl. 5/1/62 at White Sulphur Springs, Greenbrier Co., WV. MIA at Droop Mountain, WV 11/6/63. Was exchd. for Pvt. William W. Dean of Capt. William Thurmond's Co. of Partisan Rangers 10/1/64. Farm laborer, Fayette Co., WV 1860. Age: 30. PWR says enl. 5/10/61 and name is Tolbert P. Hedrick.

HENDRICK, VIRGINIUS: Pvt. Co. H. Lawyer, Charleston, WV 1860. Age: 30. Enl. 5/8/61 at Charleston, Kanawha Co., WV. Det. in Provost Marshall office at Giles C.H. (Pearisburg), Va. With leave at Lewisburg, WV on May-June 1863 roll. House of Delegates, Richmond, Va. on Nov.-Dec. 1863 roll.

HENDRICKSON, GEORGE W.: Pvt. Co. K. B. 9/24/45 at Craig Co., Va. Res. Nicholas Co., WV. Enl. 9/20/63 at Lewisburg, Greenbrier Co., WV. Entitled to bounty being underage (less than 18) when volunteered. Paroled at Charleston, WV 6/23/65. Age: 19. 5'10". Fair complexion, light hair, blue eyes. PWR says WIA in "last fight in the Valley of Va." Walked five days before finding a surg. Res. Atchison, KS 1870. Businessman, Atchison, KS and other KS cities, 1870. One of the most highly respected citizens of Atchison, KS. A first class soldier and gentleman. D. 6/12/1923 at Atchison, KS after a long illness. Still had bullet in him from the war. Age: 78. Bur. Mt. Vernon Cem., Atchison, KS.

HENDRICKSON, JOHN: Pvt. Co. C. Enl. 5/1/62 at White Sulphur Springs, Greenbrier Co., WV. Capt. at Lewisburg, WV 5/23/62. Age: 27. POW Camp Chase, OH. Exchd. 8/25/62. Returned to co. 3/10/63. Deserted on Nov.-Dec. 1863 roll. 6'. Dark complexion, dark hair, gray eyes. B. 10/13/33. Res. Craig Co., Va. D. 9/1/96. Bur. Trenor Cem., Rt. 42, behind Sinking Creek Store on Ralph Price farm, Craig Co., Va.

HENDRICKSON, JOHN M.: 1st Lt./2nd Lt. Co. K. Enl. 6/6/61 at Charleston, Kanawha Co., WV. Elected Brevet 2nd Lt. 5/1/62. Elected 1st Lt. 11/12/63. Promo. 12/31/64. Paroled at Charleston, WV 6/23/65. Age: 26. 5'10". Fair complexion, auburn hair, gray eyes. B. 8/26/38 Botetourt Co., Va. Res. Nicholas Co., WV 1848. Res. Fayette Co., WV 1854. Ft. Edward Collegiate School, NY 1858. Alleghany College 1860. In hosp. 5/23/62. Prov. Marshall at Lewisburg, WV 11/6/63. Promo. to Adj. Spring 1864. Postwar schoolteacher and general store manager, Tazewell Co., Va. Teacher and farmer, Atchison, KS 1870. Married 1873. Res. Oskaloosa, KS 1883. Real estate, Los Animos, Colorado. D. 3/19/1926 Atchison, KS. Brother of George W. Hendrickson.

HENLEY, CHARLES: B. 1833. D. 1907. Bur. Teays Hill Cem., St. Albans, WV.

HENLY, WOODSON: Pvt. Swann's Co. Laborer, Charleston, WV 1860. Enl. 5/15/61 at Charleston, Kanawha Co., WV. Age: 18 Co. transf. to Co. K, 59th Va. Inf. 7/25/61.

HENNESSEY, HENRY: Pvt. Co. F. Res. Monroe Co., WV. Enl. 7/6/61 at Charleston, Kanawha Co., WV. Det. to drive wagon 8/8/61. D. sometime prior to 1914. Postwar roster says name is Harry.

HENNING, GEORGE W.: Pvt. Co. F. Enl. 3/7/63 at Lewisburg, Greenbrier Co., WV. Capt. at Winchester, Va. 9/19/64. POW Pt. Lookout, MD 9/23/64. Exchd. 3/15/65. In Richmond, Va. hosp. for P.P. (?) 3/18/65. Relsd. from hosp. on 30 day furlo. 3/22/65.

HENNING, LORENZO: Pvt. Co. F. Enl. 5/1/62 at White Sulphur Springs, Greenbrier Co., WV. Capt. at Cold Harbor (Gaines Mill), Va. 6/3/64. POW Pt. Lookout, MD 6/11/64. POW Elmira, NY 7/12/64. D. at Elmira, NY of variola (smallpox) 1/24/65. Bur. grave #1610 Woodlawn Cem., Elmira, NY.

HENNING, MASON: Pvt. Co. F. Enl. 3/12/63 at Lewisburg, Greenbrier Co., WV. Capt. at Winchester, Va. 9/19/64. POW Pt. Lookout, MD 9/27/64. Exchd. 1/17/65. In Richmond, Va. hosp. 1/26-27/65.

HENNIS, GEORGE: Pvt. Co. B. Enl. 6/1/61 at Ripley, Jackson Co., WV. In hosp. at White Sulphur Springs, WV on 9/1/61 roll. NFR.

HENRY, CHARLES A.: Co. C. Admitted to Old Soldiers Home, Richmond, Va. 6/3/85 at age 60. D. 5/25/1904. Bur. Ridge Church.

HENSON, H.: Pvt. Co. H. Enl. 4/28/62 at White Sulphur Springs, Greenbrier Co., WV. Deserted to the enemy 12/11/63.

HEPLER, JABES M.: Pvt. Co. A. Enl. 4/10/62 at Brushy Ridge, Greenbrier Co., WV. Sick in Alleghany Co., Va. on Mar.-Apr. 1863 roll. Dischd. 8/18/63. Surg. cert. not on file.

HEPPENSTALL, SAMUEL: Pvt. Co. B. Enl. 10/28/63 at Lewisburg, Greenbrier Co., WV. Det. at bridge building Nov.-Dec. 1863. In Winchester, Va. hosp. with 'febris' 8/4/64. D. at Lynchburg, Va. of 'febris typhoid' 9/14/64.

HERDMAN, POLK: Pvt. Co. B. Enl. 9/22/62 at Charleston, Kanawha Co., WV. Absent wded. on Apr.-Oct. 1864 roll. Deserted and surrendered at Charleston, WV. Took oath at Charleston, WV 2/22/65. Age: 20. Farmer. 5'6". Dark complexion, dark hair, dark eyes. (This is probably James Polk Herdman).

HESLEP, JAMES M.: Pvt. Co. K. Enl. 6/6/61 at Charleston, Kanawha Co., WV. Sick at home 8/2/61. Present on Sept.-Oct. 1862 roll. Present on Jones-Imboden raid Mar.-Apr. 1863. Paroled at Charleston, WV 6/10/65. Age: 21. 5'10½". Light complexion, light hair, gray eyes. Day laborer, Fayette Co., WV 1860. Age: 17.

HESS, FRANKLIN: Corp. Co. C. Enl. 5/1/62 at White Sulphur Springs, Greenbrier Co., WV. Returned from captivity 3/10/62. Capt. at Winchester, Va. 9/19/64. POW Pt. Lookout, MD 9/24/64. Exchd. 3/15/65. Paroled at Charleston, WV 5/11/65. Age: 27. 6'1". Dark complexion, dark hair, gray eyes, dark whiskers. Blacksmith, Fayette Co., WV 1860. Age: 25. PWR says enl. 5/10/61.

HESS, JA. B.: Pvt. Co. A. Enl. 5/1/62 at White Sulphur Springs, Greenbrier Co., WV. Deserted at Lewisburg, WV 5/23/62. (First name is probably James).

HESS, JAMES W.: Pvt. Co. A. B. in Va. Enl. 4/10/62 at Brushy Ridge, Greenbrier Co., WV. WIA (slight arm) at White Sulphur Springs, WV 8/26-27/63. WIA (severe head) at Cold Harbor, Va. 6/4/64. Paroled at Lewisburg, WV 4/25/65. Age: 23. Farmer. 5'8". Fair complexion, brown hair, blue eyes.

HESS, JAMES W.: Pvt. Co. F. Res. Monroe Co., WV. Enl. 7/4/61 at Charleston, Kanawha Co., WV. Sick in hosp. at White Sulphur Springs, WV 8/20/61. Postwar roster says res. Fayette Co., WV postwar and middle initial H. D. sometime prior to 1914.

HESS, JOSEPH: Pvt. Co. A. Enl. 9/15/62 at Charleston, Kanawha Co., WV. KIA at Richmond, Va. (undated) on 10/31/64 roll.

HESS, JOSEPH A.: Pvt. Co. A. Enl. 1/31/64 at Camp Gauley, Greenbrier Co., WV. On clothing list 5/13/64. NFR.

HESS, T. H.: Co. C. Original member Jeb Stuart UCV Camp, Fayetteville, WV 1909.

HICKMAN, A.: Pvt. Co. B. Enl. 5/1/62 at White Sulphur Springs, Greenbrier Co., WV. WIA and capt. at Lewisburg, WV 5/23/62.

HICKMAN, ARCHIBALD: Pvt. Co. D. B. in Bath Co., Va. Res. Nicholas Co., WV. Enl. 7/7/63 at Camp Sam Jones. Deserted at Blue Ridge 6/?/64. Capt. at Bulltown, WV 3/20/65. Took oath and sent North 3/30/65. (An A. Hickman appears on a monthly report, First Brig., Army of Southwest Va., Brig. Gen. Echols, for April 1864 stating entry "died" — this may be a different man than Archibald Hickman). Age: 20. Farmer. 5'5". Dark complexion, light hair, blue eyes.

HICKMAN, GEORGE R.: 2nd Sgt. Co. D. Res. Nicholas Co., WV. Enl. 5/30/61 at Summersville, Nicholas Co., WV. Capt. at Lewisburg, WV 5/23/62. Age: 22. POW Camp Chase, OH 5/30/62. Exchd. 8/25/62. Deserted on Blue Ridge 6/?/64. Fair complexion, black hair, gray eyes, light moustache.

HICKMAN, WILLIAM R.: Pvt. Co. D. Enl. 11/10/62 at Red Sulphur Springs, Monroe Co., WV. D. of disease at Camp Gauley, WV 4/16/64.

HIGGINBOTHAM, EMANUEL: Pvt. Co. A. Enl. 7/8/61 at Camp Tompkins, Kanawha Co., WV. Unaccounted for as of 7/24/61. NFR.

HILL, HIRAM: Pvt. Co. K. Enl. 6/6/61 at Charleston, Kanawha Co., WV. Active member since 7/12/61. Honorary member after that time being 62 years of age.

HILL, JOHN: Pvt. Co. D. Enl. 5/31/61 at Summersville, Nicholas Co., WV AWOL 7/28/61.

HILL, JOHN J.: Pvt. Co. A. Res. Monroe Co., WV. Enl. 4/10/62 at Brushy Ridge, Greenbrier Co., WV. Capt. at Lewisburg, WV 5/23/62. Age: 20. POW Camp Chase, OH. Apparently exchd. KIA at White Sulphur Springs, WV 8/26/63. 5'6". Dark complexion, dark hair, black eyes, light whiskers.

HILL, JOHN J.: Pvt. Co. F. B. in Monroe Co., WV. Enl. 7/4/61 at Charleston, Kanawha Co. WV. D. at Talcott, WV of old age in 1883. Bur. Pence Springs Cem., Summers Co., WV.

HILL, JOHN W.: Pvt. Co. A. Res. Kanawha Co., WV. Enl. 6/5/61 at Camp Tompkins, Kanawha Co., WV. Capt. in Boone Co., WV 2/9/64. Another card says capt. in Kanawha Valley, WV. POW Camp Chase, OH 3/7/64. Transf. to Ft. Delaware 3/14/64. Took oath 6/20/65. Age: 20. Farmer. 5'4" or 5'10". Ruddy/light complexion, light hair, gray or blue eyes. Bur. Godwin Cem., near Belington, WV.

HILL, M. H.: Pvt. Co. H. Res. Craig Co., Va. Enl. 4/2/62 in Greenbrier Co., WV. Another card says enl. Apr. 1864. NFR.

HILL, WILLIAM M.: Pvt. Co. A. Res. Monroe Co., WV. Enl. 4/10/62 at Brushy Ridge, Greenbrier Co., WV. Left sick at Beverly, WV on Mar.-Apr. 1863 roll. Present on Aug.-Dec. 1863 roll. In Charlottesville, Va. hosp. for acute diarrhea 6/12-17/64. Capt. at Winchester, Va. 9/19/64. POW Pt. Lookout, MD 9/24/64. Exchd. 1/17/65. In Richmond, Va. hosp. 1/26/65. Capt. again and took oath at Charleston, WV and sent North 4/6/65. Age: 40. Farmer. 5'4". Fair complexion, fair hair, gray eyes.

HILTON, JOSHUA: Pvt. Co. H. Enl. 7/14/62 in Monroe Co., WV. Deserted 11/?/62.

HINCHEY, WILLIAM E.: Pvt. Co. E. Res. Roanoke Co., Va. Enl. 5/1/62 at White Sulphur Springs, Greenbrier Co., WV. Sick in hosp. on Nov.-Dec. 1862 roll. Det. service as blacksmith on Mar.-Apr. 1864 roll. Deserted while on det. service 10/?/64. Went to the enemy. Postwar res. Roanoke Co., Va. PWR also lists him as a W. G. Hinchee.

HINESMAN, OTTMAR L. C.: Quarter Master/2nd Lt. Co. B. Enl. 6/1/61 at Ripley, Jackson Co., WV. Appted. 2nd Lt. 10/1/61. Dropped 5/2/62.

HINKLE, A. D. K.: Pvt. Taylor's Co. Enl. 6/19/61 at Frankford, Greenbrier Co., WV. Age: 22. Co. transf. to Wise Brigade 7/10/61.

HINKLE, AMOS: Pvt. Taylor's Co. Enl. 6/19/61 at Frankford, Greenbrier Co., WV. Age: 23. Co. transf. to Wise Brigade 7/10/61.

HINKLE, NATHAN G.: Pvt. Co. D. Enl. 5/30/61 at Summersville, Nicholas Co., WV. Sick at home 7/6/61. NFR.

HINTON, PETER: Co. F. Res. Monroe Co., WV. D. sometime prior to 1914. Postwar roster.

HIPPS, JOHN M.: Co. K. Deserted 5/23/62 on Aug. 1862 broadside. Sinking Creek, Craig Co., Va.

HIVELY, W. A.: Pvt. Co. C. Date and place of enlistment not stated. Dischd. at Camp McCausland, WV 7/27/62.

HIX, JAMES M.: Pvt. Co. A. B. 5/6/32 Putnam Co., WV Enl. 5/22/61 at Charleston, Kanawha Co., WV. On special duty en route to White Sulphur Spring, WV 8/27/61. In Richmond, Va. hosp. 11/29/62. WIA (serious breast) at White Sulphur Springs, WV 8/26-27/63. On wded. furlo. on Aug.-Dec. 1863 roll. Det. service Apr. 1864. In Richmond, Va. hosp. with rubeola 6/4/64. Transf. to Chimborazo Hosp. 6/15/64. On a list of POW's, deserters, refugees at Clarksburg, WV 9/7/64. Took oath and relsd. 8/31/64. Age: 25. Farmer. 5'11". Fair complexion, sandy hair, gray eyes. D. 2/4/1911 of pneumonia at Pickens, WV. Bur. Cherry Grove Cem. near Pickens, WV. (Also spelled Hicks). Middle name is probably Mat.

HIX, JOHN: Pvt. Co. A. Enl. 4/10/62 at Brushy Ridge, Greenbrier Co., WV. Capt. at Cold Harbor (Gaines Mill), Va. 6/3/64. POW Pt. Lookout, MD 6/11/64. POW Elmira, NY 7/17/64. D. at Elmira, NY of chronic diarrhea 9/2/64. Bur. grave #134 Woodlawn Cem., Elmira, NY. (Also spelled Hicks).

HIX, MILTON H.: 1st Corp. Co. A. B. in Va. Enl. 5/22/61 at Charleston, Kanawha Co., WV. Capt. (date and place not stated). Paroled at Lewisburg, WV 4/25/65. Age: 24. Farmer. 6'. Dark complexion, black hair, dark eyes. (Also spelled Hicks).

HODGES, ALEXANDER: Pvt. Swann's Co. Enl. 6/9/61 at Charleston, Kanawha Co., WV. Age: 31. Co. transf. to Co. K, 59th Va. Inf. 7/25/61.

HOGAN, JOHN A.: Pvt. Co. C. Enl. 5/1/62 at White Sulphur Springs, Greenbrier Co., WV. Capt. at Winchester, Va. 9/19/64. POW Pt. Lookout, MD 9/24/64. D. at Pt. Lookout, MD of pneumonia 12/27/64. Bur. POW graveyard at Pt. Lookout, MD.

HOLCOMB, J. M.: Pvt. Co. H. Enl. 4/28/62 at White Sulphur Springs, Greenbrier Co., WV. Was assigned to & joined this co. 4/28/62. Capt. at Winchester, Va. 9/19/64. In Chimborazo Hosp. as paroled POW 8/31/64. In Chimborazo Hosp. for chronic diarrhea 10/7/64 to 10/26/64. D. at Staunton, Va. hosp. of disease 11/7/64. Coffin furnished.

HOLCOMB, ROBERT M.: Pvt. Co. K. Enl. 6/1/62 at White Sulphur Springs, Greenbrier Co., WV. Deserted 5/3/62 on Aug. 1862 broadside. Lick Creek, Greenbrier Co., WV. In Chimborazo Hosp. 5/1/64. WIA at Cold Harbor, Va. 6/3/64. In hosp. for amputation of left thumb 6/5-17/64. Paroled at Charleston, WV 5/11/65. Age: 21. 5'5". Light complexion, brown hair, blue eyes.

HOLCOMB SAMUEL: Pvt. Co. A. Enl. 4/10/62 at Brushy Ridge, Greenbrier Co., WV. D. in hosp. at Charleston, WV 9/20/62.

HOLLY, HARRISON: Pvt. Co. B. Enl. 5/1/62 at White Sulphur Springs, Greenbrier Co., WV. Conscript. In Richmond, Va. hosp. 5/31/64. WIA (slight left leg) at Cold Harbor, Va. 6/4/64.

HOLMAN, J. T.: Pvt. Co. G. WIA at Cold Harbor, Va. 6/3/64. Postwar roster.

HOLSTEN, HENRY: Co. F. Deserted on Aug. 1862 broadside. Craig Co., Va.

HOLSTEN, WILLIAM A.: Pvt. Co. F. Possible res. Craig Co., Va. Enl. 5/1/62 at White Sulphur Springs, Greenbrier Co., WV. In arrest on Jan.-Feb. 1863 roll. In arrest at Lewisburg, WV on Mar.-Apr. 1863 roll. In arrest on Nov.-Dec. 1863 roll. Also spelled Holston.

HOLSTINE, ANDREW JACKSON: Pvt. Co. I. Possible res. Craig Co., Va. Enl. 6/22/61 at Charleston, Kanawha Co., WV. Entitled to bounty. Capt. in Fayette Co., WV 9/13/61. POW Camp Chase, OH. Exchd. 8/25/62. MIA at White Sulphur Springs, WV 8/26-27/63. Deserted near Staunton, Va. 6/28/64. Also spelled Holstein.

HOLSTINE, VAN BUREN: Pvt. Co. I. Possible res. Craig Co., Va. Enl. 6/22/61 at Charleston, Kanawha Co., WV. WIA (serious hip) at White Sulphur Springs, WV 8/26-27/63. On wded. sick leave on Aug.-Dec. 1863 roll. Capt. at Hanover Junction, Va. 5/30/64. POW Pt. Lookout, MD. Transf. to Elmira, NY 7/9/64. Exchd. 10/29/64. Deserted and surrendered at Charleston, WV Feb. 1865. Took oath at Charleston, WV 2/14/65. Age: 21. Farmer. 5'7". Fair complexion, light hair, gray eyes. Also spelled Holstein.

HOLT, PLEASANT: Pvt. Co. G. Enl. 3/5/64 at Staunton, Va. Conscript from Camp Lee. Deserted 4/12/64.

HOLT, R. C.: Co. G. WIA at Cold Harbor (Gaines Mill), Va. 6/3/64. Postwar roster.

HOOK, PECHY F.: Pvt. Co. F. Res. Monroe Co., WV. Enl. 7/4/61 at Charleston, Kanawha Co., WV. Det. as nurse at White Sulphur Springs, WV 7/27/61. Det. to make coffins July 1862. Det. to make coffins on Sept.-Oct. 1862 roll. Postwar roster says name is Peachy. Believed still alive 1914.

HOOKER, JOHN A.: Pvt. Co. D. Enl. 5/1/62 at White Sulphur Springs, Greenbrier Co., WV. Capt. in Nicholas Co., WV while on furlo. on Nov.-Dec. 1862 roll. Taken the oath and with the enemy on Jan.-Feb. 1863 roll.

HOOKER, WILLIAM C.: Pvt. Co. D. Enl. 5/30/61 at Summersville, Nicholas Co., WV. AWOL 7/28/61.

HOPKINS, BILL: Co. I.

HOPKINS, MANLY: Co. I.

HOPKINS, THOMAS: Pvt. Co. H. Enl. 5/8/61 at Charleston, Kanawha Co., WV. Apptd. hosp. steward at hosp. at White Sulphur Springs, WV and served in that capacity 8/4-24/61.

HOSKINS, JOSEPH: Pvt. Co. B. Enl. 9/22/62 at Charleston, Kanawha Co., WV. KIA (mortal breast) at New Market, Va. 5/15/64.

HOWARD, GEORGE L.: Pvt. Co. F. Enl. 11/2/63 at Camp Beauregard. Deserted 5/8/64. Dead by 1894. Had been living in MD et. group. Listed as a Lt.

HOWARD, MOSES R.: Pvt. Co. F. B. 1845. Res. Monroe Co., WV. Enl. 7/13/61 at Camp Two Mile, Kanawha Co., WV. Sick in hosp. at White Sulphur Springs, WV 8/20/61. Sick at home on surg. cert. on Sept.-Oct. 1862 roll. Present on Nov.-Dec. 1862 roll. Sick in hosp. at Lewisburg, WV on Jan.-Feb. 1863 roll. Present on Mar.-Apr. 1863 roll. Took oath at Clarksburg, WV (must have deserted) 9/11/63. Age: 19. Farmer. 6'. Light complexion, dark hair, dark eyes. Postwar res. Belfountain, OH. Res. Cinn., OH 1875. Res. Claremont, OH 1880. Res. Newport, KY 1890. Postwar insurance salesman, Equity Life. D. ?/14/1908. Bur. Spring Grove Cem., Cinn., OH. PWR also gives Pittsburg, PA and Cleveland, OH as res. (Middle initial is M according to PWR roster).

HOWARD, ROBERT: Pvt. Co. F. Enl. 9/14/63 at Camp Price. Deserted 5/8/64. Took oath at Richmond, Va. 4/12/65. Shoemaker. Res. 1st St. between Byrd & Canal, Richmond, Va.

HUBBARD, CHARLES W.: Pvt. Co. I. Res. Craig Co., Va. Enl. 6/29/63 near Lewisburg, Greenbrier Co., WV. On arrest on Mar.-Apr. 1864 roll. Deserted at Rockfish Gap, Nelson Co., Va. 6/15/64.

HUBBARD, M.: Pvt. Co. D. Enl. 5/1/62 at White Sulphur Springs, Greenbrier Co., WV. Listed as dead on Sept.-Oct. 1862 roll.

HUBBARD, WILLIAM H.: 2nd Corp. Co. A. Enl. 7/1/61 at Camp Tompkins, Kanawha Co., WV. Served as a Pvt. until elected 2nd Corp. 7/30/61. KIA at Droop Mountain, WV 11/6/63.

HUDDLESTON, CHARLES W.: Pvt. Co. K. Enl. 6/6/61 at Charleston, Kanawha Co., WV. AWOL 9/1/61. Day laborer, Fayette Co., WV 1860. Age: 22. Bahlmann says "left."

HUDDLESTON, GEORGE WASHINGTON: Musician/drummer/Quartermaster Sgt. Co. K. Attd. Beuhring H. Jones' school for young men, Fayetteville, WV pre-war. Enl. 6/6/61 at Charleston, Kanawha Co., WV. Sick since 7/24/61 on June-Sept. 1861 roll. Present on Sept.-Oct. 1861 roll. Left sick at Lewisburg, WV on Mar.-Apr. 1863. Sick since 10/25/64 on Apr.-Oct. 1864 roll. Paroled at Charleston, WV 5/1/65. Age: 27. 5'6". Fair complexion, fair hair, blue eyes. Entered Richmond, Va. Old Soldiers Home 8/25/1915 at age 78. Dischd. 2/21/1917. Insane on furlo. Brother of Jackson Burwell Huddleston.

HUDDLESTON, JACKSON BURWELL: Pvt. Co. K. B. 7/20/40 near Kanawha Falls, WV. Attd. Beuhring H. Jones' school for young men, Fayetteville, WV pre-war. Enl. 12/1/62 at Centerville, Monroe Co., WV. Transf. from Co. C, 60th Va. Inf. 12/1/62. POW paroled at Charleston, WV 5/10/65. Age: 22. 5'8". Sandy complexion, light hair, blue eyes, sandy whiskers. D. 12/25/1916 at Oak Hill, WV. PWR says died Feb. 1873 by being kicked by a horse on his farm at Fayetteville, WV. Photo of him reportedly taken in 1861. Brother of George Washington Huddleston.

HUDDLESTON, JOB: Pvt. Co. K. Enl. 12/1/62 at Centerville, Monroe Co., WV. Transf. from Co. C, 60th Va. Inf. 12/1/62. Capt. 3/8/64 (place not stated). Listed as deserted while on furlo. 4/1/64.

HUDDLESTON, THOMAS J.: Capt. Kanawha Rangers (cavalry) Co. K. Date and place of enlistment not stated. Absent at camp on Cotton Hill, WV for unknown reason 10/26/61. Accidently shot and killed at Guyandotte, WV 11/10/61. Bur. Spring Hill Cem., Huntington, WV. See also 8th Va. Cav.

HUDSON, SAMUEL: Co. I. Listed as deserter on Aug. 1862 broadside. Botetourt Co., Va.

HUFF, DAVID J.: 2nd Lt. Co. I. Enl. 5/1/62 at White Sulphur Springs, Greenbrier Co., WV. Capt. at Winchester, Va. 9/19/64. POW Ft. Delaware 9/25/64. Took oath and relsd. 6/17/65. 6'3". Ruddy complexion, dark hair, gray eyes. Res. Botetourt Co., Va. (An additional card says W. J. Huff: WIA (slight hand) at White Sulphur Springs, WV 8/26-27-63).

HUFF, JOHN A.: Pvt. Co. D. Enl. 7/7/63 at Camp Sam Jones. Deserted at Goshen Depot, Va. 5/?/64.

HUFF, LUTHER J.: Pvt. Co. D. Enl. 5/30/61 at Summersville, Nicholas Co., WV. Present on 8/31/61 roll. NFR.

HUFFMAN, BAXLEY: Served under Patton, McDonald, and McCausland. Fought at Droop Mountain, WV and with Breckinridge in the Valley of Va. Enl. 4/16/64 in 11th Battn. Va. Reserves at New Castle, Va.

HUFFMAN, CALEB: Pvt. Co. E. Res. Craig Co., Va. Enl. 5/1/62 at White Sulphur Springs, Greenbrier Co., WV. Present on 10/31/64 roll.

HUFFMAN, D.: Pvt. Co. H. Res. Christiansburg, Va. Enl. 4/18/62 in Greenbrier Co., WV. Capt. at Cold Harbor, Va. 6/3/64. POW Pt. Lookout, MD. Transf. to Elmira, NY 7/12/64. Took oath and relsd. at Elmira, NY 7/3/65. 5'9". Fair complexion, light hair, blue eyes.

HUFFMAN, HUGH A.: Pvt. Co. G. B. ca. 1814 Craig Co., Va. Res. Craig Co., Va. Enl. 11/20/63 at New Castle, Va. Assigned to this co. 12/19/63. Deserted 12/28/63. Capt. at Winchester, Va. 9/19/64 (or Fisher's Hill, Va. 9/22/64). POW Pt. Lookout, MD 10/1/64. Exchd. 11/1/64. D. 12/20/64 of diarrhea at Broad Run, age 43.

HUFFMAN, JACOB: Capt. 12/?/63 in Craig Co., Va. POW Camp Chase, OH. D. at Camp Chase, OH 1/10/65, age 39. Was also with 30th Va. Inf. and 46th Va. Inf. Listed in UDC booklet *Craig's Share in the War Between the States*. Res. Craig Co., Va.

HUFFMAN, JAMES ALLEN: Pvt. Co. K. Res. Craig Creek, Va. Enl. 4/24/62 at White Sulphur Springs, Greenbrier Co., WV. Left sick at Lewisburg, WV on Mar.-Apr. 1863 roll. Sick at Montgomery White Sulphur Springs, Va. on Nov.-Dec. 1863 roll. D. 6/22/64 of gunshot wd. Age: 21.

HUFFMAN, JAMES HECTOR: Pvt. Co. H. B. 10/31/30. Res. Craig Co., Va. Enl. 4/28/62 at White Sulphur Springs, Greenbrier Co., WV. PWR says enl. Mar. 1862 for three years. Assigned to this co. 4/28/62. Sick in Craig Co., Va. on Nov.-Dec. 1863 roll. In hosp. June 1864. Capt. at Winchester, Va. 9/19/64. POW Pt. Lookout, MD 9/24/64. Exchd. 2/10/65 and in camp hosp. In Chimborazo Hosp. with chronic diarrhea 2/16/65. PWR says WIA (right leg). Res. Craig Co., Va. 1898 at age 63. D. 12/19/1910. Bur. Huffman Cem., intersection of Rts. 42 and 622, Arch Huffman farm, Craig Co., Va., or Bell cem., Rt. 632, Upper Johns Creek, on Bell farm.

HUFFMAN, JAMES W.: Pvt. Co. C. Res. Craig Co., Va. Cabinetmaker. Enl. 5/1/62 at White Sulphur Springs, Greenbrier Co., WV. PWR says enl. March 1862 for three years. Sick at hosp. on Sept.-Oct. 1862 roll. Under arrest on rolls for Jan.-Feb. and Mar.-Apr. 1863. Present on Nov.-Dec. 1863 roll. Res. Craig Co., Va. 1898 at age 60.

HUFFMAN, MAZARINE: Listed in UDC booklet *Craigs's Share in the War Between the States*. Res. Craig Co., Va.

HUGGARD, J.: Pvt. Co. H. Enl. 1/4/63 in Monroe Co., WV. Sick in Greenbrier Co., WV on May-June 1863 roll. KIA at White Sulphur Springs, WV 8/26/63.

HUGGARD, J. H.: Pvt. Co. E. Enl. 5/1/62 at White Sulphur Springs, Greenbrier Co., WV. Sick in hosp. on Sept.-Oct. 1862 roll. Joined from desertion on Mar.-Apr. 1863 roll. Dischd. due to disability 4/6/63.

HUGHS, CHRISTOPHER C.: Pvt. Co. C. B. ca. 1836 Craig Co., Va. Enl. 5/1/62 at White Sulphur Springs, Greenbrier Co., WV. Deserted at Camp McCausland, WV July 1862. In Winchester, Va. hosp. with debilitas and diarrhea 7/23/64. Returned to duty 7/30/64.

HUGHS, JOHN W.: Pvt. Co. C. B. ca. 1806 Craig Co., Va. Enl. 5/1/62 at White Sulphur Springs, Greenbrier Co., WV. Deserted 8/6/63.

HUGHS, ROBERT J.: Pvt. Co. C. Enl. 3/14/64 at Camp Gauley, Greenbrier Co., WV. Capt. at Winchester, Va. 9/19/64. POW Pt. Lookout, MD 9/24/64. Exchd. 3/15/65.

HUGHS, WILLIAM P.: Pvt. Co. B. Enl. 7/1/61 at Ripley, Jackson Co., WV. In Hosp. with rubeola 6/4/64 to 9/16/64. Det. as Provost Guard on Apr.-Oct. 1864 roll. Paroled at Charleston, WV 5/10/65. Age: 20. Farmer. 5'6". Fair complexion, light hair, gray eyes, light whiskers. Bur. Creston Cem., near Gunville, Mason Co., WV.

HULL, D. S.: Co. E. Res. Augusta Co., Va. In Richmond, Va. Old Soldiers Home 1903.

HULL, JOHN McKEE: Co. K. Conscripted at Kerrs Creek, Greenbrier Co., WV. 3/11/64. Age: 42. Farmer. 5'10". Dark complexion, black hair, black eyes. Brother of James B. and William H. Hull. Res: Rockbridge Co.,Va. Res. Kerrs Creek, WV 1860, farmer. Tanner, Kerrs Creek, WV 1870. Farmer and sawmill operator, Kerrs Creek, WV 1880. D. at Kerrs Creek, WV 10/15/89.

HULL, JOSHUA: Possibly served in 22nd Va. Inf. Exemption denied 3/?/62.

HULL, P. T.: Pvt. Co. D. D. at Pt. Lookout, MD.

HULL, WILLIAM CRAWFORD: Co. unknown.

HULL, WILLIAM M.: Capt. Co. K. Originally in Co. E, 27th Va. Inf. B. Lexington, Va., 1828. Res. Rockbridge Co., Va. Farmer. Age: 30, Collierstown, Va. District 1860. Hired James J. Macalin as a substitute 7/?/62. D. at Collierstown, Va. 5/10/89. Bur. Collierstown Presb. Church Cem., Va.

HUMES, DAVID P.: Pvt. Co. A. Enl. 9/15/62 at Charleston, Kanawha Co., WV. Left sick at Beverly, WV on Mar.-Apri. 1863 roll. Present on Aug.-Dec. 1863 roll. D. at Union, WV (undated). Claim filed by Clara Humes, mother 2/21/65. Bur. St. Matthews Lutheran Church Cem., New Market, Va. Another says bur. Presbyterian Church Cem., New Market, Va.

HUMES, JAMES: Pvt. Co. A. Enl. 9/1/62 at Camp Price. Absent wded. on Aug.-Dec. 1863 roll. D. at Mt. Jackson, Va. (undated). Claim filed by Clara Humes, mother 2/21/65.

HUMPHREYS, C. H.: Pvt. Co. B. Enl. 5/1/62 at White Sulphur Springs, Greenbrier Co., WV. Deserted on Aug. 1862 broadside. Deserter on Sept.-Oct. 1862 roll.

HUMPHREYS, C. L.: Pvt. Co. B. Enl. 5/1/62 at White Sulphur Springs, Greenbrier Co., WV. Conscript. Deserted 5/4/62 on Aug. 1862 broadside, White Sulphur Springs, WV. Deserter on Sept.-Oct. 1862 roll.

HUMPHREYS, CHARLES K.: Pvt. Co. D. Enl. either 5/1/62 at White Sulphur Springs, Greenbrier Co., WV or 7/21/62 at Camp McCausland, Monroe Co., WV. Deserted at Buchanan, Va. June 1864. D. 12/23/1900. Bur. Humphries Family Cem., Rt. 616, Alleghany Co., Va.

HUMPHREYS, CHRISTOPHER: Co. F. Res. Monroe Co., WV. Believed still alive 1914.

HUMPHREYS, M. N.: Co. K. PWR.

HUMPHRIES, DAVIS M.: Pvt. Co. D. Enl. 5/1/62 at White Sulphur Springs, Greenbrier Co., WV. Left Sick at Lewisburg, WV 4/13/63. Present on Nov.-Dec. 1863 roll. AWOL 4/24/64. Deserted at Buchanan, Va. June 1864.

HUMPHRIES, SAMPSON: Pvt. Co. B. Res. Jackson Co., WV. Enl. 5/1/62 at White Sulphur Springs, Greenbrier Co., WV. Conscript. Deserted 5/4/62 on Aug. 1862 broadside, White Sulphur Springs, WV. Listed as a deserter but returned to his co. on Sept.-Oct. 1862 roll. Appears on a (believed to be) later updated list of deserters. Took oath 12/31/63. Age: 28. Farmer. 6'. Dark complexion, dark hair, gray eyes. Res. Alleghany Co., Va. Aug. 1862 broadside says last name is Humphrey.

HUNT, ABIJAH: Pvt. Co. B. Res. Jackson Co., WV. Enl. 9/22/62 at Charleston, Kanawha Co., WV. Capt. at Ripley, WV by 7th WV. Cav. 1/4/65. POW Wheeling, WV. POW Cumberland, MD. POW Camp Chase, OH 2/13/65. Took oath at Camp Chase, OH 5/16/65. Age: 63. Farmer. 5'8¾". Gray hair and blue eyes.

HUNT, HENRY FISHER: Pvt. Co. B. Res. Cow Run, 1860, age 17. Res. Jackson Co., WV. Enl. 6/1/61 at Ripley, Jackson Co., WV. Capt. in either Jackson or Clay counties, WV 2/20/64. POW Camp Chase, OH 3/7/65. Transf. to Ft. Delaware. Believed to have escaped en route. Took oath at Camp Chase, OH 6/10/64. Age: 20. (or 21). Farmer. 5'8". Florid complexion, dark hair, blue eyes. D. at Camp Chase, OH of pneumonia 1/28/65. Bur. grave #936 just south of Camp Chase, OH.

HUNT, HENRY J.: Pvt. Co. C. Enl. 5/10/61.

HUNT, J. L.: Pvt. Co. B. Enl. 9/22/62 at Charleston, Kanawha Co., WV. Present on Nov.-Dec. 1863 roll. NFR.

HUNTER, MOSES: Pvt. Co. E. Enl. 3/6/63 at Lewisburg, Greenbrier Co., WV. Substitute for W. Deats. Joined from desertion. KIA at Droop Mountain, WV 11/6/63. Claim filed by Sarah J. Hunter, widow 12/30/64.

HUNTER, WILLIAM: Pvt. Swann's Co. Enl. 5/22/61 at Buffalo, Putnam Co., WV. Age: 23. Co. transf. to Co. K, 59th Va. Inf. 7/25/61.

HUSE, HARLOW: Pvt. Co. K. Probably born Fayetteville, WV. Res. Fayette Co., WV. Enl. 6/6/61 at Charleston, Kanawha Co., WV. Det. as a courier on Mar.-Apr. 1863 roll. Capt. at Lewisburg, WV 5/23/62. Age: 28. POW Camp Chase, OH. Exchd. 8/25/62. "Came home barefoot." Enl. 9/1/63 in Co. I, 8th Va. Cav. 5'8½". Dark complexion, dark hair with gray/hazel eyes, thin dark beard. Paroled at Charleston, WV 5/1/65. Age: 29. Bur. Fayetteville, WV. Miller, Fayette Co., WV 1860. Age: 25.

HUTCHESON, GEORGE W.: Pvt. Co. B. Res. Jackson Co., WV and/or Greenbrier Co., WV. Enl. 6/?/63 at Lewisburg, Greenbrier Co., WV. Capt. at Droop Mountain, WV 11/6/63. POW Camp Chase, OH 11/18/63. Age: 21. D. at Ft. Delaware. Bur. Finn's Point Nat'l. Cem., NJ. Farmer. 5'6¾". Black hair and blue eyes.

HUTCHINSON, JOSEPH F.: Pvt. Co. D. Enl. 5/30/61 at Summersville, Nicholas Co., WV. AWOL 7/28/61.

HUTCHINSON, E. P.: Listed in UDC booklet *Craig's Share in the War Between the States*. Res. Craig Co., Va.

HUTCHINSON, R. E.: Pvt. Co. C. Enl. 4/?/63 for three years. Res. Craig Co., Va. 1898 at age 50. Bur. Williams Cem., Rt. 42, on hill across from Level Green Church, Craig Co., Va.

HYLTON, JACOB S.: Pvt. Co. K. Res. Floyd Co., Va. Enl. 2/10/62 at Lewisburg, Greenbrier Co., WV. Capt. at Droop Mountain, WV 11/6/63. Age: 30. POW Camp Chase, OH 11/16/63. Transf. to Ft. Delaware 3/4/64. Bur. Jersey shore of Ft. Delaware, Finn's Point Nat'l. Cem., NJ. Farmer. 5'5¼". Florid complexion, sandy hair, blue eyes.

HYPES, ASBURY: Listed in UDC booklet *Craig's Share in the War Between the States*. Res. Craig Co., Va. This may be George A. Hypes.

HYPES, GEORGE A.: Pvt. Co. C. B. ca. 1837 Craig Co., Va. Carpenter. Res. Craig Co., Va. Enl. 5/1/62 at White Sulphur Springs Greenbrier Co., WV. Mention of a court-martial date of record 9/16/64. Capt. at Winchester, Va. 9/19/64. POW Pt. Lookout MD. Another card says capt. at Woodstock, Va. 9/23/64. Took oath 5/?/65.

HYPES, JACOB L.: Pvt. Co. D. Res. Craig Co., Va. Enl. 5/1/62 at White Sulphur Springs, Greenbrier Co., WV. Capt. at Droop Mountain, WV 11/6/63. Age: 28. POW Camp Chase, OH 11/18/63. Transf. to Ft. Delaware 3/3/64. Took oath at Ft. Delaware 6/20/65. Farmer. 5'5" or 5'6½". Florid or fair complexion, auburn or dark hair, gray or brown eyes.

HYPES, JAMES H.: Pvt. Co. C. B. ca. 1844 Craig Co., Va. Enl. 6/22/62 at Salt Sulphur Springs, Monroe Co., WV. In Winchester Va. hosp. with diarrhea 7/26-30/64. KIA at Cedar Creek, Va. 10/19/64. Death listed in Gravel Hill Christian Church Records Craig Co., Va.

HYPES, JAMES MADISON: Pvt. Co. K. B. ca. 1840 Craig Co., Va. Enl. 4/24/62 at White Sulphur Springs, Greenbrier Co., WV. PWR says enl. Mar. 1862 for three years. Deserted at Staunton, Va. 5/12/64. Res. Craig Co., Va. 1898 at age 58.

HYPES, LEWIS: Listed in UDC booklet *Craig's Share in the War Between the States*. Res. Craig Co., Va. This may be Jacob L. Hypes.

HYPES, AUGUSTUS: Pvt. Co. B. B. 1/16/41. Res. Jackson Co., WV 1860, age 19. Enl. 6/1/61 at Ripley, Jackson Co., WV. PWR says mustered 12/10/61. D. of typhoid fever early in the war at Princeton, WV.

HYRE, PERRY A.: Pvt. Co. B. Res. Jackson Co., WV. Enl. 6/1/61 at Ripley, Jackson Co., WV. In hosp. on June-Aug. 1861 roll Dischd. at Lewisburg, WV 2/27/62 for scrotal hernia which unfit him for military duty and reduced his capacity for labor one third. Age: 22. Farm laborer. Sallow complexion, light hair, blue eyes. 5'11". Entered Richmond, Va. Old Soldiers Home 10/14/1915 at age 76. D. 7/26/1916. Bur. Hollywood Cem., Richmond, Va. Registered at Home as serving in Co. F.

INGLES, GEORGE A.: Pvt. Co. F. Res. Monroe Co., WV. Enl. 7/4/61 at Charleston, Kanawha Co., WV. Det. as teamster. Capt. at Winchester, Va. 9/19/64. POW Pt. Lookout, MD 9/23/64. D. at Pt. Lookout, MD of hepatitis 12/31/64. Bur. POW graveyard at Pt. Lookout, MD.

INGLES, JAMES H.: Pvt. Co. F. Enl. 2/17/63 at Lewisburg, Greenbrier Co., WV. Det. as teamster. WIA (slight back) at White Sulphur Springs, WV 8/26-27/63. In hosp. for debility. D. in hosp. at Charlottesville, Va. of febris typhoides 10/7/64.

INGLES, JOHN P.: Pvt. Co. F. Enl. 7/4/61 at Charleston, Kanawha Co., WV. Home sick on surg. cert. since 7/27/61 on July-Aug. 1861 roll. Present on Nov.-Dec. 1862 roll. Driving brig. wagon Apr. 1864. Det. service on Mar.-Apr. 1864 roll. Postwar res Monroe Co., WV. Crippled.

INGRAHAM, CYRUS D.: Pvt. Co. A. Enl. 7/14/61 at Camp Tompkins, Kanawha Co., WV. NFR.

INGRAHAM, GRANVILLE P.: Pvt. Co. A. Enl. 5/22/61 at Poca, Putnam Co., WV. NFR.

INGRAHAM, MARSHALL N.: Pvt. Co. A. Enl. 5/22/61 at Poca, Putnam Co., WV. NFR.

INKLE, A.: Pvt. Co. F. Date and place of enlistment not stated. Det. as butcher July 1862.

JACKSON, A.: Rank unknown. Co. G. Date and place of enlistment not stated. Capt. at Winchester, Va. 9/19/64. POW Pt. Lookout, MD 9/23/64. Exchd. 2/13/65.

JACKSON, A. B.: 4th Sgt. Co. E. Res. Roane Co., WV. Enl. 6/8/61 at Charleston, Kanawha Co., WV. Left sick at White Sulphur Springs, WV 8/20/61. Apptd. 4th Sgt. 10/17/62. Capt. in Roane Co., WV or Logan Co., WV 1/13/64. Age: 26. POW Camp Chase, OH 2/2/64. Transf. to Ft. Delaware 3/17/64. Took oath at Ft. Delaware and relsd. 6/20/65. Farmer. 5'9" or 5'9½". Dark or fair complexion, dark or light hair, hazel or gray eyes.

JACKSON, C. L.: Pvt. Co. H. Enl. 9/23/62 in Monroe Co., WV. WIA at Droop Mountain, WV 11/6/63.

JACKSON, EDWARD: Pvt. Co. H. Laborer, Malden, WV 1860. Age: 17. Enl. 8/10/61 at White Sulphur Springs, Greenbrier Co., WV. Present on 8/31/61 roll. NFR.

JACKSON, EDWARD T.: 1st Lt. Co. E. Enl. 5/1/62 at White Sulphur Springs, Greenbrier Co., WV. Left sick in Charleston, WV and supposed to be a POW on Sept.-Oct. 1862 roll. Signed a parole or oath at Charleston, WV 11/5/62 at the age of 21. POW Camp Chase, OH 11/10/62. Exchd. 12/2/62. Present on Jan.-Feb. 1863 roll. WIA at Kernstown, Va. 7/24/64. In Winchester, Va. hosp. for serious wd. of right shoulder 7/26/64. In prisoner quarters at Mrs. O. Daniels, nurse 7/26/64. In Lynchburg, Va. hosp. 8/25/64. Camped near New Market, Va. 11/28/64. Wded. on Apr.-Nov. 1864 roll. Granted furlo. Dec. 1864. Merchant. 5'11". Fair complexion, dark hair, blue eyes. Res. Charleston, WV.

JACKSON, E. J.: Capt. Co. unknown. Died by 1864. Had been in a Md. vet. group.

JACKSON, GEORGE W.: Pvt. Co. G. B. Alleghany Co., Va. Res. Alleghany Co. Farmer. Enl. 5/1/62 at White Sulphur Springs, Greenbrier Co., WV. Capt. at Droop Mt., WV 11/6/63. Age: 23. PWR says enl. Greenbrier Bridge, WV Mar. 1862. POW Camp Chase, OH 11/18/63. Transf. to Ft. Delaware 3/4/64. Took oath and relsd. 6/20/65. Farmer. 6'2" or 6'2½". Fair or sallow complexion, light hair, blue eyes. Res. Healing Springs, Bath Co., Va. 1911 at age 71. Still alive 1929 and res. Alleghany Co., Va.

JACKSON, JOHN W.: Pvt. Co. G. Res. Wyoming Co., WV. Enl. 6/3/61 at Charleston, Kanawha Co., WV. Left sick at White Sulphur Springs, WV 8/20/61. Present on Sept.-Oct. 1862 roll. Capt. at Winchester, Va. 9/19/64. POW Pt. Lookout, MD. Exchd. 2/13/65. Took oath at Charleston, WV 5/9/65. Age: 25. Farmer. 5'10". Dark complexion, dark hair, blue eyes.

JACKSON, S. A.: Pvt. Co. E. Enl. 9/?/62 at Charleston, Kanawha Co., WV. KIA at Droop Mountain, WV 11/6/63.

JACKSON, SAMUEL WILLIAM: Pvt. Co. G. Res. Alleghany Co., Va. Enl. 7/28/62 at Union, Monroe Co., WV. Left sick at Lewisburg, WV 4/13/63. Dischd. for chronic rheumatism and pulmonary disease 5/23/63. Age: 30. Farmer. 5'11". Fair complexion, light hair, blue eyes. D. 7/18/1900.

JACKSON, WILLIAM A.: Lt. Col. Field and Staff. B. 1806 Fredericksburg, Va. Coal operator, Malden, WV 1860. Apptd. Lt. Col. due to reorganization Nov. 1861 (actually due to the resignation of Col. Tompkins and Lt. Col. Patton's status as a POW). Another card says apptd. Lt. Col. 12/21/61. Dropped 5/2/62. Successor: A. R. Barbee. Lt. Col. 4th Va. State Line. Had a large family including four sons. Recommended to be Commandant at Dublin Depot, Va. 10/10/63. Served in Nitre and Mining Bureau late in war. D. at son's house at Pulaski Co., Va. 1/15/75. Bur. Fredericksburg Confederate Cem.

JACKSON, WILLIAM R.: Co. F. D. sometime prior to 1914. Postwar roster.

JANANEY, JOHN: Pvt. Co. E. Enl. 5/1/62 at White Sulphur Springs, Greenbrier Co., WV. Det. as wagoner on Sept.-Oct. 1862 roll. Paroled at Charleston, WV 5/10/65. Age: 50. 5'6". Dark complexion, black hair, black eyes, brown whiskers.

JARRELL, ANDERSON: Pvt. Co. I. Res. Boone Co., WV. Enl. 6/22/61 at Charleston, Kanawha Co., WV. Deserted at Union, WV 6/10/62. In hosp. at Montgomery White Sulphur Springs, Va. 10/11/62 and found unable to perform duties of a soldier due to inflammation erysepolas. Capt. in Boone Co., WV by the 47th O.V.I. 11/15/62. Age: 22. POW Camp Chase, OH 12/2/62. Exchd. Returned to regt. on Jan.-Feb. 1863 roll. WIA (slight arm) at White Sulphur Springs, WV 8/26-27/63. Farmer. 6'. Dark complexion, brown hair, blue eyes. (Another card says: Anderson Jarrett. 5'10". Fair complexion, fair hair, gray eyes. Farmer. Deserter. Took oath at Charleston, WV 4/8/65).

JARRETT, ABRAHAM: Pvt. Co. E. Laborer, Charleston, WV 1860. Age: 17. Enl. 6/8/61 at Charleston, Kanawha Co., WV. Wded. on Mar.-Apr. 1864 roll. Capt. at Cloyd's Mountain, Va. 5/15/64. POW Camp Chase, OH 5/26/64. Requested to take oath at Camp Chase, OH 6/10/65. D. at Camp Chase, OH of smallpox 11/10/64. Bur. grave #444 south of Camp Chase, OH.

JARRETT, HARRISON: Pvt. Co. E. Laborer, Paint Creek, WV 1860. Age: 22. Enl. 7/1/61 on Sandy. Left sick at Charleston, WV on Sept.-Oct. 1862 roll.

JARRETT, JAMES: Corp. Co. E. Enl. 12/26/62 at Lewisburg, Greenbrier Co., WV. Present throughout the war.

JARRETT, S.: Pvt. Co. E. Enl. 6/8/61 at Charleston, Kanawha Co., WV. Present on 8/31/61 roll. NFR.

JARRILL, JOHN: Pvt. Co. I. Enl. 6/22/61 at Charleston, Kanawha Co., WV. Present on June-Aug. 1861 roll. NFR. (Probably spelled Jarrell).

JEFFRIES, GEORGE V.: 2nd Sgt. Taylor's Co. Enl. 6/19/61 at Frankford, Greenbrier Co., WV. Age: 22. Co. transf. to Wise Brigade 7/10/61.

JENKINS, J. W.: Co. B. Entered Richmond, Va. Old Soldiers Home 10/31/94 at age 70. At own request 5/1/95.

JENNELLE, JAMES: Pvt. Co. B. Enl. 6/?/62 at Union, Monroe Co., WV. Det. at home on Sept.-Oct. 1862 roll. D. at Montgomery White Sulphur Springs, Va. 12/24/62.

JENNINGS, (DR.), DRYDEN: Pvt. Swann's Co. Engineer, Charleston, WV 1860. Enl. 5/15/61 at Charleston, Kanawha Co., WV. Age: 29. Co. transf. to Co. K, 59th Va. Inf. 7/25/61.

JESSE, JOSEPH B.: Corp. Co. B. Enl. 6/1/61 at Ripley, Jackson Co., WV. WIA (gunshot wd. of ankle) and capt. at Lewisburg, WV 5/23/62. Present on Nov.-Dec. 1862 roll. Dischd. at Packs Ferry, WV due to his Lewisburg wd. 7/19/63. Age: 21. Student. 5'8". Fair complexion, light hair, black eyes. B. in Caroline Co., Va.

JESTER, J. B.: Co. G. WIA (thigh) at Winchester, Va. 9/19/64. In Winchester hosp. 9/19/64-11/19/64.

JOHNS, JAMES C.: 3rd Sgt./Capt. Co. B. Res. Elk Fork, WV 1860, age 24. Enl. 6/1/61 at Ripley, Jackson Co., WV. Elected 1st Lt. 5/1/62. Promo. to Capt. 5/23/62. Capt. at Lewisburg, WV 5/23/62. Age: 26. POW Camp Chase, OH. Transf. to Johnson's Island, OH 7/3/62. Exchd. 9/1/62. WIA at New Market, Va. 5/15/64. In Charlottesville, Va. hosp. for acute diarrhea 6/10-21/64. Paroled at Charleston, WV 5/10/65. 5'10". Light complexion, dark hair, blue eyes, black whiskers. Listed as commanding regt. 10/31/64 with Lt. Col. McDonald commanding brig. (Echols').

JOHNS, WILLIAM: Pvt. Co. E. Enl. 5/1/62 at White Sulphur Springs, Greenbrier Co., WV. Hosp. nurse at Pearisburg, Va. May-June 1862. Sick in hosp. on Sept.-Oct. 1862 roll. Hosp. nurse at White Sulphur Springs, WV on Nov.-Dec. 1863 roll. Exchd. to Engineer Corps for G. B. Samuels 4/14/64.

JOHNSON, CHARLES B.: Pvt. Co. C. Farmer, Fayette Co., WV 1860. Age: 28. Enl. 5/10/61.

JOHNSON, COLLINS J.: Pvt. Co. F. Enl. 7/4/61 at Charleston, Kanawha Co., WV. Capt. at Lewisburg, WV 5/23/62. POW Camp Chase, OH. Exchd. 8/25/62.

JOHNSON, DAVID: Pvt. Co. B. Enl. 5/?/63 at Lewisburg, Greenbrier Co., WV. Capt. and paroled at Charleston, WV 5/10/65. Age: 28. 5'11". Fair complexion, fair hair, blue eyes.

JOHNSON, DAVID: Pvt. Co. K. Enl. 6/6/61 at Charleston, Kanawha Co., WV. Hosp. nurse at Union, WV on Nov.-Dec. 1862 roll. KIA at Cold Harbor, Va. 6/3/64. Farm laborer, Fayette Co., WV 1860. Age: 23.

JOHNSON, GEORGE A.: Pvt. Co. H. Date and place of enlistment not stated. Capt. in Wyoming Co., WV 3/14/62. Age: 26. POW Camp Chase, OH 5/7/62. Exchd. 8/25/62. 6'. Fair complexion, blonde hair, blue eyes, almost red whiskers.

JOHNSON, HENRY: Pvt. Co. A. B. ca. 1840 (?). Enl. 4/10/62 at Brushy Ridge, Greenbrier Co., WV. Sick at Alleghany Co., Va. on Mar.-Apr. 1863 roll. Provost Guard in Alleghany Co., Va. on Aug.-Dec. 1863 roll. Det. service in conscript dept. in Alleghany Co., Va. on Feb.-Mar. 1864 roll. On Alleghany Co. pension rolls.

JOHNSON, HENRY: 4th Corp. Co. K. Enl. 6/6/61 at Charleston, Kanawha Co., WV. Conscript. Capt. at Droop Mountain, WV 11/6/63. Age: 25. POW Camp Chase, OH 11/18/63. Requested to take oath at Camp Chase, OH 6/10/64. Relsd. 5/8/65. Took oath at Camp Chase, OH 5/13/65. Farmer. 5'8½". Florid complexion, brown hair, dark eyes. Res. Fayette Co., WV.

JOHNSON, ISRAEL: Pvt. Co. K. Res. Fayette Co., WV. Farmer. Enl. 5/27/61 at Kanawha Falls, Fayette Co., WV or 6/6/61 at Charleston, Kanawha Co., WV. Absent cause unknown on June-Sept. 1861 roll. Transf. to Co. I, 8th Va. Cav. 1862. Took oath and sent North 2/17/65, Age: 19.

JOHNSON, J. H.: Co. G. B. at Guilford C.H., NC. Photographer. Res. Staunton, Va. Alive 1909, age: 69.

JOHNSON, JOHN: Jr. 2nd Lt./2nd Lt. Co. G. Enl. 5/1/62 at White Sulphur Springs, Greenbrier Co., WV. Elected Jr. 2nd Lt. 4/6/64. Elected 2nd Lt. 4/28/64. Wded. and absent from 5/15/64.

OHNSON, JOHN M.: Pvt. Co. K. Enl. 6/6/61 at Charleston, Kanawha Co., WV. Dischd. for disability 6/27/61. WIA (shot in breast) at New Market, Va. 5/15/64. PWR says died from wd.

JOHNSON, JOSEPHUS: Pvt. Co. H. B. Salem, Va. Enl. 6/20/62 in Monroe Co., WV. Det. to Ordnance Dept. at New Bern, Va. on July-Aug. 1862 roll. Det. as master gunsmith, Ordnance Dept., Wytheville, Va. 9/30/64 to 10/31/64. Det. as master armorer 10/31/64 to 11/?/64. Age: 23. 5'5½". Dark complexion, black hair, dark gray eyes.

JOHNSON, J. T.: Pvt. Co. H. Enl. 5/13/63 in Pulaski Co., Va. Det. in Ordnance Dept. at Dublin, Va. Det. as armorer 10/31/64 to 11/?/64. (Probably the brother of Josephus Johnson).

JOHNSON, LEWIS W.: Pvt. Co. G. Res. Alleghany Co., Va. Enl. 5/1/62 at White Sulphur Springs, Greenbrier Co., WV. Conscript. Deserted 10/28/62. Dischd. 11/4/62. POW on Mar.-Apr. 1863 roll.

JOHNSON, MILES: Co. K. William Bahlmann says KIA (?), shot through temple, at Lewisburg, WV 5/23/62. Day laborer, Fayette Co., WV 1860. Age: 18. PWR says enl. 1861.

JOHNSON, PHILIP: Pvt. Co. D. B. 2/1/34. Enl. 5/1/62 at White Sulphur Springs, Greenbrier Co., WV. Det. as brig. teamster Nov. 1863. On Alleghany pension rolls. D. 8/25/80. Bur. Palestine Presbyterian Chapel, Alleghany Co., Va. or cem. at Rts. 614 & 18.

JOHNSON, WILLIAM B.: Pvt. Co. H. Date and place of enlistment not stated. Capt. and paroled at Charleston, WV 5/10/65. Age: 35. 6'. Light complexion, light hair, gray eyes.

JOHNSON, WILLIAM H.: Pvt. Co. K. Enl. 9/15/62 at Fayetteville, Fayette Co., WV. In Charlottesville, Va. hosp. for acute diarrhea 6/9/64. Transf. to convalescent camp 6/18/64. Capt. and paroled at Charleston, WV 5/1/65. Age: 20. 5'7". Fair complexion, light hair, dark eyes. Also served in Capt. Thurmond's Co. Bahlmann says enl. 1861.

JOHNSON, WILLIAM RUFUS: Corp. Co. F. Res. Monroe Co., WV. Enl. 7/4/61 at Charleston, Kanawha Co., WV. Served as corp. to 7/17/61. Sick in hosp. at White Sulphur Springs, WV 8/20/61. Det. as scout for Capt. Thurmond July 1862. Det. to drive brig. wagon on Nov.-Dec. 1862 roll. Two months pay taken by sentence of court-martial on 5/1/64 to 10/31/64 roll. Capt. and took oath at Charleston, WV 4/1/65. Age: 24. Farmer. 6'4". Florid complexion, sandy hair, blue eyes. (Also listed under this name in the CSR is a Rufus J. Johnson who was det. to drive brig. wagon 12/1/62). D. sometime prior to 1914.

JOHNSON, WILLIAM STEPHEN: Co. B. Bur. Oakwood Cem., Charlottesville, Va.

JOHNSTON, ERI: Pvt. Co. G. Enl. 7/3/61 at Charleston, Kanawha Co., WV. Left sick at White Sulphur Springs, WV 8/20/61. Present on Sept.-Oct. 1862 roll. In Richmond, Va. hosp. with debility 5/26/64. Returned to duty 6/6/64. Capt. at Winchester, Va. 9/19/64. POW Pt. Lookout, MD 9/23/64. Exchd. 3/15/65. Paroled at Charleston, WV 6/22/65 and gave name as "Eli". Age: 21. 5'4". Florid complexion, light hair, gray eyes.

JOHNSTON, OSCAR F.: Pvt. Co. G. Enl. 7/3/61 at Charleston, Kanawha Co., WV. Sick in Giles Co., Va. July 1862. Sick at Pearisburg, Va. on Sept.-Oct. 1862 roll.

JOICE, JOHN: Pvt. Co. F. Res. Monroe Co., WV. Enl. 7/4/61 at Charleston, Kanawha Co., WV. Held prisoner at White Sulphur Springs, WV on charges of intoxication and not yet on trial 8/14/61. Sick at Montgomery White Sulphur Springs, Va. July 1862. Postwar roster says dischd. 1862. Believed still alive in 1914. Also spelled Joyce.

JONES, A. F.: Pvt. Co. E. Enl. 5/1/62 at White Sulphur Springs, Greenbrier Co., WV. Capt. at Fisher's Hill, Va. 9/22/64. POW Pt. Lookout, MD 10/1/64. Exchd. 3/17/65. In Farmville, Va. hosp. 3/22/65. Took oath at Charleston, WV 4/20/65. Listed as deserter. Age: 27. Farmer. 5'7". Dark complexion, dark hair, blue eyes.

JONES, ANDREW A.: Co. H. Deserted 5/3/62 on Aug. 1862 broadside. Craig Co., Va.

JONES, CHARLES: Pvt. Co. I. Res. Kanawha Co., WV. Date and place of enlistment not stated. Deserted at Camp McCausland, WV 7/4/62. Listed as deserter on Aug. 1862 broadside. Took oath at Charleston, WV 4/22/65. Age: 27. Farmer. 5'8". Dark complexion, dark hair, blue eyes. From Maryland.

JONES, EULIS: Pvt. Co. C. Enl. Apr. 1862 for three years. Living in Craig Co., Va. in 1898 at age 70.

JONES, FRANCIS MARION: Pvt. Co. I. Res. Botetourt Co., Va. Enl. 5/1/62 at White Sulphur Springs, Greenbrier Co., WV. Capt. at Droop Mountain, WV 11/6/63. Age: 45. POW Camp Chase, OH 11/18/63. Transf. to Ft. Delaware 3/4/64. Exchd. 9/18/64. In Richmond, Va. hosp. with ulcer 9/22-23/64. Blacksmith. 5'9¼". Florid complexion, dark hair, dark eyes.

JONES, GEORGE R.: Pvt. Co. H. Enl. 4/28/62 at White Sulphur Springs, Greenbrier Co., WV. Conscript. Joined this co. 4/28/62. Deserted near Staunton, Va. May 1864. Res. Craig Co., Va. 1898 at age 53. Bur. Lee Cem., Rt. 311, near Ramey Lee house, Craig Co., Va.

JONES, J. A.: Pvt. Co. E. Enl. 7/18/61 at Porters Shoals, WV. Present on 8/31/61 roll. NFR.

JONES, JAMES L.: Co. H. Deserted 5/3/62 on Aug. 1862 broadside, Craig Co., Va.

JONES, JAMES W.: Pvt. Co. G. Res. Monroe Co., WV. Enl. 7/4/61 at Charleston, Kanawha Co., WV. Deserted about 5/6/64. Postwar res. Monroe Co., WV. Believed still alive in 1914.

JONES, J. M.: Pvt. Co. E. Enl. 6/8/61 at Charleston, Kanawha Co., WV. Present on 8/31/61 roll. NFR.

JONES, JOHN FRANK: Res. Craig Co., Va. NFR.

JONES, JOSEPH W.: Pvt. Co. K. Enl. 6/6/61 at Charleston, Kanawha Co., WV. NFR. This man's full name may be Joseph Warren Jones.

JONES, LEWIS ALLEN: Pvt. Co. C. B. ca. 1844 Craig Co., Va. Enl. 5/1/62 at White Sulphur Springs, Greenbrier Co., WV. PWR says enl. Mar. 1862 for three years. Absent sick on Mar.-Apr. 1863 roll. Present on Nov.-Dec. 1863 roll. Res. Craig Co., Va. 1898 at age 54. Bur. Jones Cem., Rt. 658, John's Creek, on Robert Dunbar Farm, Criag Co., Va.

JONES, MICHAEL K.: Pvt. Swann's Co. Enl. 5/8/61 at Charleston, Kanawha Co., WV. Age: 24. Co. transf. to Co. K, 59th Va. Inf. 7/25/61.

JONES, SAMUEL G.: Pvt. Co. D. Enl. 8/?/64 at Darksville, Va. Sick at hosp. on Apr.-Oct. 1864 roll.

JONES, TURPIN: Pvt. Co. E. Enl. 9/25/62 at Charleston, Kanawha Co., WV. KIA at Droop Mountain, WV 11/6/63.

JONES, WARREN: Pvt. Co. K. Enl. 6/6/61 at Charleston, Kanawha Co., WV. WIA at Scary Creek, WV 7/17/61. Not in CSR. (This may be Joseph W. Jones). Also served as Sgt. Bahlmann says believed KIA at Lewisburg, WV 5/23/62.

JONES, WILLIAM A.: 1st Corp. Co. D. Enl. 5/30/61 at Summersville, Nicholas Co., WV. WIA (severe shoulder) at White Sulphur Springs, WV 8/26-27/63. D. 9/15/63.

JONES, WILLIAM A.: Co. D. POW killed in a train wreck at Shohola, PA en route to Elmira, NY.

JONES, WILLIAM H.: Sgt. Co. E. Enl. 6/8/61 at Charleston, Kanawha Co., WV. Left sick at White Sulphur Springs, WV 8/20/61. Sick at Giles C.H. (Pearisburg), Va. since May 1862 on July 1862 roll. Present on Sept.-Oct. 1862 roll. Capt. at Winchester, Va. 9/19/64. POW Pt. Lookout, MD 9/23/64. Exchd. 3/15/65. In Farmville, Va. hosp. and transf. 5/22/65. Paroled at Charleston, WV 5/3/65. Age: 23. Farmer. 5'8". Dark complexion, dark hair, hazel eyes.

JUSTICE, ALEXANDER: Pvt. Co. E. Enl. 6/8/61 at Charleston, Kanawha Co., WV. Absent sick on Mar.-Apr. 1863 roll.

JUSTICE, WILLIAM N.: Pvt. Co. E. Enl. 6/8/61 at Charleston, Kanawha Co., WV. NFR.

JUSTICE, WILLIAM N.: Swann's Co. Date and place of enlistment not stated. Relieved Pvt. Cotteral at White Sulphur Springs, WV 8/14/61.

KARNES, GEORGE, W. C.: Corp./Sgt. Co. B. Res. Jackson Co., WV. Enl. 5/1/62 at White Sulphur Springs, Greenbrier Co., WV. Capt. at Strasburg, Va. 10/19/64. POW Pt. Lookout, MD. Relsd. 6/28/65. Took oath at Pt. Lookout, MD 6/28/65. 5'11½". Brown hair and dark blue eyes. Res. Alleghany Co., Va.

KARNES, JAMES BABER: Pvt. Co. H. Res. Craig Co., Va. Enl. either 3/21/64 or 3/22/64 in either Monroe or Greenbrier counties of WV. Rcvd. clothing 5/13/64. Present 12/25/64. On Alleghaney Co. pension rolls.

KASEY, GEORGE W.: Pvt. Co. F. Res. Roanoke Co., Va. Enl. 5/1/62(?) at White Sulphur Springs, Greenbrier Co., WV. Sick at Lewisburg, WV on 2/28/63 roll. Present on Mar.-Apr. 1863 roll. Capt. at Winchester, Va. 9/19/64. POW Pt. Lookout, MD. Exchd. 3/10/65. Postwar res. Roanoke Co., Va. Member of Hupp-Deyerle CV Camp, Salem, Va. 1906.

KEAN, ALFRED H.: Pvt. Co. G. Enl. 5/1/62 at White Sulphur Springs, Greenbrier Co., WV. Sick in hosp. on 11/13/62 roll. Present on 12/31/62 roll. Rcvd. clothing 5/13/64. A postwar roster says died of disease at home during war. Res. Alleghany Co., Va. Age: 21 at enl.

KEAN, LEWIS M.: Pvt. Co. G. Enl. 2/7/64 at Bunch Hill, Greenbrier Co., WV. D. at Staunton, Va. of disease 8/4/64. Res. Allegheny Co., Va. PWR says enl. Mar. 1862 at Lewisburg, WV. Age: 18.

KEELING, THOMAS L.: Pvt. Co. A. B. Halifax, Va. Enl. 5/22/61 at Poca, Putnam Co., WV or 6/5/61 at Charleston, Kanawha Co., WV. Disability discharge at Lewisburg, WV 2/24/62 due to severe typhoid fever which left him deaf and dumb. Paroled at Lewisburg, WV 4/27/65. Age: 25. Farmer. 5'6". Light complexion, light hair, blue eyes.

KEENEY, LEWIS M.: Pvt. Co. B. Res. Ripley, WV, 1860. Age: 32.

KEFFER, ISAAC C.: Pvt. Co. D. Enl. 5/30/61 at Summersville, Nicholas Co., WV. Capt. in Nicholas Co., WV while on furlo. on 12/31/62 roll. "Has taken the oath and is with the enemy" on 2/28/63 roll. Present on 4/30/63 roll. Deserted at Cow Pasture Bridge, Va. 5/?/64.

KELLY, AARON: Pvt. Co. A. Res. Putnam Co., WV. Enl. 7/13/61 at Camp Tompkins, Kanawha Co., WV. Det. service as hosp. nurse from July 1862 to 12/31/62. Det. as regt. nurse on 4/30/63 roll. Capt. in Boone Co., WV 2/9/64. POW Camp Chase, OH 3/7/64. Transf. to Ft. Delaware 3/14/64. Exchd. 9/18/64. In Chimborazo Hosp. as a paroled prisoner for debility 9/22-27/64. Took oath at Charleston, WV and sent North 4/22/65. Age: 34 or 35. 6'2". Florid complexion, black hair, dark eyes.

KELLY, JOSEPH T.: Pvt. Co. E. B. 9/19/41. Enl. 5/1/62 at White Sulphur Springs, Greenbrier, WV. Capt. at Winchester, Va. 9/19/64. POW Pt. Lookout, MD 9/23/64. Exchd. 3/15/65. Paroled at Charleston, WV 5/8/65. Age: 24. 5'10". Fair complexion, brown hair, hazel eyes, sandy whiskers. Marr. 2/22/66 in Fayette Co., WV. Farmer all his life. School trustee for three years. Res. Grassy Meadows, Greenbrier Co., WV.

KELLY, MARTIN: Pvt. Taylor's Co. Enl. 6/19/61 at Frankford, Greenbrier Co., WV. Age: 30. Co. transf. to Wise Brigade 7/10/61.

KELLY, SIMEON: Pvt. Co. A. Enl. 7/13/61 at Camp Tompkins, Kanawha Co., WV. Det. service as nurse at hosp. in Lewisburg, WV on 4/30/63 roll. MWIA (severe breast) at White Sulphur Springs, WV 8/26-27/63. D. 9/16/63.

KELLY, WILLIAM A.: Pvt. Co. E. Enl. 5/1/62 at White Sulphur Springs, Greenbrier Co., WV. D. in hosp. at Lewisburg, WV 4/22/63.

KEPKER, H.: Pvt. Co. B. Enl. 7/12/63 at Camp McCausland, Monroe Co., WV. NFR.

KERR, BENNETT H.: Pvt. Co. G. B. ca 1820 Craig Co., Va. Enl. 3/21/64 at Dublin, Pulaski Co., Va. In Charlottesville, Va. hosp. for acute diarrhea 6/9-27/64. Capt. at Winchester, VA 9/19/64. POW Pt. Lookout, MD 9/23/64. Exchd. 11/1/64. At Camp Lee near Richmond, Va. 11/22/64. Res. Craig Co., Va. in 1898 at age 77. Bur. Medley-Carr Cem., Rt. 622, Happy Hollow, at Rex Medley homeplace, Craig Co., Va. Also spelled Carr or Karr.

KERSHNER, GEORGE W.: Pvt. Taylor's Co. Enl. 6/19/61 at Frankford, Greenbrier Co., WV. Age: 22. Co. transf. to Wise Brigade 7/10/61.

KERSHNER, WILLIAM H.: Pvt. Taylor's Co. Enl. 6/19/61 at Frankford, Greenbrier Co., WV. Age: 30. Co. transf. to Wise Brigade 7/10/61.

KESLOR, SMITH: Exchd. to 22nd Va. from Thurmond's Co. 11/29/63.

KESSLER, LEWIS: Pvt. Co. C. B. ca. 1830 Craig Co., Va. Enl. 1/26/64 (?) at Richmond, Va. Sick in hosp. on 10/31/64 roll. D. at Staunton, Va. 12/19/64. Age: 34. PWR says died 12/18/64.

KESSLER, PETER: PWR. B. 2/24/36. D. 3/23/1913. Bur. Ross Cem., Rt. 42, near Maywood, on Fred Keffer farm, Craig Co., Va.

KESSLER, STEWART R.: Pvt. Co. C. Enl. 12/29/63(?) at Camp Gauley. In Charlottesville, Va. hosp. for acute dysentery 6/8-17/64. Rcvd. clothing at Lexington, Va. 6/27/64. Rcvd. clothing 4 qr. 1864.

KESSLER, WILLIAM: Pvt. Co. B/Co. D. Enl. 5/?/62 at White Sulphur Springs, Greenbrier Co., WV. In Danville, Va. hosp. for rubeola 11/17/64 to 12/9/64. Capt. at Winchester, Va. 9/19/64. POW Pt. Lookout, MD. Relsd. by order of the President 3/12/65.

KESTER, M. J.: Pvt. Co. B. Res. Bar Run, WV 1880, age 27.

KEYSOR, A. J.: Sgt./Pvt. Co. G. Res. Alleghany Co., Va. Enl. 5/1/62 at White Sulphur Springs, Greenbrier Co., WV. Capt. at Lewisburg, WV 5/23/62. Age: 29. POW Camp Chase, OH. Exchd. 8/25/62. "Prisoner in hands of the enemy" on 11/13/62 roll. 6'1". Dark complexion, dark hair, blue eyes, sandy whiskers.

KEYSOR, WILLIAM H.: Pvt. Co. G. B. 6/12/27. Enl. 10/23/62 near Lewisburg, Greenbrier Co., WV. AWOL on 4/30/64 roll. Sick in hosp. on 10/31/64 roll. Also spelled Keyser. D. 2/17/1901. Bur. Emory Methodist Church, Alleghany Co., Va.

KEYSTER, JOSEPH J. J.: Pvt. Co. F. Enl. 9/1/(?) at Red Sulphur Springs, Monroe Co., WV. Capt. at Strasburg (Cedar Creek), Va. 10/19/64. POW Pt. Lookout, MD 10/23/64. Relsd. 5/13/65. Mechanic. Res. Greenbrier Co., WV.

KIEFFER, JOSEPH: Pvt. Co. H. Enl. 5/8/61 at Charleston, Kanawha Co., WV. Absent cause unknown on 8/31/61 roll. Bur. Spring Hill Cem., Charleston, WV. (Also spelled Keifer and Keiffer). Father of Valentine Keifer.

KIEFFER, VALENTINE (SR.): Pvt. Co. H. B. 2/28/28 Baden, Germany. Res. U.S. 1848. Enl. 5/8/61 at Charleston, Kanawha Co., WV. Absent cause unknown on 8/31/61 roll. D. 12/12/93. Bur. Spring Hill Cem., Charleston, WV. (Also spelled Keifer and Keiffer). Postwar liquor dealer, Charleston, WV.

KIMBERLAND, L.: Pvt. Co. G. Enl. 5/1/62 at White Sulphur Springs, Greenbrier Co., WV. Deserted 5/1/62. Listed as deserter on Aug. 1862 broadside, left 5/24/62.

KIMES, GEORGE A.: Pvt. Co. B. Mustered 1862 for three years. Res. Jackson Co., WV.

KIMES, JOHN T.: Pvt. Co. B. Mustered 1862 for three years. Res. Jackson Co., WV.

KINCADE, ANDREW B.: Pvt. Co. G. B. 1835 (?). Res. Alleghany Co., Va. Enl. 7/1/62 at Salt Sulphur Springs, Monroe Co., WV or 7/10/62 at Camp McCausland, Monroe Co., WV. Capt. at Droop Mountain, WV 11/6/63. Age: 31. POW Camp Chase, OH 11/18/63. Transf. to Ft. Delaware 2/29/64. Exchd. 9/18/64. In Chimborazo Hosp. for inflammation of eye 9/23/64 to 10/2/64. Returned to regt. on 10/31/64 roll. Farmer. 5'6". Florid complexion, auburn hair, blue eyes. (Also spelled Kincaid). On Alleghany Co. list as being wounded, (pension).

KINCADE, HENRY DICKINSON: Pvt. Co. C. B. 6/?/40. Res. Fayette Co., WV. Enl. either 7/13/62 at Salt Sulphur Springs, Monroe Co., WV or 8/13/62 at Camp McCausland, Monroe Co., WV. On scout by order of Gen. Williams on 12/31/62 roll. Capt. in Greenbrier Co., WV 4/6/64. Age: 22. Circa. Mar. 1864 became Pvt. in Co. A, Hounshell's Battn. Va. Inf. POW Camp Chase, OH 4/29/64. Exchd. 2/25/65. Record also indicates he was exchd. 4/18/64 for Pvt. Simon Thomas of Capt. John Amick's Co. of Partisan Rangers. Farmer. 5'9". Florid complexion, black hair, brown eyes. D. 2/?/89.

KINCADE, WILLIAM A.: Pvt. Co. G. Enl. either 7/1/62 at Salt Sulphur Springs, Monroe Co., WV or 7/10/62 at Camp McCausland, Monroe Co., WV. Sent home sick from Red Sulphur Springs, WV on 12/31/62 roll. Dischd. 4/12/63. (Also spelled Kincaid). On Alleghany Co. pension list as being wounded. Res. Alleghany Co., Va. Age: 18 at enl.

KINCADE, WILLIAM M.: Pvt. Co. C. Enl. 5/20/63 at Camp Stonewall. WIA (slight) at White Sulphur Springs, WV 8/26-27/63. AWOL on 10/31/64 roll. Listed as POW. Paroled at Charleston, WV 5/11/65. Age: 49. 5'10". Dark complexion, brown hair, blue eyes, dark whiskers. (Also spelled Kincaid). Farmer, Fayette Co., WV 1860.

KINCAID, CHARLES: Pvt. Co. K. Res. Kanawha Co., WV. Enl. either 4/15/62 at Lewisburg, WV or 6/7/63(?) at Meadow Bluff, Greenbrier Co., WV. Deserted 9/4/63. Capt. in Greenbrier Co., WV 10/8/63. Age: 24. POW Camp Chase, OH 10/23/63. Transf. to Rock Island, Ill. 1/14/64. Record also states he was taken prisoner Aug. 1863. Farmer. 5'9". Light complexion, dark hair, blue eyes, sandy whiskers.

KINCAID, L. B.: Pvt. Co. C. Enl. 5/10/61.

KINCAID, SAMPSON: Pvt. Co. E/Co. A. Date and place of enlistment not stated. Capt. in Fayette Co., WV 9/3/63. Age: 21. Bushwhacker. POW Camp Chase, OH 9/18/63. Transf. to Rock Island, Ill. 1/14/64. D. at Rock Island, ILL of diarrhea 9/19/64. Bur. grave #1513 at Rock Island, ILL. 6'. Light complexion, brown hair, blue eyes. Also with Capt. Thurmond's Co.

KINCAID, WILLIAM R.: Co. G. B. 1831. D. 1884. Bur. Cedar Hill Cem., Covington, Va.

KING, A.: Pvt. Co. H. Res. Greenbrier Co., WV. Enl. 4/15/62 at Lewisburg, Greenbrier Co., WV. Capt. at White Sulphur Springs, WV 8/26/63. Age: 21. POW Camp Chase, OH 9/7/63. Transf. to Rock Island, ILL 1/22/64. Exchd. 3/2/65. Farmer. 5'8". Dark complexion, dark hair, gray eyes.

KING, CHARLES L.: Pvt./Corp. Co. I. Enl. 5/1 or 2/62 at White Sulphur Springs, Greenbrier Co., WV. Present on 10/31/64 roll. NFR.

KING, DAVID: Pvt. Co. E. Enl. 6/8/61 at Charleston, Kanawha Co., WV. AWOL 7/1/61.

KING, JAMES A.: Pvt. Co. D. Enl. 5/1/62 at White Sulphur Springs, Greenbrier Co., WV. Left sick at Beverly, WV on 4/25/63 roll. Present on 12/31/63 roll. Det. service with Provost Guard in Alleghany Co., Va. 2/1/64. Present on 10/31/64 roll.

KING, WILLIAM: Pvt. Co. B. Enl. 10/?/63 at Lewisburg, Greenbrier Co., WV. Admitted to Wayside Hosp., Richmond, Va. 11/28/63. Absent sick on 1/10/64 roll.

KINGAN, J.: Pvt. Co. H. Enl. 10/13/63 at Lewisburg, Greenbrier Co., WV. Det. butcher at Dublin, Va. on 1/6/64 roll. Transf. to Edgars' Battn. 3/25/64. Det. butcher at Dublin, Va. on 12/28/64 roll.

KINNISON, CALDHILL J.: Pvt. Co. A. Farm laborer, Sissonville, WV 1860. Age: 31. Enl. 6/12/61 at Camp Tompkins, Kanawha Co., WV. Sick with leave on 7/8/61. NFR.

KINNISON, JAMES H.: Pvt. Co. A. Farm laborer, Sissonville, WV 1860. Age: 23. Enl. 5/22/61 at Poca, Putnam Co., WV. Sick with leave 7/24/61. NFR.

KIOUS, HIRAM: Pvt. Co. C. Day laborer, Fayette Co., WV 1860. Age: 25. Enl. 5/10/61.

KIRLEY, ANDREW: Pvt. Swann's Co. Enl. 5/22/61 at Charleston, Kanawha Co., WV. Age: 21. Co. transf. to Co. K, 59th Va. Inf. 7/25/61.

KNAPP, CHARLES: Pvt. Co. B. Enl. 7/12/62 at Camp McCausland, Monroe Co., WV or 7/16/62 at Union, Monroe Co., WV. In hosp. at Charlottesville, Va. for rubeola 6/10-19/64. In hosp. on 11/30/64 roll.

KNAPP, CHARLES S.: Pvt./Corp. Co. B. Res. Greenbrier Co., WV. Enl. 3/1/63 at Lewisburg, Greenbrier Co., WV. Capt. at Winchester, Va. 9/19/64. POW Pt. Lookout, MD 9/23/64. Exchd. 10/30/64. Took oath at Charleston, WV 4/4/65. Age: 20. Farmer. 5'8". Fair complexion, fair hair, gray eyes. (Also listed under the name is a Stewart Knap as well as a Pvt. Charles A. Knapp, Co. B, 22nd Va. Inf. who was age: 31. 6'11". Dark complexion, blue eyes, dark hair, dark whiskers. Paroled at Charleston, WV 5/11/65)

KNAPP, GEORGE W.: Pvt. Co. F. Date and place of enlistment not stated. Rcvd. clothing 12/3/64. Paroled at Charleston, WV 5/11/65. Age: 19. 5'9". Fair complexion, dark hair, blue eyes.

KNAPP, JAMES C.: Pvt. Co. A. Enl. 5/1/62 at White Sulphur Springs, Greenbrier Co., WV. D. in hosp. at Pearisburg, Va. 5/22/62.

KNIGHT, JAMES: Pvt. Co. E. Enl. 7/18/61 at Porter's Shoals, WV. Present on 8/31/61 roll. NFR.

KNIGHTSTEP, WILLIAM C.: Pvt. Co. B. Enl. 6/1/61 at Ripley, Jackson Co., WV. Present on 9/1/61 roll. NFR.

KNISELEY, DAVID: Pvt. Swann's Co. Enl. 5/22/61 at Charleston, Kanawha Co., WV. Age: 23. Co. transf. to Co. K, 59th Va. Inf. 7/25/61. (Also spelled Knicely).

KNOX, H. H.: Co. H. Deserted 7/27/62 on Aug. 1862 broadside. Greenbrier Co., WV.

KOONTZ, MIDDLETON H.: Pvt. Co. D. Enl. 6/15/61 at Charleston, Kanawha Co., WV. Present on 8/31/61 roll. NFR.

KRANTZ, MICHAEL: Pvt. Co. D. Enl. 5/30/61 at Summersville, Nicholas Co., WV. Sick at home 7/28/61. NFR.

LAFAN, JOHN C.: Pvt./Sgt. Co. C. Enl. 5/1/62 at White Sulphur Springs, Greenbrier Co., WV. PWR says enl. Mar. 1862 for three years. Rcvd. clothing 5/23/64. (Also spelled Lafon). Res. Craig Co., Va. 1898 at age 54.

LAIDLEY, RICHARD QUARRIER: Capt./1st Lt. Co. H. B. either 1834 or 6/2/36 Wheeling, WV. Studied medicine and practiced law. Laywer, Charleston, WV 1860 at age 23. Physician. Enl. 5/8/61 at Charleston, Kanawha Co., WV. Served as 1st Sgt. until comm. 1st Lt. 7/27/61. Elected 1st Lt. 8/1/61. Elected Capt. 5/1/62. WIA (slight) at Fayetteville, WV 9/10/62. Married during 1862 occupation of Charleston, WV. Absent until again present on 6/30/63 roll. WIA at Droop Mountain, WV 11/6/63. Retired to Invalid Corps 7/6/64. PWR says made Capt. upon Col. Patton's death 9/19/64. Postwar druggist, Charleston, WV. D. 2/12/73 of consumption at Charleston, WV from disease contracted during war. Bur. Spring Hill Cem., Charleston, WV.

LAKE, ANDREW: Pvt. Co. G. Res. Monroe Co., WV. Enl. 11/24/62 at Centerville, Monroe Co., WV. On sick leave 1/?/63. Deserted while on sick leave. Left for White Sulphur Springs, WV (time not known?) on 12/31/63 roll. Deserted while on sick leave on 10/31/64 roll. Surrendered 11/?/63. Took oath and sent North 11/15/63. Age: 32. Carpenter. 5'10". Light complexion, dark hair, blue eyes.

LAM, ISAAC: Pvt. Co. C. Enl. 9/23/64 at Dublin, Pulaski Co., Va. Present on 10/31/64 roll. Rcvd. Clothing for 4 QR 1864.

LANDERS, PETER: Rank unknown. Co. B. B. 1/10/29 Botetourt Co., Va. Date and place of enlistment not stated. Rcvd. clothing 4 QR 1864. Marr. Mary Kesler 1857. D. 8/17/1906 Botetourt Co., Va.

LANDRUM, ANDREW L.: Pvt. Taylor's Co. Enl. 6/19/61 at Frankford, Greenbrier Co., WV. Age: 21. Co. transf. to Wise Brigade 7/10/61.

LANE, A. M.: Pvt. Co. G. Enl. 7/7/61 at Charleston, Kanawha Co., WV. Transf. from Capt. Sayre's (Sarver's) Co. 7/15/61. Present on 4/30/63 roll.

LANGHORNE, ________: 1st Lt. Co. E. WIA and left service.

LANHAM, ALBERT: Pvt. Co. A. Enl. 7/2/61 at Camp Tompkins, Kanawha Co., WV. Present on 8/31/61 roll. NFR.

LANHAM, ALEXANDER: Pvt. Co. A. Enl. 7/2/61 at Camp Tompkins, Kanawha Co., WV. Unaccounted for as of 7/24/61.

LANHAM, BENJAMIN FRANKLIN: Pvt. Co. A. B. 1840 Putnam Co., WV. Res. Putnam Co., WV. Enl. 10/30/62 at Raleigh (Beckley), Co., WV. Capt. at Winchester, Va. 9/19/64. POW Pt. Lookout, MD 9/24/64. Exchd. 3/15/65. Deserted and surrendered 4/?/65. Age: 24. Farmer. 5'6". Fair complexion, fair hair, gray eyes. Res. Putnam Co., WV 1910 at age 69. Brother of William H. Lanham.

LANHAM, CHRISTOPHER COLUMBUS: Capt. at Gettysburg, PA. POW Elmira, NY. May be brother of James Lanham of Putnam Co., WV.

LANHAM, DAVID B.: Pvt. Co. A. Enl. 5/27/61 at Buffalo, Putnam Co., WV. Present on 8/31/61 roll. NFR.

LANHAM, JAMES PLEASANT: Pvt. Co. A. Res. Putnam Co., WV. Enl. 11/1/62 at Packs Ferry, Monroe Co., WV. Left sick in Maryland on 10/31/64 roll. Capt. near Washington, D.C. 7/15/64. POW Old Capitol Prison, Washington, D.C. In Lincoln Hosp., Washington, D.C. for typhoid fever 7/15/64 to 9/24/64. Age: 20. Then to Gen. Hosp. POW Elmira, NY 10/24 or 27/64. Relsd. 5/17/65. Too oath at Elmira, NY 5/17/65. 5'11½". Florid complexion, dark hair, blue eyes. Res. Putnam Co., WV 1910 at age 66. In a photo of R. E. Lee Camp #887, Charleston, WV taken Aug. 1911. Bur. Monroe Bailey Cem., Flatwoods, WV (now Cross Lanes, WV).

LANHAM, JOHN W.: Pvt. Co. A. B. Nelson Co., Va. 11/14/25. Marr. 1850 in Kanawha Co., WV and settled in Putnam Co., WV. Enl. 6/28/61 at Camp Tompkins, Kanawha Co., WV. Capt. at Scary Creek, WV 7/17/61. PWR says served one month. Postwar res. Liberty, WV.

LANHAM, RICHARD: Pvt. Co. A. Farm laborer, Tyler Mountain, WV 1860. Age: 26. Enl. 6/5/61 at Camp Tompkins, Kanawha Co., WV. Unaccounted for as of 7/24/61.

LANHAM, WILLIAM HENRY: Pvt. Co. A. B. 12/6/34 Putnam Co., WV. Res. Putnam Co., WV. Enl. 10/1/62 at Charleston, Kanawha Co., WV. WIA (head) at White Sulphur Springs, WV 8/26-27/63. Absent wded. on 12/31/63 roll. Present on 4/30/64 roll. Capt. at Winchester, Va. 9/19/64. POW Pt. Lookout, MD 9/24/65. Exchd. 3/15/65. Deserted and surrendered 4/?/65. Took oath at Charleston, WV and sent North 4/20/65. Age: 30. Farmer. 5'6". Dark complexion, dark hair, gray eyes. Postwar res. Kanawha Co., WV. Brother of Benjamin Franklin Lanham.

LARK, DANIEL: Co. K. Deserted 5/1/62 on Aug. 1862 broadside. Sinking Creek, Craig Co., Va.

LARK, JAMES: Co. K. Deserted 5/20/62 on Aug. 1862 broadside. Sinking Creek, Craig Co., Va.

LARK, SAMUEL: Pvt. Co. K. B. ca. 1841 Craig Co., Va. Enl. 9/1/62 at White Sulphur Springs, Greenbrier Co., WV. Also stated enl. 4/24/63. Deserted 5/20/62 on Aug. 1862 broadside. Marr. 12/23/62 at age 20. Deserted at Charleston, WV 10/10/62. D. 9/?/29 Roane Co., WV at age 100 years, 7 months, 5 days. Also spelled Larch.

LAVINDER, ROBERT: Pvt. Co. I. Enl. 6/22/61 at Charleston, Kanawha Co., WV. Capt. in Fayette Co., WV 9/13/61. POW Camp Chase, OH. Exchd. 8/25/62. Returned from prison on 11/11/62 roll. On sick leave on 3/4/63 roll.

LAWHON, JAMES: Pvt. Co. B. enl. either 8/1/61 in Monroe Co., WV or 7/?/62 at Union, Monroe Co., WV. On furlo. and overstayed his time on 2/28/63 roll. Present on 4/30/63 roll. In Chimborazo Hosp. with very serious left lung 6/5/64 to 7/12/64. At home wded. on 11/30/64 roll. Took oath 6/12 (?)/65 at Charleston, WV. Age: 20. 6'1". Light complexion, dark hair, blue eyes. (This is probably J. A. Lawhorn).

LAWHON, JAMES E.: Pvt. Co. B. Enl. 8/?/63 at Lewisburg, Greenbrier Co., WV. Entitled to bounty being underage when enlisted. Paroled at Charleston, WV 6/12/65. Age: 18. 5'7". Light complexion, dark hair, blue eyes. (Name is probably spelled Lawhorn).

LEACH, EDMOND K.: Pvt. Co. F. Res. Monroe Co., WV. Enl. 5/1/62 at White Sulphur Springs, Greenbrier Co., WV. Capt. at Winchester, Va. 9/19/64. POW Pt. Lookout, MD 9/23/64. Exchd. 3/15/65. In Chimborazo Hosp. 3/19-24/65. Admitted to Wayside Hosp., Richmond, Va. 3/25/65. Paroled at Charleston, WV 6/5/65. Age: 29. 5'7". Fair complexion, light hair, gray eyes. D. sometime prior to 1914.

LEACH, JAMES HARVEY: Pvt./Corp. Co. F. B. Monroe Co., WV 1842. Res. Monroe Co., WV. Enl. 7/4/61 at Charleston, Kanawha Co., WV. On special duty 8/22/61. Present on 10/31/62 roll. Promo. Sgt. 1863. WIA (slight hand) at Winchester, Va. 9/19/64. Paroled at Charleston, WV 6/12/65. Age: 23. 6'1". Dark complexion, black hair, black eyes. PWR says "shot 3 times at Col. Hays at Giles C.H., Va. 5/10/62. Said Co. F lost 18 killed and captured at Winchester, Va. 9/19/64 out of 22 who went into action." Postwar farmer, Monroe Co., WV. One term county assessor. Still alive 1899. D. sometime prior to 1914.

LEACHMAN, WILLIAM T.: Pvt. Co. B. Enl. 6/1/61 at Ripley, Jackson Co., WV. Joined by transf. order. Authorized by War Dept. to raise a co. 1/10/64. Authority revoked 2/24/64. AWOL on 4/1/64 roll. Listed as a deserter on Gen. Echols' April 1864 report.

LEE, JONATHAN CALVIN: Pvt./Musician. Co. H. B. ca. 1836 Craig Co., Va. Enl. 4/28/62 at White Sulphur Springs, Greenbrier Co., WV. (PWR says enl. Mar. 1862 for 3 years). Was assigned to and joined this co. 4/28/62. Wrote a letter from Lewisburg, WV 3/23/63. Deserted near Staunton, Va. 5/?/64. Res. Craig Co., Va. in 1898 at age 62. Bur. Lee Cem., Rt. 311, near Ramey Lee house, Craig Co., Va. A "pacifist at heart".

LEE, LEANDER S.: Pvt. Co. F. Enl. 5/1/62 at White Sulphur Springs, Greenbrier Co., WV. Deserted about 5/8/64.

LEE, WILLIAM R.: Pvt. Co. G. Enl. 7/3/61 at Charleston, Kanawha Co., WV. NFR.

LEECH, THOMAS HACKNEY: Not in CSR. Co. K. B. in Rockbridge Co., Va. ca. 1840. Served in Co. C 1st Va. Cav. Married in Rockbridge Co., Va. 1860 and remarried in Rockbridge Co., Va. 1878.

LEFFEL, JOHN ALLEN: Pvt. Co. H. B. ca. 1823 Craig Co., Va. Enl. 3/7/64 in Greenbrier Co., WV. Capt. at Winchester, Va. 9/19/64. POW Pt. Lookout, MD 9/24/64. Exchd. 3/15/65. Res. Craig Co., Va. 1898 at age 52. D. in hosp. at Staunton, Va. Bur. Thornrose Cem., Staunton, Va.

LEFFEL, SAMUEL: Son of Jacob Leffel. Res. Craig Co., Va. Photo in *Hard Times.*

LEGENMORE, JOSEPH ('FRENCH JOE'): Pvt. Co. B. B. in France. Res. Ripley, WV 1860, age 25. Enl. 6/1/61 at Ripley, Jackson Co., WV. Capt. at Ripley, WV by 2nd Va. Cav. (U.S.) 10/11/62. POW Camp Chase, OH 10/24/62. Exchd. 10/27/62. Age: 30. Stonemason. 5'11". Fair complexion, dark hair, blue eyes, black/sandy whiskers. Also spelled Legamore.

LEGG, THOMAS A.: Pvt./Corp. Co. D. Enl. 5/30/61 at Summersville, Nicholas Co., WV. AWOL 7/28/61. Present on 10/31/62 roll. Apptd. color corp. 4/1/64. Present on 10/31/64 roll. In Winchester, Va. hosp. for leg wd. (?) 1864. Middle name is Adams.

LEGG, WILLIAM H.: Pvt. Co. D. Enl. 5/30/61 at Summersville, Nicholas Co., WV. "Capt. by Yankees and carried to Gauley Bridge, WV" 8/20/61.

LEIGHTON, JOHN R.: Co. G. Enl. 10/?/64 at New Market, Va. Postwar roster. Res. Alleghany Co., Va. Age: 21. Surrendered Apr. 1865.

LEMON, TILMAN O.: Pvt./Musician. Co. B. Enl. 5/?/62 at White Sulphur Springs, Greenbrier Co., WV. Conscript. Sick in hosp. on 11/16/62 roll. Present on rolls for 2/28/63 and 11/30/64. NFR.

LEMONS, GEORGE L.: Pvt. Co. F. B. ca. 1844-45 in Va. Enl. 2/17/63(?) at Lewisburg, Greenbrier Co., WV. MWIA at Winchester, Va. 9/19/64. Very serious arm wd. Transf. to hosp. 10/15/64.

LEMONS, JOHN ALEXANDER: Pvt./Sgt. Co. F. B. 12/6/37 Sinks Grove, Monroe Co., WV. Enl. either 7/4/61 at Charleston, Kanawha Co., WV or 5/1/62 at White Sulphur Springs, Greenbrier Co., WV. Married 3/2/63. Capt. at Winchester, Va. 9/19/64. Paroled at Charleston, WV 5/11/65. Age: 27. 5'11". Light complexion, light hair, blue eyes, red whiskers. PWR says WIA twice. Postwar res. Monroe Co., WV. D. 10/31/1927 Sinks Grove, WV. Bur. Mt. Pleasant Churchyard, Sinks Grove, WV.

LEMONS, JOHN ANDERSON: Pvt. Co. F. B. ca. 1839. Res. Monroe Co., WV. Enl. 7/4/61 at Charleston, Kanawha Co., WV. Sick at home on surg. cert. 8/12/61. Present on 10/31/62 roll. KIA at Lynchburg, Va. 6/19/64.

LEMONS, LEWIS HENDERSON: Pvt. Co. F. B. 10/14/38. Res. Monroe Co., WV. Enl. 7/4/61 at Charleston, Kanawha Co., WV. AWOL 8/21/61. Present on 10/31/62 roll. On wded. leave on 12/31/63 roll. Present on 4/1/64 roll. Two mos. wages taken by court martial on 10/31/64 roll. Paroled at Charleston, WV 5/11/65. Age: 25. 6'11". Light complexion, light hair, gray eyes. Married 1866. D. 9/20/1907. Postwar res. Russell Co., Va.

LEMONS, WILLIAM HOUSTON: Pvt. Co. F. B. 6/4/46 Monroe Co., WV. Enl. 8/10 or 11/63 at Red Sulphur Springs, Monroe Co., WV. Another card says enl. in Loudoun Co., Va. Capt. at Winchester, Va. 9/19/64. POW Pt. Lookout, MD 9/26/64. Exchd. 11/1/64. Paroled at Charleston, WV 5/11/65. Age: 21. 5'6". Light complexion, light hair, blue eyes. Married 1/7/69. Postwar res. Monroe Co., WV. D. 12/26/1926. Bur. Longacre Cem., Fort Spring, WV.

LEONARD, ANTHONY: Pvt. Co. B. Enl. 6/?/62 at Union, Monroe Co., WV. Det. as cooper 11/16/62 roll. Det. as cooper at Dublin, Va. on rolls for 12/31/62, 2/28/63 and 11/30/64.

LESTER, A. B.: Pvt. Co. G. B. 9/16/35. Enl. either 7/3/61 at Charleston, Kanawha Co., WV or 1/25/63 at Union, Monroe Co., WV. WIA (slight hand) at White Sulphur Springs, WV 8/26-27/63. Returned from desertion on Gen. Echols' report for April 1864. Under arrest 4/30/64. WIA and capt. at Winchester, Va. 9/19/64. Age: 29. Musketball in left side of chest passing through left lung. Paroled at West Bldg., U.S.A. Gen. Hosp., Baltimore, MD 11/16/64. Transf. to Ft. McHenry for exchange. Admitted to Wayside Hosp. 2/25/65. Transf. to Howard's Grove Hosp., Richmond, Va. 2/26/65 to 3/6/65. D. 3/9/1907. Bur. Tuggle Cem., near Bluefield, WV.

LESTER, DAVID B.: Pvt. Co. G. Res. Wyoming Co., WV. Enl. 7/1/62 at Union, Monroe Co., WV. Had a furlo. and time expired 3/10/64. Deserter. Took oath at Charleston, WV 5/9/65. Age: 21. Farmer. 5'6". Dark complexion, dark hair, dark eyes.

LESTER, JOSEPH D.: Pvt. Co. G. Res. Wyoming Co., WV. Enl. 7/3/(?) at Charleston, Kanawha Co., WV or 5/1/62 at White Sulphur Springs, Greenbrier Co., WV. Capt. at Droop Mountain, WV 11/6/63. Age: 26. POW Camp Chase, OH 11/18/63. Transf. to Ft. Delaware 2/29/64. Relsd. 6/20/65. Farmer. 6'. Florid complexion, dark hair, blue eyes.

LETT, JAMES THOMAS: Pvt. Co. A. Enl. 6/24/61 at Camp Tompkins, Kanawha Co., WV. Unaccounted for as of 7/24/61.

LEVISAY, VALENTINE: Sgt. Co. D. Enl. 5/30/61 at Summersville, Nicholas Co., WV. Capt. at Winchester, Va. 9/19/64. POW Pt. Lookout, MD 9/24/64. Exchd. 2/18/65. Paroled at Charleston, WV 6/7/65. Age: 33. 5'6". Dark complexion, gray hair, gray eyes.

LEWIS, CALEB V.: Pvt. Co. K. Enl. 4/24/62(?) at White Sulphur Springs, Greenbrier Co., WV. Left sick at Lewisburg, WV on Mar.-Apr. 1863 roll. Rcvd. clothing 5/13/64(?). Bahlmann says transf. to 14th Va. Cav. and killed in a skirmish 1864.

LEWIS, CHARLES CAMERON: Pvt. Co. H. B. 4/15/39 in Kanawha Co., WV. Attended private school and Mercer Academy. Kanawha Rifleman, 1859. Enl. 5/8/61 at Charleston, Kanawha Co., WV. On furlo. from Gen. Wise 7/25/61 on 8/31/61 roll. Honorably dischd. following Wise's retreat to Kanawha Falls, WV. Involved in salt industry of Kanawha Valley 1861-1869. Married 10/19/64. President of Kanawha Valley Bank of Charleston, WV 1870 for 15 years. In 1885 organized wholesale grocery house of P. H Noyes & Co. and became its president and was still so engaged in 1915. Served as city treasurer. D. 9/9/1917 Charleston, WV. Bur. Spring Hill Cem., Charleston, WV. Younger brother of Joel Lewis.

LEWIS, CHARLES IRVINE: Capt. Lewis's Co. Date and place of enlistment not stated. On a report from Charleston, WV dated 7/8/61. Co. transf. to 8th Va. Cav. Raised Kanawha Rangers Co. in 1861 and elected Capt. Promo. was offered but he preferred to remain with his co. Practiced medicine postwar. Bur. Kanawha Falls, WV.

LEWIS, COLUMBUS: Pvt. Co. B. Enl. 9/?/62 at Charleston, Kanawha Co., WV or 10/?/63 at Lewisburg, Greenbrier Co., WV. KIA at New Market, Va. 5/15/64, shot in head.

LEWIS, GEORGE: Pvt. Co. K. Enl. 4/24/62(?) at White Sulphur Springs, Greenbrier Co., WV. Sick on 11/1/62 roll. At home sick on Nov.-Dec. 1862 roll. Present on rolls for 2/28/63 and Mar.-Apr. 1863. NFR. Bahlmann says transf. to 14th Va. Cav. and killed in retreat from Droop Mountain, WV 1863.

LEWIS, HENRY: Pvt. Co. K. Res. Alleghany Co., Va. Enl. 6/22/63(?) in Greenbrier Co., WV. Capt. in retreat from Droop Mountain, WV 11/6/63. Appears on a list of rebel deserters for Nov. 1863. Took oath at Charleston, WV 11/15/63 and sent North of the Ohio River. Age: 19. Carpenter. 5'11". Light complexion, dark hair, hazel eyes.

LEWIS, JAMES F.: Pvt. Co. H. B. 8/30/30 Charleston, WV. Grad. Univ. of Va. Visited California 1853. Enl. 5/8/61 at Charleston, Kanawha Co., WV. Specially det. in July 1861 as asst. in the office of the adjutant of the Wise Legion and still so engaged on 8/31/61 roll. Injured his foot on the march from Charleston, WV to Buffalo, WV in May or June 1861, caused by an old sprain, and therefore det. to clerical duties. Applied for disability discharge 11/5/61. Age: 20. Rcvd. clothing 7/28/64. Was with his brother, John Lewis, when KIA at Winchester, Va. 9/19/64. Merchant. 5'8". Florid complexion, light brown hair, blue eyes. Married 6/17/68. Postwar merchantile business, Charleston, WV. D. 1/31/1908, Charleston, WV.

LEWIS, JOEL S.: Pvt./Sgt. Co. H. Enl. 5/8/61 in Charleston, Kanawha Co., WV. Served as a Pvt. until elected 2nd Sgt. 7/27/61. Sick in Bedford Co., Va. July 1862. Undated entry says: "At home sick — discharged." PWR says served 22nd Va. Inf. throughout war except for a brief period of cav. service at which time he was capt. and sent to Camp Chase, OH. Postwar farmer. D. 1894 in Raleigh Co., WV. Older brother of Charles Cameron Lewis.

LEWIS, JOHN: Pvt./2nd Lt. Co. H. B. 1841. Enl. 6/26/61 at Charleston, Kanawha Co., WV. Elected 2nd Lt. 6/5/64. Promo. 7/?/64. KIA at Winchester, Va. 9/19/64. Brother of James F. Lewis. See also 36th Va. Inf.

LEWIS, JOHN: Pvt. Co. K. Enl. either 4/24/62 or 5/2/62 at White Sulphur Springs, Greenbrier Co., WV. On mounted guard at White Sulphur Springs, WV 8/13-14/61. Sick at Montgomery White Sulphur Springs, Va. July 1862. Capt. at Winchester, Va. 9/19/64. POW Pt. Lookout, MD 9/23/64. Exchd. 3/15/65.

LEWIS, JOHN L.: Pvt. Co. A. B. 1/6/40 Middleway, WV. D. 7/8/1915 at Leetown, WV. Bur. Masonic Cem., Middleway, Jefferson Co., WV.

LEWIS, JOHN M.: Pvt. Co. A. B. 1/6/40 at Middleway, Jefferson Co., WV. Enl. 4/10/62 at Brushy Ridge, Greenbrier Co., WV. Capt. at Droop Mountain, WV 11/6/63. (This could possibly be John L. Lewis, Co. B. 22nd Va. Inf. D. 7/18/1915 in Leetown, Va. Bur. Masonic Cem., at Middleway, Jefferson Co., WV).

LEWIS, WILLIAM S.: Co. H. Enl. 1861. Farmer. Transf. to Co. I, 8th Va. Cav. July 1862. Took oath 12/6/64. Age: 25.

L'HOMMEDIEU, CHARLES: Pvt. Co. K. Enl. 8/22 or 24/64 at Union, Monroe Co., WV. Left sick in Hightown on Mar.-Apr. 1863 roll. On Provost Guard 12/30/63. Det. as a nurse for Capt. Bahlmann 4/1/64 and 5/1/64. Det. as colonel's orderly 7/20/64. Paroled at Charleston, WV 5/31/65. Age: 44. 5'6½". Dark complexion, gray hair, blue eyes, dark whiskers.

LIKENS, JAMES E.: Pvt. Co. C. Res. Fayette Co., WV. Enl. 5/1/62(?) at White Sulphur Springs, Greenbrier Co., WV. Capt. at Lewisburg, WV 5/23/62. Age: 32. POW Camp Chase, OH 6/30/62. Exchd. 8/25/62. 5'8¾". Dark complexion, black hair, hazel eyes, black whiskers. An original member of Jeb Stuart UCV Camp, Fayette Co., WV. Gave Victor, WV as res. 1909. Also spelled Lykins. PWR says enl. 5/10/61. Farm laborer, Fayette Co., WV 1860.

LILLY, COLUMBUS: Pvt. Co. B. Res. Kanawha Co., WV. Enl. 6 or 7/?/63 at Lewisburg, Greenbrier Co., WV. Det. teamster 4/1/64. Det. brig. wagoner 11/30/64. Paroled at Lewisburg, WV 5/10/65. Age: 28. Farmer. 6'. Fair complexion, dark hair, blue eyes.

LINDOL, JOHN: Pvt. Co. F. Date and place of enlistment not stated. Sick at Buchanan, Va. July 1862. NFR.

LINN, W.: Pvt. Co. H. Enl. 5/8/61 at Charleston, Kanawha Co., WV. Transf. from Co. K, 59th Va. Inf. 3/19/62. Re-transf. to Capt. John S. Swann's Co., Edgar's Battn. 7/6/62. This may be William H. Lynn.

LINTON, JOHN B.: Pvt. Co. B. B. 8/13/44. Res. Craig Co., Va. Enl. 12/18 or 28/62 at Camp Sam Miller, Monroe Co., WV. At home wded. on 10/31/64 roll. Paroled at Charleston, WV 6/12/65. Age: 20. 5'10". Light complexion, dark hair, blue eyes. PWR says wded. in left hip. Res. Craig Co., Va. in 1889 at age 54. Bur. Bostic Cem., Rt. 602, Back Valley, Monroe Co., WV (near Paint Back).

LINTON, WILLIAM: Pvt. Co. D. B. ca. 1835 Craig Co., Va. Res. Craig Co., Va. Enl. 5/1/62 at White Sulphur Springs, Greenbrier Co., WV. Deserted at White Sulphur Springs, WV 5/1/62.

LINVILLE, SQUIRE: Sgt. Co. I. Enl. 6/29/61 at Charleston, Kanawha Co., WV. Present on 8/31/61 roll. NFR.

LIPPS, ARCH: Pvt. Co. B. Res. Greenbrier Co., WV. Enl. 5/?/62 at White Sulphur Springs, Greenbrier Co., WV. Conscript. At home wded. on 11/30/64 roll. Deserted and surrendered at Charleston, WV March 1865. Took oath at Charleston, WV and sent North 3/28/65. Age: 29. Farmer. 5'11". Fair complexion, dark hair, blue eyes.

LIPPS, JOSEPH H.: Pvt. Co. B. Enl. 5/?/62 at White Sulphur Springs, Greenbrier Co., WV. Conscript. WIA at Lewisburg, WV 5/23/62. Det. to farm 5/1/64. WIA (slight thigh) at Gardner's Farm, Va. 5/30-31/64. Another card says WIA (very serious thigh) at Cold Harbor, Va. 6/2-3/64. At Wayside Hosp. 6/2-3/64. At Chimborazo Hosp. 6/3-30/64. POW. Paroled at Charleston, WV 5/10/65. Age: 37. 5'11". Light complexion, dark hair, blue eyes, dark whiskers. D. at Rich Hollow 1/24/84. Bur. Johnson Cem., at Rich Hollow. A photo of him appears in an issue of the *Greenbrier Independent* newspaper.

LIPSCOMB, WILLIAM H.: Capt. Co. B. Lawyer, Ripley, WV 1860, age 25. Enl. 6/1/61 at Ripley, Jackson Co., WV. Mustered into state service 6/1/61. Elected Capt. 6/1/61. Mustered into C.S. service July 1861. Not re-elected 5/1/62.

LIVELY, CHRISTOPHER: Pvt. Co. H. Date and place of enlistment not stated. Rcvd. clothing 4 QR 1864. Took oath at Charleston, WV 6/12/65. Age: 36. 6'1". Light complexion, dark hair, blue eyes.

LIVESAY, GEORGE W.: Pvt. Taylor's Co. Enl. 6/19/61 at Frankford, Greenbrier Co., WV. Age: 18. Co. transf. to Wise Brigade 7/10/61.

LIVESAY, MARION F.: Pvt. Taylor's Co. Enl. 6/19/61 at Frankford, Greenbrier Co., WV. Age: 20. Co. transf. to Wise Brigade 7/10/61.

LIVESAY, ROBERT F.: Pvt. Taylor's Co. Enl. 6/19/61 at Frankford Greenbrier Co., WV. Age: 21. Co. transf. to Wise Brigade 7/10/61.

LIVISEY, VALENTINE: 2nd Sgt. Co. D. Date and place of enlistment not stated. WIA (slight arm) at White Sulphur Springs, WV 8/26-27/63. (This is probably the same man as Valentine Levisay listed above in this roll).

LOLLIS, WILLIAM E.: Pvt. Co. E. Res. Botetourt Co., Va. Enl. 5/?/62 at White Sulphur Springs, Greenbrier Co., WV. Conscript. WIA and capt. at Lewisburg, WV 5/23/62. POW Camp Chase, OH 6/30/62. Exchd. 8/25/62. Sick in Miss. hosp. on 12/31/62 and 2/28/63 rolls. Present on 4/30/63 roll. Capt. at Droop Mountain, WV 11/6/63. Age: 34. POW Camp Chase, OH 11/18/63. Relsd. by order of the President 3/10/65. Took oath 3/22/65. 5'10". Florid complexion, dark hair, gray eyes.

LONEY, CARRY: Co. I. Attended a vet. reunion at Beckley, WV 9/12/1918 and gave res. as Boone Co., WV.

LONG, O. B.: Pvt. Co. B. Enl. 7/1/62 at Union, Monroe Co., WV. Deserted at Camp McCausland, WV on 7/5/62 roll. Listed as a deserter on 11/10/62 roll.

LOONEY, ALEXANDER C.: Pvt. Co. K. B. ca. 1831 Craig Co., Va. Enl. 4/11 or 24/62 at White Sulphur Springs, Greenbrier Co., WV. Deserted 5/26/62 on Aug. 1862 broadside, Sinking Creek, Craig Co., Va. Deserted at Blue Ridge Tunnel 6/23/64. Res. Craig Co., Va. 1898 at age 55. UDC report says WIA (shoulder) at Gettysburg, PA. Therefore must have served with another unit. Bur. Wilson Cem., Paint Back, Va. Bahlmann says WIA at Lewisburg, WV 5/23/62.

LOONEY, DAVID: Rank unknown. Co. C. B. Craig Co., Va. Enl. 8/1/63 at Camp Price. D. 2/20/65 of disease in Monroe Co., WV. Claim filed by Mary Jane Looney, mother. Not in CSR.

LOONEY, JAMES: Co. C. Listed in UDC booklet *Craig's Share in the War Between the States.* Res. Craig Co., Va.

LOONEY, JOSEPH: Pvt. Co. K. Res. Craig Co., Va. Enl. 8/1/63 in Craig Co., Va. Deserted 12/25/63 but had returned on 12/30/63 roll. Deserted in Wilson Co., Va. (Nelson Co. ?) 6/25/64.

LOONEY, JOSEPH B.: Pvt. Co. K. Enl. 2/1/63 at either Lewisburg, Greenbrier Co., Wv or in Monroe Co., WV. B. ca. 1846 Craig Co., Va. Deserted 12/25/63 but returned on 12/30/63 roll. Deserted near Jackson's River Depot, Va. 5/9/64. Res. Craig Co., Va. 1898 at age 53.

LOONEY, WILLIAM: Pvt. Co. C. Res. Craig Co., Va. Enl. 5/1/62(?) at White Sulphur Springs, Greenbrier Co., WV. In Winchester, Va. hosp. for leg amputation 7/25/64. Absent wded. on 10/31/64 roll.

LOONY, DAVID: Pvt. Co. C. B. in Craig Co., Va. Enl. 8/1/63(?) at Camp Price. D. of disease in hosp. at Union, WV 2/20/64. Claim filed by Mary Jane Looney. (Undoubtedly same as David Loony; also spelled Looney).

LOUDERMILK, JAMES H.: Pvt. Taylor's Co. Enl. 6/19/61 at Frankford, Greenbrier Co., WV. Age: 19. Co. transf. to Wise Brigade 7/10/61.

LOUDERMILK, JOSEPH: Pvt. Co. F. Enl. 5/1/62(?) at White Sulphur Springs, Greenbrier Co., WV. Capt. at Lewisburg, WV 5/23/62. POW Camp Chase, OH. Exchd. 8/25/62. Present on 2/28/63 roll. AWOL on 10/31/64 roll. Paroled at Charleston, WV 5/11/65. Age: 25. 5'11". Light complexion, light hair, light eyes, light whiskers.

LOVEJOY, MILTON A.: Pvt. Co. A. Enl. 9/15/62 at Charleston, Kanawha Co., WV. Deserted in the Valley of Virginia on 10/31/64 roll.

LOVEJOY, NATHANIEL G.: Pvt. Co. A. Enl. 5/22/61 at Poca, Putnam Co., WV. Dischd. 2/20/64.

LOVEJOY, S. G.: Co. A. Attended vet. reunion in Charleston, WV in 1924 and gave address as Winter, WV.

LOVEJOY, THORNTON F.: Pvt. Co. A. Enl. 10/15/62 at Raleigh (Beckley) Raleigh Co., WV. Left sick on Cheat Mountain, WV on 4/30/63 roll. Absent wded. on 12/31/63 roll. Present on 4/30/64 roll. In Charlottesville, Va. hosp. for vulv. sclo. 6/25-27/64. POW. Paroled at Staunton, Va. 5/1/65. Age: 39. 5'9". Dark complexion, black hair, black eyes.

LOVING, THEODOSIUS V.: Pvt. Co. D/Co. H. Enl. 5/30/61 at Summersville, Nicholas Co., WV. Transf. from Co. D to Co. H 5/25/62. Capt. in Nicholas Co., WV 5/13/63. WIA at Droop Mountain, WV 11/6/63. Present on 12/3/63 roll. POW. Paroled at Charleston, WV 5/15/65. Age: 23. 5'9". Fair complexion, dark hair, blue eyes, dark whiskers.

LOVING, V. B.: Pvt. Co. H. Enl. 10/8/62 in Kanawha Co., WV. Capt. in Nicholas Co., WV 5/13/63. WIA at Droop Mountain, WV 11/6/63. Present at 12/3/63 roll. Capt. at Cold Harbor (Gaines Mill), Va. 6/3/64. POW Pt. Lookout, MD. Transf. to Elmira, NY 7/17/64. Apparently died before exchange at Pt. Lookout, MD 7/20/64.

LOYWELL, MELTON C.: Pvt. Co. A. Enl. 5/22/61 at Poca, Putnam Co., WV. Unaccounted for as of 7/24/61.

LUCAS, CALVIN: Rank unknown. Co. C. Date and place of enl. not stated. Present on 3/4/63 roll. This is probably John Calvin Lucas.

LUCAS, EDWARD: Pvt. Co. C. B. ca. 1821 Craig Co., Va. Enl. 10/1/63 at Camp Beauregard. AWOL on 12/31/63 roll. Present on 4/8/64 roll. In Chimborazo Hosp. for acute rheumatism 6/1-11/64. D. at Way Hosp. at Lynchburg, Va. of cholera morbus 6/22/64. Effects turned over to Quarter Master.

LUCAS, JOHN CALVIN: Pvt. Co. C. B. 9/18/46 Craig Co., Va. Enl. 2/13 or 16/63 at Lewisburg, Greenbrier Co., WV. Substitute for John E. Lugar. In hosp. at Charlottesville, Va. for very serious left leg 9/26-28/64. Sent to Lynchburg, Va. At home wded. on 10/31/64 roll. PWR says enl. in Winter 1862 at age 15. WIA (bullet struck head) at Winchester, Va. 9/19/64. WIA (bullet in thigh) near New Market, Va. 9/23/64. Never able to return to command. Bullet cut out of thigh 9 years and 18 days later. Married 2/20/69. Res. Craig Co., Va. 1898 at age 52. D. 1/23/1911 in Craig Co., Va. Bur. Huffman Cem., Rt. 632, Herman Huffman farm, Craig Co., Va.

LUCAS, WILLIAM HENDERSON: Pvt. Co. C. Res. Craig Co., Va. Enl. 1/1/63 at Cam Sam Miller. In Chimborazo Hosp. for very serious left cheek 6/5-11/64. KIA at Cedar Creek, Va. 8/12/64. A PWR says died of gunshot wd. of left thigh at Fishers Hill, Va. 6(?)/12/64 at age 34.

LUGAR, GEORGE W.: Listed in UDC booklet *Craig's Share in the War Between the States.* Res. Craig Co., Va.

LUGAR, JAMES W.: B. 1828. Also served in 46th Va. Inf. D. 6/23/97. Bur. Ross Cem., Rt. 42, near Maywood, on Fred Keffer farm, Craig Co., Va. PWR.

LUGAR, JOHN A. ("Dusty"): Pvt. Co. K. B. 10/5/32 Craig Co., Va. Enl. 7/5/62 at Union/Camp McCausland, Monroe Co., WV. Det. service 7/8/62. Det. to scout with Capt. Thurmond July 1862. Det. as tanner on Nov.-Dec. 1862 roll. Unaccounted for as of 11/1/62. Listed as deserter. D. 7/29/1915. Bur. Carmel Cem. Gap Mills, WV.

LUGAR, JOHN E.: Pvt. Co. C. Res. Craig Co., Va. Enl. 5/1/62 at White Sulphur Springs, Greenbrier Co., WV. WIA at Lewisburg, WV 5/23/62. Sick at Montgomery White Sulphur Springs, Va. July 1862.

LUSK, A. J.: Corp./Sgt. Co. G. Enl. 7/3/61 at Charleston, Kanawha Co., WV. Had a furlo. and time had expired on 11/13/62 roll. NFR.

LUSK, A. P.: Pvt. Co. G. Enl. 7/7/61 at Charleston, Kanawha Co., WV. Transf. from Capt. Sarver's Co. 7/25/61. NFR.

LYNCH, JOSEPH: Sgt. Co. E. Lumberman, Clendenin, WV 1860. Age: 38. Enl. 6/8/61 at Charleston, Kanawha Co., WV. Served as a 4th Sgt. until elected 2nd Sgt. 7/3/61. NFR.

LYMAN, JOHN: Pvt. Swann's Co. Enl. 5/8/61 at Charleston, Kanawha Co., WV. Age: 20. Co. transf. to Co. K, 59th Va. Inf. 7/25/61.

LYNN, THOMAS ESTILL: Pvt. Co. H. B. 3/26/42. D. 11/13/1930 at Richmond, Va. Confed. Home. Bur. Spring Hill Cem., Charleston, WV.

LYNN, WILLIAM HARRISON ("TIP"): Pvt. Swann's Co./Co. H. B. 7/16/40 Kanawha Co., WV. Enl. 5/8/61 at Charleston, Kanawha Co., WV. Age: 21. Co. transf. to Co. K, 59th Va. Inf. 7/25/61. D. 6/2/1916 in Charleston, WV. Bur. Spring Hill Cem., Charleston, WV. Probably same man as W. Linn.

LYONS, JOHN A.: Pvt. Co. E. Res. Calhoun Co., WV. Enl. 6/18/61 at Charleston, Kanawha Co., WV. Capt. in Calhoun Co., WV 1/20/64. Age: 23. POW Camp Chase, OH 2/15/64. Transf. to Ft. Delaware 3/14/64. Exchd. 2/27/65. In Chimborazo Hosp. for abcess 3/4-10/65. Paroled at Charleston, WV 5/12/65. (Gave age as 22 on oath). Farmer. 6'. Light complexion, brown hair, black or gray eyes. Pension in Benton Co., Ark. D. 3/15/1903.

LYPES, ALLEN LEFFEL: Pvt. Co. C. B. ca. 1847 Craig Co., Va. Enl. 2/29/64 at Camp Gauley, Greenbrier Co., WV. KIA at Lynchburg, Va. 6/18/64. Age: 19.

LYPES, SAMUEL: Pvt. Co. I. Res. Craig Co., Va. Enl. 5/1/62 at White Sulphur Springs, Greenbrier Co., WV. Det. by Gen. Heth on 12/30/62 roll. Carpenter at Narrows, Va. on 5/1/63 roll. Det. wagon maker 8/31/63 to 12/31/63. Det. as a wagon maker in Quarter Master Dept. at Narrows, Va. 3/1/64 to 6/30/64. Det. service on 10/31/64 roll. Gave his age as 37 in Jan. 1865. Wheelwright/carpenter.

MACOMB, JAMES G.: Pvt. Co. D. Enl. 5/30/61 at Summersville, Nicholas Co., WV. Present on 8/31/61 roll. (This is probably spelled Malcolm). NFR.

MADISON, L. D.: Pvt. Co. C. Enl. 5/10/61.

MAHAN, WILLIAM L.: Pvt. Co. H/Co. G. Res. Charleston, WV 1860. Age: 19. Enl. 5/8/61 at Charleston, Kanawha Co., WV. Det. for Q.M. Dept. 6/8/61 and still so engaged on 8/31/61 roll. Det. as a teamster in A.Q.M. Dept. at Lewisburg, WV on 5/24/62 roll. Surrendered and paroled at Appomattox, Va. 4/9/65.

MAHONEY, JOHN C.: Pvt. Co. B. Enl. 5 or 6/?/62 at White Sulphur Springs, Greenbrier Co., WV. Capt. at Lewisburg, WV 5/23/62. POW Camp Chase, OH. Exchd. 8/25/62. Det. service 12/31/62. Det. service as blacksmith on 2/28/63 and 5/1/64 rolls.

MALONE, WILLIAM D.: Pvt. Co. H. Clerk, Charleston, WV 1860. Age: 18. Enl. 5/8/61 at Charleston, Kanawha Co., WV. Transf. to Co. G, 8th Va. Cav. 7/11/63 (or 6/11/63). Det. service Signal Corps 10/31/64. B. in Kanawha Co., WV 9/18/41. PWR says enl. at age 18. Acting aide to Breckenridge at close of war. Noted for bravery but refused promotion from the ranks. Married 1870. Res. Ashland, KY early 1870's then res. Grayson, KY. Res. and businessman, Catlettsburg, KY ca. 1916. D. 1/29/1928. Bur. Grayson, KY.

MANKEY, J. A.: Co. I. B. 4/1/36. Enl. at age 17. Fought at New Market, Va. Postwar res. Eagle Rock, Va. Res. Rockbridge Co., Va. until 1935, then res. Buchanan, Va., which would have made him the oldest living vet. in Rockbridge Co., Va. D. 6/12/1937 at Clifton Forge, Va.

MANKEY, WILLIAM A.: Pvt. Co. I. Res. Botetourt Co., Va. Enl. 4/1/64 at Camp Gauley, Greenbrier Co., WV. Deserted in Botetourt Co., Va. 6/20/64. Capt. 4/8/65. Took oath at Clarksburg, WV 4/10/65.

MANN, ARCHIBALD H. ("ARCH"): Pvt. Co. G. B. 6/19/34. Res. Alleghany Co., Va. Enl. 5/1/62 at White Sulphur Springs, Greenbrier Co., WV. Sick in hosp. on 11/13/62 roll. Sent home from Red Sulphur Springs, WV on sick furlo. on 12/31/62 roll. Present on 2/28/63 roll. Capt. in Alleghany Co., Va. 11/10/63. Age: 29. POW Camp Chase, OH 11/19/63. Transf. to Ft. Delaware 3/14/64. Relsd. 6/7/65. Took oath at Ft. Delaware. Farmer. 6'. Florid complexion, dark hair, gray eyes. D. 2/5/1900. Bur. Falling Spring Presbyterian Church Cem., Falling Spring, Va.

MANN, H. C.: Co. G. Enl. 2/?/62 for 4 months. WIA at Cold Harbor (Gaines Mill), Va. 6/3/64. Postwar roster.

MANN, JAMES W.: Pvt. Co. G. Enl. 5/1/62 at White Sulphur Springs, Greenbrier Co., WV. Sick in hosp. 11/13/62. Present 12/31/62. In Danville, Va. hosp. for vulvius sclopeticum hand 6/16-21/64. Rcvd. clothing at Columbia, SC hosp. 8/1/64. Capt. at Winchester, Va. 9/19/64. POW Pt. Lookout, MD 9/23/64. Exchd. 3/15/65. Res. Alleghany Co., Va. Age: 21 at enl.

MANN, JOHN L.: Pvt. Co. G. B. 2/6/39. Enl. 7/10/62 at Union/Camp McCausland, Monroe Co., WV. Sick in hosp. on 11/13/62 roll. Present on 12/31/62 roll. Nurse in hosp. but since returned on 10/31/64. D. 12/24/1918. Bur. Falling Spring Presbyterian Church Cem., Falling Spring, Va. Res. Alleghany Co., Va.

MANN, ROBERT S.: Pvt. Co. H. B. Alleghany Co., Va. Res. Alleghany Co., Va. Enl. 6/25/62 at Salt Sulphur Springs, Monroe Co., WV. On sick furlo. on rolls for July-Aug. 1862 to 2/28/63. Age listed as 30 on pay voucher dated 3/12/63. Dischd. for paralyzed bladder 3/12/65. Farmer. 6'2¾". Dark complexion, dark hair, gray eyes.

MANSPILE, RICHARD G.: Pvt. Co. A. B. ca. 1813 Craig Co., Va. Enl. 3/18/63 at Lewisburg, Greenbrier Co., WV. Left sick at Camp Bartow, WV on 4/30/63 roll. Det. hosp. nurse on 12/31/63 roll. Deserted 3/1/64. Also spelled Mankspile.

MARKHAM, BENJAMIN D.: Pvt. Co. A. Farm manager, Coalsmouth, WV 1860. Age: 27. Enl. 5/22/61 at Poca, Putnam Co., WV. Orderly for Lt. Col. Barbee July 1862. Orderly for Barbee on 11/1/62 roll. Dischd. 4/10/63.

MARKHAM, BEVERLY O.: Pvt. Co. A. Enl. 5/22/61 at Poca, Putnam Co., WV. WIA (musketball) at Scary Creek, WV 7/17/61. Capt. at Winchester, Va. 9/19/64. POW Pt. Lookout, MD 9/24/64. Exchd. 3/15/65. Transf. from Farmville, Va. hosp. 3/22/65. Paroled at Charleston, WV 5/10/65. Age: 30. 5'7". Dark complexion, black hair, black eyes, black whiskers.

MARKHAM, SAMUEL P.: Corp. Co. A. Enl. 5/22/61 at Poca, Putnam Co., WV. Unaccounted for as of 7/24/61.

MARRS, F. MARION: Pvt. Co. C. Res. Fayette Co., WV. Enl. 5/1/62(?) at White Sulphur Springs, Greenbrier Co., WV. Det. to work at Gauley Bridge, WV on Sept.-Oct. 1862 roll. Det. at Narrows, Va. on 12/31/62 roll. Det. as sappor(?) and miner 3/4/63 and 6/11/63. Day laborer, Fayette Co., WV 1860. Age: 25.

MARTIN, ALLEN M.: Pvt. Co. D. Res. Summerville, WV. Enl. 10/8/62 at Charleston, Kanawha Co., WV. Capt. at Cold Harbor (Gaines Mill), Va. 6/3/64. POW Pt. Lookout, MD. Transf. to Elmira, NY 7/12/64. Relsd. 6/21/65. 5'11". Florid complexion, auburn hair, blue eyes.

MARTIN, CHARLES: Pvt. Swann's Co. Enl. 5/22/61 at Charleston, Kanawha Co., WV. Age: 25. Co. transf. to Co. K, 59th Va. Inf. 7/25/61.

MARTIN, E.: Rank unknown. Co. I. Date and place of enlistment not stated. Rcvd. clothing 5/13/64(?). (This name is filed under Edward R. Martin in the CSR).

MARTIN, EDWARD R.: Pvt./Sgt. Co. D. Enl. 5/30/61 at Summersville, Nicholas Co., WV. D. in hosp. at Union, WV 2/8/63. Effects turned over to the Quarter Master.

MARTIN, JAMES: Pvt. Swann's Co. Enl. 5/25/61 at Charleston, Kanawha Co., WV. Age: 18. Co. transf. to Co. K, 59th Va. Inf. 7/25/61.

MARTIN, JOHN: Pvt. Co. K. Enl. 9/18/64 at Winchester, Va. Capt. at Fishers Hill, Va. 9/21/64. (Other dates given for capture include 9/22/64 and 9/23/64). POW Pt. Lookout, MD 10/1/64. Exchd. 11/1/64. In Richmond, Va. hosp. 2/26/65 to 3/6 or 8/65.

MARTIN, JOHN: Pvt. Co. A. Date and place of enlistment not stated. Capt. in Boone Co., WV either 2/9/64 or 8/16/64. POW Ft. Delaware. D. at Ft. Delaware of smallpox 5/8/64. Bur. Finn's Point Nat. Cem., NJ

MARTIN, JOSEPH: Pvt. Co. K. Enl. 9/1/62 at Red Sulphur Springs, Monroe Co., WV. Under arrest and dropped as a deserter but on trial by court-martial — acquitted. WIA and capt. at Cedar Creek, Va. 10/19/64. Age: 25. Serious shell wd. of right side of face passing through the right molar and right (illegible) maxillary bone. Postwar account: Shell exploded and piece struck him on the left side of the face. In Baltimore, MD hosp. POW Ft. McHenry about 1/5/65. Sent to Pt. Lookout, MD for exchange 2/20/65. Postwar res. Green Sulphur Dist., Summers Co., WV. Farmer. Still alive there in 1908.

MARTIN, THOMAS: Pvt. Co. B. Enl. 8/?/62 at Union, Monroe Co., WV. WIA at Lewisburg, WV 5/23/62. WIA at Fayetteville, WV 9/10/62. Det. as courier on 4/30/63 roll. Teamster 1/1-30/64.

MARTIN, WILLIAM: Pvt. Co. A. Res. Kanawha Co., WV. Enl. 5/22/61 at Poca, Putnam Co., WV. Unaccounted for 7/24/61. Present on 11/1/62 roll. Deserted in face of the enemy 12/13/63. (Rcvd clothing 5/13/64(?). Took oath and sent North 12/31/63. Age: 23. Farmer. 5'11". Fair complexion, dark hair, hazel eyes.

MARTIN, WILLIAM G.: Pvt. Co. E. Enl. 6/8/61 at Charleston, Kanawha Co., WV or 5/4/63 at Weston, WV. Paroled at Charleston, WV 5/11/65. Age: 21. 5'10". Light complexion, brown hair, blue eyes.

MARTIN, WILLIAM H.: Pvt. Co. D. Res. Kanawha Co., WV. Enl. 4/2/63 at Lewisburg, Greenbrier Co., WV. Deserted Nov. 1863. POW. Took oath at Ft. Delaware and sent North 11/15/63. Age: 17. Relsd. 6/20/65. Farmer. 5'10". Fair complexion, light hair, blue eyes.

MARTIN, WILLIAM J.: Pvt. Co. A. Enl. 6/24/61 at Camp Tompkins, Kanawha Co., WV. Unaccounted for 7/24/61. Present on 11/1/62 roll. Capt. in Boone Co., WV 11/14/63. POW Ft. Delaware. D. at Ft. Delaware of smallpox 7/26/64. Bur. Jersey shore of Ft. Delaware.

MARTIN, WILLIAM TUCKER: Pvt. Co. E. Enl. 7/17 or 22/62 at Camp McCausland, Monroe Co., WV. Det. brig. wagoner on 4/1/64 roll. Paroled at Charleston, WV 6/12/65. Age: 33. 6'2". Dark complexion, black eyes, dark whiskers.

MASON, D. S.: Pvt. Co. H. Date and place of enlistment not stated. Appears on an undated list of POW's paroled at Columbia, SC.

MASON, THOMAS L.: Sgt./1st Lt. Co. A. Enl. 5/22/61 at Poca, Putnam Co., WV. Elected 1st Sgt. 7/18/61. Elected 1st Lt. 5/1/62. KIA at Lewisburg, WV 5/23/62.

MASON, WILLIAM M.: Pvt. Co. A. Enl. 5/22/61 at Poca, Putnam Co., WV. Det. service 2/28/63. Left sick at Lewisburg, WV on 4/30/63 roll. Present on 12/31/63 roll. In Chimborazo Hosp. for very serious right shoulder 6/4-28/64. Killed at Leetown on 10/31/64 roll. Bur. Edgehill Cem., Charlestown, WV. Also spelled Maison.

MASSIE, H. W.: Co. G. Postwar roster. Pvt. Enl. 10/?/64 at New Market, Va. Age: 21. Res. Alleghany Co., Va. Surrendered Apr. 1865.

MASTERSON, THOMAS V.: 2nd Lt. Co. C. Enl. 6/6/61. Place not stated. Elected 2nd Lt. 6/6/61. Not re-elected. Farmer, Fayette Co., WV 1860. Age: 32.

MASTON JOHN A.: Pvt. Co. B. Res. Greenbrier Co., WV. Enl. 7/12/62 at Camp McCausland, Monroe Co., WV or 7/16/62 at Union, Monroe Co., WV. Deserted. Took oath at Charleston, WV 3/28/65. Age: 35. Farmer. 5'4" or 5'6". Fair complexion, hazel eyes, gray hair.

MATE, ROBERT: Pvt. Co. B. Res. Ripley, WV, 1860, age 38, blacksmith.

MATHENY, WILLIAM: Pvt. Swann's Co./Co. H. Enl. 5/25/61 at Charleston, Kanawha Co., WV. Age: 19. Transf. from Co. K, 59th Va. Inf. to Co. H, 22nd Va. Inf. 3/19/62. Deserted near Staunton, Va. May 1864.

MATTHEWS, JOHN W.: Pvt./Sgt. Co. H. Enl. 5/8/61 at Charleston, Kanawha Co., WV. Either capt. at Mechanicsville, Va. 5/20/64 or capt. at Cold Harbor, Va. 6/3/64. POW Pt. Lookout, MD. Transf. to Elmira, NY 7/9/64. Serious cut of temple in train wreck at Shohola, PA 7/15/64 en route to Elmira, NY. Exchd. 10/29/64. Paroled at Charleston, WV 5/26/65. Age: 24. 5'7". Dark complexion, black hair, black eyes, black whiskers.

MAYES, A.: Pvt. Co. H. Enl. 10/8/62 in Kanawha Co., WV. D. at Lewisburg, WV 4/6/63. (Also spelled Mays).

MAYES, BENJAMIN F.: Pvt. Co. H. Res. Kanawha Co., WV. B. near Charleston, WV 10/28/38. Farmer, mercantile business. Enl. 10/1/62 in Kanawha Co., WV. Capt. at Fishers Hill (Strasburg), Va. 9/22/64. PWR says enl. 1863. POW Pt. Lookout, MD 10/1/64. Took oath and relsd. at Ft. Delaware 6/29/65. 5'11". Dark complexion, dark brown hair, blue eyes. "Fought in 19 engagements." Used much as a sharpshooter. Postwar mercantile business, Charleston, WV. Member Camp R. E. Lee, UCV. D. 10/?/1916. Bur. Spring Hill Cem., Charleston, WV.

MAYES, F. M.: Pvt. Co. H. Enl. 10/1/62 in Kanawha Co., WV. Present on 10/31/62 roll. NFR.

MAYNOR, COLUMBUS: Pvt. Co. K. Enl. 6/6/61 at Charleston, Kanawha Co., WV. Absent cause unknown on 9/4/61. NFR. Bahlmann says "left."

MAYNOR, WILLIAM C.: Pvt. Co. K. Enl. 6/6/61 at Charleston, Kanawha Co., WV. Absent cause unknown on 9/4/61 roll. NFR. Bahlmann says "left."

MAYS, JAMES HENRY: Pvt./Sgt. Co. F. B. 2/15/42 in Amherst Co., Va. Moved to Greenbrier Co., WV. ca. 1849-50 and later moved to Monroe (now Summers) Co., WV near Pence Springs. Attended "subscription schools". First married Cleopatra Hatcher, who died, and then her cousin, Mary Elizabeth Hatcher. Enl. 7/4/61 at Charleston, Kanawha Co., WV. WIA at Cold Harbor, Va. 6/3/64. In Chimborazo Hosp. for very serious hand fracture caused by minie ball. Stayed 18 days. Transf. to Lynchburg, Va. hosp. At home wded. in Monroe Co., WV on 10/31/64 roll. Spent remainder of war with co. doing det. service in Monroe Co., WV. After war moved to Charleston, WV and soon thereafter to Kentuck (Fletcher), Jackson Co., WV. Was a Baptist. Attended a vet. reunion at Charleston, WV 1924 and gave res. as Fletcher, WV. D. in Kentuck, WV 12/4/1928 of a complication of diseases. Bur. Cleek Cem., Kentuck, WV. Wrote his war memoirs in 1922 which were not published until 1970 by his son and titled *Four Years For Old Virginia.*

MAYS, JESSE D.: Pvt. Co. D. Enl. 5/1 or 30/61 at Summersville, Nicholas Co., WV. MWIA at Lewisburg, WV 5/23/62. D. 6/8/62.

MAYS, J. J.: Co. H. Bur. Presb. Church Cem., New Market, Va. PWR.

MAYS, JOHN: Pvt./Sgt. Co. I. Res. Kanawha Co., WV. Enl. 8/22/61 at Meadow Bluff, Greenbrier Co., WV. Capt. at Droop Mountain, WV 11/6/63. Age: 37. POW Camp Chase, OH 11/18/63. Transf. to Ft. Delaware 2/29/64. Relsd. and took oath at Ft. Delaware 6/30/65. Farmer. 6'. Florid complexion, dark hair, brown eyes.

MAYS, JONATHAN: 1st Lt. Taylor's Co. Enl. 6/19/61 at Frankford, Greenbrier Co., WV. Age: 30. Co. transf. to Wise Brigade 7/10/61.

MAYS, JOSEPH M.: Pvt. Taylor's Co. Enl. 6/19/61 at Frankford, Greenbrier Co., WV. Age: 20. Co. transf. to Wise Brigade 7/10/61. Postwar roster says also served in Co. F and postwar res. Kanawha Co., WV.

MAYS, PIERCE W.: Pvt. Co. A. B. in Amherst Co., Va. Enl. 5/22 or 24/61 at Poca, Putnam Co., WV. Sick at White Sulphur Springs, WV 8/20/61. In hosp. at Charlottesville, Va. for paratitis 1/2-31/62. Dischd. for constititional disability at Packs Ferry, WV 11/3/62. Age: 22. Farmer. 6'. Fair complexion, light hair, blue eyes.

McCALEB, BRUCE: Co. G. Postwar roster. Pvt. Enl. 10/?/64 at New Market, Va. Age: 22. Res. Alleghany Co., Va. Surrendered Apr. 1865.

McCALEB, JOSEPH THOMAS: Co. F. B. 1/14/27 Craig Co., Va. D. 1/19/1919 Alleghany Co., Va. PWR.

McCALLESTER, A. C.: Pvt. Co. G. Enl. 7/10/62 at Camp McCausland, Monroe Co., WV. NFR.

McCAY, WILLIAM H.: Pvt. Co. D. Enl. 5/30/61 at Summersville, Nicholas Co., WV. AWOL 7/28/61.

McCLANAHAN, JOHN W.: Pvt. Co. A. Laborer, Sissonville, WV 1860. Age: 24. Res. Kanawha Co., WV. Enl. 5/22/61 at Poca, Putnam Co., WV. On special duty and cut off by Federal army 7/24/61. Present on 11/1/62 roll. Deserted and surrendered at Charleston, WV March 1865. Took oath at Charleston, WV 3/20/65. Age: 30. Farmer. 5'8". Fair complexion, fair hair, blue eyes.

McCLANAHAN, R. W.: Pvt. Co. C. B. ca. 1834 Craig Co., Va. Enl. 5/1/62(?) at White Sulphur Springs, Greenbrier Co., WV. Det. as hosp. cook on rolls from Sept.-Oct. 1862 to 4/18/64. Capt. at Winchester, Va. 9/19/64. POW Pt. Lookout, MD 9/24/64. Exchd. 3/15/65. Paroled at Charleston, WV 4/24/65.

McCLANAHAN, WILLIAM S.: 3rd Lt. Co. A. Laborer, Sissonville, WV. 1860. Age: 26. Enl. 5/22/61 at Poca, Putnam Co., WV. Sick at White Sulphur Springs, WV 8/20/61. Elected 3rd Lt. 5/22/61. Promo. to 2nd Lt. 5/23/62. MWIA at Droop Mountain, WV 11/6/63. D. 11/6 or 7/63.

McCLELLAND, ROBERT: Pvt. Co. H. Enl. 5/8/61 at Charleston, Kanawha Co., WV. Present at Camp Sam Miller in Jan. 1863. Then listed as POW. Paroled at Charleston, WV 4/24/65. Age: 29. 5'11". Dark complexion, brown hair, dark eyes, dark whiskers.

McCLINTIC, JAMES S.: Pvt. Taylor's Co. Enl. 6/19/61 at Frankford, Greenbrier Co., WV. Age: 21. Co. transf. to Wise Brigade 7/10/61.

McCLINTIC, MOSES: 2nd Lt. Taylor's Co. Enl. 6/19/61 at Frankford, Greenbrier Co., WV. Age: 26. Co. transf. to Wise Brigade 7/10/61.

McCLOUD, AUGUSTUS: Pvt. Swann's Co. Enl. 5/22/61 at Charleston, Kanawha Co., WV. Age not stated. Co. transf. to Co. K, 59th Va. Inf. 7/25/61.

McCLUNG, ADONIRAM JUDSON: Pvt. Co. C. B. Mt. Lookout, Nicholas Co., WV 3/12/42. Enl. 4/7/63(?) at Lewisburg, Greenbrier Co., WV. Re-enlisted in the co. 4/7/63. Transf. to Co. F, 36th Va. Inf. 4/7/63. Paroled 3/18/65. Age: 22. 5'11¾". Gray eyes and fair hair. Postwar res. Mt. Lookout, WV. D. 3/24/1934 Gatewood, WV. Bur. Mt. Pleasant Cem., Mt. Lookout, WV. Also called Adam and Adorian.

McCLUNG, F. M.: Pvt. Co. C. Enl. 5/10/61. Farm laborer, Fayette Co., WV 1860. Age: 23.

McCLUNG, JOSEPH: Pvt. Taylor's Co. Enl. 6/19/61 at Frankford, Greenbrier Co., WV. Age: 24. Co. transf. to Wise Brigade 7/10/61.

McCLUNG, MICHAEL B.: 2nd Lt. Co. D. Enl. 5/30/61 at Summersville, Nicholas Co., WV. Elected 2nd Lt. 6/6/61. Resgnd. 9/15/61.

McCLUNG, MILICE: Pvt. Co. D. Enl. 5/30/61 at Summersville, Nicholas Co., WV. AWOL 11/5/62. Reported as captured.

McCLUNG, W. A.: Co. C. D. 8/23/64 at Ft. Delaware, Del. Bur. Finn's Point Nat. Cem., NJ.

McCLUNG, WILLIAM: Pvt. Taylor's Co. Enl. 6/19/61 at Frankford, Greenbrier Co., WV. Age: 25. Co. transf. to Wise Brigade 7/10/61.

McCONIHAY, JAMES D.: Pvt. Swann's Co./Co. H. Res. Brownstown (Marmet), Kanawha Co., WV. Enl. 6/8/61 at Charleston, WV in Co. E. On mounted guard at White Sulphur Springs, WV 8/13-14/61. Transf. to Co. H sometime prior to July-Aug. 1862 roll. Capt. at White Sulphur Springs, WV 8/26/63. Age: 34. (Listed as sick in Greenbrier Co., WV on 6/30/63 roll). POW Camp Chase, OH 9/7/63. Transf. to Rock Island, Ill. 1/22/64. Took oath at Rock Island, ILL 6/10/65. Merchant. 5'9½". Dark complexion, dark hair, hazel eyes.

McCORKLE, WILLIAM G.: Corp. Co. F. Enl. 7/4/61 at Charleston, Kanawha Co., WV. Served as a corp. until 7/20/61. Sick at home on surg. cert. 8/20/61. Postwar res. Talcott Dist., Summers Co., WV. Believed still alive 1914. Postwar roster says middle initial is H.

McCORMAC, SAMUEL J.: Pvt. Co. A. Res. Putnam Co., WV. Enl. 10/1/62 at Charleston, Kanawha Co., WV. Wded. on 12/31/63 roll. Present on 4/30/64 roll. In Farmville, Va. hosp. for rubeola 6/3/64 to 9/7/64. Capt. at Winchester, Va. 9/19/64. POW Pt. Lookout, MD 9/24/64. Exchd. 3/15/65. Listed as deserter. Took oath at Charleston, WV 5/2/65. Age: 25. Farmer. 5'8". Dark complexion, dark hair, blue eyes. Also spelled McCormick and McCormmac.

McCORMICK, DAVID: Pvt. Swann's Co. Enl. 5/8/61 at Charleston, Kanawha Co., WV. Age: 27. Co. transf. to Co. K, 59th Va. Inf. 7/25/61.

McCORMICK, J. W.: 2nd Lt./1st Lt. Co. F. Enl. 5/1/62(?) at White Sulphur Springs, Greenbrier Co., WV. Elected 2nd Lt. 5/1 or 2/62. At home sick at Potts Creek, Va. on 10/31/62 roll. Promo. to 1st Lt. 2/4/63. Still present on 2/28/63 roll.

McCOY, DAVID: Corp./2nd Lt. Co. B. B. 5/17/34 Tyler Co., WV. Res. Jackson Co., WV 1860 at age 26. Enl. 6/1/61 at Ripley, Jackson Co., WV. Elected 2nd Lt. either 7/26/63 or 8/20/63. Capt. in Jackson Co., WV while on furlo. 1/22/64. Age: 29. POW Camp Chase, OH 2/6/64. Transf. to Ft. Delaware 3/25/64. Took oath at Ft. Delaware and relsd. 6/16/65. Farmer. 6'. Fair/ruddy complexion, sandy/auburn hair, blue eyes. Farmer, Putnam Co. ,WV 1866. Sheriff, Putnam Co., WV 1888 and 1891.

McCOY, FRAZIER: Pvt. Co. C. Enl. 5/10/61.

McCRASKY, W.: Pvt. Co. E. Enl. 6/8/61 at Charleston, Kanawha Co., WV. AWOL 7/1/61.

McCUNE, GILBERT: Pvt. Co. B. Res. Calhoun Co., WV. Enl. 4/1/64 at Lewisburg, Greenbrier Co., WV. Listed as a deserter in Mar. 1865. Surrendered at Charleston, WV Mar. 1865. Took oath at Charleston, WV 3/20/65. Age: 23. Farmer. 5'7". Fair complexion, dark hair, gray eyes.

McCUTCHEN, ROBERT W.: Pvt. Co. D. Enl. 5/30/61 at Summersville, Nicholas Co., WV. AWOL 11/5/62. Reported as captured. Went home on a furlo. and did not return. Recaptured by 22nd Va. scouts near Cross Lanes, WV 10/30/62. Court-martialed for desertion. Sentenced to be shot.

McCUTCHEN, THOMAS: Pvt. Co. D. Enl. 6/8/63 at Camp Jackson. Sick on 12/31/63 roll. NFR.

McCUTCHEN, WILLIAM T.: Pvt. Co. D. Enl. 5/30/61 at Summersville, Nicholas Co., WV. On mounted guard at White Sulphur Springs, WV 8/13-14/61. Deserted on 4/30/64 roll.

McCUTCHEON, GEORGE: Pvt. Co. C. Enl. 5/1/62 at White Sulphur Springs, Greenbrier Co., WV. Capt. at Lewisburg, WV 5/23/62. WIA (severe leg) at White Sulphur Springs, WV 8/26-27/63. Was capt. either 11/10/63 in Fayette Co., WV or 11/16/63 in Kanawha Co., WV by Gen. Duffie. Another source says capt. at Droop Mountain, WV 11/6/63 although that is unlikely). POW Camp Chase OH 1/28/63. Transf. to Ft. Delaware 2/29/64. Exchd. 9/30/64. At Camp Lee, Richmond, Va. for paroled prisoners 10/12/64. PWR says enl. 5/10/61. Farmer, Fayette Co., WV 1860. Age: 33.

McCUTCHEON, G. S.: Chaplain. B. 2/28/18 Greenbrier Co., WV. Enl. 11/?/62. Capt. at Big Sewell Mountain, WV 1/8/63. (listed in a Wheeling, WV prison 1/18/62). Paroled. Age: 44. 5'½'. Fair complexion, sandy hair, blue eyes. Was also involved with 3rd State Line and Swann's Co. Also served with Crawford's Co., 14th Va. Cav. D. 8/25/1915 Wirt Co., WV. Bur. Foster Cem., Wirt Co., WV.

McCUTCHEON, GRANVILLE S.: Pvt. Co. C. Court-martial record dated 4/3/63. Military court sentence 4/30/63. Possibly same man as G. S. McCutcheon.

McDANIEL, G. W.: Pvt. Co. C. Date and place of enlistment not stated. WIA (slight arm) at White Sulphur Springs, WV 8/26-27/63.

McDANIEL, WILLIAM M.: Pvt. Co. C. B. ca. 1841 Craig Co., Va. Enl. 6/11/63 at Camp Jackson. Absent due to wounds recvd. at White Sulphur Springs, WV 8/26-27/63 on 8/?/63 roll. Recvd. clothing at Montgomery White Sulphur Springs, Va. hosp. on list dated 2/10/64. Res. Craig Co., Va. 1898 at age 52.

McDERMOTT, ADDISON: Pvt. Co. C. Enl. 3/5/64 at New Castle, Va. D. of disease in Lynchburg, Va. hosp. 7/6 or 10/64. Effects turned over to Q.M. Also spelled Adison.

McDERMOTT, CHARLES: Pvt./Corp. Co. D. Enl. 5/30/61 at Summersville, Nicholas Co., WV. KIA at Lewisburg, WV 5/23/62.

McDERMOTT, JOSEPH M.: Corp. Co. D. Enl. 5/30/61 at Summersville, Nicholas Co., WV, Sick from 10/31/61 to 11/26/61 on 11/28/62 roll. Death claim filed 5/13/62 by Jacob Brillheart, atty.

McDEVITTE, JOHN HENRY: Pvt./Sgt. Co. G. B. 8/31/43 Sinks Grove, WV. Possible res. Alleghany Co., Va. Enl. 5/1/62 at White Sulphur Springs, Greenbrier Co., WV. Det. to work on roads on 11/13/62 roll. Sent home sick 12/18/62. Promo. for saving flag in battle at Droop Mountain, WV 11/6/63. Apptd. corp. 12/1/63. Rcvd. clothing 5/13/64. Dischd. at Christiansburg, Va. Postwar res. Gap Mills, WV. Res. Stevensburg, Va. 1913. Res. Culpepper, Va. 1919. D. 11/1/1931 of illness near Culpepper, Va. at age 88.

McDONALD, E. K.: Pvt. Co. K. Enl. 10/26/64 at New Market, Va. Rcvd. clothing 4 QR 1864. In Charlottesville, Va. hosp. with chronic diarrhea 12/20/64 to 1/16/65.

McDONALD, FLOYD: 1st Lt. Co. G. Res. Wyoming Co., WV. 1860, store merchant, age 29. Wyoming Co. School Commissioner 1856-61. Resgnd. 3rd term to enlist. Enl. 7/3/61 at Charleston, Kanawha Co., WV. Elected 1st Lt. 7/1 or 8/61. WIA (sabre cut on arm) at Cross Lanes, WV 8/26/61. PWR says commanded co. at Cross Lanes in absence of officers Duncan and Cook. T. F. Bailey saw McDonald with his arm bandaged, riding his horse among the troops, encouraging them to "Give 'em Hell". Not re-elected. PWR says died during war. PWR says 2nd Lt.

McDONALD, GABRIEL C.: Surgeon. Field and Staff. B. Campbell Co., Va. Grad. Randolph Macon College of Va., medical dept. 1851. Married 1859. Enl. 4/22/61 at Covington, Va. as a Pvt. in Co. A, 27th Va. Inf. at the age of 31. Occupation: Doctor. Apptd. surgeon 4/?/61. Apptd. surgeon of 22nd Va. Inf. 5/16/61. Apptd. major and surgeon of 22nd Va. Inf. 7/1/61. Promo. to Division surgeon July 1864. Served as surgeon in both Echols Brigade and Wharton's Division. Relsd. as Wharton's surgeon 2/11/65 and apptd. to Echols Brig. (CSR also lists him as George McDonald). Elected to Va. legislature during war. Grad. Jefferson Medical College, Philadelphia, PA 1870. V.P. State Medical Society. Res. Monroe Co., WV 1872. Medicine.

McDONALD, JOHN CLAIRBORNE: Lt. Col./2nd Lt./Capt. Field & Staff, Co. K. B. 3/14/31 Montgomery Co., Va. Res. Giles Co., Va. Merchant. Then res. Wyoming Co., WV. Fayette Co., WV patrol Jan. 1861. Helped R. A. Bailey recruit Co. K. Enl. 6/6/61 at Charleston, Kanawha Co., WV. Elected 2nd Lt. 6/6/61. Promo. to major 11/12/63. Elected Capt. of Co. K 12/24/61. Bahlmann says re-elected Capt. 5/1/62 becoming Senior Capt. To Lt. Col. 5/15/64. In Winchester, Va. hosp. for erysipelas 7/27-31/64. Absent on surg. cert. 7/25/64. POW paroled at Charleston, WV 5/25/65. Age: 34. His reported promo. to Col. of the 22nd Va. Inf. not found although he was recommended for the position by Gen. Wharton 12/26/64 and 3/23/65. Postwar res. Hinton, WV. Building contractor. Res. Craig Co., Va. being interested in development of the county. Attended vet. reunion at Union, WV 8/26/97 and gave res. at New Castle, Va. Res. Craig Co., Va. 1898 at age 60. D. 12/22/1914 at Craig City, Va. Bur. McDonald Cem., Craig City at Craig City, Va. Bur. McDonald Cem., Craig City, near Va. highways shop.

McDONALD, JOSEPH: Pvt. Co. K. Enl. 1862. Bahlmann says dischd.

McDONALD, SAMUEL: Pvt. Co. K. Enl. 5/1/62(?) at White Sulphur Springs, Greenbrier Co., WV. Sick in Montgomery White Sulphur Springs, Va. hosp. on Mar.-Apr. 1863 roll. Present on 4/1/64 roll. KIA at Cold Harbor, Va. 6/3/64. Brother of Lt. Col. John C. McDonald.

McDONALD, WILLIAM: Pvt. Co. K. Enl. 9/15/63 at White Sulphur Spring, Greenbrier Co., WV. Sick at Montgomery White Sulphur Springs, Va. hosp. 4/1/64. Det. guard at hosp. on 5/1/64 roll. Present on 10/31/64 roll.

McDONALD, JAMES D.: Pvt. Co. A. Enl. 10/1/62 at Charleston, Kanawha Co., WV. Deserted in face of the enemy 12/13/63.

McDOWELL, J. WILLIAM: Still alive in 1920. Res. Alleghany Co., Va.

McFADDEN, WILLIAM: Pvt. Co. I. Enl. 6/22/61 at Charleston, Kanawha Co., WV. Present on 8/31/61 roll. NFR.

McFARLAND, HENRY DEVOL: Pvt. Co. H. B. 8/28/38 Charleston, WV. Res. Kanawha Co., WV. Enl. 5/8/61 at Charleston, Kanawha Co., WV. Det. as a drillmaster in Edgar's Battn. July-Aug. 1862. Det. in Signal Corps, Breckenridge's Division, Narrows, Va. on rolls dated 12/3/63 to 12/25/64. Promo. Sgt. Jan. 1865. Listed as POW. Paroled at Charleston, WV 5/3/65 and at Lewisburg, WV 5/10 or 20/65. Age: 25. Clerk. 6'. Fair/florid complexion, light brown hair, gray/hazel eyes, light whiskers. In photo of R. E. Lee UCV Camp #887 taken Aug. 1911 at Charleston, WV. Commissary at Camp. Wartime letters appear in book, *Dear Annie*.

McGINNIS, WILLIAM J.: Pvt. Co. B. Res. Ripley, WV, 1860, age 24. Blacksmith.

McGRAW, THOMPSON: Pvt. Co. C. Enl. 5/1/62(?) at White Sulphur Springs, Greenbrier Co., WV. Deserted on Blue Ridge in June 1864. Farm laborer, Fayette Co., WV 1860. Age: 21. PWR says enl. 5/10/61.

McGRAW, WILLIAM H.: Pvt. Co. C. Enl. 7/8/62 at Camp McCausland, Monroe Co., WV. On leave by order of Gen. Williams on 12/31/62 roll. Present on 3/4/63 roll. KIA at White Sulphur Springs, WV 8/26-27/63. Farm laborer, Fayette Co., WV 1860. Age: 29.

McGUIRE, A.: Co. F. Listed as deserter on Aug. 1862 broadside. Roanoke Co., Va.

McKENNY, JAMES C.: Pvt. Swann's Co. Enl. 5/28/61 at Charleston, Kanawha Co., WV. Age: 27. Co. transf. to Co. K, 59th Va. Inf. 7/25/61.

McKINEY, ISAAC FRENCH: Pvt./3rd Sgt. Co. G. B. 9/12/40 Princeton, WV. Enl. 7/3/61 at Charleston, Kanawha Co., WV. Sick at White Sulphur Springs, WV 8/20/61. Present on 11/13/62 roll. Absent with leave on 4/30/63 roll. D. 11/7/1904 at Elgood, Va. Also spelled McKinney.

McKINEY, JAMES H. W.: Pvt. Co. G. Enl. 5/1/62 at White Sulphur Springs, Greenbrier Co., WV. Left sick at White Sulphur Springs, WV 8/20/61. MWIA (gunshot wd. of thigh) and capt. at Lewisburg, WV 5/23/62. D. at U.S. Military Hosp., Gallipolis, OH 6/19/62. (Wded. and in the hands of the enemy on 11/13/62 roll?).

McKINN, PATRICK: Pvt. Taylor's Co. Enl. 6/19/61 at Frankford, Greenbrier Co., WV. Age: 46. Co. transf. to Wise Brigade 7/10/61.

McKINSEY, ALEXANDER: Pvt. Co. K. B. ca. 1833. Res. Craig Co., Va. Date and place of enlistment not stated. D. at Montgomery White Sulphur Springs, Va. 7/1/62.

McKNIGHT, GEORGE: Co. G. Bahlmann says MWIA at Lewisburg, WV 5/23/62. Died more from consumption than wounds. Had light, curly hair.

McKOWN, JAMES: Pvt. Co. B. Res. Mud Run, WV 1860, age 31.

McLAUGHLIN, EDWARD: Pvt. Co. A. Enl. 5/22/61 at Poca, Putnam Co., WV. Unaccounted for as of 7/24/61.

McLAUGHLIN, JOHN: Pvt. Co.E. Enl. 6/8/61 at Charleston, Kanawha Co., WV. Elected as a Lt. in Floyd's command 10/30/62. Left 9/16/62.

McLAUGHLIN, NOAH: Pvt. Co. A. Enl. 5/22/61 at Poca, Putnam Co., WV. Unaccounted for as of 7/24/61.

McMAHON, JOSEPH: Corp. Co. B. Enl. 6/6/61 at Ripley, Jackson Co., WV. Present on 9/1/61 roll. NFR.

McMAHON, WILLIAM B.: Pvt. Co. B. Res. Big Sandy Creek, WV 1860, age 48. Doctor.

McMILLAN, JAMES W.: Pvt. Taylor's Co. Enl. 6/19/61 at Frankford, Greenbrier Co., WV. Age: 20. Co. transf. to Wise Brigade 7/10/61.

McMILLAN, LEVI: Pvt. Taylor's Co. Enl. 6/19/61 at Frankford, Greenbrier Co., WV. Age: 19. Co. transf. to Wise Brigade 7/10/61.

McMILLAN, ROBERT: Exchd. to/from Co. A, 14th Va. Cav. for John Aquilla Miles 9/20/62. PWR.

McMILLIAN, JAMES: Pvt. Co. K. Res. Nicholas Co., WV. Enl. 10/15/63(?) in Greenbrier Co., WV. Capt. at Droop Mountain, WV 11/6/63. Age: 26. POW Camp Chase, OH 11/18/63. In Camp Chase, OH hosp. for smallpox 2/8/64 to 3/9/64. Took oath at Camp Chase, OH 6/10/64. Listed as a conscript deserter. Enl. in U.S. Navy 7/10 or 21/64. Farmer. 6'1½". Florid complexion, dark hair, gray eyes.

McMILTY JOHN: Pvt. Swann's Co. Enl. 5/25/61 at Charleston, Kanawha Co., WV. Age: 24. Co. trans. to Co. K, 59th Va. Inf. 7/25/61.

McMULLEN, A. J.: Pvt. Co. H. Enl. 9/3/62 in Monroe Co., WV. Det. as a scout on 1/1/64 roll. Present on 4/1/64 roll. POW. Paroled at Lewisburg, WV 5/10/65. Age: 27. 5'9". Fair complexion, dark hair, dark eyes, moustache.

McMULLEN, JOHN L.: Pvt. Co. H. Enl. 5/8/61 at Charleston, Kanawha Co., WV. Det. in Brig. Gen. Williams' office on rolls dated July-Aug. 1862 and Sept.-Oct. 1862. Present on 12/31/62 roll. Det. in Signal Corps at Union, WV on 6/30'63 roll. Det. in Signal Corps at Narrows, Va. 4/1/64. Signal operator.

McMULLIN, HUGH: Pvt. Swann's Co. Enl. 5/8/61 at Charleston, Kanawha Co., WV. Age: 20. Co. transf. to Co. K, 59th Va. Inf. 7/25/61.

McNEELY, ANDREW D.: Pvt. Co. I. Enl. 6/22/61 at Charleston, Kanawha Co., WV. AWOL 7/26/61.

McNEELY, JOHN W.: Pvt. Co. I. Res. Boone Co., WV 1860, age 17. Logan Co., WV. Enl. 6/22/61 at Charleston, Kanawha Co., WV. Deserted 9/18/61.

McNEELY, LORENZO D.: Pvt. Co. I. Res. Boone Co., WV 1860, age 20. Logan Co., WV. Enl. 6/22/61 at Charleston, Kanawha Co., WV. Deserted 9/18/61.

McNEILL, JAMES M.: 1st Lt./Capt. Co. D. B. 5/9/23. Res. Nicholas Co., WV. Elected Capt. 5/1/62. Capt. at Droop Mountain, WV 11/6/63. Age: 40. POW Camp Chase, OH 11/18/63. Transf. to Ft. Delaware 3/25/64. Relsd. 6/12/65. Carpenter. 6'½" or 6'1". Dark complexion, gray hair, blue eyes. D. 3/26/1911. Bur. McNeill Cem., near Marlinton, WV.

McNUTT, JAMES R.: Pvt./2nd Lt. Co. C. Res. Craig Co., Va. Enl. either 1/10/64 at Narrows, Va. or 3/24/64 at Camp Gauley, Greenbrier Co., WV. Originally in Capt. Rowan's Co., 60th Va. Inf. Exchd. to 22nd Va. Inf. for R. B. Wiley 3/25/64. Elected 2nd Lt. 6/5/64. KIA 10/19/64 (probably at Cedar Creek, Va.) PWR says KIA at Cold Harbor, Va. 6/1/64 of gunshot wd. to head and gives age as 21.

McNUTT, JOHN: Pvt. Tyree's Co. Enl. 6/25/61. Place of enlistment not stated. On mounted guard at White Sulphur Springs, WV 8/13-14/61. Paid to 3/9/62.

McPHERSON, JAMES R.: Co. G. B. 10/6/43. D. 9/10/1912. Bur. Cedar Hill Cem., Covington, Va. Enl. 10/?/64 at New Market, Va. Age: 21. Res. Alleghany Co., Va. Surrendered Apr. 1865.

McPHERSON, WILLIAM HARRISON: Pvt. Co. G. Res. Alleghany Co., Va. Enl. 7/20/62 at Union, Monroe Co., WV. Det. as regt. blacksmith on 11/13/62 roll. Det. as regt. blacksmith at Lewisburg, WV on 4/30/63 roll. Regt. blacksmith on 4/1/64 roll. Det. with ordnance train 10/31/64. Blacksmith. UDC William Watts Chapter. Age: 22 at enl. Surrendered Apr. 1865 at Staunton, Va.

McQUEEN, A.: Pvt./Sgt. Co. H/Co. E. B. in Nova Scotia, Canada. Enl. 6/8/61 at Charleston, Kanawha Co., WV. Enl. in Co. E and transf. to Co. H prior to July 1862. Det. as a recruiting officer on 6/30/63 roll. Det. to raise a co. on roll dated 12/3/63. Apptd. corp. 12/19/63. Det. to raise a co. on 1/1/64 roll. Authority to raise a co. revoked 2/24/64. In Winchester, Va. hosp. for debilitas 7/31/64. On a list of POW's paroled at Lewisburg, WV 4/25/65. Age: 25. Mechanic. 5'9". Ruddy complexion, dark hair, blue eyes.

MEADOWS, ANDERSON: Pvt. Co. G. Enl. 10/1/62 at Charleston, Kanawha Co., WV. Left sick at Lewisburg, WV on 4/30/63 roll. WIA at Droop Mountain, WV 11/6/63. Absent wded. on 12/31/63 roll. Paroled at Charleston, WV 6/22/65. Age: 39. 5'8". Dark complexion, black hair, hazel eyes.

MEADOWS, DAVID: Pvt. Co. G. Enl. 7/3/61 at Charleston, Kanawha Co., WV. Left his co. at Clifton (Pratt), WV 7/24/61. AWOL.

MEADOWS, ELIJAH: Pvt./Corp. Co. G. Enl. 7/3/61 at Charleston, Kanawha Co., WV. Left sick at White Sulphur Springs, WV 8/20/61. Sick at Montgomery White Sulphur Springs, Va. July 1862. Present on 11/13/62 roll. WIA 10/19/64 (probably at Cedar Creek, Va.). On a list of POW's paroled at Charleston, WV 5/6/65. Age: 24. 5'10". Fair complexion, light hair, blue eyes, moustache.

MEADOWS, JOHN W.: Pvt. Co. D. Date and place of enrollment not stated. In Farmville, Va. hosp. for rubeola 3/16-28/65. Paroled at Charleston, WV 6/24/65. Age: 18. 5'8". Dark complexion, black hair, black eyes.

MEADOWS, VALENTINE: Pvt. Co. G. Enl. 10/20/62 at Fayetteville, Fayette Co., WV. In Danville, Va. hosp. for leg 6/17-21/64. Capt. at Winchester, Va. 9/19/64. POW Pt. Lookout, MD 9/23/64. Exchd. 3/15/65. In Chimborazo Hosp. for bronchitis 3/19-23/65. In Wayside Hosp., Richmond, Va. 3/23/65. Also in Capt. Thurmond's Co.

MEDLIN, DANIEL P.: Pvt. Co. I. Res. Boone Co., WV. Enl. 6/22/61 at Charleston, Kanawha Co., WV. Capt. at Winchester, Va. 9/19/64. POW Pt. Lookout, MD 9/24/64. Exchd. 3/15/65. Took oath at Charleston, WV 5/2/65. Age: 29. Farmer. 5'10". Fair complexion, fair hair, gray eyes.

MEDLIN, GEORGE W.: Pvt./Corp. Co. I. Res. Boone Co., WV. Enl. 6/22/61 at Charleston, Kanawha Co., WV. Capt. at Winchester, Va. 9/19/64. POW Pt. Lookout, MD 9/24/64. Exchd. 3/15/65. Took oath at Charleston, WV 5/2/65. Age: 26. Farmer. 5'10". Dark complexion, dark hair, hazel eyes.

MEEKS, JOHN W.: 1st Sgt./Sgt. Co. E. Enl. 5/1/62 at White Sulphur Springs, Greenbrier Co., WV. Capt. by Capt. King's Union Home Guard and paroled same day 9/20/62. Deserted and capt. 5/1/65. Reported 4/30/65 at Bulltown, WV. Age: 33. Stonecutter. 5'11". Dark complexion, dark hair, blue eyes.

MELTON, CHARLES A.: Pvt. Co. A. Res. Kanawha Co., WV. Enl. 5/20/63 at Dublin, Pulaski Co., Va. Det. service in Q.M. Dept. at Dublin, Va. 12/31/63. Deserted and surrendered. Took oath at Charleston, WV 4/17/65. Age: 27. Carpenter/wheelwright. 6'1". Fair complexion, fair hair, gray eyes.

MELTON, ROBERT T.: Corp./Sgt. Co. A. Enl. 5/22/61 at Poca, Putnam Co., WV. Elected 2nd Sgt. 6/6/61. Unaccounted for 7/24/61. Present on 11/1/62 roll. Apptd. 5th Sgt. 11/24/63. Capt. in Boone Co., WV 2/9/64 or 3/2/64. Age: 25. POW Camp Chase, OH 3/7/64. Transf. to Ft. Delaware 3/14/64. Relsd. and took oath at Ft. Delaware 6/7/65. Farmer. 6'1". Light complexion, dark hair, hazel eyes.

MELTON, SAMUEL H.: Pvt. Co. A. Enl. 7/18/61 at Camp Tompkins, Kanawha Co., WV. Unaccounted for as of 7/25/61.

MELTON, WILSON: Pvt. Co. A. Enl. 7/18/61 at Camp Tompkins, Kanawha Co., WV. Unaccounted for 7/25/61. Deserted at Charleston, WV 10/9/62.

MERIDETH, EVAN A.: Pvt. Co. F. Res. Monroe Co., WV. Enl. 7/4/61 at Charleston, Kanawha Co., WV or 5/1/62 at White Sulphur Springs, Greenbrier Co., WV. In Charlottesville, Va. hosp. for very serious left hip 9/25/64. At home wded. on 10/31/64 roll. POW paroled at Charleston, WV 5/11/65. Age: 25. 6'2". Fair complexion, fair hair, blue eyes. Postwar roster says died from wd. recvd. at Cedar Creek, Va. 10/19/64.

MERIDETH, JAMES: Pvt. Co. F. Res. Monroe Co., WV. Enl. 7/4/61 at Charleston, Kanawha Co., WV. Sick at home on surg. cert. 7/27/61. On sick furlo. July 1862. D. 1872.

MERIDETH, JUNIOR M.: Pvt. Co. F. Res. Monroe Co., WV. Enl. 7/4/61 at Charleston, Kanawha Co., WV. D. in Monroe Co., WV 8/17/63. Claim presented to Junior Meridith, Sr. Postwar roster says died 1861 and name is Junius.

MERIDETH, L. D.: Pvt. Co. F. Enl. 4/13/63(?) at Lewisburg, Greenbrier Co., WV. Absent wded. on 4/30/64 roll. POW paroled at Charleston, WV 5/11/65. Age: 37. 5'7". Dark complexion, black hair, black eyes, black whiskers.

MERRILL, GEORGE W.: Pvt. Co. B. Enl. 5/?/62 at White Sulphur Springs, Greenbrier Co., WV. Capt. at Winchester, Va. 9/19/64 and/or at Strasburg, Va. 10/19/64. POW Pt. Lookout, MD. Relsd. 1/25 or 26/65 by order of the President.

METTS, GILES: Pvt. Co. C. Farm laborer, Fayette Co., WV 1860. Age: 20. Also spelled Metz.

MICHELL, EDW.: Co. K. Deserted 5/15/62 near Salt Sulphur Springs, WV on Aug. 1862 broadside.

MILAM, BENJAMIN: Pvt. Co. A. Enl. 5/22/61 at Poca, Putnam Co., WV. Present on 8/31/61 roll. NFR.

MILAM, DUDLY: Pvt./Corp. Co. G. Enl. 7/3/61 at Charleston, Kanawha Co., WV. Under arrest on 4/30/64 roll. AWOL from Co. B, Ward's Battn. C.S. prisoners at Lynchburg, Va. July 1864. Deserted from guard at Lynchburg, Va. on 10/31/64 roll.

MILAM, JOHN B.: Pvt./Corp. Co. A. Enl. 5/22/61 at Poca, Putnam Co., WV. Cut off from his co. by Federal army 7/24/61. Deserted at Charleston, WV 10/9/62.

MILAM, MILTON: Pvt. Swann's Co. Enl. 5/17/61 at Charleston, Kanawha Co., WV. Age: 22. Co. transf. to Co. K, 59th Va. Inf. 7/25/61.

MILAM, WILLIAM: Pvt. Co. A. Enl. 9/15/62 at Charleston, Kanawha Co., WV. Deserted at Charleston, WV 10/9/62.

MILAM, WILLIAM B.: Pvt./Corp. Co. A. Enl. 7/13/61 at Camp Tompkins, Kanawha Co., WV. Unaccounted for 7/24/61. Sick on 11/1/62 roll.

MILAM, WILLIAM THOMAS: Pvt. Co. A. Enl. 5/30/61 at Charleston, Kanawha Co., WV. Unaccounted for as of 7/24/61.

MILBEE, JOHN: Pvt. Co. A. Res. Putnam Co., WV. Doctor. Enl. 5/22/61 at Poca, Putnam Co., WV. Capt. in Boone Co., WV either 2/9/64 or 3/2/64. Age: 24. POW Camp Chase, OH 3/7/64. Transf. To Ft. Delaware 3/14/64. Relsd. and took oath at Ft. Delaware 6/7/65. In Charlottesville, Va. hosp. for acute diarrhea 6/12-17/64. Farmer. 5'7". Light complexion, dark hair, dark eyes. Attended vet. reunion in Charleston, WV in 1924 and gave address as Charleston, WV.

MILBEE, MARTIN: Pvt. Co. A. Enl. 9/15/62 at Charleston, Kanawha Co., WV. WIA (severe leg) at Cold Harbor, Va. 6/2-3/64. In Chimborazo Hosp. for 'very serious" 6/15-28/64.

MILBEE, WILLIAM: Pvt. Swann's Co./Co. A. Enl. 5/25/61 at Charleston, Kanawha Co., WV. Joined Co. A by transf. 12/31/63. Paroled at Charleston, WV 6/2/65. Age: 24 or 28. 5'8". Fair copmlexion, brown hair, blue eyes.

MILBY, JAMES: Pvt. Swann's Co. Enl. 5/8/61 at Charleston, Kanawha Co., WV. Age: 24. Co. transf. to Co. K, 59th Va. Inf. 7/25/61.

MILBY, JAMES L.: Pvt. Co. A. B. 2/9/43 Nicholas Co., WV. Res. Buffalo, WV 1854. Farmer on Poca River. Attended local schools and helped on the farm until age 14. Grad. King's College (medicine), Cinn, OH 1857. Practiced medicine until war. Enl. 6/4/61 at Camp Tompkins, Kanawha Co., WV. Sick at White Sulphur Springs, WV 8/20/61. Traveled from White Sulphur Springs, WV to Narrows, Va. to rejoin regt. 5/4-12/62. Left sick at Lewisburg, WV on 4/30/63 roll. Present on 12/31/63 roll. POW paroled at Lewisburg, WV 4/25/65. Age: 21. Farmer. 5'8". Light complexion, dark hair, hazel eyes. PWR says slightly wded. in war and capt. Feb. 1864. POW until war end at Ft. Delaware. Asst. surg. during much of the service. Postwar medicine Putnam Co., WV 4 years. Medicine, Sissonville, WV 8 years. Res. Elk District then res. Charleston, WV. Practiced medicine and surgery for 38 years. Correct spelling probably Milbee.

MILES, JOHN AQUILLA: Transf. to Co. A, 14th Va. Cav. 9/20/62 in exchange for Robert McMillan. Also served in 60th Va. Inf. This may be Aquilla Myles.

MILES, JOHN W.: Pvt. Co. K. Enl. 6/6/61 at Charleston, Kanawha Co., WV. Sick 8/2/61. NFR.

MILLEGAN, JAMES D.: Pvt. Co. D. Enl. 5/1/62 at White Sulphur Springs, Greenbrier Co., WV. Sick in hosp. on 10/31/62 roll. NFR.

MILLER, ANDERSON: Pvt. Co. I. Res. Boone Co., WV 1860, age 25. Logan, WV. Enl. either 6/?/61 at Charleston, Kanawha Co., WV or 5/2/62 at White Sulphur Springs, Greenbrier Co., WV or 9/10/62 in Kanawha Co., WV. Surrendered at Charleston, WV Feb. 1865. Age 30. Took oath at Charleston, WV and sent North 2/9/65. Farmer. 5'9". Dark complexion, dark hair, brown eyes.

MILLER, CORNELIUS P.: Pvt. Co. I. Enl. 5/1/62 at White Sulphur Springs, Greenbrier Co., WV. Listed as deserter on Aug. 1862 broadside. Botetourt Co., Va. Absent sick on 10/3/64 roll.

MILLER, DANIEL: Pvt. Co. I. Enl. 6/22/61 at Charleston, Kanawha Co., WV. Under arrest on 4/30/64 roll. In Charlottesville, Va. hosp. for catarrhus (?) 11/10-22/64. Deserted and surrendered at Charleston, WV Mar. 1865. Age: 23 or 25. Took oath at Charleston, WV and relsd. 3/2/65. Farmer. 6'. Fair complexion, fair hair, hazel eyes.

MILLER, D. G.: Pvt. Co. E. Enl. 3/18/64 at Camp Gauley, Greenbrier Co., WV. Recvd. clothing 4 QR 1864. NFR. PWR says remained with co. until disbanded at Christiansburg, Va.

MILLER, DRYDEN: Pvt. Co. I. Res. Boone Co., WV 1860, age 21. Logan Co., WV. Enl. 6/22/61 at Charleston, Kanawha Co., WV. AWOL 7/26/61. Present on 4/1/64 roll. Under arrest on 4/30/64 roll. Deserted and surrendered at Charleston, WV Mar. 1865. Age: 25. Took oath at Charleston, WV 3/2/65. Farmer. 5'11". Fair complexion, light hair, gray eyes.

MILLER, ELIJAH: Pvt./Fifer. Co. I. Res. Boone Co., WV 1860, age 23. Enl. 6/29/61 at Charleston, Kanawha Co., WV. Sick at Giles C.H. (Pearisburg), Va. July 1862. KIA at White Sulphur Springs, WV 8/26/63. (Also called Eli).

MILLER, GEORGE H.: 1st Sgt. Co. E. Enl. 6/8/61 at Charleston, Kanawha Co., WV. Served as 3rd Sgt. until elected 1st Sgt. 6 or 7/3/61. NFR. PWR says promo. to Capt. and assigned to another regt. Captured and POW until end of war.

MILLER, GEORGE H.: Pvt. Co. H/Co. K. Enl. 6/8/61 at Charleston, Kanawha Co., WV. Promo. to 1st Lt. in Jackson's regt. 3/26/62. Transf. to Co. K from Co. H 7/18 or 21/62.

MILLER, HEATH JONES: Pvt./2nd Sgt. Co. E. Enl. 6/8/61 at Charleston, Kanawha Co., WV. KIA in a skirmish at Totopotomoy, Va. 5/30/64. (Also listed as J. H. Miller in CSR).

MILLER, ISAAC L.: Pvt. Co. I. Res. Boone Co., WV 1860, age 25. Enl. 6/22/61 at Charleston, Kanawha Co., WV. Capt. in Fayette Co., WV 9/13/61. POW Camp Chase, OH. Exchd. 8/25/62. At home in Boone Co., WV on sick furlo. on 12/31/62 roll.

MILLER, JAMES H. JR.: Pvt. Co. C. Farmer, Fayette Co., WV 1860. Age: 24. Enl. 5/10/61.

MILLER, JAMES O.: Pvt./2nd Lt. Co. F. Res. Monroe Co., WV. Enl. 7/4/61 at Charleston, Kanawha Co., WV. Elected 2nd Lt. 2/14/63. WIA (slight hand) at White Sulphur Springs, WV 8/26-27/63. Capt. at Winchester, Va. 9/19/64. Listed as wded. on 9/29/64 roll. POW Ft. Delaware. Relsd. 6/17/65. 5'7". Fair complexion, light hair, blue eyes. D. in Monroe Co., WV sometime prior to 1914.

MILLER, JAMES S.: Pvt./Corp. Co. D. Enl. 5/30/61 at Summersville, Nicholas Co., WV. Left sick in Greenbrier Co., WV 8/22/61. At Giles Co. hosp. July 1862.

MILLER, JOHN S.: Pvt. Co. C. Farm laborer, Fayette Co., WV 1860. Age: 20. Enl. 5/10/61.

MILLER, ROBERT L.: Pvt. Co. C. Enl. 5/10/61.

MILLER, SAMUEL AUGUSTINE: Pvt./A.Q.M. Field & Staff Co. I. B. 10/16/20 Mt. Jackson, Va. Son of Reuben Moore and Atlantic Ocean (Walton) Miller. Clerk and financial agent for cattle king, Beale Steenburgen. Grad. Gettysburg College, PA. Went to England, Scotland, and returned by way of Mexico. Enjoined father in Mason Co., WV business. Moved to Charleston, WV ca. 1840-41 and studied law with George W. Summers until 1844, when he entered law partnership with Summers until 1851. Married Helen Quarrier 7/27/45. Induced to accept financial agency of Ruffner, Donnally & Co. salt firm 1851. Chief legal counsel and president of the co. When firm ceased he went into law partnership with G. B. Newman ("Miller & Newman"). Lawyer, Charleston, WV 1860, age 40. Enl. 6/22/61 at Charleston, Kanawha Co., WV. Apptd. A.Q.M. 8/?/61. Apptd. Capt. A.Q.M. 9/11/61. Date of acceptance 9/25/61. Date of confirmation 12/24/61. A.D.C. of McCausland's (2nd) Brig. 5/13/62. Apptd. Brig. Q.M. (Patton's) 2nd Brig., 22nd Va. Inf. 11/12/62. Resgnd. 2/27/63. Elected to Confederate Congress on 3/1/63 roll taken at Lewisburg, WV. Filled unexpired term of Gen. A. G. Jenkins in Confederate Congress. Re-elected 1864 and served until end of war. Fled to Canada briefly. Pardoned by President Johnson. Disbarred as an ex-Confederate 1865-70. Returned home ca. Sept. 1865. Law partnership in 1870's with brother-in-law William A. Quarrier. WV House of Delegates 1874. WV Legislature 1874-75. Law partnership with ex-governor E. W. Wilson 1876, then with D. C. Gallaher. D. 11/17/90. Bur. Spring Hill Cem., Charleston, WV.

MILLER, SAMUEL J.: Co. I. Listed as deserter on Aug. 1862 broadside. Botetourt Co., Va.

MILLER, THOMAS: Pvt./2nd Lt. Co. E. Enl. 6/8/61 at Charleston, Kanawha Co., WV. Appted. 2nd Lt. 5/2/62. KIA at Winchester, Va. 9/19/64.

MILLER, THOMAS, JR.: Pvt. Co. E. Enl. 7/20/61 on Sandy. AWOL 7/1/61.

MILLER, W. A.: Pvt. Co. C. Enl. 5/10/61.

MILLER, WARREN: Pvt. Co. I. Enl. 6/22/61 at Charleston, Kanawha Co., WV. Present on 8/27/61 roll. NFR.

MILLER, WILLIAM: Pvt. Co. I. Enl. 8/22/61 at Meadow Bluff, Greenbrier Co., WV. Present on 8/31/61 roll. NFR.

MILTON, BENJAMIN A.: Pvt. Co. E. Enl. 6/8/61 at Charleston, Kanawha Co., WV. Transf. to 36th Va. Inf. 8/8/61. (Listed as Benjamin A. Melton on 36th Va. Inf. roster). Originally enl. in Beckett's Cav. Co., 36th Va. Inf. 5/16/61 at Buffalo, WV. Transf. to Sweeney's Co.

MINICK, J.: Co. F. Listed as deserter on Aug. 1862 broadside. Monroe Co., WV.

MINNER, JAMES J.: Pvt. Co. D. Enl. 5/1/62 at White Sulphur Springs. Greenbrier Co., WV. Det. to feed and care for brig. horses and stock as of June 1863 and still so engaged on 4/12/64. In Chimborazo Hosp. 5/7/64. In Monroe Co., WV keeping horses on 10/31/64 roll. POW. Paroled at Charleston, WV 5/10/65. Age: 32. 5'11". Light complexion, light hair, blue eyes, gray whiskers.

MITCHELL, JOHN L.: Pvt. Swann's Co. Enl. 6/27/61 at Charleston, Kanawha Co., WV. Age: 19. Co. transf. to Co. K, 59th Va. Inf. 7/25/61.

MODENA, T. H.: Pvt. Co. I. Date and place of enlistment not stated. POW. Paroled at Columbia, Va. 5/8/65. (Also spelled Modenia).

MONROE, DAVID: Corp./1st Lt. Co. E. Enl. 7/17/61 at Charleston, Kanawha Co., WV. Elected 1st Lt. 12/24/61. Dropped 5/2/62.

MONROE, JOHN: Pvt. Co. B. Enl. 8/?/63 at Lewisburg, Greenbrier Co., WV. Present on 1/10/64 roll. NFR.

MONROE, WILLIAM A.: Co. D. B. Alleghany Co., Va. Entered Richmond, Va. Old Soldiers Home 12/16/1912 at age 91. D. 2/10/1913. Bur. Hollywood Cem., Richmond, Va.

MONTGOMERY, SAMUEL A.: Co. K. B. 10/10/43. Served in Bolivar Mills Co., Militia. Exempt 3/?/62 for disease of ankle. Res. Rockbridge Co., Va.

MOOKLAR, ATWELL TEBBS: Lt. Tyler Co., WV (?).

MOONEY, WILLIAM J.: Pvt. Co. A. Enl. 11/1/62 at Raleigh (Beckley), Raleigh Co., WV. Present on 4/30/63 roll. NFR.

MOORE, ANDREW: 1st Lt./Capt. Co. H. B. 7/16/32 "Mt. Jay", Botetourt Co., Va. Attended Washington College 1853. LLD 1854. Lawyer in Georgia, Texas and Kanawha Co., WV. Enl. 5/22/61 at Charleston, Kanawha Co., WV. Resgnd. 6/28/61. Also served in Imboden's Cav. Surrendered at Charlotte, NC as an aide on Gen. D. H. Hill's staff. Farmer. Res. Pulaski Co., Va. for 35 years. D. 8/29/1913 Lexington, Va. Bur. New Dublin Presbyterian Church Cem., Pulaski Co., Va.

MOORE, DAVID B.: Pvt. Co. E. Fought at Carnifex Ferry, Droop Mountain, Gettysburg, Culpepper, and Valley of Va. B. 1838 Bath Co., Va. Came to WV with parents at age 12. Married twice and had one son. Loved horses and recalled viewing "Traveller" on Sewell Mountain, WV in 1861. In 1927 was one of the last 6 surviving members of Jeb Stuart UCV Camp, Fayette Co., WV and gave res. as Russellville, WV.

MOORE, ENOCH: Pvt. Co. F. Enl. 5/1/62 at White Sulphur Springs, Greenbrier Co., WV. Det. to drive brig. teams on rolls for 10/31/62 and 2/28/63. Working in government shop on rolls for Mar.-Apr. 1863 and 12/31/63. Det. service on 10/31/64 roll. In Charlottesville, Va. hosp. for acute diarrhea 10/12-24/64.

MOORE, MALZEY: Pvt. Co. E. Enl. 5/18/63 at Camp Stonewall. Deserted 11/8/63.

MOORE, MELICE: Pvt. Co. E. Enl. 6/8/61 at Charleston, Kanawha Co., WV. AWOL 7/24/61.

MOORE, P. H.: Major. Field & Staff. Comm. Maj. 6/13/61. In charge of Q.M. Dept. at Charleston, WV. Resgnd. 8/1 or 13/61.

MOORE, THOMAS: Pvt. Co. D. Date and place of enlistment not stated. POW paroled at Charleston, WV 5/11/65. Age: 39. 5'5" Dark complexion, black hair, dark eyes, dark whiskers.

MOORE, THOMAS: Co. G. Postwar roster.

MOREHEAD, CHARLES R.: Pvt. Taylor's Co. Enl. 6/19/61 at Frankford, Greenbrier Co., WV. Age: 23. Co. transf. to Wise Brigade 7/10/61.

MOREHEAD, MILTON: Pvt. Taylor's Co. Enl. 6/19/61 at Frankford, Greenbrier Co., WV. Age: 20. Co. transf. to Wise Brigade 7/10/61.

MORGAN, HARVEY G.: Pvt. Co. G. Enl. 7/7/61 at Charleston, Kanawha Co., WV. Absent wded. on 12/31/63 roll and 4/1/64 roll. Present on 4/30/64 roll. Transf. to Co. E, 8th Va. Cav. 5/6/64.

MORGAN, JAMES B.: 1st Lt. Co. A.Enl. 5/22/61 at Poca, Putnam Co., WV. Elected 1st Lt. 5/22/61. Comm. burned on *Julia Maffitt* 7/24/61. Resgnd. Sept. 1861.

MORGAN, JOHN JR.: Sgt. Co. A. B. 10/10/40 Shenandoah Co., Va. Res. Putnam Co., WV 1846. Enl. 5/22/61 at Poca, Putnam Co., WV. Dischd. in consequence of ill health (of family) 5/27/61. Postwar res. Poca, WV. Married 1875. Farmer. Wrote book *The Last Dollar*, telling war experience. D. ca. 1898 at age 58 of kidney disease. Bur. Morgan family cem., east of Morgan farm, Morgan's Landing, Putnam Co., WV.

MORGAN, JONATHAN: Pvt. Co. G. Enl. 7/3/61 at Charleston, Kanawha Co., WV. Left his co. at Charleston, WV 7/5/61. AWOL.

MORGAN, SILAS: Pvt. Swann's Co. Enl. 5/8/61 at Charleston, Kanawha Co., WV. Age: 25. Co. transf. to Co. K, 59th Va. Inf. 7/25/61.

MORRIS, CHARLES B.: Pvt. Co. A. Enl. 7/4/61 at Camp Tompkins, Kanawha Co., WV. Unaccounted for as of 7/24/61.

MORRIS, FENELL: Pvt. Co. I. Enl. 6/29/61 at Charleston, Kanawha Co., WV. D. by being shot in Boone Co., WV on Big Coal River 8/8/61.

MORRIS, GEORGE H.: Pvt./Corp. Co. A. B. 1837 in Va. Enl. 5/22/61 at Poca, Putnam Co., WV. Paroled at both Lewisburg, WV 4/25/65 and Charleston, WV 5/11/65. Age: 25. Farmer. 5'6" or 6'. Fair complexion, light hair, dark/gray eyes. D. 1908. Bur. Mt. Nebo Cem. near Winfield, Marion (?) Co., WV. (Probably Putnam Co., WV).

MORRIS, JAMES M.: Pvt. Co. D. Enl. 5/30/61 at Summersville, Nicholas Co., WV. Present on 8/31/61 roll. NFR.

MORRIS, JOHN A.: Pvt. Co. A. Enl. 6/7/62 in Greenbrier Co., WV or 7/1/63 at Camp Jones. Teamster 1/30/63 to 2/1/64. Paroled at Lewisburg, WV 5/10/65. Age: 19. Farmer. 5'9". Fair complexion, dark hair, blue eyes. Res. Kanawha Co., WV.

MORRIS, WILLIAM S. S.: Pvt./Corp. Co. A. Enl. 5/22/61 at Poca, Putnam Co., WV. KIA at Droop Mountain, WV 11/6/63.

MORRISON, JAMES H.: Pvt. Co. D. Enl. 8/19/61 at White Sulphur Springs, Greenbrier Co., WV. Transf. from Capt. Rigg's co. by order of Col. Tompkins 8/19/61. Teamster from 4/1/63 to 7/31/63 and left Lewisburg, WV 4/13/63. Present on 12/31/63 roll. Paroled at Charleston, WV 5/10/65. Age: 20. 5'7". Fair complexion, brown hair, blue eyes, light whiskers. (Also listed as Harvey Morrison in CSR).

MOSELEY, J. ROBERT F.: Corp. Co. K. Res. Loup Creek, Fayette Co., WV. Enl. 6/6/61 at Charleston, Kanawha Co., WV. Dischd. for fractured ankle 12/12/61. A PWR says fought in all battles up to and including Winchester, Va. 9/19/64. Called "Captain". Farmer, Fayette Co., WV 1860. Age: 39.

MOSELY, JOSHUA: Pvt. Co. E. Enl. 5/1/62 at White Sulphur Springs, Greenbrier Co., WV. Listed as deserter on Aug. 1862 broadside. Deserted on Sept.-Oct. 1862 roll.

MOULS, JAMES: Pvt. Co. F. Enl. 5/1/62 at White Sulphur Springs, Greenbrier Co., WV. Sick at Dublin Depot, Va. July 1862. D. 5/13/64. Lydia Mouls, widow.

MULLINS, NATHANIEL: Pvt. Co. G. Enl. 7/3/61 at Charleston, Kanawha Co., WV. Present on 9/31/61 roll. NFR.

MUNCY, VAN B.: Pvt. Co. D. Enl. 6/3/61 at Kanawha Falls, Fayette Co., WV. Left sick at White Sulphur Springs, WV 8/20/61. MWIA at Lewisburg, WV 5/23/62. D. 10/15/62. Res. Fayette Co., WV 1860. Age: 20.

MURRAY, PATRICK: Pvt./Corp. Co. K. B. in Va. Enl. 6/6/61 at Charleston, Kanawha Co., WV. Corp. 6/24/61. WIA (shot through thigh) and capt. at Lewisburg, WV 5/23/62. POW Camp Chase, OH 7/22/62. Exchd. 8/25/62. At Montgomery White Sulphur Springs, Va. for wd. on 11/1/62 roll. Present on 2/28/63 roll. Det. as courier on Mar.-Apr. 1863 roll. Det. as hosp. nurse on 12/30/63 roll. WIA (shot in neck) at New Market, Va. 5/15/64. Recovered. Paroled at Lewisburg, WV 4/27/65 and Charleston, WV 5/11/65. Age: 25. Farmer. 5'5". Fair or ruddy complexion, black hair, dark brown/hazel eyes, dark whiskers.

MYERS, H. C.: Pvt. Co. H. Enl. 4/2/62 in Greenbrier Co., WV. Present on Apr. 1864 roll. NFR.

MYLES, AQUILA: 4th Sgt. Taylor's Co. Enl. 6/19/61 at Frankford, Greenbrier Co., WV. Age: 20. Co. transf. to Wise Brigade 7/10/61. Probably John Aquila Miles.

MYLES, JOHN T.: Pvt. Co. K. Enl. 1861. Bahlmann says transf. to Capt. Thurmond's Co.

MYRES, ABRAHAM S.: Pvt. Co. G. Res. Giles Co., Va. Enl. 5/1/62 at White Sulphur Springs, Greenbrier Co., WV. Deserted 5/24/62. Enl. 6/23/62 at Meadow Bluff, Greenbrier Co., WV. Listed as deserter on Aug. 1862 broadside. Left 5/24/62. WIA (very serious knee joint & femur) and capt. at Winchester, Va. 9/19/64. D. 10/13/64. Bur. Winchester Cem., Winchester, Va. #1359. Also spelled Myers.

NACE, J. C.: Pvt. Co. E. Enl. either 3/30/64 at Dublin, Pulaski Co., Va. or 4/10/64 at White Sulphur Springs, Greenbrier Co., WV. Conscript. In Charlottsville, Va. hosp. for ulcers 9/1-7/64. NFR.

NACE, JOHN C.: Pvt. Co. I. Enl. 6/22/61 at Charleston, Kanawha Co., WV. Sick in Giles Co., Va. on rolls for 11/11/62 and 12/31/62. NFR.

NAREY, LAREY: Pvt. Co. C. Enl. 5/1/62 at White Sulphur Springs, Greenbrier Co., WV. Recvd. clothing at Harrisonburg, Va. hosp. 9/20/64. Det. service on 10/31/64 roll. (Also Larry and Neary).

NEAL, ARCHIBALD: Pvt. Co. C. General agent, Fayette Co., WV 1860. Age: 35. Enl. 5/10/61.

NEAL, JOHN F.: Pvt. Co. F. Enl. 5/1/62 at White Sulphur Springs, Greenbrier Co., WV. POW paroled at Charleston, WV 5/10/65. Age: 23. 5'9". Light complexion, light hair, gray eyes.

NEBERGALL, ANDREW A.: Pvt. Co. D. Enl. 5/30/61 at Summersville, Nicholas Co., WV. Det. to work in government shop at Dublin, Va. on rolls from 10/31/62 through 1/28/63. Wagoner.

NEBERGALL, JAMES H.: Pvt. Co. D/Co. F. Enl. 5/30/61 at Summersville, Nicholas Co., WV. Det. wagoner on Sept.-Oct. 1862 roll. Present 1/2/63 roll. Det. on 3/4/63 roll. Regt. wagoner on 4/30/63 roll. Det. wagoner 12/31/63 roll. Teamster.

NECESSARY, JOHN W.: Pvt. Co. unknown. Res. Scott Co., Va. Date and place of enlistment not stated. Took oath at Chattanooga, TN 6/8/65. 5'10". Dark complexion, brown hair, gray eyes.

NEIGHBORS, WILSON: Pvt. Co. K. Res. Monroe Co., WV. Enl. 4/15/62(?) at Lewisburg, Greenbrier Co., WV. WIA and capt. at Lewisburg, WV 5/23/62. Age: 47. POW Camp Chase, OH 7/22/62. Exchd. 8/25/62. At Montgomery White Sulphur Springs, Va. for wd. on 11/1/62 roll. At home unable for duty on Nov.-Dec. 1862 roll. Permanent wded. furlo. granted. Capt. in Greenbrier Co., WV 5/24/64 (gave age as 50). POW Camp Chase, OH 9/18/64 or 10/6/64. Desired to take oath 10/7/64. D. at Camp Chase, OH 10/30/64. Grave #387. Farmer. 5'7½". Dark/fair complexion, sandy/light hair, blue eyes.

NEIL, ROBERT C.: Pvt./Sgt. Co. D. Enl. 5/30/61 at Summersville, Nicholas Co., WV. Apptd. 3rd Sgt. 3/1/63. Det. service 4/27/64 roll. Capt. at Winchester, Va. 9/19/64. POW Pt. Lookout, MD 9/23/65. Exchd. 3/15/65. POW paroled at Charleston, WV 5/26/65. Age: 20. 5'10". Fair complexion, brown hair, gray eyes.

NEIL, SAMUEL B.: Pvt. Co. D. Enl. 5/30/61 at Summersville, Nicholas Co., WV. Wagoner on July 1862 roll. Teamster on rolls 4/1/63 through 1/31/64. Keeping horses at Dublin, Va. on 10/31/64 roll. Recvd. clothing 4 QR 1864.

NEIL, SAMUEL JR.: Pvt. Co. D. Enl. 5/1/62 at White Sulphur Springs, Greenbrier Co., WV. Deserted on or about 5/25/62. Sick at Montgomery White Sulphur Springs, Va. July 1862. Took oath at Charleston, WV 7/15/65. Age: 28. 6'. Dark complexion, dark hair, dark eyes.

NEIL, WILLIAM H. H.: Pvt. Co. D. Res. Nicholas Co., WV. Enl. 5/1/62 at White Sulphur Springs, Greenbrier Co., WV. Capt. at Lewisburg, WV 5/23/62. Age: 21. POW Camp Chase, OH 5/30/62. Exchd. 8/25/62. Capt. in Nicholas Co., WV 11/7/62 (gave age as 28). POW Camp Chase, OH 12/2/62. Exchd. 12/2/62. Det. service 4/27/64. Recvd. clothing 5/13/64. Deserted on Blue Ridge June 1864. Farmer. 5'11". Sallow/dark complexion, dark hair, dark/blue eyes, dark whiskers. PWR says last fought at Cold Harbor, Va. Res. Lockwood, WV at age 86.

NEIL, WILLIAM S.: Pvt. Co. D. Res. Nicholas Co., WV. Enl. 5/30/61 at Summersville, Nicholas Co., WV. Sick at Montgomery White Sulphur Springs, Va. July 1862. Deserted 10/31/62. Present on 2/28/63 roll. Deserted near Hillsboro, WV 4/14/63. Capt. at Summersville, WV 4/22/63. Age: 28. POW Camp Chase, OH 5/11/63. Transf. to Johnson's Island, OH 6/14/63. Relsd. on taking oath 11/29/63. Farmer. 5'10¾". Dark complexion, light hair, dark eyes.

NELSON, GEORGE W.: Pvt. Co. I. Enl. 6/29/61 at Charleston, Kanawha Co., WV. AWOL 7/26/61. Returned from desertion 9/15/61. Capt. in Logan Co., WV 2/6/62. POW Camp Chase, OH. Exchd. 8/25/62. Recvd. clothing at Montgomery White Sulphur Springs, Va. hosp. 12/2/63. Sick at home in Boone Co., WV on rolls for 4/1/64 and 4/30/64.

NELSON, JAMES: Pvt. Co. I. Enl. 6/29/61 at Charleston, Kanawha Co., WV. Capt. at Carnifex Ferry, WV 9/11/61. POW Camp Chase, OH. Exchd. 8/25/62.

NELSON, JAMES M.: Pvt./Corp. Co. F. Res. Monroe Co., WV. Enl. 7/4/61 at Charleston, Kanawha Co., WV. Present on Mar.-Apr. 1863 roll. D. sometime prior to 1914.

NEVINS, GEORGE R.: Pvt. Co. H. Enl. 5/8/61 at Charleston, Kanawha Co., WV. Absent cause unknown on 8/31/61 roll. NFR.

NEWCOMB, SAMUEL C.: Pvt. Co. E. Date and place of enlistment not stated. NFR.

NEWHOUSE, SIMON: Pvt. Swann's Co. Enl. 5/25/61 at Charleston, Kanawha Co., WV. Age: 25. Co. transf. to Co. K, 59th Va. Inf. 7/25/61.

NEWMAN, J. GREEN: Pvt. Co. H. Enl. 1861 at Charleston, Kanawha Co.. WV. Lawyer, Charleston, WV 1860. Age: 30.

NEWMAN, JONATHAN M.: Pvt./Sgt. Co. F. Res. Monroe Co., WV Enl. 7/4/61 at Charleston, Kanawha Co., WV. Sick at home 8/8/61. Capt. in Monroe Co., WV 7/6/62. Age: 18. POW Camp Chase, OH 8/6/62. Exchd. 8/25/62. Present on 10/31/62 roll. Recvd. clothing 5/13/64. Deserted about 6/30/64. Farmer. 5'7". Fair complexion, sandy hair, blue eyes. Postwar res. McClain Co., IL. Believed still alive 1914.

NICELY, ALEX: Co. I. Listed as deserter on Aug. 1862 broadside. Botetourt Co., Va.

NICELY, EMANUEL: Pvt. Co. G. B. 6/5/34. Enl. 3/5/64 at Staunton, Va. Conscript. Deserted 4/12/64. Absent wded. on 10/31/64 roll. D. 11/8/98. Bur. Sharon Union Church Cem., Irongate, Alleghany Co., Va.

NICELY, GEORGE A.: Co. G. B. Alleghany Co., Va. Enl. 1863 in Monroe Co., WV. Filed for pension 4/20/1900, age 68. Collierstown. Farmer, Buffalo District, 1910, age 77. D. 12/20/1911 Collierstown. Bur. Collierstown Presbyterian Cem.

NICELY, GEORGE E.: Pvt. Co. G. Enl. 3/5/64 at Staunton, Va. Conscript. Deserted 4/12/64. Recvd. clothing 5/13/64. Absent wded. on 10/31/64 roll.

NICELY, PETER: Pvt. Co. G. Enl. 3/5/64 at Staunton, Va. Conscript. Deserted 4/12/64 roll. Recvd. clothing 5/13/64. Nurse in hosp. on 10/31/64 roll. On Alleghany Co., Va. pension rolls.

NICHOLAS, GEORGE W.: Pvt. Co. E. Enl. 6/8/61 at Charleston, Kanawha Co., WV. AWOL 7/1/61.

NICHOLAS, MICHAEL: Pvt. Taylor's Co. Enl. 6/19/61 at Frankford, Greenbrier Co., WV. Age: 28. Co. transf. to Wise Brigade 7/10/61.

NICHOLAS, RICHARD: Pvt. Co. D. Enl. 5/1/62 at White Sulphur Springs, Greenbrier Co., WV. At Charlottesville, Va. hosp. for very serious right foot 9/1-8/64. At home wded. on 10/31/64 roll.

NICHOLS, ADAM: Pvt. Swann's Co. Enl. 5/8/61 at Charleston, Kanawha Co., WV. Age: 23. Co. transf. to Co. K, 59th Va. Inf. 7/25/61.

NICHOLS, WILLIAM J.: Pvt. Co. I. Enl. 5/1/62 at White Sulphur Springs, Greenbrier Co., WV. Deserted at Lewisburg, WV 4/24/64.

NICKELL, C. M.: Pvt. Co. F. Res. Monroe Co., WV. Enl. 11/8/64 in Shenandoah, Va. POW paroled at Lewisburg, WV 5/10/65. Age: 26. Farmer. 5'5". Fair complexion, light hair, gray eyes.

NICKELL, ROBERT M.: Pvt. Co. F. Res. Monroe Co., WV. Enl. 7/4/61 at Charleston, Kanawha Co., WV. Present on 8/31/61 roll. D. sometime prior to 1914.

NICOL, GEORGE T.: 2nd Lt. Co. F. Res. Monroe Co., WV. Enl. 7/4/61 at Charleston, Kanawha Co., WV. Elected 2nd Lt. 7/4/61. Tendered resignation 8/20/61. D. sometime prior to 1914. Postwar roster says middle initial is W. also says name is Nickell. Another PWR says middle initials W. H.

NIDA, C. J.: Pvt. Co. G. Enl. 5/1/62 at White Sulphur Springs, Greenbrier Co., WV. Deserted 5/1/62.

NIDAY, JOHN A.: Pvt. Co. H. B. ca. 1830 Craig Co., Va. Enl. 3/10/64 in Greenbrier Co., WV. Deserted near Staunton, Va. May 1864.

NIDAY, PETER: Pvt. Co. H. Enl. 7/14/63 in Monroe Co., WV. Deserted near Staunton, Va. May 1864.

NIDY, JOHN A.: Pvt. Co. F. B. ca. 1824 Craig Co., Va. Shoemaker. Enl. 5/1/62 at White Sulphur Springs, Greenbrier Co., WV. Deserted on 7/14/62 list. Listed as deserter on Aug. 1862 broadside, Craig Co., Va. Det. to work shoe shop 12/31/63 roll. Present on 10/31/64 roll. Also spelled Niday.

NIEL, ANDREW: Pvt. Co. D. Enl. 5/30/61 at Summersville, Nicholas Co., WV. Left sick at White Sulphur Springs, WV 8/20/61. NFR.

NIEL, HARRISON W.: Pvt. Co. D. Enl. 5/30/61 at Summersville, Nicholas Co., WV. Present on 8/31/61 roll. NFR.

NIEL, JILSON H.: Pvt. Co. D. Enl. 7/29/61 at Cross Lanes, Nicholas Co., WV. Present on 8/31/61 roll. PWR says fought at Carnifex Ferry and Lewisburg. Transf. to Co. A. 14th Va. Cav. Res. Enon, WV at age 82.

NIMINGER, MILTON S.: Pvt. Co. E. B. Botetourt Co., Va. Enl. either 3/31/64 at Dublin, Pulaski Co., Va. or 4/10/64 at White Sulphur Springs, Greenbrier Co., WV. Conscript. D. at Bedford, Va. 12/14/1925. PWR says name is Nininger.

NOE, SAMUEL: Pvt. Co. E. B. 5/21/36 Roane Co., WV. Enl. 6/8/61 at Charleston, Kanawha Co., WV. Present on 4/30/63 roll. PWR says transf. to Co. G, 19th Va. Cav. and also served in 14th Va. Cav. Captured four times and escaped three times. Fought at Scary Creek, Cross Lanes, Cotton Hill, Dry Creek, Droop Mountain, Lynchburg, Winchester, Martinsburg, Monocacy, and Bunker Hill, where he was badly wded. Twice slightly wded. in war. POW Camp Chase, OH for six months. Married 1866 in Gilmer Co., WV. Postwar res. Newton, Roane Co., WV. 1884. Farmer. Two years school trustee.

NOFTSINGER, WILLIAM ABNER: Pvt. Co. B. Res. Jackson Co., WV. Mustered 1862 for 3 years.

NONLEY, G. W.: Pvt. Co. H. Enl. 7/22/62 at Camp McCausland, Monroe Co., WV. Deserted on 7/27/62 list at Camp Royal Oak. Probably George W. Nunaly.

NORMAN, JAMES: Pvt. Co. H. PWR says "didn't go."

NORMAN, JOHN G.: Pvt. Co. H. PWR says "didn't go."

NORVELL, WILLIAM GASTON: Pvt. Co. H. B. 1838 Charleston, WV. Reared and educated, Charleston, WV. Merchant, Charleston, WV, 1860, age 22. Married 1861. Enl. 5/8/61 at Charleston, Kanawha Co., WV. Absent cause unknown on 8/31/61 roll. Postwar mercantile business and farming, Charleston, WV. D. 10/7/1903 at Charleston, WV. Bur. Spring Hill Cem., Charleston, WV.

NOTTINGHAM, JAMES: Pvt. Co. D. Enl. 11/18/63 at Camp Bailey. Paroled at Charleston, WV 6/30/65. Age: 15. 5'5½". Sandy complexion, light hair, blue eyes.

NOULAND, D.: Pvt. Co. H. Enl. 6/18/62 at Salem, Va. Rejected for disability 7/15/62. Dischd. 7/16/62 at Camp McCausland, WV.

NOUNNAN, JAMES H.: Pvt. Co. H. Enl. 5/8/61 at Charleston, Kanawha Co., WV. Transf. at White Sulphur Springs, WV to Capt. Corns' cav. 8/6/61. Served in 8th Va. Cav. and 16th Va. Cav.

NOX, H. H.: Pvt. Co. H. Enl. 7/22/62 at Camp McCausland, Monroe Co., WV. NFR.

NOYES, BENJAMIN S.: Pvt. Co. H. Clerk, Charleston, WV, 1860, age 19. Res. Kanawha Co., WV. Enl. 8/28/62 in Monroe Co., WV. Deserter. Surrendered at Charleston, WV Mar. 1865. Took oath at Charleston, WV and sent North 3/17/65. Age: 21. Farmer. 5'8". Fair complexion, light hair, blue eyes. Bur. Spring Hill Cem., Charleston, WV.

NOYES, FRANKLIN L.: Pvt./Sgt. Co. H. Merchant, Charleston, WV. 1860, age 26. Enl. 5/8/61 at Charleston, Kanawha Co., WV. Clerk 6/30/62. Commissary clerk July 1862. Commissary Sgt. on 2/28/63 roll. Apptd. Commissary Sgt. 7/24/63. Det. A.C. Sgt. 12/25/64. POW and paroled at both Lewisburg, WV 5/10/65. Age: 32 or 33. Merchant. 5'5". Light complexion, light hair, blue/gray eyes, light whiskers.

NOYES, JAMES BRADFORD SR.: Pvt. Co. H. B. 3/4/37. Married 1859. Enl. 5/8/61 at Charleston, Kanawha Co., WV. Det. to nurse wded. Patton at Scary Creek, WV July 1861 for three months. Clerk and A.C.S. Sgt. at Lewisburg, WV 1/1/62 to 12/31/62. A.C.S. Sgt. promo. to Capt. in nitre bureau. Superintendent Jackson's River Cave June 1863. Superintendent 4th Nitre and Mining District 1/15/63 to 9/30/64. Professional assistant in the scientific arm of the Corps 12/31/64. D. 2/19/1904. Bur. Spring Hill Cem., Charleston, WV.

NOYES, JAMES BRADFORD JR.: Pvt. Co. H. Clerk, Charleston, WV, 1860, age 21. Enl. 5/8/61 at Charleston, Kanawha Co., WV. Absent cause unknown on 8/31/61 roll. Present on July-Aug. 1862 roll. AWOL 1/1/64. Present 4/1/64 roll. Capt. at Winchester, Va. 9/19/64. POW Pt. Lookout, MD 9/24/64. Relsd. on taking oath 2/25/65.

NOYES, JOHN S.: Pvt. Co. H. Enl. 11/3/61 at Cotton Hill, Fayette Co., WV. Capt. at Cold Harbor, Va. 6/3/64. POW Pt. Lookout, MD 6/11/64. Transf. to Elmira, NY 7/12/64. Took oath at Elmira, NY and relsd. 6/16/65. (?)8'¼". Florid complexion, light hair, blue eyes. Res. Cinn., OH.

NOYES, PHILLIP H.: Rank unknown. Co. H. Clerk, Charleston, WV 1860, age 25. B. 9/20/35 at Charleston, WV. Entered mercantile business at age 16. Enl. May 1861 at Charleston, Kanawha Co., WV. Not in CSR. Postwar: Wholesale goods and grocery business in Charleston, WV. Married 1868. D. 11/20/98 at Charleston, WV.

NOYES, WILLIAM A.: Pvt. Co. H. Clerk, Charleston, WV 1860, age 24. Enl. 5/8/61 at Charleston, Kanawha Co., WV. Absent cause unknown on 8/31/61 roll. Det. A.C.S. Dept. Sept.-Oct. 1862 roll. NFR.

NUGEN, FOUNTAIN: Pvt. Co. K. Enl. 6/6/61 at Charleston, Kanawha Co., WV. Absent cause unknown on 9/4/61 roll. NFR. Res. Fayette Co., WV 1860. Age: 14. Bahlmann says WIA in 1861 and died of typhoid fever 1862.

NUNALY, GEO. W.: Co. H. Deserted 7/27/62 on Aug. 1862 broadside. Greenbrier Co., WV. Probably same man as G. W. Nonley.

NUTTER, T. A.: Capt. Co. C. Recvd. in exchange for John A. Walker 4/1/64.

NUTTER, TURP: Pvt. Co. C. Res. Nicholas Co., WV. Enl. 3/25/64 at Camp Gauley, Greenbrier Co., WV. Deserted 4/28/64. Capt. in Nicholas Co., WV 3/18/64. Age: 19. POW Camp Chase, OH 4/4/64. Paroled at Camp Chase, OH. Exchd. 2/25/65. Farmer. 5'8". Dark complexion, black hair, brown eyes.

O'BRIEN, JOHN: Pvt. Taylor's Co. Enl. 6/19/61 at Frankford, Greenbrier Co., WV. Age: 25. Co. transf. to Wise Brigade 7/10/61.

O'BRIEN, WILLIAM P.: 3rd Sgt. Co. E. B. 3/14/34. Res. Roane Co., WV. Enl. 6/8/61 at Charleston, Kanawha Co., WV. Pvt. until apptd. 3rd Sgt. 7/3/61. Det. as a mounted guard at White Sulphur Springs, WV 8/13-14/61. Left with authority to raise a co. 9/16/62. PWR says recruited a co. for 3rd Va. Inf. and served as it's Capt. until Apr. 1864. Comm. Capt. of a co. of 19th Va. Cav. until end of war. Fought in fifteen pitched battles and skirmishes and in 1863 had a horse shot out from under him near home. Capt. Feb. 1865. POW Camp Chase, OH until end of war. Married 1868 in Clay Co., WV. Farmer. Postwar res. Newton, Roane Co., WV. Believed bur. in a cem. in Roane Co., WV now covered by Interstate 79.

O'DANIEL, OWEN: Pvt. Co. G. Enl. 7/3/61 at Charleston, Kanawha Co., WV. Death claim filed 3/21/62(?) by J. W. M., atty.

O'DANIEL, RUFUS: Pvt. Co. G. Res. Fayette Co., WV. Enl. 7/3/61 at Charleston, Kanawha Co., WV. Took oath and sent North 10/11/63. Age: 34. Farmer. Light complexion, sandy hair, hazel eyes.

O'DELL, WILLIAM: Pvt. Co. E. Enl. 10/12/62 at Cotton Hill, Fayette Co., WV. Sick in hosp. on Jan.-Feb. 1863 roll. Present on Mar.-Apr. 1863 roll. Deserter Feb. 1864. Prisoner on Mar.-Apr. 1864 roll.

OGBURN, MOSES M.: Pvt. Swann's Co. Enl. 5/15/61 at Charleston, Kanawha Co., WV. Age: 28. Co. transf. to Co. K, 59th Va. Inf. 7/25/61.

OILER, GEORGE W.: Pvt. Co. B. Enl. 5/?/62 at White Sulphur Springs, Greenbrier Co., WV. Det. wagoner at White Sulphur Springs, WV May 1862. Det. teamster Apr. 1864.

OILER, JOHN: Co. G. Postwar roster. Pvt. Enl. 10/?/64 at New Market, Va. Age: 21. Res. Alleghany Co., Va. Surrendered Apr. 1865.

OILER, JOHN H.: Pvt. Co. G. Enl. 10/?/64 at New Market, Va. Age: 34. Res. Alleghany Co., Va. Surrendered Apr. 1865.

OLD, THOMAS J.: Pvt. Co. B. B. in Botetourt Co., Va. (Amstersdam). Enl. 5/1/62 at White Sulphur Springs, Greenbrier Co., WV. Conscript. WIA and capt. at Lewisburg, WV 5/23/62. Age: 32. Dischd. due to wds. 2/28/63. D. 1/24/1908. Farmer. 6'1". Fair complexion, dark hair, blue eyes.

OLIVER, J. F. (or T.): Co. G. Enl. 2/?/62. Served as a Pvt. until close of war. Postwar roster.

OLIVER, JOHN HILL: Co. G. B. 2/12/23 Alleghany Co., Va. Enl. 10/?/64 at New Market, Va. Postwar farmer, Covington, Va. D. 9/19 or 20/1902 at Covington, Va. Bur. Cedar Hill Cem., Covington, Va.

OLIVER, T. J.: Co. G. Enl. 8/?/64 at Nottoway, Va. for 8 mo. WIA at Five Forks, Va. Postwar roster.

O'NEAL, DANIEL: Pvt. Co. G. Enl. 10/2/62 at Raleigh (Beckley), Raleigh Co., WV. At Greenbrier Bridge, WV on Mar.-Apr. 1863 roll. NFR.

O'NEAL, JAMES C.: Pvt. Co. K. Enl. 6/6/61 at Charleston, Kanawha Co., WV. Sick since 8/2/61 on June-Sept. 1861 roll. NFR. Bahlmann says died in fall of 1861 or winter of 1861-62. B. in North Carolina. Farmer, Fayette Co., WV 1860. Age: 39.

O'NEIL, JOHN: Pvt. Co. A. Enl. 10/5/6(?) at Camp Beauregard. Joined by transf. on 8/31/63 to 12/31/63 roll. Deserted 2/24/64.

OSBORNE, MATTHEW: Pvt. Co. B. Res. Greenbrier Co., WV. Enl. 5/?/62 at White Sulphur Springs, Va. July 1862. Surrendered at Charleston, WV and took oath there 3/28/65. Age: 28. Farmer. 5'5". Fair complexion, fair hair, gray eyes.

OTT, ADDISON: Pvt. Co. A. Enl. 5/1/62 at White Sulphur Springs, Greenbrier Co., WV. KIA at Lewisburg, WV 5/23/62.

OTT, JEWIT: Pvt. Co. A. Enl. 5/1/62 at White Sulphur Springs, Greenbrier Co., WV. KIA at Lewisburg, WV 5/23/62.

OVERSTREET, STARK: Pvt. Co. H. Laborer, Charleston, WV 1860. Age: 19. Enl. 5/29/61 at Charleston, Kanawha Co., WV. WIA (slight finger) at Cold Harbor, Va. 6/2-3/64. In hosp. for amputation of right thumb 6/5-28/64. Capt. at Winchester, Va. 9/19/64. POW Pt. Lookout, MD 9/23/64. Relsd. 2/25/65. (At Cold Harbor his name is given as A. Overstreet). Bur. Higginbotham Cem., Charleston, WV.

OWEN, JOHN MERCER: Lt. Co. E. UDC application.

OWEN, JOHN S.: Pvt. Co. A. Res. Kanawha Co., WV. Date and place of enlistment not stated. Capt. in Boone Co., WV 11/14/63. Bushwhacker and horse thief. POW Camp Chase, OH. Transf. to Ft. Delaware. Took oath at Ft. Delaware 6/20/65. 6'. Light complexion, brown hair, gray eyes.

OWENS, J. S.: Pvt. Co. K. Date and place of enlistment not stated. D. at Montgomery White Sulphur Springs, Va. 7/3/62. Claim filed by John M. Rowan, atty. 1/31/63.

OXIER, LORENZO D.: Pvt. Co. I. Enl. 2/22/64 in Boone Co., WV. WIA (slight finger) at Cold Harbor, Va. 6/3/64. In Winchester, Va. hosp. for rheumatism 8/2-4/64. Capt. at Winchester, Va., 9/19/64. POW Pt. Lookout, MD. Took oath at Pt. Lookout 5/2/65. Age: 17. Farmer. 5'6". Fair complexion, fair hair, gray eyes.

OXSHIRE, ________: Co. I. In Winchester, Va. hosp. for rheumatism 1864.

PACK, JOSEPH B.: Pvt. Co. H. Res. Monroe Co., WV. Enl. 5/1/62 at White Sulphur Springs, Greenbrier Co., WV. Present throughout until transf. to Co. A, Edgar's Battn. 3/20/63. Believed still alive 1914.

PALMER, JOHN: Pvt. Co. B. Res. Franklin Co., Va. Date and place of enlistment not stated. Appears on a list of deserters at Charleston, WV and took oath 2/20/65. Age: 39. Farmer. 5'7¾". Dark complexion, light hair, gray eyes.

PARCELL, JOHN PETER: Hale's Artillery. B. 9/14/29 Henry Co., Va. Res. Charleston, WV 1845. Postwar saltmaker/gardening, Malden, WV.

PARDEE, STEPHEN: Pvt. Co. I. Date and place of enlistment not stated. Appears on a list of deserters and surrendered at Charleston, WV Mar. 1865. Took oath and sent North. Farmer.

PATRICK, ALFRED SPICER: Surgeon/Pvt. Co. H. B. 1831 Charleston, WV. Grad. Marietta College 1849. Marietta College, M.D. 1853. Doctor, Mason Co., WV. Res. Charleston, WV 1860. Reportedly first Pvt. to be mustered into Confederate service in the Kanawha Valley. Enl. 5/8/61 at Charleston, Kanawha Co., WV. Fought at Scary Creek, WV as Pvt. 7/17/61. Det. asst. surg. for hosp. duty 7/23/61 and still so engaged on July-Aug. 1861 roll. Asst. surg. at White Sulphur Springs, WV hosp. Nov. 1861. Married 1862. Det. asst. surg. winter 1863-64. Promo. from asst. surg. to regt. surg. Nov. 1864. Apptd. Surg. 2/11/65. Paroled at Lewisburg, WV 4/26/65 and at Charleston, WV 5/10/65. Age: 33. 6'1½". Light complexion, dark hair, blue eyes. Doctor, Charleston, WV until 1872. Doctor, Lewisburg, WV until 1889. Then doctor at Charleston, WV. Member of Greenbrier Co. and Charleston Medical and Surgical Societies. Member Medical Society of WV. D. 1906. Bur. Spring Hill Cem., Charleston, WV.

PATRICK, JOHN M.: Pvt. Co. H. B. in NY. Res. Malden, WV 1860. Age: 17. Enl. 7/10/62 at White Sulphur Springs, Greenbrier Co., WV. Transf. from Jackson's Battery. Earliest dated roll is July-Aug. 1862. Capt. at Cold Harbor (Gaines Mill), Va. 6/3/64. POW Pt. Lookout, MD. Transf. to Elmira, NY 7/12/64. Killed in a train wreck at Shohola, PA 7/15/64 en route to Elmira, NY. Brother of Alfred Spicer Patrick.

PATRICK, S. H.: Pvt. Co. H. Enl. 10/8/62 at Charleston, Kanawha Co., WV. Det. to shuck corn on Nov.-Dec. 1863 roll. Present on Apr. 1864 roll. Capt. at Winchester, Va. 9/19/64. POW Pt. Lookout, MD 9/24/64. Relsd. and took oath 2/25/65.

PATTERSON, JOHN M.: Co. K. B. Rockbridge Co., Va. Farmer, age 34, Lexington Dist., 1860. Originally in Rockbridge Rangers Home Guards. Enl. 1861 at Lexington, Va. Exempt March 1862 for consumption. Absent sick from 1/20/65 to end of war. Farmer, age 45, Buffalo Dist., 1870. Laborer, age 77, filed for pension, 1902. D. Roanoke, Va. 5/23/1903. Bur. Roanoke, Va.

PATTON, GEORGE SMITH: Col./Lt. Col./Capt. Field & Staff/Co. H. B. 6/26/33 Fredericksburg, Va. Fourth son of John M. Patton. Moved to Richmond, Va. and educated there. V.M.I. 1849-52. Teacher, Richmond, Va. 1853-54. Studied law. Married 1855. After admission to Bar, 1856 he moved to Charleston, WV to practice law with Thomas Lee Brown. Organized and led Kanawha Riflemen 1856-61. Enl. 5/22/61 at Charleston, Kanawha Co., WV. Elected Capt. 5/22/61. Apptd. Lt. Col. 7/7/61. WIA (severe left shoulder) at Scary Creek, WV 7/17/61. Left as an exchange prisoner in Kanawha Valley during Gen. Wise's retreat July 1861. Offered a teaching job at V.M.I. if unable to return to duty. Returned to regt. as Col. in Apr. 1862. WIA (slight) at Giles C.H. 5/10/62. Often commanded Echols' Brig. during Gen. Echols' frequent periods of absence. MWIA (leg) at Winchester, Va. 9/19/64. D. 9/25/64 at John J. Williams' house, Winchester, Va. Bur. Mt. Hebron Cem., Winchester, Va. Patton was grandfather of Gen. George S. Patton of World War II fame. Records indicate Patton was apptd. Col. 3/3/63 to rank from 11/23/61 and accepted 1/15/63.

PATTON, JAMES FRENCH: Adj. B. 9/19/43 Richmond, Va. Son of John M. Patton. Brother of George S. Patton. Attd. Hanover Academy 1857. Res. Culpepper, Va. 1859. Entered Confederate service 1861 in 22nd Va. Inf. "Served through the war, receiving several wounds, one of which narrowly escaped being fatal." Postwar Judge, Pittsylvania Co., Va. Later res. Monroe Co., WV and county prosecutor. WV Supreme Court Judge 1881. D. 3/30/82 or 4/6/82 of "chest pain" complications.

PAUL, ARMISTEAD: Chief Cook. Colored. Co. H. Enl. 5/8/61 at Charleston, Kanawha Co., WV. Absent on furlo. on Nov.-Dec. 1863 roll. Present on Apr. 1864 roll. Paroled at Charleston, WV 5/11/65. Age: 40. Farmer. 5'6". Black complexion, black hair, dark eyes. Apparently mis-filed and should be under William Armistead.

PAULEY, JAMES: Pvt. Swann's Co. Enl. 4/25/61 at Charleston, Kanawha Co., WV. Age: 27. Deserted on June 1861 roll.

PAULEY, JOHN: Pvt. Swann's Co. Enl. 4/25/61 at Charleston, Kanawha Co., WV. Age: 26. Deserted on June 1861 roll.

PAULEY, SAMUEL: Pvt. Co. I. Res. Boone Co., WV. Enl. 6/19/61 at Charleston, Kanawha Co., WV. Court-martial records on file dated circa. 8/15/64. Deserted 1865. Surrendered Feb. 1865. Took oath at Charlston, WV 2/9/65 and sent North. Age: 22. Farmer. Sandy complexion, sandy hair, blue eyes.

PAXTON, ABNER JOHN: Pvt. Co. K. B. Botetourt Co., Va. 1817. Res. Craig Co., Va. At Camp Lee, Richmond, Va. 9/1/64. Enl. 9/18/64 at Winchester, Va. Capt. at Fisher's Hill, Va. 9/21/64. POW Pt. Lookout, MD 10/1/64. Exchd. 1/17/65. Deserted 1/24/65. Also served in 14th Va. Cav. D. in WV 1898 (?). PWR says B. in Craig Co., Va.

PAXTON, CEPHAS S.: Co. K. B. 4/28/43 Craig Co., Va. Enl. 2/10/63 at Lewisburg, Greenbrier Co., WV. Deserted 5/1/62 on Aug. 1862 broadside, New Castle, Va. Deserted at Camp Gauley, WV 4/25/64. Res. Craig Co., Va. Brother of Abner Paxton. Bahlmann says enl. 1862. Nickname is 'Ceif' and middle name is Shelburne. D. 1/1/1918 at Witcher's Creek, WV.

PAXTON, G. W.: Pvt. Co. G. Enl. 5/1/62 at White Sulphur Springs, Greenbrier Co., WV. Deserted 5/1/62. Listed as deserter Aug. 1862 broadside. Left 5/24/62.

PAXTON, MILES A.: Pvt. Co. K. B. ca. 1842 Craig Co., Va. Res. Craig Co., Va. Enl. 4/24/62 at White Sulphur Springs, Greenbrier Co., WV. Capt. at Winchester, Va. 9/19/64. POW Pt. Lookout, MD 9/23/64. Exchd. 3/15/65.

PAXTON, THOMAS B.: Son of Billy Paxton. Listed in UDC booklet *Craig's Share in the War Between the States*. Res. Craig Co., Va.

PAXTON, THOMAS B., SR.: Listed in UDC booklet *Craig's Share in the War Between the States*. Res. Craig Co., Va.

PAYETTE, JOHN: Pvt. Co. A. Date and place of enlistment not stated. Deserted and surrendered Apr. 1865 and sent North. Took oath at Charleston, WV 4/24/65. Age: 22. Farmer. 5'10". Fair complexion, fair hair, gray eyes.

PAYNE, JAMES: Pvt. Co. K. Enl. 6/6/61 at Charleston, Kanawha Co., WV. Absent cause unknown on June-Sept. 1861 roll. NFR.

PAYNTER, FRANCIS M.: 1st Lt. Co. K. Enl. 6/6/61 at Charleston, Kanawha Co., WV. 1st Lt. 12/24/61. Absent sick since 8/23/61 on June-Sept. 1861 roll. NFR. Farm laborer, Fayette Co., WV 1860. Age: 24. Bahlmann says died 1862.

PAYNTER, HIRAM HILL: Sgt./3rd Corp. Co. K. Enl. 6/6/61 at Charleston, Kanawha Co., WV. 3rd Corp. on Nov.-Dec. 1863 roll. Absent sick 9/10/64 but returned on Mar.-Oct. 1864 roll. Paroled at Charleston, WV 5/24/65. Age: 23. 5'10". Dark complexion, brown hair, gray eyes. Res. Putnam Co., WV 1910 at age 69. PWR says spelled Painter.

PAYNTER, M.: 1st Lt. Co. K. Date and place of enlistment not stated. Elected 1st Lt. Dec. 1861. D. 1/10/62. Successor: William F. Bahlmann. (This is probably Francis M. Paynter).

PEALE, GEORGE A.: Pvt. Swann's Co./Co. H. Enl. 5/8/61 at Charleston, Kanawha Co., WV. Age: 24. Co. transf. to Co. K, 59th Va. Inf. 7/25/61. Transf. to Co. H, 22nd Va. Inf. 3/19/62. Retransf. to Swann's Co., Edgar's Battn. 7/6/62.

PEALE, HARVEY: Pvt. Swann's Co. Enl. 6/2/61 at Charleston, Kanawha Co., WV. Age: 19. Co. transf. to Co. K, 59th Va. Inf. 7/25/61.

PEARSON, JAMES S.: Pvt. Co. I. Enl. 5/29/61 at Charleston, Kanawha Co., WV. AWOL 7/26/61.

PEASE, GEORGE W.: 4th Sgt. Swann's Co. Clerk, Charleston, WV 1860. Enl. 5/8/61 at Charleston, Kanawha Co., WV. Age: 27. Co. transf. to Co. K, 59th Va. Inf. 7/25/61. Also spelled Peas.

PEIRCY, T. H.: Pvt. Co. D. Date and place of enlistment not stated. Captured and paroled at Burksville Junction, Va. 4/14-17/65.

PEIRCY, W. J.: Pvt. Co. D. Date and place of enlistment not stated. Appears on a list of paroled prisoners at Burksville Junction, Va. 4/14-17/65.

PERDUE, JAMES: Pvt. Co. E. Enl. 5/1/62 at White Sulphur Springs, Greenbrier Co., WV. In Chimborazo Hosp. 6/17/64. Capt. at Winchester, Va. 9/19/64. POW Pt. Lookout, MD 9/23/64. Exchd. 3/15/65.

PERFATER, JOHN E.: Pvt. Co. G./Co. H. Enl. 7/3/61 at Charleston, Kanawha Co., WV. Absent on furlo. from 7/24/61 until fit for duty on surg. cert.

PERKINS, JOSEPH A.: Shoemaker. D. 1/13/1912 at Blue Sulphur Springs, WV. Age: about 82. Bur. M. E. Church on Snake Run.

PERKINS, THOMAS H.: Pvt. Co. C. Enl. 5/1/62 at White Sulphur Springs, Greenbrier Co., WV. Prisoner charged with disobedience on orders and had not had a trial as of 8/14/61 at White Sulphur Springs, WV. Res. Monroe Co., WV. PWR says enl. 5/10/61.

PERROW, CHARLES HENRY: B. in Nelson Co., Va. 11/21/37. In 22nd Va. Inf. & 8th Va. Cav. D. in Grainger Co., TN 4/7/1904. Bur. New Gray Cem., Knoxville, TN.

PERRY, EDWARD R.: Pvt. Co. B. Res. Greenbrier Co., WV. Date and place of enlistment not stated. Deserted. Took oath 11/15/63 and sent north of the Ohio River. Age: 42. Carpenter. 6'. Light complexion, brown hair, gray eyes.

PERRY, LEWIS M.: Pvt. Co. B. Enl. 6/1/61 at Ripley, Jackson Co., WV. Deserted 11/16/64.

PERRY, ROBERT: Pvt. Co. A. Date and place of enlistment not stated. Paroled POW. Paroled at Charleston, WV 5/11/64. Age: 21. 6'2". Light complexion, brown hair, blue eyes.

PERRY, ROBERT E.: Pvt. Co. I. Enl. 6/29/61 at Charleston, Kanawha Co., WV. AWOL 7/26/61.

PERRY, WILLIAM GEORGE: Corp. Co. F. Res. Monroe Co., WV. Enl. 7/4/61 at Charleston, Kanawha Co., WV. At home sick on surg. cert. 8/10/61. Postwar res. Ronceverte, WV. Postwar roster says name is George W. Perry. D. sometime prior to 1914.

PERSINGER, ANDERSON: Pvt. Co. A. Res. Kanawha Co., WV. Enl. 6/11/61 at Camp Tompkins, Kanawha Co., WV. Present throughout. Took oath at Charleston, WV and sent North 4/20 or 22/65. Age: 21. Farmer. 5'10". Dark complexion, dark hair, hazel eyes. Res. Putnam Co., WV 1910 at age 66.

PERSINGER, CHARLES A.: Pvt. Co. B. Res. Alleghany Co., Va. Enl. 7/12/62 at Camp McCausland (Union), Monroe Co., WV. Deserted on Sept.-Oct. 1862 roll. Surrendered Dec. 1863. Took oath and sent North 12/15/63. Age: 25. Farmer. 5'9". Dark complexion, black hair, blue eyes.

PERSINGER, CHARLES O.: Pvt. Co. D. B. ca. 1841(?). Res. Oakum Depot. Enl. 5/1/62 at White Sulphur Springs, Greenbrier Co., WV. Capt. at Cold Harbor (Gaines Mill), Va. 5/31/64. POW Pt. Lookout, MD. Transf. to Elmira, NY 7/9/64. Took oath at Elmira, NY and relsd. 6/30/65. 5'11". Fair complexion, dark hair, blue eyes. On Alleghany Co., Va. pension rolls.

PERSINGER, GEORGE W.: Pvt. Co. C. B. ca. 1825 Craig Co., Va. Enl. 5/28/63 at Camp Stonewall. Sick at home on Nov.-Dec. 1863 roll. Present on Apr. 1864 roll. Deserted 5/10/64.

PERSINGER, HARVEY J.: Pvt. Co. D. Enl. 7/8/63 at Camp Sam Jones, Greenbrier Co., WV. At home wded. on Apr.-Oct. 1864 roll.

PERSINGER, JOHN: Pvt. Co. D. B. 1835. Res. Alleghany Co., Va. Enl. 5/1/62 at White Sulphur Springs, Greenbrier Co., WV. Deserted Dec. 1863. Took oath and sent North 12/15/63. Age: 22. Farmer. 5'11". Dark complexion, brown hair, brown eyes.

PERSINGER, JOSEPH W. H.: Pvt. Co. A. Res. Alleghany Co., Va. Enl. 4/10/62 at Brushy Ridge, Greenbrier Co., WV. Deserted at Packs Ferry, WV 10/17/62. Took oath and sent North 12/15/63. Age: 30. Farmer. 5'9". Dark complexion, brown hair, brown eyes.

PETERS, CHRISTOPHER: Pvt. Co. D. Enl. 6/3/61 at Clifton (Pratt), Kanawha Co., WV. Present on 8/31/61 roll. NFR.

PETERS, E. C.: 1st Lt. Co. C. Enl. 5/1/62 at White Sulphur Springs, Greenbrier Co., WV. Elected 1st Lt. 5/1/62. Comm. 1st Lt. 5/2/62. AWOL on Nov.-Dec. 1862 roll. Present on Jan.-Feb. 1863 roll. Present under arrest on Nov.-Dec. 1863 roll. Cashiered Mar. 1864. Successor: W. A. Tyree.

PETRY, WASHINGTON: Pvt. Co. I. Enl. 6/29/61 at Charleston, Kanawha Co., WV. Present on 8/31/61 roll. NFR.

PETTRY, JOHN: Co. K. Enl. 6/6/61 at Charleston, Kanawha Co., WV. Accidentally wded. 7/24/61. NFR.

PEYTON, WILLIAM H.: Pvt. Co. I. Enl. in Co. B, 36th Va. Inf. 6/1/61 at Peytona, WV. Age: 16. Transf. to 22nd Va. Inf. 8/1/62. Enl. 4/15/62 in Greenbrier Co., WV. (?). Deserted at Lewisburg, WV 1/24/64. Bur. Spring Hill Cem., Charleston, WV.

PHELPS, CHARLES W.: Pvt. Co. B. Res. Sandy, WV 1860, age 18. Enl. 6/1/61 at Ripley, Jackson Co., WV. Served as 2nd Lt. in Co. G, 10th Va. Cav.

PHELPS, EDWARD: Co. B. Res. Jackson Co., WV.

PHILLIPS, DABNEY M.: Pvt. Co. H. B. 3/26/28 Lynchburg, Va. Res. Craig Co., Va. Enl. 7/4/62 at Charleston, Kanawha Co., WV. Assigned to and joined this co. at Camp McCausland, WV 8/4/62. Det. to Maj. Gen. Sam Jones' office on Nov.-Dec. 1862 roll. Dischd. for pulmonary disease on Mar.-Apr. 1863 roll. Age: 35. Clerk. 6'2". Fair complexion, dark hair, blue eyes. D. 5/7/90. Bur. West View Cem., Old Rt. 42, New Castle, Va.

PHILLIPS, JOHN F.: Drummer. Co. A. Enl. 5/22/61 at Poca, Putnam Co., WV. Det. to go to Richmond, Va. with ten Yankee prisoners 8/28/61. Absent on furlo. on Jan.-Feb. 1863 roll. Present on Mar.-Apr. 1983 roll. Det. service on Aug.-Dec. 1863 roll. Present on Apr. 1864 roll. Capt. and paroled at Lewisburg, WV 4/26/65. Age: 55. Carpenter. 5'9". Fair complexion, dark hair, dark eyes.

PICKENS, BENJAMIN F.: Pvt. Co. B. Enl. 9/22/62 at Charleston, Kanawha Co., WV. Capt. and paroled at Charleston, WV 5/12/65. Age: 20. Farmer. 5'2". Fair complexion, light hair, gray eyes.

PICKENS, JOHN P.: Co. B. Res. Jackson Co., WV. Enl. 6/1/61 at Ripley, Jackson Co., WV. Present until deserted Feb. 1865. Took oath at Charleston, WV 2/16/65. Age: 24. Farmer. 5'6". Fair complexion, light hair, gray eyes.

PIERSON, BENJAMIN: Pvt. Co. C. Enl. 5/10/61.

PILDS, D. M.: Pvt. Co. G. Enl. 5/1/62 at White Sulphur Springs, Greenbrier Co., WV. Listed as deserter on Aug. 1862 broadside Left 5/24/64. Deserted 5/1/62.

PILES, ADAM: Pvt. Co. D. Enl. 5/1/62 at White Sulphur Springs, Greenbrier Co., WV. Deserted 5/1/62. Listed as deserter on Aug. 1862 broadside. Residence unknown.

PIPKIN, J. J.: Co. M(?). D. at Ft. Delaware 4/12/64. Bur. Finn's Point Nat. Cem., NJ. PWR says a Pvt. in Co. A and name is Pipkins.

PLATT, NICHOLAS P. J.: 3rd Corp. Co. K. Enl. 6/6/61 at Charleston, Kanawha Co., WV. Dischd. for disability on surg. cert. 6/20/61.

PLIMALL, V. S.: Pvt. Co. G. Enl. 5/1/62 at White Sulphur Springs, Greenbrier Co., WV. Deserted 5/1/62. Listed as deserter on Aug. 1862 broadside. Left 5/24/62.

PLYMALE, GEORGE W.: Pvt. Co. D. Res. Alleghany Co., Va. Enl. 5/1/62 at White Sulphur Springs, Greenbrier Co., WV. Listed as deserter on Aug. 1862 broadside. Capt. at Droop Mountain, WV 11/6/63. Age: 21. POW Camp Chase, OH 11/18/63. Transf. to Ft. Delaware 2/29/64. Relsd. 6/7/65. Farmer. 5'9½". Florid complexion, dark hair, gray eyes.

POAGUE, JOHN E.: Co. K. Age: 18, Lexington Dist., 1860. Res. Rockbridge Co., Va. PWR.

POLLOCK, JOHN F.: Pvt. Taylor's Co. Enl. 6/19/61 at Frankford, Greenbrier Co., WV. Age: 20. Co. transf. to Wise Brigade 7/10/61.

POPE, WILLIAM H.: Pvt. Co. A. PWR.

PORTIER, ISADORE: Pvt./Buglar. Swann's Co. B. in France. Master blacksmith, Charleston, WV 1860. Enl. 5/8/61 at Charleston, Kanawha Co., WV. Age: 29. Co. transf. to Co. K, 59th Va. Inf. 7/25/61. Bur. Spring Hill Cem., Charleston, WV.

POTEET, GEORGE: Pvt. Co. K. Enl. 6/6/61 at Charleston, Kanawha Co., WV. Capt. at Winchester, Va. 9/19/64. POW Pt. Lookout, MD. Exchd. 3/15/65. In Richmond, Va. hosp. 3/18/65. Also served with Capt. Thurmond's Co.

POTTER, JOHN S.: Pvt. Co. C. Res. Fayette Co., WV. Enl. 6/5/61 at Charleston, Kanawha Co., WV. Court-martial record dated 9/29/63. In Charlottesville, Va. hosp. with rubeola 6/9-17/64. Deserted 6/30/64. Took oath and sent North 4/4/65. Age: 22. Farmer. 5'2". Dark complexion, black hair, black eyes.

POWEL, WASHINGTON A. J.: Pvt. Co. D. Enl. 5/30/61 at Summersville, Nicholas Co., WV. KIA at Lewisburg, WV 5/23/62.

POWELL, HENRY W.: Pvt. Co. D. Enl. 5/30/61 at Summersville, Nicholas Co., WV. AWOL 7/28/61.

POWELL, MARTIN VAN B.: Pvt. Co. D. Enl. 5/30/61 at Summersville, Nicholas Co., WV. MWIA in a skirmish on Big Sewell Mountain, WV 12/12/63. Died same day.

POWELL, THOMAS A.: Pvt. Co. D. Enl. 9/13/63 at Camp Price. Capt. at Winchester, Va. 9/19/64. POW Pt. Lookout, MD 9/24/64. Exchd. 1/17/65. Paroled at Charleston, WV 5/10/65. Age: 19. 5'9½". Fair complexion, light hair, gray eyes. Alive in Oklahoma 1911.

POWER, J. A.: Co. D. Alive in Oklahoma 1911.

PRICE, ABRAHAM J.: Pvt. Co. B. Age: 16 in 1860. Enrolled 1861 for 4 years. Res. Jackson Co., WV.

PRICE, BENJAMIN: Corp. Co. I. Enl. 6/29/61 at Charleston, Kanawha Co., WV. AWOL 7/26/61.

PRICE, DICKSON: Pvt. Co. A. Enl. 5/1/62 at White Sulphur Springs, Greenbrier Co., WV. D. at a private house in Monroe Co., WV 9/1/62.

PRICE, JAMES POLK: Pvt. Co. E. Enl. 10/1/62 at Charleston, Kanawha Co., WV. Left sick in Charleston, WV Oct. 1862. Bur. Spring Hill Cem., Charleston, WV.

PRICE, JAMES W.: Pvt. Taylor's Co. Enl. 6/19/61 at Frankford, Greenbrier Co., WV. Age: 23. Co. transf. to Wise Brigade 7/10/61.

PRICE, JOHN: Pvt. Taylor's Co. Enl. 6/19/61 at Frankford, Greenbrier Co., WV. Age: 30. Co. transf. to Wise Brigade 7/10/61.

PRICE, MARSHALL: 5th Sgt. Co. K. Res. Fayette Co., WV. Enl. 6/6/61 at Charleston, Kanawha Co., WV. Served as a Pvt. until elected 5th Sgt. 6/24/61. Absent sick 9/1/61. Left sick in Fayette Co., WV on retreat from Kanawha on Nov.-Dec. 1862 roll. Sick on Jan.-Feb. 1863 roll and Mar.-Apr. 1863 roll. Capt. in Fayette Co., WV 4/22/63. Age: 42. POW Camp Chase, OH 5/11/63. Transf. to Johnson's Island, OH. Transf. to Ft. McHenry, MD 7/21/63 en route to City Point, Va. for exchange. Exchd. 8/4/63. Present on Nov.-Dec. 1863 roll. Tailor. 5'6¼". Dark complexion, dark hair, gray eyes.

PRICE, VINCENT: Pvt. Co. E. Res. Roane Co., WV. Enl. 6/18/61 at Charleston, Kanawha Co., WV. Capt. in Roane Co., WV 1/18/64. Age: 26. POW Camp Chase, OH 3/4/64. Transf. to Ft. Delaware 3/14/64. Took oath at Ft. Delaware and relsd. 6/20/65. 5'8". Fair complexion, dark hair, gray eyes.

PRICE, WILLIAM: Pvt. Taylor's Co. Enl. 6/19/61 at Frankford, Greenbrier Co., WV. Age: 31. Co. transf. to Wise Brigade 7/10/61.

PROGLER, JULES L.: 1st Lt. Co. B. B. in Switzerland. Res. Jacksion Co., WV 1860, Superintendent of Progler Hotel. Enl. 6/1/61 at Ripley, Jackson Co., WV. Elected 1st Lt. 6/1/61. Resgnd. 8/1/61. PWR says name is Julius.

PRUDEN, KENSEY: Pvt. Co. B. Res. Jackson Co., WV. Enl. 7/1/61 at Ripley, Jackson Co., WV. Det. as shoemaker at Salem, Va. Jan. 1863 through Mar. 1863. Det. as boatman Nov. 1863. Det. as boatman at Narrows of New River, Va. Dec. 1863 through June 1864. Capt. at Winchester, Va. 9/19/64. POW Pt. Lookout, MD 9/24/64. Exchd. 3/15/65. Took oath and sent North 4/12/65. Age: 27. Shoemaker. 5'7". Dark complexion, dark hair, dark eyes.

PUCKETT, J. A.: Co. G. Enl. 1864 for one year. Postwar roster.

PUGH, JOHN T.: Pvt. Co. F. Res. Monroe Co., WV. Enl. 8/6/61 at White Sulphur Springs, Greenbrier Co., WV. WIA (severe shoulder) at White Sulphur Springs, WV 8/27/63. PWR says KIA at White Sulphur Springs, WV 8/26-27/63. Also spelled name Pew.

PURCELL, WILLIAM: Pvt. Swann's Co. Enl. 6/20/61 at Charleston, Kanawha Co. ,WV. Age: 28. Co. transf. to Co. K, 59th Va. Inf. 7/25/61. (Probably William Parcell, brother of John Peter Parcell).

PURSLEY, WILLIAM E.: Corp. Co. I. Enl. 10/15/63 at Lewisburg, Greenbrier Co., WV. Capt. at Winchester, Va. 9/19/64. POW Pt. Lookout, MD 9/24/64. Exchd. 3/15/65.

QUARRIER, CARL S.: Pvt. Co. H. PWR.

QUARRIER, CHARLES C.: Pvt./Sgt. Co. H. B. in Va. Clerk, Charleston, WV 1860, age 16. Enl. either 6/5/61 at Coalsmouth (St. Albans), Kanawha Co., WV or 5/1 or 8/61 at Charlreston, Kanawha Co., WV. WIA (very serious flesh wd. right leg below knee joint caused by round ball) and capt. at Winchester, Va. 9/19/64. POW Pt. Lookout, MD and paroled 10/18/64. In Baltimore, MD Gen. Hosp. 10/25/64. Exchd. 10/31/64. Paroled at both Lewisburg, WV 4/25/65 and at Charleston, WV 5/10/65. Age: 22. Farmer. 5'10". Fair/light complexion, light/brown hair, gray/blue eyes, light whiskers. D. after the war from his wound.

QUARRIER, JOEL SHREWSBURY: Pvt. Co. H. Gentleman, Charleston, WV 1860, age 28. Enl. 5/8/61 at Charleston, Kanawha Co., WV. Dischd. at White Sulphur Springs, WV for disability 8/15/61. Deputy before and after the war. Kanawha Co., WV clerk, 1872. Bur. Spring Hill Cem., Charleston, WV.

QUARRIER, MONROE: Pvt./Sgt. Maj. Field & Staff/Co. H. B. Charleston, WV. Enl. 5/8/61 at Charleston, Kanawha Co., WV. Apptd. Sgt. Maj. 9/1/62. WIA (slight chin) at White Sulphur Springs, WV 8/26-27/63. WIA (gunshot) at New Market, Va. 5/15/64. Honorably dischd. from service due to wd. Age: 26. Carpenter. 6'2". Dark complexion, black hair, gray eyes. Postwar res. Lokota, WV. Entered Richmond, Va. Old Soldiers Home 9/22/1926 at age 81. D. 4/13/1928. Bur. Hollywood Cem., Richmond, Va.

QUARRIER, WILLIAM A.: Pvt./Lt. Hale's Artillery/Co. H. B. 10/1/28 in Kanawha Co., WV. Attended Mercer Academy. Lawyer, Charleston, WV 1860, age 31. Enl. 1861 at Charleston, Kanawha Co., WV. Served as a Lt. in Hale's Artillery and fought as such at Scary Creek, WV 7/17/61. Resgnd. from Hale's Artillery 8/21/61 and was in Commissary Dept. at Saltville, Va. until close of war. Postwar: returned to Charleston, WV and formed law partnership with Samuel A. Miller, and later by himself. Ran for U.S. Senate but lost. D. 9/10 or 11/88 after a short illness at Charleston, WV. Bur. Spring Hill Cem., Charleston, WV.

QUICK, WILLIAM: Pvt. Co. C. Enl. 5/1/62 at White Sulphur Springs, Greenbrier Co., WV. In Chimborazo Hosp. 6/30/64. Sent to Lynchburg, Va. 7/9/64. At home wded. on 10/31/64 roll. Paroled at Charleston, WV 6/9/65. Age: 25. 5'7". Dark complexion, dark hair, black eyes.

QUIGLEY, WILLIAM E.: Swann's Co. Laborer, Charleston, WV 1860. Enl. 5/25/61 at Charleston, Kanawha Co., WV. Age: 23. Co. transf. to Co. K, 59th Va. Inf. 7/25/61.

RADER, ARCHIBALD S.: Corp. Taylor's Co. Enl. 6/19/61 at Frankford, Greenbrier Co., WV. Age: 23. Co. transf. to Wise Brigade 7/10/61.

RADER, A. S.: Pvt. Co. B. Res. Jackson Co., WV. Enl. 5/?/62 at White Sulphur Springs, Greenbrier Co., WV. WIA and capt. at Lewisburg, WV 5/23/62. D. of consumption at U.S. hosp. at Gallipolis, OH 7/17/62. Eve Rader, widow. Confederate records state died in northern prison 7/17/62.

RADER, MICHAEL: Pvt. Co. D. Enl. 5/30/61 at Summersville, Nicholas Co., WV. AWOL 7/28/61.

RADER, ROBINSON: Pvt. Co. D. Enl. 5/30/61 at Summersville, Nicholas Co., WV. Enlisted but never joined the co. in camp. AWOL.

RAINS, CHRISTOPHER: Pvt. Co. F. Enl. 7/5/62 at Camp Royal Oak. Sick at Lewisburg, WV on Mar.-Apr. 1863 roll. Dischd. on surg. cert. 7/31/63.

RAMSEY, W. T.: Pvt. Co. B. Enl. 5/?/62 at White Sulphur Springs, Greenbrier Co., WV. Present on 4/30/63 roll. NFR.

RAND, HENRY W.: Sgt./1st Lt. Co. H. Clerk, Charleston, WV 1860. Age: 24. Res. Kanawha Co., WV. Enl. 5/8/61 at Charleston, Kanawha Co., WV. Served as a Pvt. until elected Corp. 6/6/61. Elected 4th Sgt. 7/27/61. Elected 2nd Lt. 5/1/62. Promo. 8/26/63. Capt. at Winchester, Va. 9/19/64. POW Ft. Delaware 9/25/64. Capt. Jan. 1864 at head of Big Sycamore Creek, Clay Co., WV. Escaped and returned to regt. 5'10¾". Fair complexion, dark hair, brown eyes. Res. Kanawha Co., WV. D. prior to 1901. Bur. Spring Hill Cem., Charleston, WV.

RAND, NOYES ("PLUS"): Pvt./Corp./1st Lt./Adj. Maj. Field & Staff Co. H. B. 1840 Kanawha Co., WV. Washington College. Clerk, Charleston, WV 1860, age 19. Enl. 5/8/61 at Charleston, Kanawha Co., WV. Pvt. until elected Corp. 6/6/61. Apptd. Adj. 10/7 or 8/61. Capt. at Lewisburg, WV 5/23/62. POW Camp Chase, OH 5/30/62. Transf. to Johnson's Island, OH 7/3/62. Sent to Vicksburg, MISS for exchange 9/1/62. Listed as Lt. on staff, 1st Brig., Wharton's Div., 7/23/64 on report filed by Col. Patton. Paroled at Lewisburg, WV 4/26/65. Age: 25. Merchant. 5'6". Dark complexion, dark hair, hazel eyes. Veterans Grave Registration Card says served in Co. A. D. 3/19/1911 El. Paso, TX. Bur. Spring Hill Cem., Charleston, WV. PWR says wded. twice. Displayed gallantry in his first battle. Also known as "Rand".

RAPP, J. T.: Pvt. Co. A. Date and place of enlistment not stated. POW paroled at Staunton, Va. 5/18/65. Age: 23(?). 5'11". Fair complexion, dark hair, blue eyes.

RATLIFF, OWEN: Pvt. Co. D. Res. Nicholas Co., WV. Enl. 5/30/61 at Summersville, Nicholas Co., WV. Capt. either in Greenbrier Co., WV or Monroe Co., WV 5/23/62 (Probably capt. at Lewisburg, WV battle). Age: 18. POW Camp Chase, OH 8/29/62. Transf. to Johnson's Island, OH 9/6/62. Exchd. 11/22/62. Sick at Lewisburg, WV 4/13/63. Present on 12/31/63 roll. Deserted on Blue Ridge on 10/31/64 roll. Farmer. 5'7". Fair complexion, light hair, gray eyes.

RAY, CHESLEY: Pvt. Co. K. B. in Va. Enl. 3/14/64 at Camp Gauley, Greenbrier Co., WV. Conscript. In confinement on 5/1/64 roll. Capt. at Cold Harbor, Va. 6/3/64. Age: 22. POW Pt. Lookout, MD. Wished to take oath and go to Ohio where he had a brother. Transf. to Elmira, NY 7/12/64. D. at Elmira, NY of chronic diarrhea 8/4/64. Bur. Elmira, NY, Woodlawn Cem., grave 22. POW record lists him as Chas. Ray.

RAY, JOHN: Pvt. Swann's Co. Enl. 5/8/61 at Charleston, Kanawha Co., WV. Age: 21. Co. transf. to Co. K, 59th Va. Inf. 7/25/61.

RAY, WILLIAM T.: Pvt. Swann's Co. Enl. 5/8/61 at Charleston, Kanawha Co. WV. Age: 23. Co. transf. to Co. K, 59th Va. Inf. 7/25/61. Postwar res. Poca Dist., WV. Bur. Spring Hill Cem., Charleston, WV.

RAYHILL, WILLIAM H.: Pvt. Co. K. B. 8/16/39. Res. Craig Co., Va. Enl. 4/24/62 at White Sulphur Springs, Greenbrier Co., WV. Present on 11/1/62 roll. PWR says transf. to Co. A, 60th va. Inf. D. 6/14/1913 at Paint Back, Va. at age 73, 9 months, 28 days. Bur. Ruble Cem., Rt. 603, near Helen Baker farm, Craig Co., Va.

READ, FREDERICK N.: Pvt./1st Sgt. Co. H. Lawyer, Charleston, WV 1860. Age: 25. Enl. 5/8/61 at Charleston, Kanawha Co., WV. Served as Corp. until elected 1st Sgt. 7/27/61. NFR. Census says name is N. F. Read.

REECE, JOHN A.: Musician. Co. C. Enl. 5/1/62 at White Sulphur Springs, Greenbrier Co., WV. POW paroled at Charleston, WV 5/13/65. Age: 36. 6'. Sandy complexion, brown hair, blue eyes, sandy whiskers.

REED, GEORGE W.: Pvt. Co. B. Enl. 8/?/63 at New River. Det. as Provost Guard at Lewisburg, WV on rolls from 4/1/64 to 11/30/64.

REED, GEORGE W.: Pvt. Co. F. Enl. 8/6/61 at White Sulphur Springs, Greenbrier Co., WV. On mounted guard at White Sulphur Springs, WV 8/13-14/61. AWOL 8/23/61. Postwar roster says deserted and killed.

REED, GEORGE W.: Pvt. Co. I. Res. Botetourt Co., Va. Enl. 5/2/62 at White Sulphur Springs, Greenbrier Co., WV. Sick at Bonsacks, Va. July 1862. Deserted at Lewisburg, WV 4/12/64. Took oath at Charleston, WV 4/9/65. Age: 37. Farmer. 5'7". Fair complexion, dark hair, gray eyes.

REED, JAMES H.: Pvt. Co. F. Res. Monroe Co., WV. Enl. 7/4/61 at Charleston, Kanawha Co., WV. Left sick at White Sulphur Springs, WV 8/20/61. Present on 10/31/62 and Mar.-Apr. 1863 rolls. Deserted. Took oath at Charleston, WV and sent north to Ohio 10/11/63. Age: 21. Farmer. 5'11". Light complexion, light hair, blue eyes.

REED, SAMUEL A.: Pvt./Corp. Co. B. Enl. 6/1/61 at Ripley, Jackson Co., WV. In White Sulphur Springs, WV hosp. on 9/1/61 roll. Present on 12/31/62 roll. In Montgomery White Sulphur Springs, Va. hosp. for wd. on 1/10/64 roll. Capt. at Winchester, Va. 9/19/64. POW Pt. Lookout, MD 9/23/64. Exchd. 3/15/65.

REESE, M.: Pvt. Co. B. Res. Jackson Co., WV. POW. D. at Elmira, NY 9/23/64. Bur. Woodlawn Cem., Elmira, NY.

REESE, MORRIS H.: Fifer. Co. A. Enl. 5/22/61 at Poca, Putnam Co., WV. Present on 8/31/61 roll. NFR.

REID, GASTON C.: Pvt. Co. F. Res. Monroe Co., WV. Enl. 10/20/63 at Camp Beauregard. Deserted about 5/8/64. Surrendered Mar. 1865. Took oath at Charleston, WV and sent north 3/11/65. Age: 19. Farmer. 6'. Dark complexion, black hair, hazel eyes.

REID, THOMAS: Pvt. Co. A. Enl. 7/9/62 at Salt Sulphur Springs, Monroe Co., WV. Deserted at Camp McCausland, WV 7/20/62. Aug. 1862 broadside says deserted 7/21/62. Monroe Co., WV.

REMLY, ALLEN: Pvt. Co. D. Enl. 6/13/63 at Camp Jackson. In Charlottesville, Va. hosp. for debilitas 6/12-17/64.

REMLY, AUSTON: Pvt. Co. D. Enl. 4/2/64 at Camp Gauley, Greenbrier Co., WV. Capt. at Winchester, Va. 9/19/64. POW Pt. Lookout, MD 9/24/64. Exchd. 3/15/65.

REMLY, HENRY H.: Pvt. Co. D. Enl. 5/30/61 at Summersville, Nicholas Co., WV. Present on 8/31/61 roll. NFR.

REMLY, MASON: Pvt. Co. D. B. in Greenbrier Co., WV. Res. Nicholas Co., WV. Enl. 5/30/61 at Summersville, Nicholas Co., WV. Left sick at Lewisburg, WV 8/1/61. Capt. at Lewisburg, WV 5/23/62. POW Camp Chase, OH 5/30/62. Exchd. 8/25/62. Present on 10/31/62 roll. Deserted at Cow Pasture Bridge May 1864. Took oath at Clarksburg, WV and sent north 3/30/65. Age: 23. Farmer. 5'9". Fair complexion, dark hair, brown eyes.

RENSFORD, G.: Co. A. In hosp. at Winchester, Va. for ersyplises 1864.

RETTINGER, JOHN: Pvt. Co. G. Enl. 5/1/62 at White Sulphur Springs, Greenbrier Co., WV. Deserted 5/1/62. Listed as deserter on Aug. 1862 broadside. Left 5/24/62. Also spelled Retinger.

REYNOLDS, ANDREW JACKSON: Pvt./Musician. Co. C. Res. Craig Co., Va. Enl. 5/1/62 at White Sulphur Springs, Greenbrier Co., WV. Sick at Montgomery White Sulphur Springs, Va. July 1862. Sick at hosp. on Sept.-Oct. 1862 roll. Present on 12/31/62 roll. D. of disease 12/31/63. Bur. Ross Cem., Rt. 42, near Maywood, on Fred Keffer farm, Craig Co., Va. Brother of William H. ("Bill Dooley") Reynolds.

REYNOLDS, ARCHIBALD A.: Pvt. Co. I. B. 7/26/25 at Craig Co., Va. Enl. 8/3/64 at Dublin, Pulaski Co., Va. Recvd. clothing 4 QR 1864. Res. Craig Co., Va. 1898 at age 73. D. 3/23/1902. Bur. Caldwell Cem., Rt. 624, on Holland Caldwell Farm, Craig Co., Va.

REYNOLDS, C. H.: Pvt. Co. H. Enl. 9/15/62 in Kanawha Co., WV. Left sick within enemy lines in Kanawha Co., WV on Sept.-Oct. 1862 roll. Deserted and in the enemy lines on 1/1/64 roll.

REYNOLDS, CHARLES G.: Pvt. Co. H. Enl. 6/1/63 at Lewisburg, Greenbrier Co., WV. Capt. at Winchester, Va. 9/19/64. POW Pt. Lookout, MD 9/24/64. Exchd. 11/1/64. POW and paroled at Charleston, WV 5/11/65. Age: 19. 5'8". Light complexion, light hair, blue eyes, light whiskers. Bur. Spring Hill Cem., Charleston, WV.

REYNOLDS, DAVID J.: Pvt. Co. F. Res. Craig Co., Va. Enl. 5/1/62 at White Sulphur Springs, Greenbrier Co., WV. Conscript. Capt. at Cold Harbor, Va. 5/31/64. POW Pt. Lookout, MD. Transf. to Elmira, NY 7/9/64. Desired to take oath. D. at Elmira, NY of chronic diarrhea 5/4/65. Bur. Elmira, NY, Woodlawn Cem., grave 2816.

REYNOLDS, FENTON M.: Pvt. Co. H. B. in Kanawha Co., WV. Enl. 5/8/61 at Charleston, Kanawha Co., WV. Dischd. for disability 7/23/63. Age: 27. 5'11". Dark hair and hazel eyes.

REYNOLDS, FLOYD: Pvt./Sgt. Co. C. b. 2/13/44 Craig Co., Va. Enl. 5/1/62 at White Sulphur Springs, Greenbrier Co., WV. In hosp. on Sept.-Oct. 1862 roll. Det. as wood chopper for hosp. at Union, WV on 12/31/62 roll. Present on 3/4/63 roll and still present on 10/31/64 roll. Res. Craig Co., Va. 1898 at age 54. D. 11/22/1922. Bur. Ross Cem., Rt. 42, near Maywood, on Fred Keffer farm, Craig Co., Va.

REYNOLDS, HENRY: Pvt. Co. K. Res. Craig Co., Va. Enl. 2/10 or 11/62 at Princeton, Mercer Co., WV. Present on 10/31/64 roll.

REYNOLDS, H. W.: Pvt. Co. H. PWR.

REYNOLDS, JAMES A.: Pvt. Co. K. Res. Craig Co., Va. Enl. 9/1/62 at Red Sulphur Springs, Monroe Co., WV. Deserted at Droop Mountain, WV 11/6/63. Sick at home in Craig Co., Va. on 12/30/63 roll. Took oath at Charleston, WV 12/13/64. Age: 31. Farmer. 5'10". Dark complexion, dark hair, gray eyes. Bahlmann says enl. 1863 and transf. from 60th Va. Inf.

REYNOLDS, JOHN: Pvt. Co. F. Enl. 6/1/63 at Camp Stonewall. Present on 10/31/64 roll.

REYNOLDS, JOHN ALEXANDER: Pvt. Co. C. B. 1/24/45 near Summersville, WV. Enl. 5/1/62 at White Sulphur Springs, Greenbrier Co., WV. AWOL on Sept.-Oct. 1862 roll. Volunteered in co. at Lewisburg, WV 3/27/63. Present on 10/31/64 roll. Res. Craig Co., Va. 1898 at age 53. D. 2/25/1932 at age 87 years, 1 month, 1 day. PWR says enl. 1863 and served till close of war.

REYNOLDS, JOHN D.: Pvt. Co. D. B. 5/22/42. Res. Craig Co., Va., then res. Nicholas Co., WV. Enl. 5/30/61 at Summersville, Nicholas Co., WV. AWOL 7/28/61. Present on 10/31/62 roll. Deserted and surrendered Apr. 1865. Took oath at Charleston, WV and sent North 4/22/65. Age: 23. Farmer. 5'10". Fair complexion, fair hair, gray eyes. Res. Paint Back, Va. 1898 at age 56. D. 11/14/1926. Bur. Bryan Cem., Rt. 311, Paint Back, Va.

REYNOLDS, JOHN EDWARD: Pvt. Co. C. B. 9/15/29. Res. Craig Co., Va. Enl. 5/1/62 at White Sulphur Springs, Greenbrier Co., WV. Present on 10/31/64 roll. D. 9/14/70. Bur. Leffel-Walker Cem., Rt. 42 West, at Fish Hatchery, Craig Co., Va.

REYNOLDS, JOHN FLEMIN: Pvt. Co. I. B. 1/16/29. Res. Craig Co., Va. Enl. 4/1/64 at Lewisburg, Greenbrier Co., WV. Capt. at Winchester, Va. 9/19/64. POW Pt. Lookout, MD 9/24/64. D. at Pt. Lookout, MD of acute dysentery 1/12/65. Bur. Pt. Lookout, MD grave 796.

REYNOLDS, JOHN LANDON: Res. Craig Co., Va. PWR.

REYNOLDS, JONATHAN N.: Co. E. POW D. at Elmira, NY 11/28/64. Bur. Woodlawn Cem., Elmira, NY.

REYNOLDS, JOSEPH E. (FISHER(?): Pvt. Co. C. B. ca. 1832. Res. Craig Co., Va. Enl. 5/1/62 at White Sulphur Springs, Greenbrier Co., WV. On sick furlo. 10/31/64. Res. Craig Co., Va. 1898 at age 64. D. 7/8/1916 at age 84. Bur. Ross Cem., Rt. 42, near Maywood, on Fred Keffer farm, Craig Co., Va.

REYNOLDS, M. M.: Pvt. Co. C. Farmer, Fayette Co., WV 1860. Age: 31. Enl. 5/10/61.

REYNOLDS, ROSWELL L.: Pvt. Co. H. Enl. 6/1/61 at Charleston, Kanawha Co., WV. Transf. from Capt. Hale's Artillery Co. at White Sulphur Springs, WV 8/17/61. NFR.

REYNOLDS, SHELTON: Pvt. Co. C. Enl. 8/21/62 at Lynch Mill. Absent cause unknown on Sept.-Oct. 1862 roll. NFR.

REYNOLDS, SILAS: Pvt. Swann's Co./Co. H. Enl. 5/22/61 at Charleston, Kanawha Co., WV. Age: 17. Co. transf. to Co. K, 59th Va. Inf. 7/25/61. Transf. to Co. H, 22nd Va. Inf. 3/23/62. Det. wagoner 1st Brig. on 1/1/64 roll. Capt. at Winchester, Va. 9/19/64. POW Pt. Lookout, MD 9/24/64. Exchd. 10/30/64. POW paroled at Charleston, WV 5/12/65. 5'7". Fair complexion, light hair, gray eyes, light whiskers. This is probably Silas T. Reynolds.

REYNOLDS, W.: Pvt. Co. I. B. in Va. Res. Botetourt Co., Va. Enl. 3/?/62 in Botetourt Co., Va. Sick at Bonsacks, Va. July 1862. Deserted. Capt. at Beverly, WV 7/15/64. Gave himself up. Took oath at Clarksburg, WV and relsd. 7/18/64. Age: 21. Farmer. 5'9". Fair complexion, light hair, gray eyes. Possible prewar res. Craig Co., Va.

REYNOLDS, W. A.: Pvt. Co. C. Date and place of enlistment not stated. Listed as deserter on a list dated 7/6/62 at Camp McCausland, WV.

REYNOLDS, WILLIAM B.: Pvt. Co. F. Enl. 7/4/61 at Charleston, Kanawha Co., WV. Apptd. 1st Corp. 8/29/61. Deserted on Mar.-Apr. 1863 roll. Possibly the same William Reynolds listed as deserter on Aug. 1862 broadside. Craig Co., Va.

REYNOLDS, WILLIAM CLARK: Pvt. Co. H. B. 6/12/36. Salt merchant. Lexington, MO. 1860. Enl. 5/8/61 at Charleston, Kanawha Co., WV. Det. in Brig. Gen. Echols' office on 6/30/63 roll. Promo. to Ord. Sgt. Engineers Corp. 8/6/63. Kept a diary of his first year of service in the war. D. 3/16/70.

REYNOLDS, WILLIAM H. ("BILL DOOLEY"): Pvt. Co. C. B. ca. 1830 Craig Co., Va. Enl. 5/1/62 at White Sulphur Springs, Greenbrier Co., WV. On sick furlo on rolls from 4/8/64 to 10/31/64. Postwar roster says deserted. Res. Craig Co., Va. 1898 at age 66. Bur. Ross Cem., Rt. 42, near Maywood, on Fred Keffer farm, Craig Co., Va.

REYNOLDS, WILLIAM McDOWELL: Pvt. Co. D. Res. Craig Co., Va. Enl. 5/1/62 at White Sulphur Springs, Greenbrier Co., WV. Confined at Giles C.H. (Pearisburg), Va. July 1862. Serving sentence of general court-martial on rolls from 10/31/62 to 4/30/63. AWOL 12/31/63.

RHODES, CHARLES A.: Pvt. Co. B. Enl. 6/1/61 at Ripley, Jackson Co., WV. In hosp. on 9/1/61 roll. NFR.

RHODES, CHRISTOPHER: Pvt. Co. D. Res. Greenbrier Co., WV. Enl. 5/1/62 at White Sulphur Springs, Greenbrier Co., WV. WIA and capt. at Lewisburg, WV 5/23/62. Age: 20. Dischd. from U.S. hosp. at Gallipolis, OH 7/2/62. POW Camp Chase, OH 7/8/62. Exchd. 8/25/62. Present on 4/30/63. Sick at hosp. 12/13/63. Capt. at Winchester, Va. 9/19/64. POW Pt. Lookout, MD 9/24/64. Exchd. 3/15/65. Farmer. 5'11". Fair complexion, dark hair, dark eyes.

RICE, J. H.: Pvt. Co. H. Enl. 6/18/62 at Salt Sulphur Springs, Monroe Co., WV. Det. as hosp. steward 8/30/62. NFR.

RICHARDSON, ANDREW: Pvt. Co. K. Res. Greenbrier Co., WV. Enl. 2/17/63 at Lewisburg, Greenbrier Co., WV. Capt. at Meadow Bluff, WV 10/15/63. Age: 35. POW Camp Chase, OH 12/7/63. Transf. to Ft. Delaware 2/29/64. D. at Ft. Delaware of inflammation of pleuris 3/29/64. Bur. Jersey shore of Ft. Delaware. Farmer. 5'9". Light complexion, light hair, blue eyes.

RICHARDSON, R.: Pvt. Co. H. Enl. 4/28/62 at White Sulphur Springs, Greenbrier Co., WV. KIA in a skirmish 5/16/62.

RICKETTS, WILLIAM H.: Pvt./Corp. Co. A. Enl. 6/29/61 at Camp Tompkins, Kanawha Co., WV. Apptd. 1st Corp. 11/24/63. Present on 10/31/64 roll.

RIDDLE, THOMAS EVANS: Co. I. B. 4/16/46 near Nashville, TN. In postwar years claimed service in various regiments, including 12 Tenn. Inf., and finally settled on 22nd Va. Inf. Claims he enlisted in 1863 and served until end of war. Postwar res. Wichita Falls, TX. Farmer and mason. D. 4/2/1954 of heart failure. Evidence indicates he probably did not serve in 22nd Va. Inf. His full story can be found in *Blue & Gray* magazine Apr.-May 1985. A photo of him appears with the article.

RIFFEE, AMOS D.: Pvt. Co. B. Res. Ripley, Jackson Co., WV. Enl. 7/1/62 at Ripley, Jackson Co., WV. In hosp. at White Sulphur Springs, WV on 9/1/61 roll. Present on 11/16/62 roll. Capt. in Roane Co., WV 11/15/64. Age: 18. POW Camp Chase, OH 3/4/64. Transf. to Ft. Delaware 3/14/64. Relsd. 6/21/65. Farmer. 6'1". Florid/sallow complexion, dark hair, hazel eyes. Possible res. Roane Co., WV.

RIFFEY, N.: Pvt. Co. F. Date and place of enlistment not stated. Sick at Bonsacks, Va. July 1862. NFR.

RIFFLE, MARTIN V. B.: Pvt. Co. E. Enl. 10/4/62 at Little Birch, Nicholas Co., WV. At Farmville, Va. and transf. 3/22/65. POW paroled at Charleston, WV 4/29/65. Age: 30. 5'11". Dark complexion, dark hair, blue eyes, sandy whiskers.

RIFFLE, WILLIAM A.: Pvt./Corp. Co. A. Enl. 6/8/61 at Charleston, Kanawha Co., WV. Capt. by Capt. King's Union Home Guards and paroled by his same day 9/20/62. Capt. at Winchester, Va. 9/19/64. POW Pt. Lookout, MD 9/23/64. Exchd. 3/15/65. POW paroled at both Harpers Ferry, WV and Charleston, WV 4/27/65. Age: 32. 6'. Dark complexion, dark hair, blue eyes, brown whiskers.

RIGGS, JOHN: Pvt. Swann's Co. Clerk, Charleston, WV 1860. Enl. 5/8/61 at Charleston, Kanawha Co., WV. Age: 20. Co. transf. to Co. K, 59th Va. Inf. 7/25/61.

RILING, WILLIAM: Pvt. Co. E. Enl. 6/8/61 at Charleston, Kanawha Co., WV. Left sick at White Sulphur Springs, WV 8/20/61. Hosp. nurse since May 1862 on July 1862 roll. Steward in hosp. on Sept.-Oct. 1862 roll. POW on 1/2/63 roll.

RIPPETO, JAMES P.: Pvt. Co. A. Enl. 11/1/62 at Packs Ferry, Monroe Co., WV. D. in camp 12/15/62.

RIPPETOE, ANDREW J.: Pvt. Co. A. Res. Putnam Co., WV. Enl. 5/22/61 at Poca, Putnam Co., WV. Measles at White Sulphur Springs, WV 8/20/61. Present on 11/1/62 roll. Capt. in Boone Co., WV 2/9/64. Age: 26. POW Camp Chase, OH 3/7/64. Transf. to Ft. Delaware 3/14/64. Relsd. 6/21/65. Farmer. 5'10". Florid complexion, dark hair, dark eyes. Possibly James A. Rippetoe.

RIPPETOE, JOHN J.: Pvt. Co. A. Enl. 5/22/61 at Poca, Putnam Co., WV. Served as pvt. until elected 3rd Corp. 7/30/61. On a mounted guard at White Sulphur Springs, WV 8/13-14/61. POW paroled at Staunton, Va. 5/15/65. Age: 31. 5'6". Dark complexion, black hair, hazel eyes.

RIPPETOE, LAFAYETTE: Pvt. Co. A. Enl. 11/1/61 at Packs Ferry, Monroe Co., WV. D. in hosp. at Red Sulphur Springs, WV 12/4/62.

RITTER, JOHN H.: 2nd Sgt. Swann's Co. Boatbuilder, Charleston, WV 1860, age 30. Enl. 5/8/61 at Charleston, Kanawha Co., WV. Age: 31. Co. transf. to Co. K, 59th Va. Inf. 7/25/61. Bur. Spring Hill Cem., Charleston, WV.

ROACH, ELIJAH: Pvt. B. Jackson Co., WV. Enrolled 1861 for 4 years.

ROACH, MONTRAVILLE P.: Corp./Capt. Co. G. Res. Wyoming Co., WV in 1860, farm laborer at age 19. Enl. 7/3/61 at Charleston, Kanawha Co., WV. Elected 2nd Lt. 1/24/62. Elected 1st Lt. 3/1/62. Promo. Capt. 11/22/64. Deserter. POW. Took oath at Charleston, WV 5/9/65. Age: 24. Farmer. 5'11". Dark complexion, black hair, blue eyes. Listed as commanding regt. 12/31/64 while Lt. Col. McDonald commanded the brig. (Echols').

ROACH, WILLIAM H.: Sgt. Co. G. Res. Oceana, WV 1850, constable, sheriff, farmer, trader, carpenter. Rooming business at his house 1857 and ran saloon at Oceana, WV. Elected justice member of the court 1861. Enl. 7/3/61 at Charleston, Kanawha Co., WV. Left sick at White Sulphur Springs, WV 8/20/61. Present on 11/13/62 roll. Transf. to Co. E, 17th Va. Cav. on 4/30/64 roll. Paroled at Charleston, WV 6/22/65. Age: 26. 5'10". Fair complexion, light hair, gray eyes. Postwar sheriff, Oceana, WV. Jailor, and surveyor of roads 1878. Blacksmith and wagon repair shop postwar and still so engaged in 1889.

ROBBERSON, JAMES: Pvt. Co. E. Enl. 7/16/61 at Charleston, Kanawha Co., WV. AWOL 7/24/61.

ROBERSON, ANDREW J.: Pvt. Co. A. Enl. 6/5/61 at Camp Tompkins, Kanawha Co., WV. Cut off from co. by Federal army 7/24/61.

ROBERSON, WILLIAM T.: Pvt. Co. C. Enl. 5/23/63(?) at Camp Stonewall. WIA (slight foot) at White Sulphur Springs, WV 8/26/63. Absent due to wd. on 12/31/63 roll. Present on 4/8/64 roll. At home wded. on 10/31/64 roll. POW paroled at Charleston, WV 5/11/65. Age: 19. 5'11". Dark complexion, black hair, black eyes, black whiskers.

ROBERTS, CHARLES: Pvt./Corp. Co. E. Enl. 7/17/61 at Charleston, Kanawha Co., WV. Left sick in Greenbrier Co., WV 8/10/61. Present on Sept.-Oct. 1862 roll. Apptd. Corp. 12/1/62. POW on 10/31/64 roll. At Camp Lee, Richmond, Va. 2/18/65.

ROBERTS, JOHN H.: Pvt. Co. C. B. 12/24/24 Nelson Co., Va. Married 1859. Enl. 1026/64 at Richmond, Va. Recvd. clothing 4 QR 1864. PWR says first served as drummer boy in Confederate Army. Served in Co. D, 1st Va. Cav. Served in Co. C, 22nd Va. Inf. Served in Co. F, 34th Va. Inf. Dentistry, 1867. D. 2/15/98 at Tornado, WV.

ROBERTS, JOSEPH B.: Pvt./Q.M. Sgt. Co. A. Res. Kanawha Co., WV. Enl. 5/22/61 at Poca, Putnam Co., WV. Apptd. Q.M. Sgt. 11/1/62. Capt. in Boone Co., WV 2/9/64. Age: 22. POW Camp Chase, OH 3/7/64. Transf. to Ft. Delaware 3/14/64. Took oath at Ft. Delaware and relsd. 6/7/65. Farmer. 5'11". Light complexion, dark/brown hair, blue eyes.

ROBERTS, THOMAS ALGERNON: Sgt./Capt. & Asst. Q.M. Field & Staff, Co. A. B. in Kanawha Co., WV. Enl. 5/22/61 at Poca, Putnam Co. Det. Asst. Q.M. 6/3/61. Apptd. Brig. A.Q.M. 4/13/63, confirmation 4/13/63, to rank from 11/1/62. Accepted 5/20/63. POW paroled at Charleston, WV 5/29/65. PWR says resided in Roanoke Co., Va. Member Hupp-Deyerle CV Camp, Salem, Va. 1920. D. 1931. Bur. East Hill Cem., Salem, Va.

ROBINSON, JOHN: Pvt. Co. D. Enl. 8/19/61 at White Sulphur Springs, Greenbrier Co., WV. Transf. from Capt. Riggs' Co. by order of Col. Tompkins 8/19/61. NFR.

ROBINSON, JOHN P.: Pvt. Co. F. Enl. 5/1/62 at White Sulphur Springs, Greenbrier Co., WV. Recvd. clothing at Liberty (Bedford), Va. hosp. 8/19/64. At home wded. in Bedford Co., Va. on 10/31/64 roll.

ROBINSON, ROBERT B.: Pvt. Co. D. Enl. 5/1/62 at White Sulphur Springs, Greenbrier Co., WV. Sick in Alleghany Co., Va. since 12/22/63 on rolls from 12/31/63 to 10/31/64. Recvd. clothing at Staunton, Va. hosp. 10/14/64 and 11/11/64 and 12/12/64.

ROBINSON, WILLIAM S.: Co. K. B. 1843. Shoemaker. Farm hand, Buffalo District, Va. 1870, age 35. Carpenter, Buffalo District, Va. 1900, age 66. Filed for pension 1900. Res. Rockbridge Baths, Va. 1928. D. 2/29/1929 Collierstown, Va. Bur. Collierstown Presbyterian Church Cem.

ROBISON, JOHN: Co. K. Bur. "Our Soldiers Cem.", Mount Jackson, Va.

ROBTS, JOHN R.: Pvt. Swann's Co. Enl. 6/9/61 at Charleston, Kanawha Co., WV. Age: 17. Co. transf. to Co. K, 59th Va. Inf. 7/25/61.

ROCK, GEORGE W.: Pvt. Co. C. Res. Craig Co., Va. Enl. 5/1/62 at White Sulphur Springs, Greenbrier Co., WV. Det. to cook in hosp. on Sept.-Oct. 1862 roll. Det. wagoner on 12/31/62 roll. Present on 3/4/63 roll. In Charlottesville, Va. hosp. for acute dysentery 6/8-17/64. In Charlottesville, Va. hosp. for chronic rheumatism 7/26/64. Transf. to Lynchburg, Va. 7/26/64. Recvd. clothing Staunton, Va. hosp. 12/13/64.

RODGERS, SAMUEL: Chaplain. Field & Staff. Res. Maryland. Enl. 1864 in Greenbrier Co., WV. Apptd. Chaplain Feb. 1864. POW paroled at Lewisburg, WV 4/26/65. Age: 40. Minister. 6'. Light complexion, dark hair, hazel eyes.

RODGERS, WILLIAM: Pvt. Co. K. Enl. 6/6/61 at Charleston, Kanawha Co., WV. Absent cause unknown on 9/4/61. NFR. Farm laborer, Fayette Co., WV 1860. Age: 17. Bahlmann says "left."

ROFFE, THOMAS I.: Pvt. Co. H. Res. Cabell Co., WV. Enl. 6/25/61 at Coalsmouth (St. Albans), WV. Capt. in Fayette Co., WV 9/13/61. POW Camp Chase, OH. Transf. to Camp Chase, OH 5/24/62. Exchd. 8/25/62. Present on 12/31/62 roll. Capt. in Cabell Co., WV 1/20/64. Age: 24. POW Camp Chase, OH 2/15/64. Transf. to Ft. Delaware 3/14/64. Relsd. 6/21/65. Farmer. 6'. Light complexion, brown hair, hazel/blue eyes. Drew pension in Independence Co., Ark. 1913.

ROFFE, WILLIAM D.: Pvt. Co. H. Enl. 6/25/61 at Coalsmouth (St. Albans), Kanawha Co., WV. Sick since 8/24/61 on 8/31/61 roll. Capt. in Fayette Co., WV 9/13/61 or 10/4/61. POW Camp Chase, OH 10/4/61. Transf. to Johnson's Island, OH 4/17/62 or 9/6/62. Exchd. 11/22/62. Present on 12/31/62 roll. Capt. at Winchester, Va. 9/19/64. POW Pt. Lookout, MD 9/24/64. Exchd. 11/1/64. Paroled at Charleston, WV 6/20/65. Age: 24. 6'2". Fair complexion, light hair, blue eyes, light whiskers.

ROGERS, JOHN H.: Pvt. Co. D. Enl. 5/30/61 at Summersville, Nicholas Co., WV. AWOL 7/28/61.

ROGERS, TYRANCE: Pvt. Co. D. Res. Clay Co., WV. Enl. 2/3/64 at Camp Gauley, Greenbrier Co., WV. Conscript. Deserted. POW. Surrendered at Charleston, WV Mar. 1865. Took oath at Charleston, WV and sent north 3/9/65. Age: 33. Farmer. 5'10". Dark complexion, dark hair, black eyes.

ROLLINS, JOSEPH V.: Pvt. Co. A. B. May 1842 in Monroe Co., WV. Enl. 8/10/61 in 166th Va. Militia. Enl. 4/10/62 at Brushy Ridge, Greenbrier Co., WV. Left sick at Lewisburg, WV on 4/30/63 roll. Deserted in face of the enemy 12/13/63. Took oath and sent north 12/31/63. Age: 21. Farmer. 6'2". Dark complexion, dark hair, hazel eyes. Res. Monroe Co., WV. PWR says WIA (face) at Lewisburg, WV. Postwar minister and farmer. D. 7/3/1925 Beard's Fork, WV. Bur. Settle Cem., Robson, WV.

ROLLYSON, JOHN: Pvt. Co. K. Enl. 5/28/62 at Salt Sulphur Springs, Monroe Co., WV. Sick in hosp. at Pearisburg, Va. on 12/30/63 roll. Present on 4/1/64. WIA at Cedar Creek, Va. 10/19/64. Bahlman says died from leg amputation.

RONK, ANDREW J.: Pvt. Co. A. Enl. 5/22/61 at Poca, Putnam Co., WV. Unaccounted for as of 7/25/61.

RONK, D.: Pvt. Co. F. Date and place of enlistment not stated. Sick at Dublin, Va. July 1862. NFR.

RONK, JACOB D.: Pvt. Co. A. Enl. 5/22/61 at Poca, Putnam Co., WV. Unaccounted for 7/25/61. Present on 11/1/62 roll. WIA (slight hand) at White Sulphur Springs, WV 8/26-27/63. Deserted 2/24/64. Capt. 3/2/64.

RONK, JAMES M.: Pvt./Sgt. Co. A. Enl. 5/22/61 at Poca, Putnam Co., WV. Served as Pvt. until elected 3rd Sgt. 7/30/61. Capt. at Winchester, Va. 9/19/64. POW Pt. Lookout, MD 9/24/64. Relsd. 5/13/65. Transportation provided to Putnem Co., WV.

RONK, SAMUEL T.: Pvt. Co. A. Enl. 7/1/63 at Camp Sam Jones. Teamster on 12/31/63 roll. Det. service on 4/1/64 roll. Present on 10/31/64 roll. Paroled at Charleston, WV 6/17/65. Age: 18. 5'6". Sandy complexion, light hair, gray eyes.

ROOKER, SAMUEL: Pvt. Co. I. Enl. 4/15/62 in Greenbrier Co., WV. Deserted at Centreville, Monroe Co., WV 12/10/62.

RORER, JOHN H.: Pvt. Co. K. Enl. 6/6/61 at Charleston, Kanawha Co., WV. Deserted 10/20/62.

RORER, PETER F.: Pvt. Co. K. Enl. 6/6/61 at Charleston, Kanawha Co., WV. Absent cause unknown 9/4/61 roll. NFR.

ROSE, EDWARD: Pvt. Co. G. Enl. 7/3/61 at Charleston, Kanawha Co., WV. Left his co. at Clifton (Pratt) WV 7/24/61. AWOL.

ROSE, FRANKLIN: Pvt. Co. G. Enl. 7/3/61 at Charleston, Kanawha Co., WV. Left his co. at Camp Two Mile, WV 7/16/61. AWOL.

ROSE, MARSHALL: Pvt. Co. E. Enl. 10/12/62 at Cotton Hill, Fayette Co., WV. POW paroled at Charleston, WV 5/12/65. Age: 20, 5'9", fair complexion, brown hair, gray eyes.

ROSE, WILLIAM: Pvt. Co. A. PWR.

ROSE, WILLIAM: Pvt. Co. E. Enl. 10/12/62 at Cotton Hill, Fayette Co., WV. KIA at Droop Mountain, WV 11/6/63.

ROSE, WILLIAM SETH: Pvt. Co. C. B. 6/1/20 at Potts Creek near Covington, Va. Building contractor, built mill dams. Enl. 5/1/62 at White Sulphur Springs, Greenbrier Co., WV. Sick at Montgomery White Sulphur Springs, Va. hosp. July 1862. Listed as deserter on Aug. 1862 broadside. Det. wagoner on Sept.-Oct. 1862 roll. Capt. at Winchester, Va. 9/19/64. POW Pt. Lookout, MD 9/24/64. Exchd. 5/14/65. Transportation provided to Kanawha Co., WV. Farmer. Res. Kanawha Co., WV. D. 3/23/1906 Ripley, WV. Bur. Mt. Calvary Cem., Mill Creek, near Ripley, WV.

ROSS, JOHN T.: Pvt. Co. C. Enl. 5/1/62 at White Sulphur Springs, Greenbrier Co., WV. Elected Lt. in Capt. Thurmond's Co. Pay due from 4/30/62. Bur. Spring Hill Cem., Charleston, WV. Farmer, Fayette Co., WV 1860. Age: 26. PWR says enl. 5/10/61 and also served as Corp.

ROSS, JOSEPH E.: Pvt. Co. I. Enl. 8/22/61 at Meadow Bluff, Greenbrier Co., WV. Present on 6/29/61 to 8/31/61 roll. NFR.

ROWAN, ALLAN CAPERTON: 2nd Lt. Co. C. B. 1841. Res. Craig Co., Va. Capt. in 108th Regt. Va. Militia. Enl. 5/1/62 at White Sulphur Springs, Greenbrier Co., WV. Elected 2nd Lt. 5/1/62. 3rd Lt. July 1862. Resgnd. commission 8/27/63. Elected 2nd Lt. 2/23/64. KIA at Cold Harbor, Va. 6/3/64. Claim presented to Frances Rowan. Craig Court records say d. at Cold Harbor, Va. 6/1/64 of gunshot wd. to neck.

ROWAN, JAMES W.: Pvt. Co. C. Enl. 3/13/63 at Lewisburg, Greenbrier Co., WV. WIA (slight arm) at White Sulphur Springs, WV 8/26-27/63. In Chimborazo Hosp. for rubeola 6/1-11/64. Paroled at Charleston, WV 6/12/65. Age: 20. 5'10". Dark complexion, dark hair, blue eyes. D. 7/2/1937 Gap Mills, WV at age 92 making him 3rd last CS vet in Monroe Co., WV.

ROY, CHRISTOPHER C.: 1st Lt. Swann's Co. Cooper, Charleston, WV 1860, age 22 (?). Enl. 5/8/61 at Charleston, Kanawha Co., WV. Age: 30. Co. transf. to Co. K, 59th Va. Inf. 7/25/61.

ROY, LARKIN C., JR.: Pvt. Swann's Co. Cooper, Charleston, WV 1860, age 19. Enl. 5/8/61 at Charleston, Kanawha Co., WV. Co. transf. to Co. K, 59th Va. Inf. 7/25/61. Believed bur. Spring Hill Cem., Charleston, WV.

ROY, LARKIN C., SR.: Pvt. Swann's Co. Cooper, Charleston, WV 1860, age 51(?). Enl. 5/8/61 at Charleston, Kanawha Co., WV. Age: 50. Co. transf. to Co. K, 59th Va. Inf. 7/25/61. Believed bur. Spring Hill Cem., Charleston, WV.

RUBY, EDWARD: Pvt. Co. H. Enl. 5/8/61 at Charleston, Kanawha Co., WV. Present on 8/31/61 roll. NFR.

RUBY, JOHN CHRISTOPHER: Pvt./A.C.S. Field & Staff/Co. H. B. 1838 Gallia Co., OH. Res. Charleston, WV 1856, grocery business. Clerk, Charleston, WV 1860, age 31. Enl. 5/8/61 at Charleston, Kanawha Co., WV. Det. Commissary Sgt. 7/5/61. Apptd. A.C.S. Aug. 1861. Listed as Capt. A.C.S., 1st Brig., Wharton's Div. 7/22/64 on report filed by Col. Patton. In a 1922 interview John McCausland said Ruby became a Capt. late in the war. Twice Maj. Postwar grocery business, Charleston, WV. Mayor, Charleston, WV 1875-77. D. ca. 1881 at age 52. Bur. Spring Hill Cem., Charleston, WV.

RUFFNER, ANDREW LEWIS: Pvt. Co. H. B. at Charleston, WV 9/11/41. Enl. 5/8/61 at Charleston, Kanawha Co., WV. Age: 19. Clerk, Aug. 1863. Transf. to Co. I, 8th Va. Cav. 8/6/63. Postwar mercantile business with his brother in Charleston, WV. Paroled Mercer Co., WV 4/25/65. Farmer. Age: 23. 6'10". Brown hair and hazel eyes. Also served as a Maj. Brother of Merrdith Ruffner. Cousin of David, Henry, and Joel Ruffner.

RUFFNER, CHARLES: Pvt. Co. H. Not in CSR. Possibly same Charles Ruffner B. 2/24/01 and D. 3/10/81 Barboursville, WV. Bur. Barboursville Cem., Cabell Co., WV.

RUFFNER, DAVID LEWIS: Pvt./Capt. Co. H. B. 1/10/29 Lexington, Va. Grad. Washington College 1856. AB Lawyer, Kanawha Co., WV. Schoolteacher, Charleston, WV 1860, age 31. Civil engineer and lawyer in partnership with William A. Quarrier in Kanawha Co., WV prior to the war. Enl. 5/8/61 at Charleston, Kanawha Co., WV. Served as a Pvt. until elected 2nd Lt. 5/22/61. Adj. 6/2/61. Comm. Capt. 7/27/61. Not re-elected 5/1/62. Served as Acting Asst. Adj. Gen. on the staff of Gen. John S. Williams July 1863. (A PWR says with the rank of Col). Postwar res. Charleston, WV. Surveyed Route for C & O Railroad alongside the Kanawha River. Civil engineer, lawyer, editor. Still alive 11/25/95. D. 4/24/97. Charleston, WV. Bur. Spring Hill Cem., Charleston, WV. Middle name also spelled Louis.

RUFFNER, HENRY DANIEL: 3rd Lt. Co. H. B. 8/15/34. Farm manager, Charleston, WV 1860, age 24. Enl. 5/8/61 at Charleston, Kanawha Co., WV. Tendered resignation 7/24/61. Resgnd. as 2nd Lt. 8/8 or 15/61. Absent cause unknown since resignation. Capt., Co. G, 19th Va. Cav. 3/12/63. Maj., 46th Va. Cav. Battn. 2/22/64. Maj., 26th Va. Cav. Jan. 1865. Paroled at Charleston, WV 5/10/65. Postwar silver mine operator, Colorado. Res. Charleston, WV. D. 7/25/1925 Rosebud, Florida. Considered last of original Kanawha Riflemen to die. Bur. Spring Hill Cem., Charleston, WV. Also listed as "Col". Cousin of Joel, Meredith, and Andrew Ruffner.

RUFFNER, JOEL, JR.: 1st Lt. Co. E. Attending Univ. of Va. 1860, res. Charleston, WV. age 19. Enl. 6/8/61 at Charleston, Kanawha Co., WV. Enlisted while still attending school. Assisted in raising Co. E. Served as 1st Sgt. until elected 1st Lt. 7/1/61. D. at White Sulphur Springs, WV of typhoid fever 10/12/61 or 11/4/61. Claim presented by S. A. Miller, atty. Son of Joel Ruffner. Cousin of Henry, Meredith, and Andrew Ruffner.

RUFFNER, MAUP: Pvt. Co. H. Not in CSR. Another PWR says first name is Mayre. NFR.

RUFFNER, MEREDITH PAYSON: Pvt. Co. H. B. 7/22/44 Kanawha Co., WV. Mt. Ovis Academy, Charleston, WV and Marietta College, OH. Enl. 5/8/61 at Charleston, Kanawha Co., WV. Age: 16. Served 3 years. WIA (slight) in 1862 Kanawha Valley Campaign. Det. to raise a co. on Sept.-Oct. 1862 roll. Honorably discharged for disability caused by typhoid fever. Postwar mercantile business with his brother in Charleston, WV. Married 1885. D. at Charleston, WV of diabetis 3/2/1911. Bur. Spring Hill Cem., Charleston, WV. Brother of Andrew Ruffner. Cousin of David L. and Joel Ruffner. Donated ground for Ruffner Memorial Presbyterian Church.

RULE, ANTHONY L.: Pvt. Co. C. B. 2/9/31. Enl. 5/1/62 at White Sulphur Springs, Greenbrier Co., WV. Wagoner July 1862. Regt. wagoner on 3/4/64 roll. PWR says fought at Carnifex Ferry, White Sulphur Springs, Droop Mountain, and surrendered at Appomattox. D. 1/12/1912 at Clifton Forge, Va. Bur. Crown Hill Cem., Clifton Forge, Va. See also 14th Va. Cav. Farmer, Fayette Co., WV 1860. Age: 29. PWR says enl. 5/10/61.

RULE, CARY A.: Pvt. Co. C. Enl. 11/13/62 at Camp Fair View. WIA (severe arm) at White Sulphur Springs, WV 8/26-27/63. WIA (severe finger) at Cold Harbor, Va. 6/2-3/64. POW paroled at Charleston, WV 5/10/65. Age: 20. 5'7". Dark complexion, brown hair, black eyes. Farm laborer, Fayette Co., WV 1860. Age: 17. PWR says enl. 5/10/61.

RUNDLE, JONATHAN THEODORE: Pvt. Co. H. B. 4/12/36 Wheeling, WV. Family from England. Moved to Cinn., OH then Charleston, WV. Publisher of *Kanawha Valley Star.* Later served in 36th Va. Inf. Not in CSR. Postwar res. Nashville, TN. D. 1/6/1915.

RUSK, JAMES W.: Pvt. Co. F. B. ca. 1838 Craig Co., Va. Res. Craig Co., Va. Enl. 8/1 or 10/62 at Camp Royal Oak. At home sick in Craig Co., Va. Enl. 8/1 or 10/62 at Camp Royal Oak. At home sick in Craig Co., Va. on Mar.-Apr. 1863 roll. Capt. at Droop Mountain, WV 11/6/63. Age: 25. POW Camp Chase, OH 11/18/63. Transf. to Ft. Delaware 2/29/64. Took oath,at Ft. Delaware and relsd. 5/11/65. Shoemaker. 5'11½". Fair complexion, fair hair, blue eyes.

RUSK, JOHN H.: Pvt. Co. F. B. ca. 1835 Craig Co., Va. Res. Craig Co., Va. Enl. 5/1/62 at White Sulphur Springs, Greenbrier Co., WV. Listed as deserter on Aug. 1862 broadside. Craig Co., Va. Under arrest on 10/31/62 roll. Present on 12/31/62 roll. Capt. at Droop Mountain, WV 11/6/63. Age: 27. POW Camp Chase, OH 11/18/63. Transf. to Ft. Delaware 2/29/64. In Chimborazo Hosp. for chronic diarrhea 9/22-27/64. Deserter. Took oath at Charleston, WV 12/13/64. Farmer. 5'11½". Florid complexion, light hair, dark eyes.

RUST, BENJAMIN D.: Pvt./Field & Staff. Co. B. Enl. 6/1/61 at Ripley, Jackson Co., WV. Apptd. Ordnance Sgt. 5/2/62.

RUST, GEORGE B.: 2nd Lt. Co. A. B. 5/30/26 Putnam Co., Wv. Farmer, Tyler Mountain, WV 1860. Enl. 5/22/61. Place of enlistment not stated. Elected 2nd Lt. 5/22/61. Resgnd. 7/12/61.

RUST, JAMES W.: Pvt. Co. A. B. 3/20/44 Kanawha Co., WV. Student, Coalsmouth (St. Albans), WV 1860, age 16. Enl. 5/22/61 at Poca, Putnam Co., WV. In Charlottesville, Va. hosp. for typhoid fever (paratitis) 1/2-23/62. KIA at Fayetteville, WV 9/10/62. Age: 18, 3 mos. 20 days. Bur. Rust Cem., Sattes (near Nitro), WV. Before departing Charleston, WV in 1861 he strolled with his aunt and picked the spot he wished to be buried if killed, and that is where he lies.

RUTLEDGE, EDWARD J.: Pvt./Ord. Sgt. Co. D. Enl. 5/30/61 at Summersville, Nicholas Co., WV. Dropped as 2nd Lt. 5/2/62. KIA at Cedar Creek, Va. 10/19/64.

RUTLEDGE, JAMES H.: Pvt. Swann's Co. Enl. 6/1/61 at Charleston, Kanawha Co., WV. Age: 25. Co. transf. to Co. K, 59th Va. Inf. 7/25/61.

RUTLEDGE, MICHAEL: Pvt. Co. I. Enl. 4/15/62 at Lewisburg, Greenbrier Co., WV. Sick at Giles C.H. (Pearisburg), Va. July 1862. Present on 10/31/64 roll.

RUTLEDGE, WILLIAM E.: Pvt. Co. K. Enl. 4/24/62 at White Sulphur Springs, Greenbrier Co., WV. Det. shoemaker from 2/28/63 roll to Mar.-Apr. 1863 roll. Sick at home in Monroe Co., WV since 8/25/63 on 12/30/63 roll. Took oath at Charleston, WV 12/13/64. Age: 36. Farmer. 6'. Dark complexion, dark hair, gray eyes.

SADLER, WILLIAM T.: Pvt. Co. D. Enl. 5/1/62 at White Sulphur Springs, Greenbrier Co., WV. Sick at Montgomery White Sulphur Springs, Va. July 1862. Present on 10/31/62 roll. At hosp. wded. on 10/31/64 roll. Paroled at Charleston, WV 5/11/65. Age: 34. 5'11". Light complexion, dark hair, gray eyes, sandy whiskers. Res. Pocahontas Co., WV.

SALE, JAMES E.: Pvt. Co. B. Enl. 4/?/64 at Camp Gauley, Greenbrier Co., WV. Det. light duty on 4/30/64 roll.

SAMMONS, CHARLES B.: Pvt. Co. A. Res. Lewisburg, WV. Enl. 5/1/62 at White Sulphur Springs, Greenbrier Co., WV. Det. service on 12/30/63 roll. Present on 4/30/64 roll. Capt. near Washington, D.C. 7/13/64. POW Elmira, NY 7/15/64. Took oath at Elmira, NY and relsd. 6/21/65. 5'11". Florid complexion, auburn hair, blue eyes.

SAMS, A. A.: Pvt. Co. B. Enl. 5/1/62 at White Sulphur Springs, Greenbrier Co., WV. Conscript. Deserted 5/23/62 near White Sulphur Springs, WV. Deserted on 11/16/62 roll.

SAMUELS, GREEN B.: Pvt. Co. E. Res. of Va. Enl. either 5/1/62 at Narrows, Va. or 4/1/64 at Camp Gauley, Greenbrier Co., WV. Exchd. from Engineer Corps for Wm. Johns 4/1/64. POW paroled at Lewisburg, WV 4/27/65. Age: 40. (Parole says he was mustered at Charleston, WV in 1861 by Fife). Civil engineer. 5'8". Ruddy complexion, dark hair, hazel eyes.

SANDIGE, WILLIAM H. ("BUD"): Sgt. Co. K. Enl. 6/6/61 at Charleston, Kanawha Co., WV. Absent sick since 9/1/61 on 9/4/61 roll. KIA (shot through heart) at Lewisburg, WV 5/23/62. Farmer, Fayette Co., WV 1860. Age: 20.

SANGER, WILLIAM H.: Pvt. Co. K. Enl. 6/6/61 at Charleston, Kanawha Co., WV. Absent cause unknown on 9/4/61 roll. NFR. Bahlmann says was a Dunkard and did not march off with the co. Middle name is Henry.

SARVER, B. B.: Co. K. Deserted 5/3/62 on Aug. 1862 broadside. Craig Co., Va.

SARVER, D. L.: Listed in UDC booklet *Craig's Share In The War Between The State.* Res. Craig Co., Va.

SARVER, FLEMING A.: Pvt. Co. D. B. ca. 1834 Craig Co., Va. Res. Craig Co., Va. Enl. 2/27/63 at Lewisburg, Greenbrier Co., WV. Capt. at Droop Mountain, WV 11/6/63. Age: 30. POW Camp Chase, OH 11/18/63. Transf. to Ft. Delaware 2/29/64. Took oath at Ft. Delaware and relsd. 6/21/65. Farmer. 6'2". Florid complexion, fair hair, blue eyes.

SARVER, GEORGE W.: Listed in UDC booklet *Craig's Share In The War Between The States.* Res. Craig Co., Va.

SARVER, HENRY BENTON: Pvt. Co. C. B. 3/15/46 Craig Co., Va. Enl. 5/1/62 at White Sulphur Springs, Greenbrier Co., WV. Age: 18. Capt. at Winchester, Va. 9/19/64. POW Pt. Lookout, MD 9/24/64. Took oath at Pt. Lookout, MD and relsd. 6/20/65. In Farmville, Va. hosp. for rubeola 3/16/65. Transf. to Lynchburg, Va. hosp. 4/6/65. PWR says honorably dischd. Lynchburg, Va. 5'8¼". Light complexion, brown hair, blue eyes. Res. Craig Co., Va. 1898 at age 71. D. 12/30/1935 Estacada, Oregon. Bur. Sarver Cem., near Maywood, on Sarver homeplace, Craig Co., Va.

SARVER, JAMES JR.: Pvt. Co. K. B. 8/16/33. Enl. 4/24/62 at White Sulphur Springs, Greenbrier Co., WV. Det. brig. teamster on rolls from 12/30/63 to 5/1/64. Present on 10/31/64 roll. PWR says served 3 years and 15 days. Postwar res. Upper Craig Creek, Va. Res. Craig Co., Va. 1898 at age 66. D. 8/16/33 at age 91 years, 10 months, 20 days.

SARVER, JAMES SR.: Pvt. Co. C. Res. Craig Co., Va. Enl. 5/1/62 at White Sulphur Springs, Greenbrier Co., WV. Under arrest on 3/4/63 roll. Present on 6/11/63 roll. In Winchester, Va. hosp. 7/23-28/64. PWR says enl. 4/?/61 for 1/6 year. Res. Craig Co., Va. 1898 at age 71. Also says served in Co. F.

SARVER, J. G.: Pvt. Co. C. Enl. 4/?/63 for 2 years. Res. Craig Co., Va. 1898 at age 68.

SARVER, JOHN: Rank unknown. Co. F. D. Oct. 1903. Pension applications for Giles & Bland counties, Va.

SARVER, M. JACKSON: Pvt. Co. G. B. 1/21/44. Res. Craig Co., Va. Enl. 1/1/64 at Dublin, Pulaski Co., Va. Deserted at Salem, Va. on 10/31/64 roll. Res. Craig Co., Va. 1898 at age 54. D. 1/1/1929. Bur. Sarver Cem., Rt. 621, on Frank Walker farm, Craig's Creek, Craig Co., Va.

SATIS, ALBERT: Pvt. Swann's Co. Farm manager, Charleston, WV 1860. Enl. 6/1/61 at Charleston, Kanawha Co., WV. Age: 18. Co. transf. to Co. K, 59th Va. Inf. 7/25/61. Also spelled Sattes.

SAUNDERS, GEORGE W.: Pvt. Swann's Co. Enl. 5/25/61 at Charleston, Kanawha Co., WV. Age: 40. Co. transf. to Co. K, 59th Va. Inf. 7/25/61. Bur. Spring Hill Cem., Charleston, WV.

SAUNDERS, WILLIAM: Pvt. Co. A. Enl. 5/22/61 at Poca, Putnam Co., WV. Sick at hosp. at White Sulphur Springs, WV 8/20/61. In Charlottesville, Va. hosp. for icterus 1/2/62 to 2/5/62. KIA at Lewisburg, WV 5/23/62.

SAVILLE, W. A.: Rank unknown. Co. E. Joined at New Market, Va. and served seven months. B. in Rockbridge Co., Va. Living in Giles Co., Va. Oct. 1904. Age: 78. Pension applications for Giles & Bland counties, Va.

SAWYERS, ALFRED C.: Pvt. Co. G. B. 1844 Alleghany Co., Va. Res. Alleghany Co., Va. Enl. 5/1/62 at White Sulphur Spring, Greenbrier Co., WV. Sick in hosp. on 11/13/62 roll. Present on 12/31/62 roll. Det. as guard at Greenbrier River on 4/30/63 roll. Capt. at White Sulphur Springs, WV 8/26/63. POW Camp Chase, OH 9/7/63. Transf. to Rock Island, IL 1/22/64. Transf. for exchange 2/15/65. In Howard's Grove Hosp., Richmond, Va. 2/15/65 to 3/6/65. Farmer. 5'6¾". Fair complexion, brown hair, blue eyes. Postwar res. Grassy Meadows, WV. Entered Richmond, Va. Old Soldiers Home 11/27/1922 at age 80. D. either 2/8/1925 or 2/28/1926. Bur. Hollywood Cem., Richmond, Va.

SAWYERS, WELLINGTON: Co. G. Enl. 3/?/62. Postwar roster.

SAYRE, CHARLES SR.: Pvt. Co. B. Res. Jackson Co., WV. Served under Breckinridge at New Market and Cold Harbor.

SCHULTZ, JACOB P.: Pvt. Co. F. Date and place of enlistment not stated. Recvd. clothing 4 QR 1864. Paroled at Charleston, WV 6/14/65. Age: 40. 5'10". Fair complexion, light hair, blue eyes.

SCOTT, CHARLES: Co. H. D. at Dublin, Va. 1863 of pneumonia. PWR.

SCOTT, CROCKETT: Pvt. Co. D. Enl. 5/30/61 at Summersville, Nicholas Co., WV. Present on 8/31/61 roll. Transf. to Co. I, 8th Va. Cav. either July or Aug. 1863.

SCOTT, HEZEKIAH: Pvt. Swann's Co. Enl. 5/25/61 at Charleston, Kanawha Co., WV. Age: 20. Deserted on 6/30/61 roll.

SCOTT, LEWIS C.: Pvt. Co. F. Res. Craig Co., Va. Enl. 3/11 or 12/64 at Camp Gauley, Greenbrier Co., WV. Still present on 10/31/64 roll. Res. Craig Co., Va. 1898 at age 50.

SCOTT, THOMAS L.: Pvt. Taylor's Co./Co. K. Farmer, age 20, 4th Dist., 1860. Enl. ?/19/61 at Frankford, Greenbrier Co., WV. Age: 20. Co. transf. to Wise Brigade 7/10/61. Exempt Mar. 1862 for wasted testicles. Exempt 2/8/64. Collierstown, Va., age 24. Farmer. Dark complexion, blue eyes, dark hair. 5'10". Farmer, age 31, Buffalo Dist. 1870 census. Merchant and farmer, Collierstown, Va. 1880.

SEBERT, JACOB FRY: Rank unknown. Co. K. B. 1833 Pocahontas Co., WV. Res. Rockbridge Co., Va. 1862. Date and place of enlistment not stated. Recvd. clothing 4 Qr 1864. Exempt as miller 2/8/64. Age: 30. 5'10". Fair complexion, light hair, blue eyes. Farmer, Kerr's Creek District 1870 at age 36. Res. Collierstown. Cattle dealer and farmer, Monmouth, 1880. D. 11/?/81 near Monmouth. Bur. New Monmouth Presbyterian Church Cem. Postwar photo in J. F. Miley Collection, #1616, Rockbridge Historical Society, Washington & Lee Library. Also spelled Seebert.

SEIDERS, THOMAS L.: Corp. Co. A. Laborer, Tyler Mountain, WV 1860. Age: 25. Enl. 5/22/61 at Poca, Putnam Co., WV. Resgnd. as 4th Sgt. and became a Pvt. 7/30/61. Absent with sick leave on 8/31/61 roll. NFR.

SELDOMRIDGE, PETER F.: Pvt. Co. F. Enl. 5/1/62 at White Sulphur Springs, Greenbrier Co., WV. Sick at home near Frankford, WV on Mar.-Apr. 1863 roll. NFR.

"SEMI-DADDY": Co. K. A nickname. Enl. 1861 as a Pvt. Bahlmann says died in a northern POW camp.

SHAMBLIN, JESSE: Pvt. Co. E. Res. Osborne's Mills, Kanawha Co., WV. Enl. 6/8/61 at Charleston, Kanawha Co., WV. Returned home following Wise's retreat July 1861. Re-enl. in 22nd Va. 5/1/62. Conscript. Promo. to 1st Corp. WIA (throat) and capt. at Droop Mountain, WV 11/6/63. Remained with sick at Lewisburg, WV until Federals left. Was capt. and paroled by Gen. Duffie. In hosp. wded. at Union, WV on 12/31/63 roll. Deserted at Charleston, WV Mar. 1864. Capt. in Kanawha Co., WV 4/8 or 18/64. Age: 18. POW Camp Chase, OH 5/10 or 11/64. Desired to take oath at Camp Chase, OH 6/10/64. Relsd. 2/8 or 28/65. Age: 22. Farmer. 5'10". Dark complexion, dark hair, brown eyes.

SHANK, G. W.: Pvt. Co. A. Enl. 3/24/64 at Dublin, Pulaski Co., Va. Conscript. NFR.

SHANNON, BENJAMIN F.: Pvt. Co. E. Enl. 10/12/62 at Cotton Hill, Fayette Co., WV. In Stuart Hosp., Richmond, Va. for vulvous sclopet 6/6-15/64. Transf. to Chimborazo Hosp. and there 6/15-28/64. Capt. in Clay Co., WV 2/12/65. Age: 22. POW Camp Chase, OH 2/21/65. Took oath at Camp Chase, OH 6/13/65. Farmer. 5'8". Fair complexion, dark hair, gray eyes. Res. Clay Co., WV.

SHANNON, NEWSON: Sgt./Pvt. Co. G. Son of Layne Shannon. Res. Wyoming Co., WV. Enl. 10/1/62 at Charleston, Kanawha Co., WV. D. at Red Sulphur Springs, WV 12/4 or 5/62.

SHAUNACY, PATRICK: Pvt. Taylor's Co. Enl. 6/19/61 at Frankford, Greenbrier Co., WV. Age: 24. Co. transf. to Wise Brigade 7/10/61.

SHAVER, JAMES E.: Pvt. Co. D. Enl. 10/8/62 at Charleston, Kanawha Co., WV. POW paroled at Charleston, WV 5/15/65. Age: 31. 5'10". Dark complexion, black hair, black eyes.

SHAVER, LEWIS: Pvt. Co. D. Enl. 6/3/61 at Kanawha Falls, Fayette Co., WV. D. 10/18/61 on report dated 10/26/61 from Cotton Hill, WV. Farm laborer, Fayette Co., WV 1860. Age: 22.

SHAVER, WILLIAM C.: Pvt. Co. D. Enl. 5/30/61 at Summersville, Nicholas Co., WV. Present on 8/31/61 roll. NFR.

SHAW, DAVID F.: Pvt. Co. C. B. ca. 1823 Craig Co., Va. Distiller. Enl. 5/1/62 at White Sulphur Springs, Greenbrier Co., WV. Sick at Red Sulphur Springs, WV on Sept.-Oct. 1862 roll. Recvd. clothing 5/13/64.

SHAWVER, WILLIAM CAVENDISH: Pvt. Co. C. B. 5/1/33 at "The Meadows" of Greenbrier Co., WV. Son of Robert and Mary Janes Collison Shawver. Married 12/25/55. Had three sons. Farmer. Enl. 5/1/62 at White Sulphur Springs, Greenbrier Co., WV. Capt. at Lewisburg, WV 5/23/62. POW Camp Chase, OH 6/20/62. Exchd. 8/25/62. Returned home ca. May 1864. D. 3/17/1901 at Corliss, WV. Another source says died at Meadows 1901. Bur. Shawver Cem., Corliss, WV.

SHEELOR, DAVID: Pvt. Co. E. Enl. 5/1/62 at White Sulphur Springs, Greenbrier Co., WV. Det. on Sept.-Oct. 1862 roll. Present on 1/2/63 roll to 4/30/63 roll. Paid 5/1/63 to 7/8/63. NFR.

SHELTON, WILLIAM H.: Pvt. Co. D. Enl. 7/29/61 at Cross Lanes, Nicholas Co., WV. Transf. from Capt. Lewis' Co. Sick at home on 8/31/61 roll. Transf. to Co. I, 8th Va. Cav. Aug. 1863.

SHELTON, WINSTON: Capt. Co. D. B. 1815. Farmer, Nicholas Co., WV 1850, age 35. Enl. 5/30/61 at Summersville, Nicholas Co., WV. Elected Capt. 6/1/61. Not re-elected 5/1/62. Dropped 5/2/62.

SHEPHERD, BEVERLY K.: Pvt. Co. F. Enl. 7/13/61 at Camp Two Mile, Kanawha Co., WV. Sick since 8/8/61 on 8/31/61 roll. Sick at home on surg. cert. on 10/31/62 roll. Present on 12/31/62 roll. Recvd. clothing at Liberty (Bedford), Va. hosp. 12/7/64. POW paroled at Charleston, WV 5/11/65. Age: 33. 5'5". Fair complexion, light hair, gray eyes, light whiskers. D. sometime prior to 1914.

SHEPHERD, JOHN T.: Pvt. Co. F. Res. Monroe Co., WV. Enl. 7/4/61 at Charleston, Kanawha Co., WV. Sick since 8/22/61 on surg. cert. on 8/31/61 roll. D. sometime prior to 1914.

SHIELDS, SAMEUL: Pvt. Co. G. Enl. 7/3/61 at Charleston, Kanawha Co., WV. Left sick at White Sulphur Springs, WV 8/20/61. Present on 11/13/62 roll.

SHIELDS, WILLIAM H.: 1st Sgt. Taylor's Co. Enl. 6/19/61 at Frankford, Greenbrier Co., WV. Age: 27. Co. transf. to Wise Brigade 7/10/61.

SHIRES, J. T.: Pvt. Co. F. Date and place of enlistment not stated. Dischd. at Camp McCausland, WV 7/10/62. Recvd. clothing 4 QR 1864. Paroled at Charleston, WV 6/12/65. Age: 38. 5'8". Dark complexion, black hair, black eyes.

SHIRES, WILLIAM: Pvt. Co. F. Enl. 6 or 7/5/62 at Camp McCausland, Monroe Co., WV. Present until sick with leave on 12/31/62 roll. Granted an extended furlo. due to chronic rheumatism of right leg 9/19/63. Sick at home in Monroe Co., WV on 10/31/64 roll.

SHORT, JOHN J.: Co. A. Enl. 5/22/61 at Poca, Putnam Co., WV. KIA at Droop Mountain, WV 11/6/63.

SHREWSBURY, ANDREW PARKS: Pvt. Co. H. B. at Shrewsbury, WV. V.M.I. 1854. Stayed a short time due to being roughed up by hazing. Res. Kanawha Co., WV. Enl. 5/8/61 at Charleston, Kanawha Co., WV. In Charlottesville, Va. hosp. for acute diarrhea 6/9-16/64. Capt. at Winchester, Va. 9/19/64. POW Pt. Lookout, MD 9/24/64. Exchd. 3/15/65. Deserted and surrendered Mar. 1865. Took oath at Charleston, WV and sent north 4/25/65. Age: 28. Farmer. 5'8". Fair complexion, brown hair, blue eyes. PWR says his war imprisonment broke his spirit.

SHREWSBURY, JOEL: Pvt. Co. H. Writer in clerk's office, Charleston, WV 1860, age 21. Enl. 5/8/61 at Charleston, Kanawha Co., WV. Dischd. at Pleasant Hill for disability 8/24/61.

SHREWSBURY, JOHN: Pvt. Co. H. Enl. 6/26/61 at Coalsmouth (St. Albans), Kanawha Co., WV. Absent cause unknown on 8/31/61 roll. NFR.

SHUMATE, BALLARD M.: Pvt. Co. F. Res. Monroe Co., WV. Enl. 7/4/61 at Charleston, Kanawha Co., WV. Present throughout. Paroled at Charleston, WV 6/12/65. Age: 24. 5'10". Dark complexion, black hair, gray eyes. D. in Mercer Co., WV sometime prior to 1914.

SHUMATE, BALLARD PRESTON: Pvt./2nd Lt. Co. F. B. 12/10/42 Blue City, Va. Another source says born at Pearisburg, Va. Res. near Mercer Salt Works, WV ca. 1849. Res. Glen Lyn, Va. 1858. Farm. laborer. Enl. 7/4/61 at Charleston, Kanawha Co., WV. Apptd. 4th Corp. 8/?/61. Elected 5/1/62. Nurse to Capt. Shumate July 1862. Promo. to 2nd Lt. 2/4/63. WIA (severe thigh) at White Sulphur Springs, WV 8/26-27/63. Permanently disabled due to wd. On wded. leave on rolls from 12/31/63 to 4/25/64. At home wded. in Monroe Co., WV on 10/31/64 roll. Retired 11/25/64 and/or 1/21/65. Invalid Corps. 1/27/65. Comm. Provost Marshall for southwestern counties of Va. Paroled at Charleston, WV 6/13/65. Age: 22. 5'5". Fair complexion, auburn hair, blue eyes, light whiskers. Res. Monroe Co., WV. Res. Pipestem District, Mercer (now Summers) Co., WV 1865. Mercantile business 1871. Elected Deputy Sheriff 1872. Resgnd. 1874. Pipestem Postmaster 7/22/79. Elected WV House of Delegates 1892. Ran again for WV House of Delegates and lost 1894. Elected to WV Legislature 1898. Notary Public 10/7/1903. Secretary, Board of Education, Pipestem District for 31 years. Res. Summers Co., WV 1914. D. 3/11/1915 at Athens, WV of cancer. Bur. Pipestem Cem., Pipestem, WV. Photo appears in Miller's *History of Summers County*.

SHUMATE, JAMES H.: Corp./Capt. Co. F. Enl. 7/4/61 at Charleston, Kanawha Co., WV. Elected 2nd Lt. 8/22/61. Comm. not recvd. Elected Capt. 5/1/62. WIA (gunshot wd. of knee joint) at Lewisburg, WV 5/23/62. Sick at Montgomery Whtie Sulphur Springs, Va. July 1862. Resgnd. 2/4/63 due to wd. Paroled at Charleston, WV 6/19/65. Age: 34. 5'11". Light complexion, brown hair, gray eyes.

SHUMATE, LEWIS A.: Pvt./Corp. Co. F. Enl. 7/4/61 at Charleston, Kanawha Co., WV. Present throughout. Paroled at Charleston, WV 6/12/65. Age: 25. 5'6". Dark complexion, black hair, gray eyes. Postwar res. Summers Co., WV.

SHUMATE, MATTHEW: Pvt. Co. G. B. in Fayette Co., WV. Res. Wyoming Co., WV. Enl. 7/3/61 at Charleston, Kanawha Co., WV. Surg. cert. furlo. 7/25/61 until deemed fit for service. Present on 11/13/62 roll. Dischd. at Camp Sam Jones for disability and deformity of the knees 7/8/63. Deserted Dec. 1863. Took oath and sent North 12/31/63. Age: 21. Farmer. 5'6". Light complexion, light hair, blue eyes.

SHUMATE, RUFUS A.: Pvt. Co. F. Enl. 7/4/61 at Charleston, Kanawha Co., WV. Sick at home since 8/8/61 on 8/31/61 roll. Present on 10/31/62 roll. Exchd. to Bryan's Battery on Mar.-Apr. 1863 roll. Postwar res. Mercer Co., WV.

SIAS, ANDREW S.: Pvt. Co. I. Res. Boone Co., WV 1860, age 20. Mercer Co., WV. Enl. 6/29/61 at Charleston, WV. AWOL 7/26/61.

SIAS, EVI: Pvt. Co. I. B. 1835 Fayette Co., WV. Res. Boone Co., WV 1860, age 25. Enl. 6/29/61 at Charleston, Kanawha Co., WV. AWOL 7/26/61. Postwar Lincoln Co., WV Justice of Peace. Reelected 1880. Schoolteacher, Boone Co., WV. Secretary, Lincoln Co., WV Board of Education. Farmer, Spurlocksville, Lincoln Co., WV.

SIMMONS, DAVID: Pvt. Co. I. Enl. 4/4/64 at Dublin, Pulaski Co., Va. Capt. at Cold Harbor (Gaines Farm), Va. 6/3/64. POW Pt. Lookout, MD 8/22/64. D. at Pt. Lookout, MD.

SIMMONS, GEORGE: Pvt. Co. I. Enl. 5/1/62 at White Sulphur Springs, Greenbrier Co., WV. Sick at Giles C.H. (Pearisburg), Va. July 1862. Sick at Beverly, WV on 4/30/63 roll. Present on 8/31/63 to 12/31/63 roll. Capt. at Winchester, Va. 9/19/64. POW Pt. Lookout, MD 9/24/64. Exchd. 3/15/65. D. near Zion's Hill, Botetourt Co., Va. 10/6/1924 at age 82.

SIMMONS, JAMES ALEXANDER: Pvt. Co. D. Enl. 3/17/64 at Camp Gauley, Greenbrier Co., WV. On furlo. since 3/24/64 on 4/12/64 roll. Present on 4/30/64 roll. Capt. at Winchester, Va. 9/19/64. POW Pt. Lookout, MD 9/24/64. Exchd. 2/13/65. At Camp Lee, Richmond, Va. 2/18/65. D. 2/24/1907.

SIMMONS, NATHANIEL: Pvt. Co. I. Res. Fincastle, Botetourt Co., Va. Enl. 2/4/64 at Camp Gauley, Greenbrier Co., WV. Capt. at Winchester, Va. 9/19/64. POW Pt. Lookout, MD 9/24/64. Exchd. 3/15/65. At Richmond, Va. hosp. 3/18/65. Transf. to Chimborazo Hosp. 3/19/65.

SIMMONS, WILLIAM R.: Pvt./Corp. Co. K. Enl. 6/6/61 at Charleston, Kanawha Co., WV. Helped in White Sulphur Springs, WV hosp. for 25¢ per day 11/1-15/61. WIA at Lewisburg, WV 5/23/62. Sent to Montgomery White Sulphur Springs, Va. for wd. d still there July 1862. Bahlmann says married a Dunkard girl and joined the Dunkard church.

SIMPSON, MARTIN F.: 3rd Lt./2nd Lt. Co. I. Res. Botetourt Co., Va. Enl. 5/2/62 at White Sulphur Springs, Greenbrier Co., WV. Elected 2nd Lt. 5/1/62. Resgnd. 3/1/63. On surg. report at Lewisburg, WV dated 3/23/63 surg. recommended that Simpson suffered from chronic dysentery which disqualified him from infantry service. Suggested cav. service. Transf. to Cav. service 3/24/63. Reported to U.S. officials at Beverly, WV 10/26/63. Took oath and relsd. 10/29/63. Listed as C.S. deserter on report dated 11/5/64. Age: 31. Carpenter. 5'8". Fair complexion, light hair, blue eyes.

SIMS, HENRY EDMUND: Pvt. Co. A. Enl. 5/22/61 at Poca, Putham Co., WV. Unaccounted for 7/25/61. Present on 11/1/62 roll. Deserted and surrendered Apr. 1865. Took oath and sent north 4/22/65. Age: 23. Farmer. 6'1". Dark complexion, dark hair, dark eyes. Res. Putnam Co., WV. Bur. Simms Cem., Scary Creek, WV. Also spelled Sims.

SIMS, MILTON: Pvt. Co. C. Farmer, Fayette Co., WV 1860. Age: 36. Enl. 5/10/61.

SIMS, WILLIAM: Pvt./Corp. Co. B. Enl. 6/1/61 at Ripley, Jackson Co., WV. Teamster 4/1/63 to 1/31/64. Sick at hosp. on 1/10/64 roll.

SIMS, WILLIAM H.: Pvt. Co. A. Enl. 6/12/61 at Camp Tompkins, Kanawha Co., WV. Teamster on rolls from 11/1/62 to 2/28/63. Teamster at Lewisburg, WV with wagons on 4/30/63 roll. Teamster on 12/31/63 roll. Det. service on 4/1/64 roll. Det. service in Q.M. Dept. on 4/30/63 roll. Present on 10/31/64 roll. POW paroled at Charleston, WV 5/10/65. Age: 22. 5'6". Light complexion, light hair, blue eyes, light whiskers.

SINGLETON, ALBERT R.: Pvt. Co. K. Res. Campbell Co., KY. Enl. 6/26/61 at Coalsmouth (St. Albans). Kanawha Co., WV. Capt. at Lewisburg, WV 5/23/62. POW Camp Chase, OH. Exchd. 8/25/62. Present on 12/31/62 roll. Capt. at Cold Harbor (Gaines Farm), Va. 6/3/64. POW Pt. Lookout, MD 6/14/64. Transf. to Elmira, NY 7/12/64. Both legs broken in a train wreck near Shohola, PA 7/15/64 en route to Elmira, NY. Took oath at Elmira, NY and relesd. 5/3/65. Desired to go to Newport, Campbell Co., KY where he had family. 5'11½". Fair complexion, brown hair, hazel eyes. Met with William F. Bahlmann and John P. Donaldson postwar in Cinn., OH.

SINK, JOHN D.: Pvt. Swann's Co. Laborer, Charleston, WV 1860. Enl. 5/8/61 at Charleston, Kanawha Co., WV. Age: 28. Co. transf. to Co. K, 59th Va. Inf. 7/25/61.

SITES, JOHN W.: Co. unknown. Bur. in a Nat. Cem. in MO.

SIZEMORE, JOHN: Pvt. Co. G. Enl. 7/3/61 at Charleston, Kanawha Co., WV. Left his co. at Camp Two Mile, WV 7/6/61. AWOL.

SIZEMORE, LEVI: Pvt. Co. G. Enl. 7/7/61 at Charleston, Kanawha Co., WV. Present on 9/3/61 roll. NFR.

SKAGGS, CYRUS: Pvt. Co. C. Farmer, Fayette Co., WV 1860. Age: 37. Enl. 5/10/61.

SKAGGS, HENRY H.: Pvt. Co. C. B. Ansted, WV. Res. Cross Roads, Va. (Probably Cross Lanes, WV). Enl. 5/1/62 at White Sulphur Springs, Greenbrier Co., WV. WIA at Droop Mountain, WV 11/6/63. Absent wded. on rolls from 12/31/63 to 10/31/64. Recvd. clothing at Montgomery White Sulphur Springs, Va. 2/9/64. Extended furlo. at Montgomery White Sulphur Springs, Va. for amputation of right leg 5/19/64. Original member of Jeb Stuart UCV camp, Fayette Co., WV. Res. Victor, WV. School teacher, Sugar Creek, WV. Farmer and cattleman. D. 1914. Bur. Sugar Creek, Fayette Co., WV. Day laborer, Fayette Co., WV 1860. Age: 21. PWR says enl. 5/10/61.

SKAGGS, HENRY A.: Sgt. Taylor's Co. Enl. 6/19/61 at Frankford, Greenbrier Co., WV. Age: 20. Co. transf. to Wise Brigade 7/10/61.

SKAGGS, HIRAM: Pvt. Co. C. Farmer, Fayette Co., WV 1860. Age: 40. Enl. 5/10/61.

SKAGGS, ISIAH: Pvt. Co. C. Day laborer, Fayette Co., WV 1860. Age: 18. Enl. 5/10/61.

SKAGGS, JOSEPH B.: Pvt. Co. C. Res. Fayette Co., WV. Enl. 6/5/61 at Charleston, Kanawha Co., WV. MIA & capt. at Droop Mountain, WV 11/6/63. Age: 28. POW Camp Chase, OH 11/18/63. Transf. to Ft. Delaware 2/29/64. D. at Ft. Delaware of chronic diarrhea 7/6/64. Bur. Jersey shore of Ft. Delaware, Finn's Point Nat. Cem., NJ. Farmer. 5'11". Florid complexion, black hair, dark eyes.

SKAGGS, MELITUS H.: Pvt. Co. C. Res. Fayette Co., WV. Enl. 5/1/62 at White Sulphur Springs, Greenbrier Co., WV. Det. in government shop on 12/31/63 roll. Det. brig. shoe shop on 4/18/64 roll. Capt. at Winchester, Va. 9/19/64. POW Harpers Ferry, WV 9/27/64. Exchd. 3/15/65. At Camp Lee, Richmond, Va. 3/19/65. Attended vet. reunion Beckley, WV 9/12/1918. Day laborer, Fayette Co., WV 1860. Age: 18. PWR says enl. 5/10/61.

SKAGGS, R. N.: Pvt. Co. C. Enl. 5/10/61.

SKAGGS, THOMAS A.: Pvt. Co. C. Res. Fayette Co., WV. Enl. 5/1/62 at White Sulphur Springs, Greenbrier Co., WV. Still present on 10/31/64 roll. Deserted and surrendered Mar. 1865. Took oath at Charleston, WV and sent north 4/7/65. Age: 20 or 24. Farmer. 5'8". Dark complexion, fair hair, dark eyes. Day laborer, Fayette Co., WV 1860. Age: 19. PWR says enl. 5/10/61.

SKAGGS, THOMAS L.: Pvt. Co. C. B. in Monroe Co., WV. Enl. 6/5/61 at Mountain Cove, Fayette Co., WV. Dischd. for disability at Camp Floyd 4/10/62 for advanced age and consequent infirmity. Age: 54. Farmer. 5'9½". Fair complexion, dark brown hair, dark eyes. PWR says enl. 5/10/61.

SKELTON, ANDREW: Pvt. Co. K. Enl. 1862.

SLACK, ANDREW: Pvt. Swann's Co. Enl. 6/24/61 at Charleston, Kanawha Co., WV. Age: 16. Co. transf. to Co. K, 59th Va. Inf. 7/25/61.

SLACK, JAMES H.: Pvt. Swann's Co. Enl. 6/3/61 at Charleston, Kanawha Co., WV. Co. transf. to Co. K, 59th Va. Inf. 7/25/61.

SLAUGHTER, ANDREW B.: Pvt. Swann's Co. Res. Malden, WV. 1860. Enl. 6/27/61 at Charleston, Kanawha Co., WV. Age: 19. Co. transf. to Co. K, 59th Va. Inf. 7/25/61.

SLAUGHTER, JOHN W.: Pvt. Swann's Co./Co. H. Clerk, Malden, WV 1860. Age: 22. Enl. 5/8/61 at Charleston, Kanawha Co., WV. Co. transf. to Co. K, 59th Va. Inf. 7/25/61. Transf. to Co. H, 22nd Va. Inf. 3/23/62. Re-transf. to Co. K, 59th Va. Inf., Edgar's Battn. 7/6/62.

SLOAN, M.: Co. G. Postwar roster. On Buchanan, Va. pension list.

SLOAN, THOMAS: Pvt. Co. I. Enl. 3/1/64 in Greenbrier Co., WV. Deserted 3/8/64.

SMILEY, JAMES: Co. G. PWR. Enl. as a Pvt. Mar. 1862 at Lewisburg, WV. Age: 30. Res. Alleghany Co., Va. Surrendered Apr. 1865.

SMILEY, MARTIN L.: Sgt./Capt. Co. K. Enl. 6/6/61 at Charleston, Kanawha Co., WV. Elected Dec. 1861 and 5/1/62. Bahlmann says re-elected 2nd Lt. 5/1/62. Absent with leave since 2/18/63 on 2/28/63 roll. Present on Mar.-Apr. 1863 roll. WIA at Droop Mountain, WV 11/6/63. Promo. to 2nd Lt. 11/12/63. In hosp. for Droop Mountain wd. on 12/31/63 and 5/1/64 rolls. Capt. at Winchester, Va. 9/19/64. POW Ft. Delaware 9/25/64. Trans. to Pt. Lookout, MD 10/6/64. Exchd. 10/11/64. Promo. 1st Lt. 12/31/64. Promo. Capt. 12/31/64. Surrendered with Gen. Jones' forces in Florida 4/26/65. POW paroled at Madison, Florida 5/19/65. PWR says elected 2nd Lt. 12/?/61 and 1st Lt. 11/11/62.

SMITH, A. H.: Pvt. Co. G. Res. Craig Co., Va. Enl. 5/1/62 at White Sulphur Springs, Greenbrier Co., WV. Deserted 5/1/62.

SMITH, ASA W.: Pvt. Co. A. Res. Kanawha Co., WV. Enl. 9/15/62 at Charleston, Kanawha Co., WV. Absent wded. on 12/31/63 roll. Present on 4/30/64 roll. POW paroled at Staunton, Va. 4/30/65. Age: 31. 5'9". Fair complexion, fair hair, hazel eyes.

SMITH, BALLARD J.: Pvt. Co. I. Res. Alleghany Co., Va. Enl. 4/15/62 in Greenbrier Co., WV. Sick at home in Alleghany Co., Va. on 4/30/63 roll. Capt. at Droop Mountain, WV 11/6/63. Age: 32. POW Camp Chase, OH 11/18/63. Transf. to Ft. Delaware 2/29/64. Exchd. 10/30/64. Farmer. 5'11". Dark complexion, black hair, dark eyes. PWR says Co. G.

SMITH, CHARLES M.: Chaplain. Field & Staff. Date and place of enlistment not stated. Apptd. 5/?/61. Resgnd. date not stated. Apptd. 3/4/63. Transfer letter dated 3/4/63. Resgnd. 11/1/63.

SMITH, CYRUS: Pvt. Taylor's Co. Enl. 6/19/61 at Frankford, Greenbrier Co., WV. Age: 20. Co. transf. to Wise Brigade 7/10/61.

SMITH, DAVID S.: Pvt. Co. C. Date and place of enlistment not stated. Recvd. clothing 4 QR 1864. Paroled at Charleston, WV 6/12/65. Age: 30. 5'8". Fair complexion, auburn hair, blue eyes.

SMITH, DAVIDSON: Pvt. Co. G. Enl. 7/7/61 at Charleston, Kanawha Co., WV. Transf. from Capt. Sarver's Co. 7/25/61. Had a furlo. and not reported since 7/1/63. Away with leave 8/11-21/61. WIA (severe shoulder) at White Sulphur Springs, WV 8/26-27/63. Did not return to hosp. on 12/31/63 roll. Deserted.

SMITH, EDWARD: Pvt. Co. A. Enl. 7/26/62 at Salt Sulphur Springs, Monroe Co., WV. Deserted 10/22/62. Dischd. 11/20/62.

SMITH, EDWIN: Pvt. Co. I. Enl. 6/29/61 at Charleston, Kanawha Co., WV. Present on rolls from 6/29/61 to 8/31/61. NFR.

SMITH, HARVEY W.: Pvt. Co. K. B. ca. 1834 Craig Co., Va. Enl. 4/24/61 at White Sulphur Springs, Greenbrier Co., WV. MIA and capt. at Droop Mountain, WV 11/6/63.

SMITH, ISAAC NOYES SR.: Maj. Field & Staff/Co. H. B. 4/6/31 Charleston, WV. Only son of Col. Benjamin and Roxalana Noyes Smith. Grad. Washington College 1849 A.B. Lawyer, Charleston, WV, 1860, age 29. Law partnership with father. Virginia Legislature 1860. Married 1860. Enl. 1861 at Charleston, Kanawha Co., WV. Order to be commissioned 7/16/61. Apptd. Maj. 8/?/61. Tendered resignation 10/26/61 due to his great dislike of Gen. John B. Floyd, who accused Smith of being loyal to the Union. Rsgnd. 11/?/61. Accepted 11/6/61. Given a safe pass through Federal lines 2/6/62. Smith kept a diary from early September 1861 until 11/30/61 which was confiscated from him by Gen. Jacob D. Cox when captured and now in the possession of Oberlin College, Oberlin, OH. The diary was printed in *West Virginia History* Vol. XXVII, No. 3, Apr. 1966. Smith commanded the regt. during Col. Tompkins' resignation until 11/10/61 when William A. Jackson was apptd. Lt. Col. of the regt. Due to Floyd's accusations, though, Smith was like a "prisoner in command". Law partnership with E. B. Wright, Charleston, WV 1865. Law partnership, "Smith & McKnight". President of a bank. Secretary, Kanawha Salt Co. D. 10/6/83. Bur. Spring Hill Cem., Charleston, WV.

SMITH, JACOB LUGAR: Pvt. Co. K. B. 6/15/30. Res. Craig Co., Va. Enl. 4/21/62 at White Sulphur Springs, Greenbrier Co., WV. Still present on Mar.-Apr. 1863 roll. D. 9/23/98. Bur. Smith Cem., Rts. 18 and 607, Potts Creek, Craig Co., Va. Bahlmann says exchd. to 60th Va. Inf. 1863.

SMITH, JACOB P.: Pvt. Co. A. Enl. 7/22/62 at Salt Sulphur Springs, Monroe Co., WV. Deserted at Lewisburg, WV 4/12/63.

SMITH, JAMES W.: Pvt. Co. G. Res. Alleghany Co., Va. Enl. 7/1 or 29/62 at Union, Monroe Co., WV. Conscript. Deserted on the retreat from Charleston, WV 10/28/62. Gave himself up to Col. Siber's cav. Age: 23. 5'11". Sandy hair, blue eyes, medium size. Said he "fired his musket once at Fayetteville battle (9/10/62) then the musketeers were ordered to fall back to a lying down position behind the hill."

SMITH, JOAB: Pvt. Co. B. Schoolteacher. Evans, WV 1860, age 50. Enl. 6/1/61 at Ripley, Jackson Co., WV. In White Sulphur Springs, WV hosp. with dysentery on 9/1/61 roll. Wagon master 10/7-27/61. Left as nurse at White Sulphur Springs, WV Aug. 1861. Returned to Floyd's army with convalescents 9/12/61. Returned to White Sulphur Springs, WV with the sick 9/23/61. Det. charge of one ward of 60 men who were suffering from various maladies, being in filth and destitute of every necessity of life, and all imploring relief which Smith could not give. Commissary Sgt. Age: 51. D. at White Sulphur Springs, WV hosp. 10/22 or 27/61. Smith kept a diary used as a basis for a magazine article in *West Virginia Review*, Nov. 1930. First name may be Joel.

SMITH, JOHN: Pvt. Co. A. Date and place of enlistment unknown. KIA at Beverly, WV. Bur. Mt. Iser Cem., Beverly, WV. PWR.

SMITH, JOHN: Pvt. Co. A. Res. Putnam Co., WV. Enl. 5/22/61 at Poca, Putnam Co., WV. Measles at White Sulphur Springs, WV 8/20/61. Capt. at Lewisburg, WV 5/23/62. Age: 34½. POW dischd. from hosp. at Gallipolis, OH 7/2/62. POW Camp Chase, OH 7/8/62. Relsd. on taking oath 8/25/62. Present on 12/31/62 roll. KIA at White Sulphur Springs, WV 8/26/63. Farmer. 5'8". Dark complexion, dark hair, blue eyes.

SMITH, JOHN W.: Rank unknown. Co. A. Bur. Smith Cem., Lydia Smith farm, Arnettsville, Monongalia Co., WV. This may be same John Smith listed above. PWR.

SMITH, JOSEPH: Pvt. Co. B. Lawyer, Ripley, WV 1860, age 44.

SMITH, JOSEPH: Corp. Co. F. Enl. 7/4/61 at Charleston, Kanawha Co., WV. Served as corp. from 7/17/61 to 8/22/61. Present on 8/31/61 roll. Postwar roster says dischd.

SMITH, J. S. F.: Pvt. Co. H. Enl. 3/10/62 at Lewisburg, Greenbrier Co., WV. Det. in Adj. office. Acting Brig. Q.M. Sgt. July 1862. Det. A.Q.M. Sgt. on rolls for July-Aug. 1862 and Sept.-Oct. 1862 and 12/31/62. Dischd. 1/9 or 16/63. Half-brother of Maj. R. A. Bailey.

SMITH, MAJOR W.: Pvt. Co. K. Enl. 4/24/62 at White Sulphur Springs, Greenbrier Co., WV. Left sick at Lewisburg, WV on Mar.-Apr. 1863 roll. Disability discharge at Camp Sam Jones for "hemostylis of long standing" 7/2/63. Age: 27. 6'½". Dark complexion, dark hair, blue eyes. Res. Monroe Co., WV. PWR says MIA 1862.

SMITH, N. A.: Pvt. Co. K. Enl. 4/24/62 at White Sulphur Springs, Greenbrier Co., WV. Present on 11/1/63 roll. NFR.

SMITH, PETER: Pvt. Co. G. B. 1/10/34. Enl. 2/22/64 at Orange Court House, Va. PWR says enl. Mar. 1862 at Lewisburg, WV. Conscript. Sick in hosp. but since returned on 10/31/64 roll. D. 9/10/1918 Healing Spring, Va. Bur. Smith family cem., along Rt. 20, approximately 2 miles north of Mt. Pleasant U. Church, Covington, Va.

SMITH, RICHARD F.: Pvt. Co. G. Res. Alleghany Co., Va. Enl. 5/1/62 at White Sulphur Springs, Greenbrier Co., WV. Left sick at Beverly, WV on 4/30/63 roll. Capt. at White Sulphur Springs, WV 8/26/63. Age: 26. POW Camp Chase, OH 9/7/63. Transf. to Rock Island, IL 2/22/64. Took oath at Rock Island, IL 10/24/64. Farmer. 5'11½". Fair complexion, auburn hair, gray eyes.

SMITH, R. T.: Co. K. Deserted 5/3/62 on Aug. 1862 broadside. Craig Co., Va.

SMITH, SAMPSON: Pvt. Co. G. Probable res. Covington, Va. Date and place of enlistment not stated. Wrote a letter from camp near Fishersville, Va. 12/29/64. In Farmville, Va. hosp. for rubeola 3/16/65. Transf. to Lynchburg, Va. 4/6/65. On a list of POW's surrendered by Gen. Lee at Lynchburg, Va. Apr. 1865. PWR says enl. Mar. 1862 at Lewisburg, WV. Age: 24. Res. Alleghany Co., Va.

SMITH, THOMAS: Adjutant. Field & Staff/Co. H. Lawyer. Enl. 1861 at Charleston, Kanawha Co., WV. Appears on a morning report from Charleston, WV dated 7/8/61. At start of war was a Pvt. in Co. H. Apptd. Maj. in 36th Va. Inf. 7/16/61. Promo. Lt. Col. of 36th Va. Inf. 3/30/64. WIA at Cloyd's Mountain, Va. 5/9/64. Promo. to Col. of 36th Va. Inf. 5/18/64. In postwar period was not permitted to practice law in Charleston, WV and then practiced law in Warrenton, Va. Apptd. Chief Justice of U.S. District Court in New Mexico 1893. B. 8/25/36. Son of Gov. "Extra Billy" Smith. D. at Warrenton, Va. 6/29/1918.

SMITH, THOMAS: Pvt. Co. A. Enl. 6/25/61 at Salt Sulphur Springs, Monroe Co., WV. WIA (slight) at Fayetteville, WV 9/10/62. Absent wded. on 11/1/62 roll. Deserted on or about 4/25/63.

SMITH, THOMAS: Pvt. Co. B. Enl. 9/22/62 at Charleston, Kanawha Co., WV. Sick at hosp. on 4/30/63 roll. D. of wounds (probably inflicted at Droop Mountain, WV 11/6/63) at Lewisburg, WV 11/6/63. Effects turned over to Q.M.

SMITH, THOMPSON L.: Chaplain Field & Staff. Date place of enlistment not stated. Gave a sermon to regt. at White Sulphur Springs, WV 8/18/61. Assignment dated 2/3/62. Possibly apptd. 14th Va. Cav. Jan. 1862. Chaplain at St. Luke's Hosp., Chicago, IL 1895.

SMITH, WILLIAM: Pvt. Co. G. Enl. 5/1/62 at White Sulphur Springs, Greenbrier Co., WV. Joined from desertion 7/12/62. Left sick at Lewisburg, WV on 4/30/63 roll. Present on 12/31/63 roll. Absent sick on 4/1/64 roll. Dischd. on surg. cert. 4/15 or 18/64 at Camp Gauley, WV for chronic disease of kidneys. On Alleghany Co., Va. pension rolls. Age: 33 at enl. Res. Alleghany Co., Va.

SMITH, W. H.: Co. K. Deserted 5/3/62 on Aug. 1862 broadside. Craig Co., Va.

SMITH, WILLIAM M.: Pvt. Swann's Co. Res. Craig Co., Va. Enl. 5/22/61 at Charleston, Kanawha Co., WV. Age: 19. Co. transf. to Co. K, 59th Va. Inf. 7/25/61.

SMITHERS, DAVID SHREWSBURY: Pvt. Co. H. B. 10/4 or 8/37 on Dickinson farm, Malden, WV. Farmer, Belle, WV. Enl. 5/8/61 at Charleston, Kanawha Co., WV. Det. teamster 7/9/61 and still so engaged on 8/31/61 roll. Brig. wagoner July 1862. Det. teamster on rolls from July-Aug. 1862 to 2/28/63. Det. teamster at Montgomery White Sulphur Springs, Va. on rolls for 4/30/63 and 6/30/63. Employed as a car conductor from July 1863 to Mar. 1864. Employed as a car conductor at Montgomery White Sulphur Springs, Va. 3/1-27/64. Det. in A.Q.M. Dept., Dublin, Va. on 12/3/63 and 1/1/64 and 4/1/64 rolls. WIA (contusion of left leg) and capt. at Winchester, Va. 9/19/64. At Winchester, Va. field hosp. 9/20/64. Transf. to Baltimore, MD hosp. 10/10/64. POW Pt. Lookout, MD 10/30/64. Exchd. 10/30/64. Returned to regt. at Abington, Va. Paroled at Charleston, WV 6/7/65. Age: 27. Car conductor. 5'10". Light complexion, light hair, blue eyes, dark whiskers. Postwar farmer, Kanawha Co., WV. D. 10 or 12/18/1903 at Simmons Creek, WV of heart failure. Bur. Spring Hill Cem., Charleston, WV.

SMITHERS, SAMUEL W.: Pvt. Co. H. Farmer, Shrewsbury, WV 1860. Age: 21. Enl. 5/8/61 at Charleston, Kanawha Co., WV. Absent cause unknown on 8/31/61 roll. Bur. Spring Hill Cem., Charleston, WV.

SMITHSON, EPHRIAM B.: Pvt. Co. B. D. at Pt. Lookout, MD.

SMOOT, CREED F.: Pvt. Co. G. Enl. 5/1/62 at White Sulphur Springs, Greenbrier Co., WV. Det. as Provost Guard at Greenbrier River on 4/30/63 roll. Present on 12/31/63 roll. Deserted at Salem, Va. on 10/31/64 roll. PWR says enl. May 1861 at Charleston, WV. Age: 18. Res. Wyoming Co., WV. Surrendered Apr. 1865 at Staunton, Va.

SMOOTH, GABRIEL: 1st Sgt. Co. G. B. in Bath Co., Va. Res. Wyoming Co., WV. Enl. 7/3/61 at Charleston, Kanawha Co., WV. Capt. at Wyoming Court House, Wyoming Co., WV by the 44th O.V.I. while home on furlo. 3/14/62. POW Camp Chase, OH 5/7/62. Exchd. 8/25/62. Dischd. at Lewisburg, WV 2/17/63 being over the age required for military service. Age: 45. On a list of POW rebel deserters. Took oath at Charleston, WV 5/10/65. Farmer. 5'8". Dark/fair complexion, sandy/dark/gray hair, blue eyes, turning gray whiskers.

SNEAD, AUSTEN: Pvt. Co. G. Enl. 5/1/62 at White Sulphur Springs, Greenbrier Co., WV. Deserted 5/1/62. Listed as deserter on Aug. 1862 broadside. Left 5/24/62. Also spelled Austin.

SNIDOW, CHRISTIAN N.: Pension application Giles and Bland counties, Va. D. 6/14/86 Giles Co., Va. A log rolled over him.

SNYDER, DAVID HALL: Sgt. Co. H. B. 2/?/25 Charleston, WV. Postmaster 10 or 12 years prior to the war. Enl. 5/8/61 at Charleston, Kanawha Co., WV. Det. as postmaster 5/8/61. Apptd. to Gen. Henry A. Wise's staff but resigned due to ill health which ultimately caused his death. D. 2/11/91. Believed bur. Spring Hill Cem., Charleston, WV. Another source says died 8/19/66.

SNYDER, JOHN B.: Pvt. Swann's Co. Enl. 5/22/61 at Charleston, Kanawha Co., WV. Age: 22. Co. transf. to Co. K, 59th Va. Inf. 7/25/61.

SNYDER, SOLOMON: Pvt. Co. I. Enl. 10/15/63 at Lewisburg, Greenbrier Co., WV. Deserted at Lewisburg, WV 4/24/64.

SNYDER, W. B.: Pvt. Co. H.

SOLESBURY, ALLEN P.: Pvt. Co. G. Enl. 7/3/61 at Charleston, Kanawha Co., WV. Left sick at White Sulphur Springs, WV 8/20/61. Absent on furlo. on 11/13/62 roll. Present on 12/31/62 roll. Had a furlo. and had not reported back on 10/31/64 roll.

SOLESBURY, LAFAYETTE: Pvt. Co. G. Enl. 9/20/62 at Charleston, Kanawha Co., WV. Furlo. expired 3/10/63. AWOL on 4/30/63 roll.

SOLESBURY, MARTIN V.: Pvt. Co. G. Enl. 1/1/62 at Union, Monroe Co., WV. Deserted at Salem, Va. on 10/31/64 roll.

SOUTHALL, ALEX F.: Pvt. Co. A. Fnl. 5/22/61 at Poca, Putnam Co., WV. Teamster for regt. on 8/31/61 roll. NFR.

SOUTHALL, ALEX F.: Pvt. Co. B. Enl. 7/?/62 at Union, Monroe Co., WV. AWOL on 11/16/62 roll.

SOUTHALL, WILLIAM: Pvt. Co. B. Enl. 6/1/61 at Ripley, Jackson Co., WV. Present on 9/1/61 roll. NFR.

SOVLER, BARTROT: Pvt. Co. D. Date and place of enlistment not stated. On a list of prisoners paroled at Burkeville Junction, Va. 4/14-17/65.

SPANGLER, JACOB: Pvt. Co. I. Enl. 6/29/61 at Charleston, Kanawha Co., WV. Returned from desertion 9/15/62. Det. to work on boats at Central Depot on 3/4/63 roll. Det. to build boats at New River on 4/30/63 roll. Det. building boats at Narrows of New River on rolls 4/1/64 through 10/31/64.

SPARKS, EDWARD W.: Pvt. Co. D. B. 5/8/41 Gallis Co., OH. Enl. 5/30/61 at Summersville, Nicholas Co., WV. AWOL 7/28/61. Married 1862. PWR says served 1 year then served with State Guard until dischd. at expiration of term 3/24/64. Res. Spencer, WV 1876, blacksmith.

SPARR, JOHN V.: Pvt. Co. C. Farmer, Fayette Co., WV 1860. Age: 25. Enl. 5/10/61.

SPENCER, ABRAHAM D.: Pvt. Co. D. Enl. 5/30/61 at Summersville, Nicholas Co., WV. AWOL 7/28/61.

SPENCER, ELIAS C.: Pvt. Co. D. Enl. 5/30/61 at Summersville, Nicholas Co., WV. Left sick at White Sulphur Springs, WV 8/20/61. NFR.

SPENCER, HENRY H.: Corp. Co. D. Enl. 5/30/61 at Summersville, Nicholas Co., WV. AWOL 7/28/61.

SPENCER, HENRY O.: Sgt. Swann's Co. Enl. 5/8/61 at Charleston, Kanawha Co., WV. Age: 21. Co. transf. to Co. K, 59th Va. Inf. 7/25/61.

SPENCER, WILLIAM T.: Pvt. Swann's Co. Enl. 5/8/61 at Charleston, Kanawha Co., WV. Age: 23. Co. transf. to Co. K, 59th Va. Inf. 7/25/61.

SPESSARD, J. P.: Pvt. Co. H. B. 1831 Craig Co., Va. Enl. 4/28/62 at White Sulphur Springs, Greenbrier Co., WV. Absent on recruiting service on 2/28/63 roll. Present on 4/30/63 roll. On scout on 4/1/64 roll. WIA (gunshot wd. of right foot-metatarsol region-toes by roundball) at Winchester, Va. 9/19/64. Capt. at Harrisonburg, Va. 9/25/64. Transf. to Gen. Hosp. at Baltimore, MD 10/12/64. Transf. to Gen. Hosp. 10/17/64. POW Pt. Lookout, MD 10/18/64. Exchd. 10/30/64. First name is either Jacob or John. D. 1906.

SPROAT, JAMES: Pvt. Co. B. Enl. 6/1/61 at Ripley, Jackson Co., WV. Dischd. at Camp McCausland, WV 7/16/62.

SPURR, CHARLES: 1st Sgt. Swann's Co. Enl. 5/8/61 at Charleston, Kanawha Co., WV. Age: 20. Co. transf. to Co. K, 59th Va. Inf. 7/25/61.

STALNAKER, ASBURY: Pvt. Swann's Co. Enl. 5/8/61 at Charleston, Kanawha Co., WV. Age: 21. Co. transf. to Co. K, 59th Va. Inf. 7/25/61.

STALNAKER, WELLINGTON: Corp. Swann's Co. Boatman, Charleston, WV 1860. Enl. at Charleston, Kanawha Co., WV. Date not stated. Muster in date from 4/25/61. Age: 27. Co. transf. to Co. K, 59th Va. Inf. 7/25/61.

STANLEY, REUBEN: Pvt. Swann's Co. Enl. 5/25/61 at Charleston, Kanawha Co., WV. Age: 22. Co. transf. to Co. K, 59th Va. Inf. 7/25/61.

STARCHER, EDWARD: Pvt. Co. B. Res. Jackson Co., WV 1850. Enl. 9/22/62 at Charleston, Kanawha Co., WV. In Winchester, Va. Gen. Hosp. for very serious hand 7/25/64. Sent to Gen. Hosp. 8/3/64. At hosp. wded. on 11/30/64 roll. POW paroled at Charleston, WV 5/12/65. Age: 20. 5'7". Fair complexion, light hair, blue eyes.

STARCHER, JAMES: Pvt. Co. B. Res. Jackson Co., WV 1850, age 17. Enl. 9/22/62 at Charleston, Kanawha Co., WV. Capt. at Winchester, Va. 9/19/64. POW Pt. Lookout, MD 9/23/64. D. at Pt. Lookout, MD of chronic diarrhea 11/4/64. Bur. Pt. Lookout, MD #628 POW graveyard.

STARK, WILLIAM: Pvt. Swann's Co. Apprentice carpenter, Charleston, WV 1860. Enl. 5/8/61 at Charleston, Kanawha Co., WV. Age: 18. Co. transf. to Co. K, 59th Va. Inf. 7/25/61.

STARKS, ERASTUS P.: Pvt. Co. K. B. ca. 1844 Craig Co., Va. Enl. 4/24/62 at White Sulphur Springs, Greenbrier Co., WV. Still present on 10/31/64 roll.

ST. CLAIR, VIRGIL M.: Pvt. Co. B. Enl. 5/?/62 at White Sulphur Springs, Greenbrier Co., WV. Sick in Botetourt Co., Va. July 1862. At hosp. on 11/16/62 roll. Det. at hosp. on 12/31/62 roll. Det. as nurse at hosp. on rolls dated 2/28/63 and 4/30/63. At home sick on rolls from 1/10/64 through 5/1/64. Deserted on 11/30/64 roll.

STEEL, JEFFREY: Pvt. Co. C. B. 3/22/31 Giles Co., Va. First settler on Rich Creek (Ramont), Fayette Co., WV ca 1860. Built first school house at Rich Creek. Taught school in his house at Ramont. Date and place of enlistment not stated. Capt. in Fayette Co., WV by 47th O.V.I. 4/24/62. Age: 31. Charged with aiding and abetting the enemy and harboring and feeding rebel scouts. POW Camp Chase, OH 5/7/62. Exchd. 8/25/62. 5'8". Fair complexion, black hair, hazel eyes, sandy whiskers. Res. Fayette Co., WV. D. 5/?/1914 Rich Creek, WV of cancer. Bur. Steele Cem., Rich Creek, WV. Also spelled Steele.

STEEL, WALDEN M.: Pvt. Co. F. Enl. 7/4/61 at Charleston, Kanawha Co., WV. Sick at White Sulphur Springs, WV hosp. 8/20/61. Present on 10/31/62 roll. POW paroled at Charleston, WV 5/11/65. Age: 28. 5'6". Dark complexion, dark hair, hazel eyes, dark whiskers. Went to Missouri postwar.

STEPHENSON, JOHN: Pvt./Corp. Co. I. B. in Monroe Co., WV. Res. Boone Co., WV 1860, age 18. Enl. 6/29/61 at Charleston, Kanawha Co., WV. Deserted at Charleston, WV 10/18/62.

STEPHENSON, JOSEPH: Pvt. Swann's Co. Enl. 5/25/61 at Charleston, Kanawha Co., WV. Age: 25. Co. transf. to Co. K, 59th Va. Inf. 7/25/61.

STEPHENSON, T.: Co. I. Attended vet. reunion at Charleston, WV 1924 and gave res. as Charleston, WV.

STEWARD, WILLIAM: Pvt. Co. B. Enl. 7/12 or 16/62 at Union, Monroe Co., WV. Deserted on 11/16/62 roll.

STEWART, JOHN ANDERSON: Pvt. B. 5/20/46 Milboro Springs, Va. Enl. 1864 in Bath Co., Va. in 3rd (Christman's) Battn. Va. Reserves. Latter in 22nd Va. Inf. Guard at Libby Prison, Richmond, Va. 1865. Postwar farmer and carpenter, Mingo Co., WV. D. 7/19/1935.

STEWART, WILLIAM H.: Pvt. Co. G. Res. Monroe Co., WV. Enl. 5/1/62 at White Sulphur Springs, Greenbrier Co., WV. Left sick at White Sulphur Springs, WV 8/20/61. Deserted 10/28/62. Believed still alive in 1914.

STEWART, WILLIAM M.: Pvt. Co. F. Enl. 7/4/61 at Charleston, Kanawha Co., WV. AWOL 8/8/61.

STIFF, WILLIAM H.: Pvt. Co. A. Enl. 5/22/61 at Poca, Putnam Co., WV. Deserted near Charleston, WV 7/20/61.

STILL, LEWIS A.: Pvt. Co. F. Enl. 7/4/61 at Charleston, Kanawha Co., WV. Sick at home 8/8/61 on 8/31/61 roll. NFR.

STINE, CARRY L.: Pvt. Co. I. Res. Botetourt Co., Va. Enl. 5/1/62 at White Sulphur Springs, Greenbrier Co., WV. Capt. at Droop Mountain, WV 11/6/63. Age: 29. POW Camp Chase, OH 11/18/63. Transf. to Ft. Delaware 2/29/64. Relsd. 6/21/65. Farmer. 5'11". Dark complexion, black hair, gray eyes.

STINNETT, JAMES P.: Pvt./Sgt. Co. K. Enl. 6/6/61 at Charleston, Kanawha Co., WV. Accidentally wded. 6/20/61. Present on 11/1/62 roll. Left sick at Hightown on Mar.-Apr. 1863 roll. Present on 12/30/63 roll. Capt. at Winchester, Va. 9/19/64. POW Pt. Lookout, MD 9/23/64. Exchd. 3/15/65. In Chimborazo Hosp. with chronic diarrhea 3/19-27/65. Paroled at Charleston, WV 6/2/65. Age: 27. Fair complexion, dark hair, gray eyes.

STITES, JOHN J. (DR.): Co. K. Doctor, Fayette Co., WV 1860. Age: 33. Enl. 1861. Surgeon of Militia Brig. Bahlmann says left or resgnd.

STOGDEN, ROBERT A.: Pvt. Co. A. Enl. 6/13/61 at Camp Tompkins, Kanawha Co., WV. Unaccounted for as of 7/24/61.

STONE, JOHN A.: Pvt. Co. G. Enl. 3/25/64 at Bunches Hill, Greenbrier Co., WV. WIA (flesh wd. of finger) at Cold Harbor, Va. 6/2-3/64. In Chimborazo Hosp. for gunshot wd. of right hand 6/2-12/64. Had a furlo. and reported as gone north on 10/31/64 roll.

STONE, WILLIAM: Drummer/2nd Lt. Co. I. Enl. 6/29/61 at Charleston, Kanawha Co., WV. Elected 2nd Lt. 3/1/62. Not re-elected 5/1/62. Dropped 5/2/62.

STONE, WILLIAM P.: Pvt. Co. G. Res. Alleghany Co., Va. Enl. 5/1/62 at White Sulphur Springs, Greenbrier Co., WV. Conscript. Deserted 5/24/62. Listed as deserter on Aug. 1862 broadside. Left 5/24/62. Capt. at Gauley Bridge, WV by 91st O.V.I. 5/28/63. Age: 21. POW Camp Chase, OH 6/16/63. Transf. to Ft. Delaware 7/14/63. Farmer. 5'10". Dark complexion, brown hair, blue eyes.

STONEMAN, DANIEL O. H.: Pvt. Co. D. Enl. 5/30/61 at Summersville, Nicholas Co., WV. Sick at Montgomery White Sulphur Springs, Va. July 1862. Still present on 10/31/64 roll.

STRATTON, JAMES S.: Co. H. B. 6/12/40 Kanawha Co., WV. Clerk, Charleston, WV 1860. Age: 20. Transf. to CS Cav.

STREALEY, ABRAHAM H.: Pvt. Co. A. B. in Va. Enl. 4/1/62 at Lewisburg, Greenbrier Co., WV. Det. brig. blacksmith on 12/31/63 roll. Present on 10/31/64 roll. POW paroled at Lewisburg, WV 4/25/65. Age: 29. Blacksmith. 5'11". Dark complexion, dark hair, dark eyes.

STRIBLING, PETE E.: Pvt. Co. H. Enl. 9/3/62 in Monroe Co., WV. Absent wded. on Sept.-Oct. 1862 roll. Present on 12/31/62 roll. WIA (shot in chest, ball entering left side and exiting on right of spine) at Droop Mountain, WV 11/6/63. In Pearisburg, Va. hosp. for "vul. sclopet" and died there of disease caused by Droop Mountain wd. 6/17/64.

STRICKLER, ALLEN: Pvt. Co. A. Res. Putnam Co., WV. Date and place of enlistment not stated. Capt. in Boone Co., WV 2/9/64. Age: 18. POW Camp Chase, OH 3/7/64. Transf. to Ft. Delaware 3/14/64. D. at Ft. Delaware of chronic dysentery 9/3/64. Bur. Jersey shore of Ft. Delaware, Finn's Pt. Nat. Cem., NJ. Farmer. 5'8". Florid complexion, dark hair, dark eyes.

STRICKLIN, HENRY: Pvt. Swann's Co. Enl. 5/25/61 at Charleston, Kanawha Co., WV. Age: 24. Co. transf. to Co. K, 59th Va. Inf. 7/25/61.

STULL, A. J.: Pvt. Co. B. Enl. 7/12/62 at Camp McCausland, Monroe Co., WV. Det. to work in Salt Petre Cave July 1862.

STULL, CHARLES MEADE: Pvt. Co. B. Enl. 5/?/62 at White Sulphur Springs, Greenbrier Co., WV. Joined from desertion at Camp McCausland, WV 7/5/62. AWOL on 11/1/62 roll. Present on 12/31/62 roll. Substituted A. J. Brown in his stead on 4/30/63 roll. D. 1910.

STULL, GEORGE L.: Pvt. Co. A. Res. Alleghany Co., Va. Enl. 4/10/62 at Brushy Ridge, Greenbrier Co., WV. Capt. at Lewisburg, WV 5/23/62. Age: 23. Dischd. from Gen. Hosp. at Gallipolis, OH 7/2/62. POW Camp Chase, OH 7/8/62. Exchd. 8/25/62. Capt. at Beverly, WV 11/20 or 30/63. Conscript and deserter. POW Camp Chase, OH 12/7/63. Relsd. 5/8/65. Took oath at Camp Chase, OH 5/13/65. Farmer. 5'7½". Dark complexion, dark hair, hazel eyes.

STULL, JOHN J.: Pvt. Co. B. Enl. 7/25/62 at Union, Monroe Co., WV. AWOL on 11/16/62 roll. Present on 12/31/62 roll. Substituted Adam Bush on 2/28/63 roll.

STULL, LEWIS: Co. F. Listed as deserter on Aug. 1862 broadside. Greenbrier Co., WV.

STULL, JOSEPH P.: Corp. Co. D. Enl. 5/1/62 at White Sulphur Springs, Greenbrier Co., WV. Capt. at Winchester, Va. 9/19/64. POW Pt. Lookout, MD 9/24/64. Exchd. 3/15/65.

STULL, PERRY T.: Pvt. Co. D. Enl. 6/10/61 in Co. F, 36th Va. Inf. at Summerville, Nicholas Co., WV. Age: 18. Sick Apr. 1862. Present June 1862. Enl. 5/1/62 at White Sulphur Springs, Greenbrier Co., WV in Co. D, 22nd Va. Sick at hosp. at Union, Monroe Co., WV on 2/28/63 roll. Present on 4/30/63 roll. WIA (slight neck) at White Sulphur Springs, WV 8/26-27/63. Capt. at Winchester, Va. 9/19/64. POW Pt. Lookout, MD 9/24/64. At Pt. Lookout, MD hosp. for chronic diarrhea 2/4/65 to 3/17/65.

SUMMERS, A. E.: Pvt. Co. H. PWR.

SUMMERS, GEORGE W.: Pvt. Co. H. Enl. 1861. Transf. to Co. E, 8th Va. Cav. PWR.

SUMMERS, L.: Pvt. Co. H. PWR.

SUMMERS, WILLIAM B.: Pvt. Co. D. Enl. 5/30/61 at Summersville, Nicholas Co., WV. Left sick at White Sulphur Springs, WV 8/20/61. NFR.

SUMMERS, WILLIAM SYDNEY: Pvt. Co. H. B. 8/27/34 in Kanawha Co., WV. Mercer Academy and Univ. of Va. Salt business. Lawyer, Charleston, WV 1860, age 23. Enl. 5/8/61 at Charleston, Kanawha Co., WV. Transf. to Va. Cav. 6/1/62. Postwar lawyer and civil engineer, Charleston, WV. Putnam Co., WV 1891. D. 1894. Bur. Spring Hill Cem., Charleston, WV.

SURBAUGH, G. WASHINGTON C.: Apptd. in 22nd Va. Inf. 4/10/63. Also in 14th Va. Cav.

SUTFIN, DAVID: Pvt. Co. I. Enl. 6/29/61 at Charleston, Kanawha Co., WV. Under arrest on 4/30/64 roll. On clothing list 5/13/64.

SUTPHIN, WILLIAM: Co. I.

SUTTLE, WILLIAM D.: Pvt. Taylor's Co. Enl. 6/19/61 at Frankford, Greenbrier Co., WV. Age: 26. Co. transf. to Wise Brigade 7/10/61.

SUTVAN, E.: Pvt. Co. F. Date and place of enlistment not stated. Sick at Bonsacks Depot, Va. July 1862.

SWANN, JOHN STERLING: Capt. Swann's Co. B. 10/5/22. Lawyer, Charleston, WV 1860, age 35. Enl. 5/8/61 at Charleston, Kanawha Co., WV. Age: 37. Co. transf. to Co. K, 59th Va. Inf. 7/25/61. PWR says also served in Co. H, 22nd Va. Inf. Law partnership with brother, Thomas Belt Swann, at Charleston, WV 1893. Considered one of the best reference lawyers of the Charleston, WV Bar. D. 12/5/1903. Bur. Spring Hill Cem., Charleston, WV.

SWANN, THOMAS BELT: Capt. Co. E. B. 9/12/24. Educated Amelia Academy, Va. Schoolteacher, Orange C.H., Va. Studied law and became attorney. Moved to Charleston, WV with his brother, John Sterling Swann, to practice law 3/18/49. Enl. 6/8/61 at Charleston, Kanawha Co., WV. Elected Capt. 6/8/61. Not re-elected 5/1/62. Dropped 5/1/62. Volunteer aide Gen. Heth's staff 1862. Lt. Col. 3rd Va. State Line. Lt. Col. Swann's Cav. Battn. Married Mary Tompkins, aunt of Gen. U.S. Grant shortly after the war. Law partnership with brother, Charleston, WV 1893. Age: 69. D. 5/13/97. Bur. Spring Hill Cem., Charleston, WV.

SWORD, ELI: Pvt. Co. E. Enl. 6/8/61 at Charleston, Kanawha Co., WV. At Winchester, Va. hosp. for very serious leg and hip 7/25/64 to 8/1/64. Absent wded. on 10/31/64 roll. Took oath 12/26/64. Age: 27. Farmer. 6'. Fair complexion, light hair, blue eyes.

SWORD, J. T.: Pvt. Co. E. Enl. 6/8/61 at Charleston, Kanawha Co., WV. Left sick in Kanawha Co., WV 7/1/61. NFR.

SWORD, THOMAS J.: Pvt. Co. E. B. in Va. Enl. 10/6/62 at Charleston, Kanawha Co., WV. Sick at Giles C.H. (Pearisburg), Va. hosp. on 3/4/64 roll. Absent sick on 4/30/63 roll. Employed as nurse at Giles C.H. (Pearisburg), Va. hosp. 7/1/63 and still so det. on 12/31/63 roll. Present on 4/1/64 roll. In Chimborazo Hosp. for chronic diarrhea 5/25/64. Transf. to Danville, Va. hosp. 6/6/64. In Danville, Va. for chronic diarrhea 6/7-21/64. Paroled at Lewisburg, WV 4/25/65. Age: 33. Farmer. 6'. Dark complexion, dark hair, dark eyes.

SYDENSTRICKER, DAVID: Pvt. Taylor's Co. Enl. 6/19/61 at Frankford, Greenbrier Co., WV. Age: 26. Co. transf. to Wise Brigade 7/10/61.

SYDENSTRICKER, JOHN N.: Pvt. Taylor's Co. Enl. 6/19/61 at Frankford, Greenbrier Co., WV. Age: 24. Co. transf. to Wise Brigade 7/10/61.

TAYLOR, A. G.: Pvt. Co. C. Farmer, Fayette Co., WV 1860. Age: 32. Enl. 5/10/61.

TAYLOR, A. P. ("PENT"): Mentioned in an obituary in the *New Castle Record*. Postwar res. Craig Co., Va.

TAYLOR, C.: Co. G. Enl. 1861 for 4 years. Postwar roster.

TAYLOR, DAVID H.: Pvt. Co. D. Enl. 5/30/61 at Summersville, Nicholas Co., WV. On mounted guard at White Sulphur Springs, WV 8/13-14/61. D. at Lewisburg, WV hosp. 3/29/63.

TAYLOR, ELI: Pvt. Co. E. Enl. 11/1/63 at Camp Beauregard. Capt. at Winchester, Va. 9/19/64. POW Pt. Lookout, MD 9/23/64. Exchd. 3/17/65. Also on a list of Pt. Lookout, MD prisoners relsd. on taking oath.

TAYLOR, GEORGE ALLEN: Pvt. Co. C. B. ca. 1832 Craig Co., Va. Enl. 5/1/62 at White Sulphur Springs, Greenbrier Co., WV. Wagoner July 1862. Regt. wagoner on 3/4/63 roll. Teamster 4/1/63 to 9/30/63. Left at Lewisburg, WV as regt. wagoner on 6/1/63 roll. Teamster/regt. wagoner 12/1/63 to 1/31/64. Present on 4/1/64 roll. D. in Staunton, Va. Gen. Hosp. 8/15/64. Effects turned over to Q.M.

TAYLOR, HOR: Pvt. Co. D. Enl. 5/30/61 at Summersville, Nicholas Co., WV. Left at White Sulphur Springs, WV hosp. Capt. 8/20/61.

TAYLOR, JACOB N.: Capt. Taylor's Co. Enl. 6/19/61 at Frankford, Greenbrier Co., WV. Age: 23. Co. transf. to Wise Brigade 7/10/61.

TAYLOR, JAMES A.: Pvt. Co. C. Res. Fayette Co., WV. Enl. 5/1/62 at White Sulphur Springs, Greenbrier Co., WV. Wagoner July 1862. Reg. wagoner on 3/4/63 roll. Regt. wagoner on 6/11/63 roll. Regt. wagoner on 12/31/63 roll. Teamster 12/1/63 to 1/31/64. Present on 4/18/64 roll. Deserted 7/17/64. Capt. at Frederick, MD 7/14/64. Stated that he enl. as as teamster for one year and was afterwards conscripted until deserting 7/14/64. POW Elmira, NY 7/23/64. Desired to take oath at Elmira, NY 9/15/64. Desired to go to his home in Fayette Co., WV. Took oath at Elmira, NY and relsd. 5/17/65. 6'½". Dark complexion, dark hair, hazel eyes. PWR says enl. 5/10/61.

TAYLOR, JOHN J.: Pvt. Co. A. Res. Putnam Co., WV. Enl. 5/22/61 at Poca, Putnam Co., WV. Capt. at Lewisburg, WV 5/23/62. Ae: 23. POW Camp Chase, OH 7/14/62. Exchd. 8/25/62. Sick in hosp. on 12/31/62 roll. Sick on 2/28/63 roll. Sick at Montgomery White Sulphur Springs, Va. on 4/30/63 roll. Teamster at Montgomery White Sulphur Springs, Va. hosp. 6/3/63 to 12/21/63. Wagon master on medical board cert. at Montgomery White Sulphur Springs, Va. 4/30/64. Det. service in Medical Dept. 4/30/64 roll. Employed as forage agent in Q.M. Dept. at Montgomery White Sulphur Springs, Va. 1/26/65 to 2/28/65. Farmer. 5'10". Fair complexion, black hair, gray eyes.

TAYLOR, JOHN J.: Pvt. Co. D. Res. Nicholas Co., WV. Enl. 5/30/61 at Summersville, Nicholas Co., WV. Deserted 10/30/62. Capt. at Summersville, WV by 3rd Va. Regt. 12/26/62. Age: 22. POW Camp Chase, OH 1/4/63. Exchd. 3/28/63. Reported 4/14/63. Left sick at Beverly, WV 4/25/63. WIA (slight thigh) at White Sulphur Springs, WV 8/26-27/63. Absent wded. on 12/31/63 roll. Capt. at Summersville, WV 2/20/64. POW Camp Chase, OH 3/9/64. Transf. to Ft. Delaware 3/14/64. Took oath at Ft. Delaware and relsd. 6/22/65. Farmer. 6'2". Fair complexion, auburn hair, gray eyes.

TAYLOR, JOHN W.: Sgt. Co. F. Res. of Monroe Co., WV. Enl. 7/4/61 at Charleston, Kanawha Co., WV. Sick at White Sulphur Springs, WV 8/20/61. MWIA at Lewisburg, WV 5/23/62. D. 5/31/62.

TAYLOR, ROBERT T.: Pvt. Co. D. Enl. 5/30/61 at Summersville, Nicholas Co., WV. Left at Camp Bartow, WV to guard commissary stores, etc. 4/23/63. Present on 12/31/63 roll. Capt. at Winchester, Va. 9/19/64. POW Pt. Lookout, MD 9/24/64. Exchd. 3/15/65. At Camp Lee, Richmond, Va. 3/19/65.

TAYLOR, SOLOMON C.: Pvt. Co. D. Enl. 5/30/61 at Summersville, Nicholas Co., WV. Present on 8/31/61 roll. Died. Claim filed 2/28/63 by Mason Mathews, atty.

TAYLOR, WILLIAM: Pvt. Co. B. Res. Jackson Co., WV. Enl. 2/?/62 at Lewisburg, Greenbrier Co., WV. WIA (slight head) at White Sulphur Springs, WV 8/26-27/63. In Winchester, Va. Gen. Hosp. for very serious gunshot wd. on hand 7/25/64 to 8/4/64. At Charlottesville, Va. hosp. for very serious right thigh 8/2-14/64. At hosp. wded. on 11/30/64 roll. POW paroled at Charleston, WV 5/12/65. Age: 19. 6'3". Light complexion, light hair, blue eyes. Census says name is Traylor.

TAYLOR, WILLIAM J.: Pvt. Co. K. Res. Greenbrier Co., WV. Enl. 4/24/62 at White Sulphur Springs, Greenbrier Co., WV. WIA (shot through shoulder) and capt. at Lewisburg, WV 5/23/62. Age: 20. Dischd. from Gallipolis, OH Gen. Hosp. 7/2/62. POW Camp Chase, OH 7/8/62. Exchd. 8/25/62. At White Sulphur Springs, WV due to his Lewisburg, WV wd. on 11/1/62 roll. At home due to wd. on Nov.-Dec. 1862 and 2/28/63 rolls. Present on 12/30/63 roll. Absent sick on May 1864 roll. Capt. in Greenbrier Co., WV 5/24/64. POW Camp Chase, OH 6/11/64. D. at Camp Chase, OH of pneumonia 8/24/64. Bur. grave #218 at Camp Chase, OH. Farmer/bricklayer. 5'9¾". Fair complexion, brown hair, hazel eyes.

TAYLOR, WILLIAM P.: Pvt. Co. C. Enl. 5/10/61.

TAYLOR, WILLIAM T.: Pvt. Co. D. Enl. 5/30/61 at Summersville, Nicholas Co., WV. KIA at Lewisburg, WV 5/23/62.

TEAYS, STEPHEN T.: Pvt. Co. H. B. 10/24/39 St. Albans, WV. Laborer, Upper Falls of Coal River, 1860, age 20. Enl. 5/8/61 at Charleston, Kanawha Co., WV. Capt. at Winchester, Va. 9/19/64. POW Pt. Lookout, MD 9/24/64. Took oath and relsd. 2/25/65. Postwar 4 years in Missouri. Merchant, St. Albans, WV. City treasurer 1896. D. 1901. Bur. Teays Hills Cem., St. Albans, WV.

TEEL, GEORGE: Co. K. Enl. 1864. Served 3 months. Member Hupp-Deyerle CV Camp, Salem, Va. 1907.

TEMPLETON, JAMES A. (DR.): Asst. Surgeon. Field & Staff. Grad. Univ. of Va. 1855. Doctor. Four years Confederate service. Served 22nd Va. Inf., 36th Va. Inf., and present in that regt. July 1862 until transf. 45th Va. Inf. 11/25/63. Also served 4th and 23rd battns. Res. Neilsville, Wisconsin 1887-89. D. 12/17/87 Summersville (?), WV of feeble health. Bur. M. E. Church South graveyard.

TENTHORG, DANIEL: Pvt. Co. B. Res. Jackson Co., WV. Enl. 6/1/61 at Ripley, Jackson Co., WV. Present on 9/1/61 roll. NFR. PWR says name is Tenthory.

TERRY, JOSEPH P.: Pvt. Co. G. B. 1831. Enl. 7/3/61 at Charleston, Kanawha Co., WV. Det. to drive teams 4/30/64. Present on 10/31/64 roll. D. 1891. Bur. cem. at Rts. 614 & 18, Alleghany Co., Va.

TEUBERT, JOHN: Pvt. Co. C. Master carpenter, Fayette Co., WV 1860. Age: 44. Enl. 5/10/61.

THACKER, ALEX L.: Pvt. Co. A. Enl. 5/22/61 at Poca, Putnam Co., WV. Capt. at Winchester, Va. 9/19/64. POW Pt. Lookout, MD. Exchd. 3/15/65. Deserted. Surrendered Apr. 1865. Took oath at Charleston, WV and sent north 4/9 or 11/65. Age: 24. Farmer. 5'9". Dark complexion, black hair, dark eyes.

THACKER, JAMES: Co. A. Postwar roster

THAYER, WILLIAM TAYLOR: Pvt. Co. H. B. 7/22/31 Charleston, WV. With brother, Otis A. Thayer, discovered, operated, and owned coal mines in Kanawha Valley and New River area. Enl. 1861 at Charleston, Kanawha Co., WV. Transf. to Hale's Art. Postwar coal business. D. 5/5/1901 Charleston, WV.

THOMAS, BALLARD L.: Pvt. Co. C. B. ca. 1844 Craig Co., Va. Enl. 5/1/62 at White Sulphur Springs, Greenbrier Co., WV. Sick at home on rolls for Sept.-Oct. 1862 and 12/31/62. Present on 3/4/63 to 4/18/64 roll. On clothing list 5/13/64. Postwar res. Craig Co., Va. Twin brother of H. Benton Thomas. Bur. on Zane Jones farm, Rt. 42, near Sinking Creek store, Craig Co., Va.

THOMAS, CHARLES B.: Co. G. B. ca. 1845. On Alleghany Co., Va. pension rolls.

THOMAS, FATHER: Farmer and hunter. Ansted, WV. D. 1876. Bur. Terry Cem.

THOMAS, H. BENTON: Pvt. Co. C. B. ca. 1844 Giles Co., Va. Enl. 2/28/63 at Lewisburg, Greenbrier Co., WV. Dischd. for disability - mental incapacity - Camp Smith, Monroe Co., WV 8/9/63. Age: 18. Farmer. 5'3". Fair complexion, fair hair, blue eyes. Postwar res. Craig Co., Va. Twin brother of Ballard L. Thomas. Bur. Clover Hollow Cem., Marshall Humphrey farm, Craig Co., Va.

THOMAS, JAMES: Pvt. Co. A. Enl. 7/2/61 at Camp Tompkins, Kanawha Co., WV. Unaccounted for as of 7/24/61.

THOMAS, JAMES M.: Pvt. Swann's Co. Enl. 5/14/61 at Charleston, Kanawha Co., WV. Age: 26. Co. transf. to Co. K, 59th Va. Inf. 7/25/61.

THOMAS, JAMES W.: Pvt. Co. B. Enl. 5/?/62 at White Sulphur Springs, Greenbrier Co., WV. Conscript. POW paroled at Lewisburg, WV 4/26/65 and at Charleston, WV 5/10/65. Gave age as 26 at Charleston and 27 at Lewisburg. Tailor. 5'10" or 6'. Fair complexion, dark/light hair, dark/blue eyes, black whiskers.

THOMAS, J. H.: Pvt. Co. E. Enl. 6/8/61 at Charleston, Kanawha Co., WV. Left sick in Kanawha Co., WV 7/1/61. NFR.

THOMAS, JOHN M.: Pvt. Co. A. B. in Va. Enl. 6/5/61 at Camp Tompkins, Kanawha Co., WV. Unaccounted for 7/24/61. Deserted at Charleston, WV 10/9/62. Capt. at Antietam, MD 9/18/62. Conscript. Deserted and capt. in Franklin Co. 1/15/64.

THOMAS, JOHN S.: Pvt. Co. C. Enl. 6/5/61 at Charleston, Kanawha Co., WV. Left in hands of the enemy 9/9/61. Absent on furlo. on 12/31/63 roll. Present on 4/18/63 roll. MIA and capt. at Winchester, Va. 9/19/64. POW Pt. Lookout, MD. Exchd. 3/15/65. Transf. from Farmville, Va. hosp. 3/22/65. POW paroled at Charleston, WV 5/11/65. Age: 30. 5'7". Light complexion, dark hair, hazel eyes, dark whiskers.

THOMAS, JOSEPH B.: Pvt. Co. I. Res. Alleghany Co., Va. Enl. 5/1/62 at White Sulphur Springs, Greenbrier Co., WV. Capt. at Lewisburg, WV 5/23/62. Age: 33. POW Camp Chase, OH 6/30/62. On furlo. on 3/4/63 roll. Present on 4/30/63 roll. Serving sentence on G.C.M. at Dublin, Va. on 8/31/63 to 12/31/63, 4/1/64, and 10/31/64 roll. 5'7". Dark complexion, black hair, hazel eyes, black whiskers.

THOMAS, SIMON: Pvt. Co. C. Enl. 3/29/64 at Camp Gauley, Greenbrier Co., WV. Exchd. for Peter W. Deal of Co. C, 60th Va. Inf. 4/24/64.

THOMAS, W. H.: Pvt. Co. H. Enl. 11/1/62 at Red Sulphur Springs, Monroe Co., WV. Admitted to Chimborazo Hosp. 6/1/64. In Richmond, Va. hosp. 6/2/64. Absent sick on 12/25/64 roll. POW paroled at Charleston, WV 5/26/65. Age: 19. 5'10". Dark complexion, brown hair, blue eyes.

THOMAS, WILLIAM B.: Pvt. Co. C. Enl. 2/28/63 at Lewisburg, Greenbrier Co., WV. Present on 6/11/63 roll. NFR.

THOMPSON, ALLEN R.: Pvt. Swann's Co. Enl. 5/22/61 at Charleston, Kanawha Co., WV. Age: 30. Co. transf. to Co. K, 59th Va. Inf. 7/25/61.

THOMPSON, CAMERON LEWIS ("CAM"): Pvt. Co. H. B. 4/22/42 at Coalsmouth (St. Albans), WV. Son of Capt. B. S. Thompson of the Kanawha Militia. Grad. of Greenbrier Academy. Enl. 4/17/61 or 5/8/61 at Charleston, Kanawha Co., WV. Age: 18. On a mounted guard at White Sulphur Springs, WV 8/13-14/61. Present on 8/31/61 roll. Served as Capt. on staff of Gen. Terry 4/9/65. Was promo. to Capt. for gallentry at Baker's Creek, MISS. Captured at Vicksburg, MISS. Surrendered with Gen. Lee at Appomattox, Va. Salesman, hardware store, Cinn., OH 1865. Merchant, Mayfield, KY 1872. Moved to Hinton, WV due to ill health 2/1/72. Studied law. Started newspaper Dec. 1873. Publisher, *Mountain Herald* 6/1/74 - 10/?/85. Publisher, *Independent Herald.* Nominee for WV State Senate. Huntington, WV publisher, *Huntington Advertiser,* 1885, and publisher *Huntington Daily Advertiser* 1889. Postmaster, Huntington, WV 1888, and Charleston, WV 1893. WV State Auditors Office 1892. Insurance Dept. of State of WV. Huntington, WV real estate. Member of CV Camp 1911. Wrote an article on the Battle of Scary Creek, WV (7/17/61) in the June, 1918 *Confederate Veteran.* Attended at vet. reunion Beckley, WV 9/12/1918.

THOMPSON, CHARLES BLAINE: Pvt. Co. G. B. ca. 1845. Enl. 3/19/64 at Bunches Hill, Greenbrier Co., WV. Sick in Alleghany Co., Va. on 4/30/64 roll. Deserted 5/8/64. Recvd. clothing at "Castle Thunder" 5/13/64. Prisoner at "Castle Thunder" 6/28/64 and 9/2/64 lists. Postwar roster says enl. 1862 for 3 years and WIA at Winchester, Va. 9/19/64. D. 6/24/1915.

THOMPSON, C. R.: Co. G. Enl. 1862. Served 3 years. WIA at Winchester, Va. 9/19/64. On Alleghany Co. pension list.

THOMPSON, FRANCIS THORNTON: Pvt. Co. H. Enl. 5/8/61 at Charleston, Kanawha Co., WV. On a mounted guard at White Sulphur Springs, WV 8/13-14/61. Present on 8/31/61 roll. NFR.

THOMPSON, JOHILE: Pvt. Co. I. Enl. 6/29/61 at Charleston, Kanawha Co., WV. Present on 8/31/61 roll. NFR.

THOMPSON, JOHN KOONTZ: Capt./1st Lt. Co. A. B. 12/30/42 Luray, Va. Son of John J. Thompson. Res. Pt. Pleasant, WV ca. 1849. Educated in schools in Pt. Pleasant, WV. Res. Raymond City ca. 1851. V.M.I. 1859. Left V.M.I. Apr. 1861 to serve as drill master at Richmond, Va. Said he mustered in at Richmond, Va. in 1861 by Lt. Gen. T. J. Jackson. Transf. to WV. Arrived at Camp Tompkins, WV 7/3/61 and became officer in charge of instruction. Apptd. Adj. 8/10/61. WIA (face) at Carnifex Ferry, WV 9/10/61. Elected 1st Lt. 10/10/61. Promo. Capt. 10/10/61. Elected Capt. 5/1/62. WIA (right eye shot out) and capt. at Lewisburg, WV 5/23/62. Listed as POW 6/3/62. Paroled. Absent on furlo. on 12/31/62 roll. Present on 2/28/63 roll. WIA (shot in face) at White Sulphur Springs, WV 8/26-27/63. WIA (shot through neck) at Droop Mountain, WV 11/6/63. Absent on court-martial at Lewisburg, WV 12/31/63 roll. On sick furlo. on 4/1/64 roll. Present on 4/30/64 roll. WIA (severe side - shot through left lung) at Cold Harbor, Va. 6/2-3/64. At Stuart Hosp., Richmond, Va. for vulvus sclopet 6/4-7/64. Transf. to private quarters. Apparently fought at Winchester, Va. 9/19/64 and led remnants of regt. from field. On surg. cert. 9/28/64 to 11/28/64. Absent sick on 10/31/64 roll. AWOL at Fisherville, Va. 12/31/64. Returned to regt. Jan. 1865. Dischd. at Appomattox, Va. POW paroled at Lewisburg, WV 4/26/65. Age: 22. 5'10½". Light complexion, light hair, blue eyes. Res. Putnam Co., WV. Returned to farm at Poca, WV. Could not decide whether to practice medicine or law and became a farmer. WV Legislature 1880. U.S. Marshall 1897. U.S. Marshall southern judicial district 1901-05. Putnam Co., WV Tax Commission (2 terms). Refused to serve 3rd term. Served on WV school board. Retired to farm at Poca, WV. Stricken with rheumatism 1917. Unable to care for farm, sold it and moved to Pt. Pleasant, WV 1919 to be cared for by nephew, Dr. Hugh Barbee. Almost bedridden last 10 years of life due to war wounds. Bruise caused by spent ball in face caused most annoyance in last 6 years. PWR says WIA 6 times and "brave to the point of recklessness." In later years he would show his war wounds by sticking his finger in bullet holes. Never married. D. 1/3/1925 at Pt. Pleasant, WV of senility at age 82. Bur. Lone Oak Cem., Pt. Pleasant, WV.

THOMPSON, JOHN M.: Pvt. Swann's Co. Enl. 5/22/61 at Charleston, Kanawha Co., WV. Age: 24. Co. transf. to Co. K, 59th Va. Inf. 7/25/61.

THOMPSON, WILLIAM H.: Drummer. Co. A. Son of R. N. B. Thompson. Enl. 5/22/61 at Poca, Putnam Co., WV. Unaccounted for as of 7/24/61 at which time deemed absent with leave. PWR says fought at Scary Creek, with Gen. Early in 1864, and with Longstreet in the winter campaign at Norfolk, Va. Grad. Wytheville College, 1867. Studied law under James W. Hogue of Putnam Co., WV. Lawyer, Barboursville, WV Aug. 1870. Prosecuting Attorney, Cabell Co., WV 1876. Re-elected prosecutor 1880. Nominated WV State Democratic party 1884. Elected WV State Treasurer. Re-elected 1888.

THORNE, GEORGE A.: Pvt./Sgt. Co. D. Enl. 5/30/61 at Summersville, Nicholas Co., WV. WIA at Lynchburg, Va. At Charlottesville, Va. hosp. for very serious right side 9/1-26/64. Transf. to Lynchburg, Va. POW paroled at Charleston, WV 5/29/65. Age: 31. 6'2". Dark complexion, black hair, black eyes, black whiskers.

THORNE, J. F.: Co. C. Attended vet. reunion at Charleston, WV 1924 and gave res. Spring Hill, WV.

THRASHER, M. F.: Pvt. Co. H. B. ca. 1832 Botetourt Co., Va. Enl. 5/30/62 at Salt Sulphur Springs, Monroe Co., WV. Assigned to and joined this co. 5/30/62. At Camp Price, Greenbrier Co., WV 9/15/63. Dischd. for disability (hernia) 9/19/63. Age: 38. Farmer. 5'7". (Illegible) complexion, dark hair, black eyes. Postwar res. Craig Co., Va. First name is probably Monk.

THRASHER, ROBERT J.: Pvt. Co. I. B. in Botetourt Co., Va. Enl. 4/28/62. Place of enl. not stated. MWIA (gunshot wd. of thigh) and capt. at Lewisburg, WV 5/23/62. D. at Gallipolis, OH hosp. of vulvus sclopet 10/1/62. Age: 29 years and 5 months. Farmer. 5'10". Dark complexion, dark hair, black eyes.

THURMOND, EDMUND WYATT: Pvt. Co. K. B. 2/5/33. Res. Fayette Co., WV. En. 6/6/61 at Charleston, Kanawha Co., WV. Present on 9/4/61 roll. William Bahlmann said Edmund stood post for 17 hours in the rain, took pneumonia, and died at Princeton, WV 12/4/61. Bahlmann bur. him there. Bahlmann said Edmund was probably the best man in the co. Brother of Robert and Richard Thurmond.

THURMOND, RICHARD CLAIBOURN: Pvt. Co. K. B. 6/22/36. Res. Fayette Co., WV. Enl. 6/6/61 at Charleston, Kanawha Co., WV. Det. scout with Capt. Thurmond July 1862. Promo. 2nd Lt. in Capt. Thurmond's Co. 9/23/62. D. 10/7/69. Brother of Edmund and Robert.

THURMOND, ROBERT GIVEN: Pvt. Co. K. B. 5/28/39. Res. Fayette Co., WV. Enl. 6/6/61 at Charleston, Kanawha Co., WV. AWOL 9/1/61. Det. scout with Capt. Thurmond's guerilla co. 10/23/62. William Bahlmann said Robert had jaundice at Princeton, WV late 1861. D. 8/20/1907. Brother of Edmund and Richard Thurmond. Bahlmann says transf. to Capt. Thurmond's Co. as a 3rd Sgt. and mustered 9/19/62.

THURSTON, NAPOLEON B.: Pvt. Swann's Co. Laborer, Charleston, WV 1860. Enl. 5/22/61 at Charleston, Kanawha Co., WV. Age: 24. Co. transf. to Co. K, 59th Va. Inf. 7/25/61.

TINCHER, CHRISTOPHER L.: Pvt. Co. A. Enl. 4/10/62 at Brushy Ridge, Greenbrier Co., WV. Absent sick on 11/1/62 roll. Sick in hosp. on 12/31/62 roll. Absent sick on 2/28/63 roll. AWOL on 4/1/64 roll.

TINCHER, JOEL: Pvt. Co. I. Enl. 5/2/62 at White Sulphur Springs, Greenbrier Co., WV. Present until 4/30/63 roll. NFR.

TINCHER, JOSEPH: Co. I. Listed as deserter on Aug. 1862 broadside. Greenbrier Co., WV.

TINCHER, LEVI JR.: Pvt. Co. K. Enl. 6/6/61 at Charleston, Kanawha Co., WV. Reportedly when out of ammunition at the battle of Carnifex Ferry (9/10/61) Col. Tompkins told Levi to throw rocks at the enemy, which he is said to have done. Deserted 10/28/62. Listed as deserted 10/23/62 on Aug. 1862 broadside. Greenbrier Co., WV. Farm laborer, Fayette Co., WV 1860. Age: 16.

TINCHER, LEVI SR.: Pvt. Co. K. Enl. 4/24/62 at White Sulphur Springs, Greenbrier Co., WV. Sick at Montgomery White Sulphur Springs, Va. July 1862. Dischd. 2/28/62.

TINCHER, WILLIS: Pvt. Co. K. Enl. 5/1/62 at White Sulphur Springs, Greenbrier Co., WV. AWOL 6/18/62. Wagoner July 1862. Dropped as a deserter but produced satisfactory cert. of sickness on 2/28/63 roll. Present on Mar.-Apr. 1863 roll. On clothing list 5/13/64.

TINGLER, GEORGE W.: Pvt. Co. A. B. 4/15/44. Res. Alleghany Co., Va. Enl. 9/1/62 at Red Sulphur Springs, Monroe Co., WV. Deserted at Pack's Ferry, WV 10/17/62. Present on 12/31/62. Capt. at Beverly, WV 11/30/63. Gave himself up to U.S. troops. Age: 19. POW Camp Chase, OH 12/7/63. Transf. to Ft. Delaware 2/28/64 or 3/4/64. Farmer. 5'9" or 5'9¾". Sandy/florid complexion, dark hair, dark/brown eyes, dark whiskers. D. 3/17/1933. Bur. at Lone Star Baptist Church, Alleghany Co., Va.

TINGLER, HENRY: Pvt. Co. G. B. ca. 1830. Enl. 5/1/62 at White Sulphur Springs, Greenbrier Co., WV. Listed as deserter on Aug. 1862 broadside. Left 5/24/62. Deserted 5/30/62. Wagoner (location unknown) July 1862. Postwar res. Craig Co., Va. Bur. Tingler Cem., Paint Back, Va.

TINGLER, JOHN C.: Pvt. Co. C. Enl. 3/8/64 at Camp Gauley, Greenbrier Co., WV. Deserted 5/10/64.

TINGLER, WILLIAM: Co. K. Deserted 5/24/62 on Aug. 1862 broadside. Craig Co., Va.

TINSLEY, ISAAC: Pvt. Swann's Co. Boatman, Upper Falls, Coal River, 1860, age 23. Enl. 5/12/61 at Charleston, Kanawha Co., WV. Age: 23. Co. transf. to Co. K, 59th Va. Inf. 7/25/61.

TINSLEY, JOSEPH: Pvt. Swann's Co. Enl. 5/28/61 at Charleston, Kanawha Co., WV. Age: 21. Co. transf. to Co. K, 59th Va. Inf. 7/25/61.

TINSLEY, WHEELER: Pvt. Co. A. D. 9/7/1905. Bur. Spring Hill Cem., Charleston, WV.

TOLER, JAMES: Co. G. Postwar roster. Pvt. Enl. May 1861 at Charleston, WV. Age: 18. Res. Wyoming Co., WV. Surrendered Apr. 1865.

TOLER, JOHN: Pvt. Co. G. Enl. 7/10/63 at Packs Ferry, Monroe Co., WV. On scout on 4/1/64 roll. Present on 4/30/64 through 10/31/64.

TOLER, WYATT H.: Pvt./Sgt. Co. G. Enl. 7/3/61 at Charleston, Kanawha Co., WV. WIA (slight face) at White Sulphur Springs, WV 8/26-27/63. Capt. at Winchester, Va. 9/19/64. POW Pt. Lookout, MD 9/23/64. Exchd. 3/15/65. Res. Wyoming Co., WV. Age: 21 at enl.

TOMLINSON, ANDREW N. ("ANDY"): Pvt. Co. F. Res. Monroe Co., WV Enl. 7/4/61 at Charleston, Kanawha Co., WV. Present on 8/31/61 roll. Postwar res. Putnam Co., WV. Postwar roster says middle initial is J. Believed still alive in 1914.

TOMPKINS, CHRISTOPHER QUARLES: Col. Field & Staff. B. 1913. Res. Mathews Co., Va. West Point 1832-36. Grad. 27 in a class of 49. Brevet 2nd Lt., 3rd U.S. Art. 7/1/36. 2nd Lt., 3rd U.S. Art. 8/31/36. Seminole War 1836-40. 1st Lt., 3rd U.S. Art. 7/7/38. Recruiting duty 1840-41. Recruiting duty Florida War 1841-42. Garrison duty Ft. McHenry 1842-43. Garrison duty Ft. Macon 1843-44. Garrison duty Ft. McHenry 1844-45. Capt., 3rd U.S. Art. 6/18/46. On voyage to Calif. 1846-47. Mexican War 1847 (garrison duty, Monterey, Calif.). Resgnd. 9/22/47. Iron manufacturer, Richmond, Va. 1847-55. Capt., Va. State Militia 1848-51. Lt. Col., Va. State Militia 1848-51. Lt. Col., Va. State Militia 1851-53. Mining agent and Superintendent of Colleries, Kanawha and Fayette counties, WV 1855-61. Comm. Col. of Va. volunteer forces in Kanawha Valley, WV 5/3/61. CSR says apptd. Col. Apr. 1861 by Gov. Letcher. Resgnd. Nov. 1861, purportedly over his disagreements with Gen. John B. Floyd. Successor: George S. Patton. Took oath at Richmond, Va. 4/29/65. Coal mining, Richmond, Va. 1862-77. D. 5/28/77 at Richmond, Va. at age 64. Bur. Hollywood Cem., Richmond, Va.

TONEY, ADAM: Pvt./Sgt. Co. I. B. in Kanawha Co., WV. Res. Boone Co., WV 1860, age 41. Enl. 6/29/61 at Charleston, Kanawha Co., WV. Deserted at Lewisburg, WV 1/26/64.

TONEY, CAREY: Pvt. Co. I. B. in Kanawha Co., WV. Res. Boone Co. WV 1860, age 17. Enl. 6/29/61 at Charleston, Kanawha Co., WV. WIA at Droop Mountain, WV 11/6/63. Absent wded. on 12/31/63 roll. Sick at home in Boone Co., WV on 4/1/64 roll. Absent wded. on 10/31/64 roll.

TONEY, FREDERICK: Pvt. Co. I. B. in Kanawha Co., WV. Res. of Boone Co., WV. 1860, age 20. Enl. 6/29/61 at Charleston, Kanawha Co., WV. Absent 7/26/61. AWOL 8/21/61. Present on 11/12/62 roll. Dischd. on surg. cert. disability 4/10/63.

TONEY, JOHN P.: Capt./1st Lt. Co. I. B. in Kanawha Co., WV. Res. Boone Co., WV, age 28, boatsman. Enl. 6/20 or 29/61 at Charleston, Kanawha Co., WV. Elected 1st Lt. 6/20/61. Elected Capt. 3/1/62. Capt. at Droop Mountain, WV 11/6/63. Age: 33. POW Camp Chase, OH 11/18/63. Transf. to Ft. Delaware 3/2/64. Relsd. 6/12/65. Boatsman. 5'10". Florid complexion, dark hair, gray eyes. Res. Boone Co., WV.

TONEY, POINDEXTER, JR.: Pvt. Co. I. B. in Kanawha Co., WV. Res. Boone Co., WV 1860, age 21. Enl. 6/22/61 at Charleston, Kanawha Co., WV or 8/29/61 at Meadow Bluff, Greenbrier Co., WV. Present on 3/4/63 roll. D. on 4/30/63 roll.

TONEY, POINDEXTER SR.: Pvt. Co. I. B. in Kanawha Co., WV. Res. Boone Co., WV age 44, tending grist mill. Enl. 6/29/61 at Charleston, Kanawha Co., WV. Present on 8/31/61 roll. NFR.

TONKIN, L. E.: Pvt. Co. E. Res. Pennsylvania. Enl. 8/2/61 at Meadow Bluff, Greenbrier Co., WV. (Also claimed have enl. at Little Levels, WV). Det. butcher on 3/4/63 roll. Regt. butcher on 4/30/63 roll. Present on 12/31/63 roll. Paroled at Lewisburg, WV 4/27/65. Age: 29. Farmer. 6'. Fair complexion, brown hair, hazel eyes.

TRAYLOR, WILLIAM: Co. B. In Winchester, Va. hosp. for hand wd. 1864.

TRENOR, FRANK M.: Pvt. Co. C. B. in Craig Co., Va. Enl. 6/18/62 at Camp McCausland, Monroe Co., WV. Present through 10/31/64 roll. Postwar res. Roanoke, Va. Member Peachy Gilmer Breckinridge Camp. Entered Richmond, Va. Old Soldiers Home 9/7/1927 at age 90. D. there 4/6/1930. Bur. Hollywood Cem., Richmond, Va.

TRENOR, RUFUS C.: B. 11/22/46. Served in regt. last two years of war. Postwar res. Craig Co., Va. D. 10/29/1925 at age 79. Bur. Ross Cem., Rt. 42, near Maywood, on Fred Keffer farm, Craig Co., Va.

TRIPLETT, E. BAINS Pvt. Co. E. Enl. 6/8/61 at Charleston, Kanawha Co., WV. Capt. by Capt. King of Union Home Guard and paroled same day 9/20/62. MWIA (very serious fracture *illegible* middle three) and capt. at Winchester, Va. 9/19/64. D. 10/20/64. Another source says D. 11/1/64. Bur. Nov. 1864 at Winchester, Va. Cem., grave #374.

TRIPLETT, HEDGEMAN: Pvt. Co. E. Res. Clay Co., WV. Enl. 8/23/63 at Camp Price. Capt. at Winchester, Va. 9/19/64. POW Pt. Lookout, MD 9/23/64. Exchd. 3/15/65. At Farmville, Va. hosp. (undated) and transf. 3/22/65. Deserter. Surrendered Apr. 1865. Took oath at Charleston, WV and sent north 4/14/65. Age: 20. 5'10". Farmer. Florid complexoin, fair hair, blue eyes.

TRIPLETT, MARSHAL: 2nd Lt./Capt. Co. E. B. 3/23/11. Enl. 6/8/61 at Charleston, Kanawha Co., WV. Elected 2nd Lt. 6/8/61. Wrote a letter of resignation from camp near New Bern, Va. 12/31/61. Lt. Col. W. A. Jackson stated in a letter from Lewisburg, WV 11/18/62 he objected to Triplett's resignation for others in his co. being promo. over him. Resgnd. 1/24/62. Another card says resgnd, Mar. 1862. Dropped as 2nd Lt. 5/2/62. Capt. in Alleghany Co., Va. 5/17/62. Bushwhacker not in regular service. POW Camp Chase, OH 5/30/62. Transf. to Johnson's Island, OH. Exchd. 11/22/62. Lt. Col. Shields recommended Triplett as Enrollment Officer for counties of southwest Va. PWR says promo. to Maj. during war. D. of paralysis 10/9/95. Bur. Triplett Cem., Cowen, WV.

TRIPLETT, WILLIAM M.: Pvt./Corp. Co. D./Co. E. Res. Clay Co., WV. Enl. 6/8/61 at Charleston, Kanawha Co., WV. Transf. from Co. E 7/19/62 to Co. D 7/22/62. Capt. at Winchester, Va. 9/19/64. POW Pt. Lookout, MD 9/23/64. POW paroled at Charleston, WV 4/12/65. Age: 30. Farmer. 5'10½". Red complexion, dark hair, gray eyes.

TROUT, ARCHIBALD W.: Pvt. Co. K. Enl. 6/6/61 at Charleston, Kanawha Co., WV. Absent cause unknown on 9/4/61 roll. NFR. Bahlmann says left and heard of in 54th Va. Inf.

TROUT, HENRY S.: Pvt./2nd Lt. Co. C. B. ca. 1831 in Craig Co., Va. Enl. 8/28/62 at Red Sulphur Springs, Monroe Co., WV. Elected 2nd Lt. Aug. 1863. Apptd. 2nd Lt. 9/5/63. Promo. 6/5/64. In Winchester, Va. Gen. Hosp. 7/30/64 to 8/4/64. Transf. to Grove Hosp. (See also 28th Va. Inf.).

TROUT, JAMES A.: Pvt. Co. H. B. 3/19/65. Res. Craig Co., Va. Enl. 4/28/62 at White Sulphur Springs, Greenbrier Co., WV. Assigned to and joined this co. 4/28/62. Absent on sick leave on Sept.-Oct. 1862 and 12/31/62 rolls. Present on 2/28/63 roll. On clothing list 5/13/64. D. 10/2/1914. Bur. Summerfield Cem., Walton, WV. Middle initial is possibly H instead of A.

TUCKER, AARON C.: Pvt. Co. A. Res. Craig Co., Va. Enl. 10/1/62 at Charleston, Kanawha Co., WV. WIA (slight head) at White Sulphur Springs, WV 8/26-27/63. Capt. at Winchester, Va. 9/19/64. POW Pt. Lookout, MD 9/24/64. Paroled at Lewisburg, WV 4/27/65. Age: 21. Farmer. 5'10". Light complexion, light hair, hazel eyes.

TUCKER, CHRISTOPHER: Pvt. Co. E. Res. Angerona, WV. Deserted from Co. K, 11th WV Inf. at Ravenswood, WV 4/1/63. Enl. 5/18/63 at Camp Stonewall, age 18. At Charlottesville, Va. hosp. for very serious left thigh 9/23/64 to 10/7/64. (Probably WIA at Winchester, Va. 9/19/64). Absent wded. on 10/31/64 roll.

TUCKER, JAMES M.: Sgt./Pvt. Co. B. Enl. 2/1/62 at Lewisburg, Greenbrier Co., WV. Sick at Montgomery White Sulphur Springs, Va. July 1862. Present on 11/16/62 roll. WIA (severe thigh) at Cold Harbor, Va. 6/3/64. In Stuart Hosp., Richmond, Va. for vulvus sclopet 6/4-15/64. Transf. to Chimborazo Hosp. At Chimborazo Hosp. for very serious lower third of thigh 9/1-26/64. Transf. to Lynchburg, Va. POW surrendered with Maj. Gen. Sam Jones' forces at Tallahassee, FL 5/10/65. Paroled at Madison, FL 5/19/65.

TUCKER, JOHN: Pvt. Co. K. B. ca. 1832. Res. Craig Co., Va. Enl. 4/24/62 at White Sulphur Springs, Greenbrier Co., WV. Deserted 10/17/62.

TUCKER, STEPHEN J.: Pvt. Co. A. Enl. 5/22/61 at Poca, Putnam Co., WV. Typhoid fever at White Sulphur Springs, WV 8/20/61. NFR.

TUCKER, WILLIAM H.: Pvt. Co. A. Enl. 7/7/61 at Camp Tompkins, Kanawha Co., WV. Unaccounted for 7/24/61. Present on 11/1/62 roll. Left sick in Lewisburg, WV on 4/30/63 roll. Present on 4/30/64 roll. Capt. at Winchester, Va. 9/19/64. POW Pt. Lookout, MD 9/24/64. Exchd. 3/15/65. At Camp Lee, Richmond, Va. 3/20/65. POW paroled at Charleston, WV 5/6/65. Age: 27. 5'7". Dark complexion, dark hair, dark eyes.

TUCKWILER, CRAWFORD: Pvt. Co. E. Enl. 5/1/62 at White Sulphur Springs, Greenbrier Co., WV. Absent missing on Sept.-Oct. 1862 roll. Absent sick on 4/30/63 roll. Present on 12/31/63 roll. Det. brig. wagoner on rolls dated 4/1/64 and 5/1/64. Det. service on 10/31/64 roll. POW paroled at Charleston, WV 5/10/65. Age: 37. 5'5". Light complexion, brown hair, black eyes, brown whiskers.

TUCKWILLER, M.: Pvt. Co. B. Enl. 10/?/63 at Lewisburg, Greenbrier Co., WV. At Charlottesville, Va. hosp. for disability 6/12/64. D. at Charlottesville, Va. hosp. of typhoid fever 6/30/64. Effects turned over to Q.M.

TULLY, JAMES M.: Pvt. Co. H. B. 1840. Res. Charleston, WV 1860, age 17, at school. Enl. 10/1/62 at Charleston, Kanawha Co., WV. POW paroled at Charleston, WV 4/26/65. Age: 26. 5'6". Fair complexion, fair hair, blue eyes. D. 1894. Bur Spring Hill Cem., Charleston, WV.

TULLY, M.: Pvt. Co. H. Enl. 10/1/62 in Kanawha Co., WV. Capt. at Cold Harbor (Gaines Mill), Va. 6/3/64. POW Pt. Lookout, MD. Trans. to Elmira, NY 7/12/64. Exchd. 3/10/65. POW paroled at Charleston, WV 4/26/65. Age: 29. 5'6". Fair complexion, fair hair, blue eyes. Possibly James M. Tully.

TUNNER, ELIJAH: Pvt. Co. E. Enl. 7/2/61 on Sandy. AWOL 7/1/61.

TUNNER, JOHN: Pvt. Co. E. B. in Jackson Co., WV. Res. Clay Co., WV. Enl. 6/8/61 at Charleston, Kanawha Co., WV. AWOL 7/1/61. Capt. in Clay Co., WV 6/20/62. POW Camp Chase, OH. Exchd. 11/22/62. Capt. in Clay Co. 12/20/64. POW Camp Chase, OH 12/30/64. Applied to take oath at Camp Chase, OH 1/1-15/65. Desired to take oath. Transf. to Chicago, IL 5/20/65. Farmer. 6'1". Fair complexion, brown hair, gray eyes.

TURLEY, CLABOURN C.: Pvt. Co. A. Enl. 6/9/61 at Camp Tompkins, Kanawha Co., WV. Unaccounted for as of 7/19/61. Deserted at Charleston, WV 10/9/62.

TURLEY, THOMAS B.: Pvt. Co. A. Res. Winfield, WV. Enl. 11/1/62 at Raleigh (Beckley), Raleigh Co., WV. Left sick at Lewisburg, WV on 4/30/63 roll. Absent wded. on 12/31/62 roll. Present on 4/30/64 roll. At Charlottesville, Va. hosp. for rubeola 6/10-17/64. Capt. at Poolesville, MD 7/16/64. POW Elmira, NY 7/25/64. Desired to take oath at Elmira, NY 9/30/64 and to join friend in Kanawha Co., WV. Took oath at Elmira, NY and relsd. 5/29/65. 5'1". Florid complexion, auburn hair, blue eyes. Res. Putnam Co., WV 1910 at age 75. Bur. family cem. at Bills Creek, Rt. 35, Putnam Co., WV, mile and ½ on left bank of woods.

TURNER, BENJAMIN FRANKLIN: Pvt./Sgt. Co. H. B. 4/?/33 Shrewsbury, WV. Attended V.M.I. 1853. Enl. 5/8/61 at Charleston, Kanawha Co., WV. Served as a Pvt. until elected Sgt. 6/6/61. Absent cause unknown on 8/31/61 roll. Sick on satisfactory circumstances 1/29/62. Transf. to Co. I, 8th Va. Cav. 5/25/62. Age: 22. Det. to raise a co. for 8th Va. Cav. in Logan Co., WV Apr. 1863. Postwar res. Texas. Killed 1866.

TURNER, CHARLES: Pvt. Co. H. Res. Coalsmouth, (St. Albans), WV. Enl. 7/10/61 at Coalsmouth (St. Albans), Kanawha Co., WV. Present on 8/31/61 roll. D. at Pearisburg, Va. of typhoid fever 12/30/61. Bur. in grave with his brother, Theodore Garnett Turner, who died on the same day, at the same place, of the same disease. Son of Col. John P. Turner. Middle name is Phillip.

TURNER, THEODORE GARNETT: Pvt. Co. H. Son of Col. John P. Turner. Res. Coalsmouth (St. Albans), WV. Enl. 7/10/61 at Charleston, Kanawha, WV. Present on 8/31/61 roll. D. 12/30/61 at Giles C.H. (Pearisburg), Va. of typhoid fever. Bur. in grave with brother, Charles Phillip Turner, who died on the same day, at the same place, of the same disease.

TYREE, A. G.: Pvt. Co. C. Enl. 11/1/62 at White Sulphur Springs (Camp Fairview), Greenbrier Co., WV. On scout by order of Gen. Williams on 12/31/62 roll. Present on 3/4/63 roll. Exchd. for Smith Keslor of Capt. Thurmond's Co. 11/29/63. Middle name is probably Gus. Farm laborer, Fayette Co., WV 1860. Age: 15. First name is Augustus.

TYREE, ALEXANDER N.: Pvt. Co. G. Enl. 5/1/62 at White Sulphur Springs, Greenbrier Co., WV. On sick leave from surg. on 11/13/62 roll. Present on 12/31/62 roll. Paroled at Charleston, WV 6/12/65. Age: 26. 5'7". Dark complexion, dark hair, gray eyes. Res. Alleghany Co., Va.

TYREE, GEORGE: Pvt. Co. C. Farmer, Fayette Co., WV 1860. Age: 40. Enl. 5/10/61.

TYREE, I. H.: Co. G. On Alleghany Co. pension list.

TYREE, JAMES: Co. G. Postwar roster. Pvt. Enl. Mar. 1862 at Lewisburg, WV. Age: 21. Res. Alleghany Co., Va. Surrendered Apr. 1865.

TYREE, JOHN M.: Pvt./3rd Sgt. Co. C. Res. Fayette Co., WV. Enl. 1861 at Charleston, Kanawha Co., WV or 5/1/62 at White Sulphur Springs, Greenbrier Co., WV. Capt. in Fayette Co., WV 5/1/64. Age: 22. POW Camp Chase, OH 6/10/64. Exchd. 3/2/65. POW paroled at Lewisburg, WV 4/26/65. Farmer. 6". Light complexion, light hair, blue eyes.

TYREE, JOHN R.: Co. G. Postwar roster. Pvt. Enl. Mar. 1862 at Lewisburg, WV. Age: 24. Res. Alleghany Co., Va. Surrendered Apr. 1865.

TYREE, JOSEPH M.: Pvt. Co. C. Enl. 1861 at Charleston, Kanawha Co., WV or 5/1/62 at White Sulphur Springs, Greenbrier Co., WV. POW paroled at Lewisburg, WV 4/26/65. Age: 23. Farmer. 6'. Dark complexion, dark hair, dark eyes. Original member of Jeb Stuart UCV camp, Fayette Co., WV. Res. Ansted, WV 1909.

TYREE, SAMUEL F.: 1st Sgt. Co. C. B. 10/6/40 Fayette Co., WV. Enl. 5/1/62 at White Sulphur Springs, Greenbrier Co., WV. On scout by order of Gen. Williams on 12/31/62 roll. On furlo. on 3/4/63 roll. Present on Mar.-Apr. 1863 roll. Later organized an independent co. of rangers and was chosen Capt. His uncle, William Tyree, was Capt. of Co. C. Postwar res. Greenbrier Co., WV. Member Camp Creigh UCV. D. 1/14/1912 Frankford, WV. Bur. Frankford Cem., Frankford, WV. He was original member of Fayette Rangers. Farm laborer, Fayette Co., WV. 1860. PWR says enl. as a Pvt. 5/10/61.

TYREE, WILLIAM: Capt. Co. C. B. 11/22/07. Enl. 6/6/61 at Charleston, Kanawha Co., WV. Elected Capt. 6/6/61. Not re-elected 5/1/62. Dropped 5/2/62. Successor: Henry B. Dickinson. D. 7/19/83. Bur. Westlake Cem., Ansted, WV. Called "Colonel" postwar. Farmer, Fayette Co., WV 1860.

TYREE, WILLIAM W.: Pvt./Musician. Co. G. B. in Bath Co., Va. Res. Healing Springs, Va. Enl. 7/1/62 at Union, Monroe Co., WV. Sent home on sick furlo. 12/8/62. Present on 4/30/63 roll. Musician for regt. 12/31/63 and 3/1/64 rolls. Present on 10/31/64 roll. PWR and pension records says Co. G. Age: 69 in 1902. Blacksmith. PWR says enl. Mar. 1862 at Lewisburg, WV. Age: 22. Res. Alleghany Co., Va. Surrendered Apr. 1865.

TYREE, WOODSON A.: 2nd Lt./1st Lt. Co. C. B. 8/18/37 Fayette Co., WV. Studied medicine 1857. Enl. 4/5/61 at Charleston, Kanawha Co., WV or 5/1/62 at White Sulphur Springs, Greenbrier Co., WV. Elected 2nd Lt. 5/1/62. Apptd. 2nd Lt. 5/2/62. WIA (neck) at Droop Mountain, WV 11/6/63. Absent wded. on 12/31/63 roll. Promo. to 1st Lt. Mar. 1864. A later card says WIA at New Market, Va. but gives the date of the Droop Mountain battle. Says the wd. (which is probably the Droop Mountain wd.) caused Tyree to be permanently disabled and was recommended for retirement by his commanding officer. (Was probably not WIA at New Market, Va). POW paroled at Lewisburg, WV 4/26/65. Age: 27. Farmer. 6'2". Dark complexion, light hair, blue eyes. PWR says severe wd. of left lung at Droop Mountain, WV. Grad. Univ. of Va. 1867 (medicine). Married 1868. Res. Dewitt Co., IL 1867-78. Res. TX 1881. Res. Thurmond, WV 1881. Member Fayette Co. Medical Soc. and Medical Soc. of IL. Still alive 1891.

UNDERWOOD, SAMUEL: Pvt. Co. G. Enl. 7/3/61 at Charleston, Kanawha Co., WV. Present on 9/31/61 roll. NFR.

VAN BIBBER, ANDREW D.: Pvt. Co. D. Enl. 5/30/61 at Summersville, Nicholas Co., WV. Present through 4/30/63 roll. NFR.

VAN BIBBER, JAMES M. R.: Sgt./Pvt. Co. D. Res. Nicholas Co., WV. Enl. 5/30/61 at Summersville, Nicholas Co., WV. On mounted guard at White Sulphur Springs, WV 8/13-14/61. Relieved by Sgt. O'Brien of T. B. Swann's Co. Capt. in Nicholas Co., WV 11/7/62. Age: 31. POW Camp Chase, OH 12/2/62. Transf. to Cairo, IL for exchange 12/2/62. Present on 2/28/63 roll. Det. orderly for Col. Patton on 12/31/62 roll. Capt. Jan. 1864 head of Big Sycamore Creek, Clay Co., WV. Escaped and returned to regt. Det. service on rolls for 4/15/64 and 4/30/64. Farmer. 6'1". Dark complexion, light hair, blue eyes, sandy whiskers. Still alive 1901.

VAN BIBBER, JOHN C.: Pvt. Co. D. Enl. 10/8/62 at Charleston, Kanawha Co., WV. Another card says enlisted at Richmond, Va. Present throughout to 4/30/63 roll. NFR.

VAN BIBBER, NATHANIEL B.: Pvt. Co. D. Enl. 10/8/62 at Charleston, Kanawha Co., WV. Deserted 10/28/62. Dischd. on surg. cert. ca. 10/6/63.

VANDERGRIFT, WILLIAM H.: Pvt. Co. H. Laborer, 1860, age 24 (?). Enl. 6/1/63 at Lewisburg, Greenbrier Co., WV. Teamster at Dublin Depot, Va. 6/1-23/63. WIA at Winchester, Va. 9/19/64. Age: 23. Absent wded. on 12/25/64 roll. Deserted and surrendered Apr. 1865. Took oath and sent north 4/20 or 22/65. Farmer. 5'9". Fair complexion, brown hair, blue eyes. Gave age as 33 on oath.

VANDINE, ISIAH: Pvt. Co. B. Res. Coalsmouth (St. Albans) WV, 1860, age 25. Enl. 6/1/61 at Ripley, Jackson Co., WV. Capt. in Jackson Co., WV by 2nd Va. Cav. 10/11/62. Age: 37. POW Camp Chase, OH 10/23/62. Transf. to Cairo, IL for exchange 11/20/62. Saddler. 6'1". Fair complexion, black hair, blue eyes, black whiskers. Res. Jackson Co., WV.

VANDINE, ISHAM C.: Pvt. Co. A/Co. B. Enl. 5/22/61 at Poca, Putnam Co., WV. Dischd. at Camp McCausland, WV 7/16/62. Deserter. POW surrendered at Charleston, WV Mar. 1865. Took oath at Charleston, WV and sent north 3/7/65. Age: 22. Farmer. 5'10". Dark complexion, dark hair, gray eyes.

VANSTAVERN, N.: Pvt. Co. H./Co. K. Date and place of enlistment not stated. On clothing list 4 QR 1864. In Farmville, Va. hosp. for chronic diarrhea 3/17-25/65. Returned to duty.

VAUGHAN, R. L.: Pvt. Co. D. Date and place of enlistment not stated. POW paroled at Richmond, Va. 5/15/65. Res. Mecklinburg Co., Va.

VAUGHAN, WILLIAM: Pvt. Co. C. Day laborer, Fayette Co., WV. Age: 18. Enl. 5/10/61.

VENABLE, JAMES W.: Pvt. Swann's Co. Carpenter, Charleston, WV 1860, age 21. Enl. 5/22/61 at Charleston, Kanawha Co., WV. Age: 23. Co. transf. to Co. K, 59th Va. Inf. 7.25/61.

VERSER, L. J.: 3rd Sgt. Co. B. B. 1845. Res. Jackson Co., WV. Enl. 1861 for 4 years.

VEST, JOHN M.: Pvt. Co. E. Res. Botetourt Co., Va. Date and place of enlistment not stated. Dischd. early in the war for physical inability. Conscripted Nov. 1864 and sent from Richmond, Va. to Shenandoah Valley and assigned to 22nd Va. Father wrote from Fincastle, Va. 12/17/64 requesting his son assist on farm. Sent to Lynchburg, Va. hosp.

VIA, JOHN: Pvt. Taylor's Co. Enl. 6/19/61 at Frankford, Greenbrier Co., WV. Age: 26. Co. transf. to Wise Brigade 7/10/61.

VICKERS, FRANCIS M.: Pvt. Swann's Co. Enl. 6/9/61 at Charleston, Kanawha Co., WV. Age: 18. Co. transf. to Co. K, 59th Va. Inf. 7/25/61.

VICKERS, GRAHAM: Pvt. Co. H. D. 4/7/1926. Bur. Spring Hill Cem., Charleston, WV. Not in CSR.

VICKERS, JOHN H.: Pvt. Swann's Co. Enl. 6/9/61 at Charleston, Kanawha Co., WV. Age: 20. Co. transf. to Co. K, 59th Va. Inf. 7/25/61.

VICKERS, WILLIAM H.: Pvt. Swann's Co. Res. Craig Co., Va. Enl. 6/3/61 at Charleston, Kanawha Co., WV. Age: 20. Co. transf. to Co. K, 59th Va. Inf. 7/25/61.

VINCENT, ROBERT A.: Pvt. Co. D. Enl. 8/4/61 at White Sulphur Springs, Greenbrier Co., WV. Transf. from Capt. Riggs' Co. by order of Col. Tompkins 8/4/61. Gone to Richmond, Va. to guard prisoners on 8/31/61 roll. NFR.

VOWLES, JAMES NEWTON: Pvt. Co. G. B. 11/20/25. D. 2/22/89. Bur. Cedar Hill Cem., Covington, Va. Also spelled Vowels. Enl. Mar. 1862 at Lewisburg, WV. Age: 25. Res. Alleghany Co., Va. Surrendered Apr. 1865.

WADDELL, J. A.: Pvt. Co. B. Res. Jackson Co., WV.

WADDELL, J. L.: Pvt. Co. B. Res. Jackson Co., WV.

WADDLE, JAMES F.: Pvt. Co. D. Enl. 3/17/64 at Camp Gauley, Greenbrier Co., WV. Present through 10/31/64 roll. Res. Bland Co., Va. Feb. 1909 at age 63.

WAGGONER, JOSEPH: Pvt. Swann's Co. Enl. 5/8/61 at Charleston, Kanawha Co., WV. Age: 23. Co. transf. to Co. K, 59th Va. Inf. 7/25/61.

WAITE, (DR.) A. M.: Co. F. D. sometime prior to 1914. Postwar roster.

WAITE, SAMUEL C.: 2nd Lt. Co. F. Enl. 7/4/61 at Charleston, Kanawha Co., WV. Elected 2nd Lt. 7/4/61. On special duty 8/22/61. Resgnd. 9/3/61. Postwar roster says killed at Charlestown, WV 1864 and lists him as 4th Sgt.

WALKER, DOCK: Listed in UDC booklet *Craig's Share In The War Between The States.* Res. Craig Co., Va.

WALKER, ELIJAH: Pvt. Co. E. Res. Craig Co., Va. Enl. 6/8/61 at Charleston, Kanawha Co., WV. Present on 8/31/61 roll. NFR. PWR says died 1862.

WALKER, GEORGE: Pvt. Co. C. Enl. 5/1/62 at White Sulphur Springs, Greenbrier Co., WV. Sick at Giles C.H. (Pearisburg), Va. July 1862. In Charlottesville, Va. hosp. for chronic diarrhea 7/26/64. Transf. to Lynchburg, Va. 7/26/64. Present on 10/31/64 roll. Listed as deserter on Aug. 1862 broadside. Craig Co., Va.

WALKER, J. A.: Pvt. Co. E. Res. Craig Co., Va. Enl. 6/8/61 at Charleston, Kanawha Co., WV. Left sick in Greenbrier Co., WV 7/30/61. NFR. PWR says died 1862. First name is James.

WALKER, JAMES M.: Pvt. Co. D. Res. Nicholas Co., WV. Enl. 5/30/61 at Summersville, Nicholas Co., WV. Left sick at White Sulphur Springs, WV 8/20/61. Present on 10/31/62 roll. Capt. in Nicholas Co., WV by Ramsey's scouts 11/5/62. Age: 20. POW Camp Chase, OH 12/1/62. Exchd. 12/2/62. Present on 4/30/63. WIA (slight thigh) at White Sulphur Springs, WV 8/26-27/63. Capt. in Jan. 1864. Paroled POW remaining in Richmond, Va. after taking oath. Farmer. 5'6½". Dark complexion, brown hair, brown eyes, dark whiskers.

WALKER, JOHN: Pvt. Co. A. Date and place of enlistment not stated. Capt. at Winchester, Va. 9/19/64. POW Pt. Lookout, MD.

WALKER, JOHN A.: Pvt. Co. C. Enl. 5/1/62 at White Sulphur Springs, Greenbrier Co., WV. On scout by order of Gen. Williams on 12/31/62 roll. Present on 3/4/63 roll. Exchd. for Capt. T. A. Nutter 4/1/64. Farmer, Fayette Co., WV 1860. Age: 30. PWR says enl. as 3rd Sgt. 5/10/61.

WALKER, JOHN THOMAS: B. 10/28/40. D. at Meadow Bluff, WV. Jan. 1862.

WALKER, MUSCO L.: Swann's Co./Co. H. Enl. 5/15/61 at Charleston, Kanawha Co., WV. Age: 24. Co. transf. to Co. K, 59th Va. Inf. 7/25/61. Transf. to Co. H, 22nd Va. 3/19/62. Re-transf. to Capt. Swann's Co., Edgar's Battn., at Camp McCausland, WV 7/6/62.

WALLACE, DANIEL: Pvt. Co. B. Enl. 9/22/62 at Charleston, Kanawha Co., WV. Present until on wded. furlo. on 11/30/64 roll. POW paroled at Charleston, WV 5/25/65. Age: 18. 5'5". Dark complexion, light hair, blue eyes.

WALLACE, JOHN: Pvt. Co. A. Enl. 5/22/61 at Poca, Putnam Co., WV. WIA (severe mouth and jaw by rifle ball) at Scary Creek, WV 7/17/61. Orderly - Col. McCausland commanding July 1862. Orderly for Gen. Williams on 11/1/62 roll. WIA (severe leg) at White Sulphur Springs, WV 8/26-27/63. Absent wded. on rolls from 12/31/63 to 10/31/64.

WALLACE, R. W.: Pvt. Co. G. D. and bur. Pt. Lookout, MD. Postwar roster.

WALLS, RICHARD: Pvt. Co. H. B. in England. Carriage-maker, Charleston, WV 1860. Age: 39. Enl. 5/8/61 at Charleston, Kanawha Co., WV. Absent cause unknown on 8/31/61 roll. NFR. PWR says "didn't go."

WALLS, WILLIAM C.: Pvt. Co. G. Enl. 7/3/61 at Charleston, Kanawha Co., WV. Absent on furlo. for 30 days from 7/24/61 to 9/31/61 roll. NFR.

WALTON, J. H.: Pvt. Co. I. Date and place of enlistment not stated. Capt. at Winchester, Va. 9/19/64. POW Pt. Lookout, MD 9/24/64.

WALTON, WILLIAM: Pvt. Co. D. Enl. 5/30/61 at Summersville, Nicholas Co., WV. AWOL 11/5/62 on rolls for 12/30/62 aned 2/28/63. NFR.

WARWICK, WILLIAM W.: Pvt. Taylor's Co. Enl. 6/19/61 at Frankford, Greenbrier Co., WV. Age: 35. Co. transf. to Wise Brigade 7/10/61.

WASHUM, DANIEL: Pvt. Co. K. Enl. 8/1/63 at Narrows, Va. Entitled to bounty being underage when enlisted. Present through 5/1/64 roll. Bahlmann says enl. 1861.

WATKINS, HENRY A.: Pvt. Co. A. Res. Putnam Co., WV. Enl. 10/1/62 at Charleston, Kanawha Co., WV. Left sick at Marlins Bottom (Marlinton), WV on 4/30/63 roll. WIA (slight arm) at White Sulphur Springs, WV 8/26-27/63. Absent wded. on 12/31/63 roll. Present on 4/1/64 roll. Capt. at Winchester, Va. 9/19/64. POW Pt. Lookout, MD 9/24/64. Exchd. 3/15/65. Deserted and surrendered Apr. 1865. Took oath at Charleston, WV and sent north 4/27/65. Age: 25. Farmer. 5'5". Fair complexion, fair hair, blue eyes. Res. Putnam Co., WV 1910 at age 70.

WATKINS, JOSEPH FRIEND (DR.): Surgeon. Co. H. Grad. Univ. of Va. 1856. Grad. Jefferson Medical College, PA 1858. Physician, Charleston, WV 1860, age 25. On Kanawha Riflemen roll 4/26/61. Enl. 1861 at Charleston, Kanawha Co., WV. Apparently, transf. to 36th Va. Inf. Physician, Charleston, WV 1870's. D. Aug. 1883. Bur. Spring Hill Cem., Charleston, WV.

WATKINS, W. C.: Pvt. Co. B. Res. Jackson Co., WV. Mustered 1862 for 3 years.

WATSON, ROBERT F.: B. ca. 1844 Staunton, Va. Moved to Lewis Co., WV ca. 1846. Enl. in 22nd Va. Inf. in Lewis Co., WV 1862 and served 4 years. Postwar mercantile business, Kanawha Co., WV. D. 3/14/1911 at age 67.

WATSON, WILLIAM J.: Pvt. Co. B. Enl. not stated. Capt. at Farmville, Va. 4/6/65. POW Pt. Lookout, MD 4/14/65. Took oath at Pt. Lookout, MD 6/22/65. Relsd. 5'7½". Dark complexion, black hair, hazel eyes. Res. Lurrenburg Co., Va. Bur. Spring Hill Cem., Charleston, WV.

WATTS, LEWIS FRANKLIN: Capt. Co. F. B. 2/22/29 in Greenbrier Co., WV. Son of Salathiel Weeks Watts and Mary ("Polly") Burr. As a teen was bonded out as an apprentice tailor at Lewisburg, WV under Jacob Huffonagle. Later joined a wagon train and walked to Galveston, Texas. Worked as a cabin boy on sailing ships. Married Virginia Nickell 6/25/56. Res. Rocky Point, WV near the Nickell farm. Had five boys and five girls. Enl. 7/4/61 at Charleston, Kanawha Co., WV. Elected Capt. 7/4/61. Not re-elected. Dropped 5/2/62. Successor: J. H. Shumate. Later joined C.S. cav. and was capt. in fall of 1864. POW Pt. Lookout. MD. Reportedly nearly starved to death in prison and while sleeping on the ground his hair froze to it. Relsd. Spring 1865. Took over the ferry at Alderson, WV in 1865 and ran it until the late 1860's or early 1870's. While so engaged he built a store and saloon at the ferry, marking him as Alderson, WV's first merchant. Purchased "The Crow Place" at Crows, Va. 1872 and lived there and operated it until killed in an accident 8/15/91. Bur. Old Stone Church Cem., Lewisburg, WV.

WEAVER, JAMES: Pvt. Taylor's Co. Enl. 6/19/61 at Frankford, Greenbrier Co., WV. Age: 24. Co. transf. to Wise Brigade 7/10/61.

WEAVER, JONATHAN: Pvt. Co. K. Enl. 6/6/61 at Charleston, Kanawha Co., WV. WIA (slight) at Scary Creek, WV 7/17/61. Absent cause unknown on 9/14/61 roll. NFR. Farm laborer, Fayette Co., WV 1860. Age: 20. Bahlmann says "left and joined Union army."

WEBB, BALLARD: Listed in UDC booklet *Craig's Share In The War Between The States.* Res. Craig Co., Va.

WEBB, HENRY J.: Pvt./Drummer. Co. K. B. 1836 Craig Co., Va. Enl. 4/24/62 at White Sulphur Springs, Greenbrier Co., WV. Sick at Montgomery White Sulphur Springs, Va. July 1862. Sick at Red Sulphur Springs, WV on 11/1/62 roll. On sick furlo. on Nov.-Dec. 1862 roll. Present on 2/28/63 roll. Left sick at Beverly, WV on Mar.-Apr. 1863 roll. Present on 12/30/63 through 10/31/64 rolls. D. 1867. Bur. Bowen Cem., Upper Craig's Creek, Craig Co., Va.

WEBB, JACKSON: Listed in UDC booklet *Craig's Share in The War Between The States.* Res. Craig Co., Va. Possibly Henry J. Webb.

WEBB, JONAS: Pvt. Co. E. Enl. 10/6/62 at Charleston, Kanawha Co., WV. Deserted on the road to Belle Grove (Cedar Creek), Va. on 10/31/64 roll.

WEBB, THOMAS D.: Pvt. Co. A. B. in Pittsylvania Co., Va. Enl. 5/22/61 at Poca, Putnam Co., WV. Slightly, ruptured when enlisted. Hernia rupture caused by extricating himself from water and mud of Scary Creek, WV during the battle there 7/17/61. Nurse in private hosp. 9/15/61 to 10/22/61. At Charlottesville, Va. hosp. for typhoid fever 1/2/62 to 2/5/62. Disability discharge for scrotal hernia caused at the battle of Scary Creek, WV and given at Lewisburg, WV 2/26/62. Age: 27. Overseer. 5'8½". Dark complexion, black hair, black eyes. Bur. Spring Hill Cem., Charleston, WV.

WEBSTER, DANIEL HENRY: Pvt. Co. F. B. in Roanoke Co., Va. Enl. 5/1/62 at White Sulphur Springs, Greenbrier Co., WV. Two months wages taken by sentence of G. C. M. on 10/31/64 roll. PWR says fought for a long time and just got tired of it. Jumped off train and left. Moved to St. Albans, WV postwar. Bur. Webster Cem., off Green Valley Rd., St. Albans, WV.

WEBSTER, JAMES H.: Pvt. Co. B. Enl. 6/1/61 at Ripley, Jackson Co., WV. Nurse at White Sulphur Springs, WV. hosp. on 9/1/61 roll. NFR.

WEEKS, CORNELIUS: Pvt. Co. K. Enl. 7/28/62 at Union, Monroe Co., WV. Present on Nov.-Dec. 1862 roll. Under arrest and in confinement for desertion on 12/30/62 roll. Present on rolls for 2/28/63 and Mar.-Apr. 1863. Under sentence of G.C.M. on rolls from 4/1/64 to 5/1/64.

WEHRLE, MEINROD: Pvt. Co. H. From Baden. Brewer, Charleston, WV 1860, age 29. Enl. 5/8/61 at Charleston, Kanawha Co., WV. Absent cause unknown on 8/31/61 roll. PWR says hung by U.S. troops. Also states name was Meinhart and Mineret. PWR says "didn't go" and "hung by Yankees for killing a Yankee who attempted to outrage (rape) his wife."

WELCH, GEORGE L.: Pvt. Co. H. Enl. 1861. Son of John Welch. Clerk, Upper Falls Coal River, 1860, age 26. Res. Kanawha Salines (Malden), WV. Brother of Levi and James Welch. Not in CSR.

WELCH, J. A.: Pvt. Co. E. Enl. 6/8/61 at Charleston, Kanawha Co., WV. Left sick at White Sulphur Springs, WV 8/20/61. Det. Q.M. Dept. 3/28/62 to 5/2/62. NFR.

WELCH, JAMES CLARK: Pvt. Co. H/Kanawha Artillery. B. 1/10/36 Kanawha Co., WV. Salesman, Upper Falls Coal River, 1860, age 23. Son of John Welch. Res. Kanawha Salines (Malden), WV. Enl. 1861 at Charleston, Kanawha Co., WV. Age: 25. Comm. 2nd Lt. 5th Regt. Artillery, 22nd Brig., 5th Div. Va. Militia 6/1/61. KIA (almost decapitated by an enemy cannon ball) at Scary Creek, WV 7/17/61. Bur. Spring Hill Cem., Charleston, WV. Brother of Levi and George L. Welch.

WELCH, LEVI: Pvt./Sgt. Co. H. B. 7/8/62 Charleston, WV. Son of John Welch. Res. Upper Falls Coal River, WV. Pupil of Dr. Henry Ruffner and later of David L. Ruffner. Res. Kanawha Salines (Malden), WV. Enl. 5/8/61 at Charleston, Kanawha Co., WV. Served as a Pvt. until elected 3rd Sgt. 7/27/61. Det. to guard prisoners sent to Richmond, Va. 8/28/61. Developed serious cough on the 1861 retreat from Cotton Hill and Dr. McDonald urged him to resign or seek indoor assignment. Dischd. due to failing health. Apptd. V.M.I. 3/1/62. Visited 22nd Va. Inf. while on furlo. and participated in the battle of White Sulphur Springs, WV 8/26-27/63 as an aide-de-camp on Col. Patton's staff. As 2nd Lt., Co. B, Cadet Corps he was det. and put in charge of one 3-inch rifled gun (of) the Cadet Art. section at New Market, Va. 5/15/64. Grad. V.M.I. 1864. Postwar res. Louisville, KY, bookkeeper for mercantile business. Res. Calif., 1869, rancher. Res. Charleston, WV Sept. 1890. D. 2/11/1901 Charleston, WV. Bur. Spring Hill Cem., Charleston, WV.

WELCH, SAMUEL: Pvt./Corp. Co. F. Enl. 7/4/61 at Charleston, Kanawha Co., WV. Apptd. Corp. 8/29/61. Postwar res. Alderson, WV.

WERTEMBAKER, WILLIAM: Pvt. Swann's Co. Enl. 5/8/61 at Charleston, Kanawha Co., WV. Age: 21 or 31 (not legible). Co. transf. to Co. K, 59th Va. Inf. 7/25/61.

WERTS, JAMES M.: Pvt. Co. F. Enl. 5/1/62 at White Sulphur Springs, Greenbrier Co., WV. Deserted about 5/12/64.

WEST, JAMES A.: Pvt. Co. A. B. Putnam Co., WV. D. Putnam Co., WV. Bur. Phillips Cem., Bills Creek, Putnam Co., WV.

WETZELL, GEORGE W.: Pvt. Company unknown. Date and place of enlistment not stated. Capt. at Lewisburg, WV 5/23/62. POW Camp Chase, OH. Exchd. 8/25/62.

WETZELL, HENRY T.: Pvt. Co. G. Res. Lewisburg, WV. Enl. 2/25/64 at Camp Holmes. Left Greenbrier Co., WV. Aug. 1863 to avoid being conscripted and went to Clarksville, NC where he worked as a saddler for about 15 months. Left. Clarksville, NC 2/18/64 in order to reach Gen. Butler. Was arrested near Weldon, NC and sent to Raleigh, NC and then to Camp Lee, Richmond, Va. Enl. at Camp Lee 2/25/64 and assigned to 22nd Va. 3/7/64. Det. in hosp. at Lynchburg, Va. 1864. Deserted on Gen. Early's march north. Capt. at Martinsburg, WV 7/14/64. Age: 23. Surrendered voluntarily. POW Camp Chase, OH 7/21/64. Relsd. 5/8/65. Saddler. 5'4½". Fair complexion, light hair, blue eyes.

WETZEL, JAMES: Co. G. Postwar roster. Also spelled Witzell.

WHEAT, CASSIUS: Pvt. Co. G. Enl. 4/12/64 at Bunches Hill, Greenbrier Co., WV. Det. in Signal Corps on 4/30/64 roll. Det. Q.M. Dept. on 10/31/64 roll. POW surrendered by Maj. Gen. Sam Jones at Tallahassee, FL 5/10/65. Paroled at Madison, FL 5/20/65. Res. Alexandria, Va.

WHEELER, JOSEPH: Pvt. Co. B. Enl. 6/1/61 at Ripley, Jackson Co., WV. In White Sulphur Springs, WV hosp. on 9/1/61 roll. Sick at Montgomery White Sulphur Springs, Va. July 1862. Present on 11/10/62 roll. Capt. in Clay Co., WV 2/21/64. Age: 21. POW Camp Chase, OH 3/7/64. Transf. to Ft. Delaware 3/14/64. Took oath at Ft. Delaware. Relsd. 6/21/65. Farmer. 6'. Dark complexion, brown hair, black eyes. Res. Mason Co., WV. Bur. Creston Cem., Gunville, Mason Co., WV.

WHEELER, THOMAS J.: Pvt. Co. B. Enl. 6/1/61 at Ripley, Jackson Co., WV. In White Sulphur Springs, WV hosp. on 9/1/61 roll. WIA (side) at New Market, Va. 5/15/64. Present on rolls from 11/10/62 through 11/30/64. Pension in Independence Co., Ark. D. 1/16/1918.

WHITE, BENJAMIN: Sgt. Swann's Co. B. 1829. Son of Arthur White and Rebecca Miller.

WHITE, ELIOTT: Pvt. Co. E. B. 1823. Son of Arthur White and Rebecca Miller. Res. Roane Co., WV. Enl. 7/2/61 on Sandy. Said he enl. due to threats and accusations and served 6 months and was then forced to re-enl. in regular service. On 2/11/64 obtained a furlo. to go home but with actual intent to turn himself over to U.S. authorities. Deserted Mar. 1864 on C.S. rolls. Capt. in Braxton Co., WV 2/18/64. POW Camp Chase, OH 3/4/64. Relsd. 2/17/65. Took oath at Camp Chase, OH twice (6/10/64 and 2/17/65). Farmer. 5'9" or 5'11". Florid complexion, dark hair, blue eyes. Bur. White Cem., Tague, Braxton Co., WV.

WHITE, JAMES: Pvt. Co. E. B. 2/6/33 Tazewell Co., Va. Son of Arthur White and Rebecca Miller. Enl. either 6/8/61 at Charleston, Kanawha Co., WV or 9/8/62 at Raleigh (Beckley), Raleigh Co., WV. POW paroled at Lewisburg, WV 4/26/65. D. 2/16/1915 probably in Roane Co., WV. Bur. White Cem., near Left Hand, Roane Co., WV.

WHITE, JAMES W.: Pvt. Co. E. B. 1835. Res. Roane Co., WV. Enl. either 9/8/62 at Raleigh (Beckley), Raleigh Co., WV or 1862 at Fayetteville, Fayette Co., WV. Capt. on Sandy Creek, WV by 11th Va. Inf. 1/3/63. Age: 25. POW Camp Chase, OH 1/2/63. Exchd. 3/28/63. POW paroled at Lewisburg, WV 4/26/65. Farmer. 5'8" or 5'10". Florid complexion, red or light hair, blue eyes.

WHITE, JASPER: Pvt. Co. E. Enl. 7/1/61 at Sandy Bridge, WV. Left sick at White Sulphur Springs, WV 8/20/61. NFR.

WHITE, JOHN A.: Pvt. Co. G. Res. Wyoming Co., WV. Enl. 10/1/62 at Union, Monroe Co., WV. Deserted and surrendered Apr. 1865. Took oath at Charleston, WV and sent north 4/27/65. Age: 19. 5'3". Dark complexion, dark hair, blue eyes.

WHITE, JOHN DARIEN: Pvt. Swann's Co. B. either 7/27/22 Jeffersonville, IND. or 1826. Res. Charleston, WV 1828(?). Comm. 1st Lt., Co. A, Charleston Battn. for 6 years. Enl. 5/8/61 at Charleston, Kanawha Co., WV. Age: 35. Comm. Sgt. on 6/30/61 roll. Co. transf. to Co. K, 59th Va. Inf. 7/25/61. Capt. at Burkesville Station, Va. 4/14/65. Mayor, Charleston, WV 1874-75. Charleston, WV City Council 7 years. Restauranteer, Charleston, WV.

WHITE, PATTON: Pvt. Co. G. Res. Mercer Co., WV. Enl. on or about 10/1/62. Place not stated. Capt. in Wyoming Co., WV 12/28/62. Age 39. POW Camp Chase, OH 1/22/62. Claims he was forced to take oath on a statement dated 1/3/63. Exchd. 3/28/63. Farmer. 5'6". Dark complexion, black hair, brown eyes.

WHITE, RALPH: Pvt. Co. E. B. 5/?/43. Son of Arthur White and Rebecca Miller. Enl. 6/8/61 at Charleston, Kanawha Co., WV. Present until paroled at Lewisburg, WV 4/26/65. Age: 22. 6'2". Fair complexion, light hair, gray eyes. Some PWR say name is Rafe.

WHITE, ROBERT: Pvt. Co. D. Date and place of enlistment not stated. Deserted and surrendered Apr. 1865. Took oath at Charleston, WV and sent north 4/10/65. Age: 24. Farmer. 5'11". Fair complexion, fair hair, blue eyes. Res. Fayette Co., WV.

WHITE, TIMOTHY: Pvt. Co. E. B. 3/?/36. Son of Arthur White and Rebecca Miller. Res. Roane Co., WV. Enl. 7/2/61 on Sandy, WV. Capt. in Roane Co., WV 2/15/64. Age: 27. POW Camp Chase, OH 3/7/64. Transf. to Ft. Delaware 3/14/64. Took oath at Ft. Delaware. Relsd. 6/21/65. Farmer. 5'9". Dark complexion, sandy hair, blue eyes. Married 1900 to Elizabeth.

WHITING, JAMES W.: Pvt. Co. H. Res. Craig Co., Va. Enl. 9/5/62 in Monroe Co., WV. Det. teamster on 12/31/62 roll. Det. regt. teamster on 2/28/63 roll. Det. brig. wagoner on 4/30/63 roll. Det. brig. teamster on 6/30/63 roll. Brig. teamster at Lewisburg, WV on 12/3/63 roll. Det. brig. teamster on 1/1/64 roll. Det. brig. wagoner on rolls from 4/1/64 through 12/30/64. Postwar res. Craig Co., Va. Res. Craig Co., Va. 1898 at age 65.

WHITMAN, ANDREW M.: Pvt. Co. D. Res. Nicholas Co., WV. Enl. 5/10/61 at Summersville, Nicholas Co., WV. Left sick at White Sulphur Springs, WV 8/20/61. Capt. in Nicholas Co., WV 2/28/62. POW Camp Chase, OH 5/7/62. Exchd. 11/22/62. Sick in Montgomery White Sulphur Springs, Va. July 1862. Name appears on a blank discharge dated Dec. 1862.

WHITT, THOMAS: Pvt. Co. G. Enl. either 2/28/64 at Bunches Hill, Greenbrier Co., WV or 3/5/64 at Staunton, Va. Transf. from Co. E, 17th Va. Cav. 2/28/64. Deserted 4/20/64.

WHITTEKER, WILLIAM A.: Pvt. Co. H. Merchant, Charleston, WV 1860. Age: 36. Enl. 5/8/61 at Charleston, WV (Kanawha Co.). Absent cause unknown on 8/31/61 roll. D. 10/23/64.

WHITTEN, JOSEPH G.: Pvt. Co. H. Enl. 5/8/61 at Charleston, Kanawha Co., WV. Det. blacksmith on 5/8/61 roll. Present on July-Aug. 1861 roll. Regt. blacksmith July 1862. Det. brig. blacksmith on 2/28/63 and 4/30/63 rolls. Det. brig. blacksmith at Salt Sulphur Springs, WV on 6/30/63 roll. Det. brig. blacksmith at Lewisburg, WV on 12/13/63 roll. Det. brig. blacksmith on rolls for 1/1/64, 4/1/64, and 4/30/64. POW paroled at Charleston, WV 5/8/65. Age: 35. 5'8½". Dark complexion, dark hair, hazel eyes, brown whiskers.

WHITTEN, LEWIS: Pvt./Sgt. Co. K. Enl. 5/28/62 at Salt Sulphur Springs, Monroe Co., WV. Sick at Red Sulphur Springs, WV on 11/1/62 roll. Det. nurse at Union, WV hosp. on Nov.-Dec. 1862 roll. Det. nurse on rolls 2/28/63 through 5/1/64. Capt. at Winchester, Va. 9/19/64. POW Pt. Lookout, MD 9/24/64. Exchd. 3/15/65.

WHITTINGTON, LAWRENCE: Pvt. Co. A. Res. 9 miles below Charleston, WV. Enl. 5/30/61 at Charleston, Kanawha Co., WV. Claimed he voted for secession because he did not know better. Claimed his name was entered on rolls without his knowledge and that he had no gun and no connection with this co. when it mustered. Said co. went into camp a month after his name was entered. He was sent for and drilled. Fought at Scary Creek, WV and was on Wise's retreat from the Kanawha Valley. Three or four weeks afterwards he deserted. Claimed to be a loyal Union man. Unaccounted for as of 7/24/61. Dischd. 11/6/62. Deserter. Was entirely willing to take oath. Age: 38. 5'6". Light hair and blue eyes.

WHITTINGTON, WILLIAM: Pvt. Co. A. Enl. 5/22/61 at Poca, Putnam Co., WV. On mounted guard at White Sulphur Springs, WV 8/13-14/61. Left sick with measles at White Sulphur Springs, WV 8/20/61. NFR.

WHITTEN, JOSEPH: Corp. Swann's Co. Enl. 6/3/61 at Charleston, Kanawha Co., WV. Age: 30. Co. transf. to Co. K, 59th Va. Inf. 7/25/61.

WICKLINE, ("MR."): Possible member of regt. Bahlmann says MWIA and capt. at Lewisburg, WV 5/23/62. Shot through foot. Res. Monroe Co., WV.

WICKLINE, JACOB: Pvt. Co. K. Enl. 12/1/63 at Lewisburg, Greenbrier Co., WV. Sick at home 2/9/64 on 10/31/64 roll. NFR.

WICKLINE, WILLIAM A.: Pvt. Co. K. Enl. 2/10/63 at Lewisburg, Greenbrier Co., WV. D. of typhoid pneumonia at Union, WV either 1/9/64 or 2/9/64. (Former date is probably correct). Bahlmann says enl. 1862.

WILEY, E. P.: Pvt. Co. C. Enl. 5/1/62 at White Sulphur Springs, Greenbrier Co., WV. WIA (slight head) at White Sulphur Springs, WV 8/26-27/63. Exchd. for Pvt. James R. McNutt of Capt. Wm. S. Rowan's Co., 60th Va. Inf. 3/25/64.

WILKINSON, AMBROSE B.: Pvt. Co. A. Enl. 5/3/61 at Charleston, Kanawha Co., WV. Present on 8/31/61 roll. Hosp. cook at Salt Sulphur Springs, WV hosp. on July 1862 roll. NFR.

WILKINSON, DAVID A.: Pvt. Co. A. Res. Kanawha Co., WV. Enl. 5/22/61 at Poca, Putnam Co., WV. Measles at White Sulphur Springs, WV 8/20/61. Present on 11/1/62 roll. Left sick at Marlin's Bottom (Marlinton), WV on 4/30/63 roll. Deserted in the face of the enemy 12/13/63. Took oath and sent north 12/31/63. Age: 30. Farmer. 5'8". Dark complexion, dark hair, hazel eyes. Res. Putnam Co., WV 1910 at age 75.

WILKINSON, H. H.: Pvt. Co. A. Date and place of enlistment not stated. On mounted guard at White Sulphur Springs, WV 8/13-14/61. NFR.

WILKINSON, H. T.: Lt. Co. C. Apparently WIA and capt. 11/27-28/63. Age: 25. Amputation in upper third of femur. Right circular transf. 12/3/63.

WILKINSON, J. A.: Pvt. Company unknown. Res. Kanawha Co., WV. Date and place of enlistment not stated. Deserted Dec. 1863. Took oath and sent North 12/31/63. Age: 23. Farmer. 5'1". Light complexion, light hair, blue eyes.

WILKINSON, JOSEPH D.: Pvt. Co. A. B. in Bedford Co., Va. Res. Putnam Co., WV. Enl. 5/30/61 at Charleston, Kanawha Co., WV. Dischd. 2/3/63 for being over the age required for military service. Age: 37. Farmer. 5'9". Dark complexion, dark hair, dark eyes.

WILKINSON, RANSOM H.: Pvt. Co. A. Enl. 5/22/61 at Poca, Putnam Co., WV. Measles at White Sulphur Springs, WV 8/20/61. Sick at Montgomery White Sulphur Springs, Va. July 1862. Present on 11/1/62 roll. Dischd. for disability 2/6/63.

WILKINSON, WILLIAM D.: Pvt. Co. A. Enl. 5/22/61 at Poca, Putnam Co., WV. On mounted guard at White Sulphur Springs, WV 8/13-14/61. Capt. at Winchester, Va. 9/19/64. POW Pt. Lookout, MD 9/23/64. POW paroled at Lewisburg, WV 4/27/65. Age: 23. Farmer. 6'. Light complexion, light hair, dark eyes.

WILKINSON, WILLIAM T.: Pvt. Co. A. Enl. 5/30/61 at Poca, Putnam Co., WV. On mounted guard at White Sulphur Springs, WV 8/3-14/61. Measles at White Sulphur Springs, WV 8/20/61. Sick at Montgomery White Sulphur Springs, Va. July 1862. Present on 11/1/62 roll. Dischd. for disability 2/6/63.

WILKINSON, ZIBA A.: Pvt. Co. A. Enl. 5/22/61 at Poca, Putnam Co., WV. Measles at White Sulphur Springs, WV 8/20/61. Present on 11/1/62 roll. Deserted in the face of the enemy 12/13/63.

WILLARD, SAMUEL H.: Pvt. Swann's Co./Co. H. Farm laborer, Charleston, WV 1860. Enl. 5/25/61 at Charleston, Kanawha Co., WV. Age: 21. Co. transf. to Co. K, 59th Va. Inf. 7/25/61. Transf. to Co. H, 22nd Va. 3/19/62. Re-transf. to Capt. Swann's Co., Edgar's Battn., Camp McCausland, WV 7/6/62.

WILLIAMS, FREDERICK: Pvt. Co. B. Res. Greenbrier Co., WV. Enl. 5/?/62 at White Sulphur Springs, Greenbrier Co., WV. Capt. at Lewisburg, WV 5/23/62. Age: 25. POW Camp Chase, OH. Took oath at Camp Chase, OH either 5/30/62 or 6/30/62. Relsd. 8/14/62. Exchd. 8/25/62. 5'8". Fair complexion, light hair, blue eyes, sandy whiskers.

WILLIAMS, J. D.: Pvt. Co. I. Date and place of enlistment not stated. Sick at Giles C.H. (Pearisburg), Va. July 1862. NFR.

WILLIAMS, JOHN D.: Pvt. Taylor's Co. Enl. 6/19/61 at Frankford, Greenbrier Co., WV. Age: 35. Co. transf. to Wise Brigade 7/10/61.

WILLIAMS, JOHN J.: Pvt. Co. D. Enl. 5/30/61 at Summersville, Nicholas Co., WV. Left sick at White Sulphur Springs, WV 8/20/61. NFR.

WILLIAMS, JOHN L.: Pvt. Co. B. in Giles Co., Va. Res. Craig Co., Va. Enl. 5/8/61 at White Sulphur Springs, Greenbrier Co., WV. Sick at hosp. on 12/31/62 roll until dischd. on surg. cert. at Camp Sam Jones for hypertrophy of the heart and epilepsy 7/2/63. Age: 25. Farmer. 5'4". Dark complexion, black hair, black eyes.

WILLIAMS, MARSHAL E.: Pvt. Co. A. Enl. 5/31/61 at Charleston, Kanawha Co., WV. On mounted guard at White Sulphur Springs, WV 8/13-14/61. Measles at White Sulphur Springs, WV 8/20/61. Paroled at Charleston, WV 6/24/63. Age: 19. 5'8". Fair complexion, light hair, hazel eyes.

WILLIAMS, R. L. D.: Co. D. On Alleghany Co. pension list.

WILLSON, JAMES M.: Pvt. Co. F. Enl. 6/5/63 at Camp Jackson. Present until under arrest on 4/1/64. NFR.

WILSON, DANIEL A.: Corp./Pvt. Co. H. Enl. 5/8/61 at Charleston, Kanawha Co., WV. Served as Pvt. until elected Corp. 6/6/61. Sick 6/12/61. Deserted on Sept.-Oct. 1862 roll. Entered Richmond, Va. Old Soldiers Home 11/18/1908 at age 71. D. 7/25/1922. Bur. Hollywood Cem., Richmond, Va. Registered as serving in Co. A.

WILSON, HARRISON: Pvt. Co. H. Enl. 1861.

WILSON, HENRY: Pvt. Co. H. B. 1/27/36. Enl. 5/8/61 at Charleston, Kanawha Co., WV. WIA at Kernstown, Va. 7/24/64. POW paroled at Charleston, WV 5/1/65. Age: 25. 5'6". Fair complexion, fair hair, blue eyes. D. 5/29/1906 of dropsy at Paw Paw Creek, WV. Bur. Brookover Cem., McCurdyville, WV.

WILSON, JAMES: Pvt. Co. K. Res. Monroe Co., WV. Date and place of enlistment not stated. Deserted and surrendered at Charleston, WV and sent north 1/23/65. Age: 34. Farmer. 6'. Sandy complexion, light hair, gray eyes.

WILSON, JAMES H.: Pvt./2nd Lt. Co. K. B. Fayette Co., WV. Enl. 6/6/61 at Charleston, Kanawha Co., WV. Elected 2nd Lt. 12/15/63. WIA at Cold Harbor, Va. 6/2-3/64. At Charlottesville, Va. hosp. for "Feb. Int. Tert." 9/1-22/64. Promo. 2nd Lt. 12/31/64. POW paroled at Charleston, WV 5/5/65. Age: 21. 5'9". Fair complexion, light hair, blue eyes. Postwar: Timber business and salesman. D. before 1928. Middle name is Hiram.

WILSON, J. C.: Pvt. Co. H. B. in Va. Enl. 1861 by Patton. Place of enlistment not stated. POW paroled at Lewisburg, WV 4/25/65. Age: 25. Student. 5'6". Light complexion, light hair, blue eyes.

WILSON, JAMES M.: Pvt. Co. K. Res. Greenbrier Co., WV. Enl. 5/2/62 at White Sulphur Springs, Greenbrier Co., WV. Capt. at Meadow Bluff, WV by Col. Crook's guards 7/25/62. WIA at Winchester, Va. 9/19/64. POW paroled at Charleston, WV (undated). Age: 34. Doctor. 5'7". Dark complexion, dark hair, gray eyes. Res. Giles Co., Va. 1904 at age 74. Pension application.

WILSON, JOHN HENRY: Co. H. In a photo of R. E. Lee Camp #887, Charleston, WV taken Aug. 1911.

WILSON, JONATHAN N.: Pvt. Co. K. Res. Greenbrier Co., WV. Enl. 4/24/62 at White Sulphur Springs, Greenbrier Co., WV. Capt. at Greenbrier Co., WV 7/25/62. POW Camp Chase, OH. Exchd. 8/25/62. Sick at White Sulphur Springs, WV on 11/1/62 roll. Sick at home on Nov.-Dec. 1862 roll. Present on 2/28/63 roll. Left sick at Lewisburg, WV on Mar.-Apr. 1863 roll. Capt. at Droop Mountain, WV 11/6/63. Age: 20. POW Camp Chase, OH 11/18/63. Transf. to Ft. Delaware 3/14/64. Sick at Giles C.H. (Pearisburg), Va. July 1862. Farmer. Fair complexion, light hair, blue eyes.

WILSON, JOSIAH B.: Co. E. B. 3/23/31 Lynchburg, Va. D. Nov. 1917 Lunenburg, Va.

WILSON, SAMUEL J. ("GRANDPAP"): Pvt./Sgt. Co. K. B. in Greenbrier Co., WV. Enl. 5/25/61 at Fayetteville, Fayette Co., WV or 6/6/61 at Charleston, Kanawha Co., WV. AWOL 9/1/61. Capt. at Lewisburg, WV 5/23/62. POW Camp Chase, OH 5/30/62 or 6/30/62. Exchd. 8/25/62. Joined by exchange 10/20/62. WIA (severe arm) at Cold Harbor, Va. 6/3/64. At Chimborazo Hosp. for very serious left arm and right forearm 6/5-19/64. At Farmville, Va. hosp. 8/30/64 to 9/30/64 for wd. at Cold Harbor, Va. (6/2/64(?) consisting of ball entering over head of vulna fracturing bone and exiting opposite side. Also wd. of the extensor muscles of arm recvd. at same time of other wd. Returned to hosp. 9/30/64. Absent wded. on 12/31/64 roll. POW paroled at Charleston, WV 5/26/65. Age: 37. Farmer. 5'8". Dark complexion, gray hair, dark eyes, brown whiskers.

WILSON, THOMAS: Pvt. Co. D. Enl. 5/30/61 at Summersville, Nicholas Co., WV. Sick at home 7/28/61. Present on 10/31/62 roll. AWOL 11/5/62. In hands of enemy on 2/28/63 roll.

WILSON, WILLIAM: Pvt. Co. I/Co. F. Date and place of enlistment not stated. Capt. in Fayette Co., WV 4/3/63. Age: 36. POW Camp Chase, OH 4/13/63. Relsd. 4/24/63. On a clothing list 5/13/64. 5'10". Light complexion, light hair, gray eyes.

WILSON, WILLIAM A.: Pvt. Co. H. Res. Cinn., OH. Enl. 9/25/62 at Charleston, Kanawha Co., WV. Sick at Lewisburg, WV on 4/30/63 roll. Present on 6/30/63 roll. Capt. at Cold Harbor (Gaines Mill), Va. 6/3/64. POW Pt. Lookout, MD. Transf. to Elmira, NY 7/12/64. Took oath at Elmira, NY and relsd. 6/16/65. 5'8½". Fair complexion, dark hair, hazel eyes. Bur. Spring Hill Cem., Charleston, WV.

WILSON, WILLIAM H.: Pvt. Co. H. Enl. 5/8/61 at Charleston, Kanawha Co., WV. Present on 8/31/61 roll. NFR.

WILSON, WILLIAM J.: Pvt. Date and place of enlistment not stated.

WILSON, WILLIAM L.: Pvt. Co. K. Enl. 4/22/62 at White Sulphur Springs, Greenbrier Co., WV. WIA (slight) in 1862 Kanawha Valley Campaign. WIA at Cold Harbor, Va. 6/3/64. At Charlottesville, Va. hosp. for vulvus sclopet 6/4/64 to 7/6/64. Absent wded. on 10/31/64 roll. POW paroled at Charleston, WV 5/11/65. Age: 23 (?). 6'. Fair complexion, light hair, light eyes, light whiskers.

WINDLE, JAMES: Co. K. Deserted 5/23/62 on Aug. 1862 broadside. Greenbrier Co., WV.

WINEBRINNER, ALEX H.: Pvt. Co. A. Enl. 5/22/61 at Poca, Putnam Co., WV. Left at Coalsmouth (St. Albans), WV with typhoid fever 7/20/61. NFR.

WINGER, SAMUEL: Pvt./Corp. Co. K. Enl. 4/24/62 at White Sulphur Springs, Greenbrier Co., WV. AWOL 6/1/64.

WINGROVE, JAMES S.: Pvt. Co. K. Enl. 6/6/61 at Charleston, Kanawha Co., WV. Absent sick 7/18/61. Dischd. at Camp McCausland, WV 7/11/62. Farm laborer, Fayette Co., WV 1860. Age: 25.

WINGROVE, JOHN: Pvt. Co. K. Enl. 6/20/63 at Camp Sam Jones, Teamster 7/1/63 to 9/30/63. Sick at Pearisburg, Va. Entitled to bounty being over conscript age when enlisted on 12/30/63 roll. Present on 4/1/64 roll. POW paroled at Charleston, WV 5/29/65. Age: 37. 5'10". Dark complexion, brown hair, blue eyes. Farm laborer, Fayette Co., WV 1860. Bahlmann says enl. 1862.

WINGROVE, SAMUEL F. ("FLINT"): Pvt. Co. K. Enl. 6/6/61 at Charleston, Kanawha Co., WV. Left sick at Lewisburg, WV on Mar.-Apr. 1863 roll. Present on 12/30/63 roll. Transf. to Co. I, 8th Va. Cav. 4/29/64. Paroled at Charleston, WV 5/29/65 Age: 33. Also nicknamed "Flintlock." Farm laborer, Fayette Co., WV 1860. Age: 33.

WINTZ, JOHN WILLIAM: Pvt. Co. D. B. 8/4/42 at Brownstown (Marmet), WV. Enl. 5/30/61 at Summersville, Nicholas Co., WV. Absent with leave being sick in Kanawha Co., WV 7/21/61. PWR says Co. E and served 5/2/61 to 8/27/61. Postwar farmer, Kanawha Co., WV. Bur. Spring Hill Cem., Charleston, WV.

WISEMAN, ALEXANDER D.: Sgt. Co. D. Enl. 5/30/61 at Summersville, Nicholas Co., WV. Present on 8/31/61 roll. Transf. to Co. F, 36th Va. Inf. Paroled 5/10/65. Age: 27. 5'11". Hazel eyes and brown hair.

WITHROW, CYRUS: Pvt. Co. C. Day laborer, Fayette Co., WV 1860. Age: 22. Enl. 5/10/61.

WITHROW, NATHANIEL W.: Pvt. Co. A. Enl. 6/9/61 at Poca, Putnam Co., WV. On mounted guard at White Sulphur Springs, WV 8/13-14/61. At Charlottesville, Va. hosp. for paratitis 1/2-31/62. WIA at Lewisburg, WV 5/23/62. Kept wded. in private home 6/1/62 to 1/10/63. At Camp Sam Miller 1/10/63. Present on 2/28/63 roll. Courier on 4/30/63 roll. Det. service on 12/31/63 roll. Det. service on 4/1/64 roll. Present on 4/30/64 roll. Det. service on 10/31/64 roll. Recvd. from Staunton, Va. at medical director's office, Richmond, Va. 12/19/64. POW paroled at Staunton, Va. 4/30/65. Age: 26. 5'7". Fair complexion, light hair, blue eyes.

WITHROW, WILLIAM THOMAS: Pvt. Co. A. Enl. 6/19/61 at Poca, Putnam Co., WV. Left sick at White Sulphur Springs, WV 8/20/61. Present on 2/28/63 roll. At Chimborazo Hosp. for acute dysentery 6/8 or 9/64 to 7/2/64. Capt. at Winchester, Va. 9/19/64. POW Pt. Lookout, MD 9/23/64. Exchd. 3/15/65.

WITZELL, JAMES: Pvt. Co. G. Enl. Mar. 1862 at Lewisburg, WV. Age: 20. Res. Alleghany Co., Va. Surrendered Apr. 1865. See also James Wetzel.

WOLFE, GEORGE: Pvt. Co. I. B. in Smyth Co., Va. Date and place of enlistment not stated. On a list of refugee's and deserters recvd. at Guyandotte, WV Apr. 1865. Age: 20. 5'10". Dark complexion, black hair, hazel eyes.

WOLFENBARGER, WILLIAM M.: Pvt. Taylor's Co. Enl. 6/19/61 at Frankford, Greenbrier Co., WV. Age: 20. Co. transf. to Wise Brigade 7/10/61. Joined Co. A, 14th Va. Cav. by transfer from 22nd Va. Inf. 10/1/62 in exchange for Nathan McMullen (?).

WOOD, ARCHIBALD D.: Co. C. Enl. 6/6/61 at Charleston, Kanawha Co., WV. D. in hosp. 1/27/62. Claim filed by James Skaggs, administrator, 2/9/63. Probable res. Monroe Co., WV.

WOOD, EZRA: Pvt. Co. C. Res. Fayette Co., WV 1860. Age: 18. Enl. 5/10/61.

WOOD, FELIX A.: Pvt. Co. C. Farmer, Fayette Co., WV 1860. Age: 27. Enl. 5/10/61.

WOOD, GEORGE L.: Musician. Co. C. Res. Fayette Co., WV. Enl. 5/1/62 at White Sulphur Springs, Greenbrier Co., WV. Capt. at Droop Mountain, WV 11/6/63. Age: 20. POW Camp Chase, OH 11/18/63. Transf. to Ft. Delaware and relsd. 6/22/65. Farmer. 5'9" or 5'11¾". Florid complexion, dark hair, gray eyes.

WOOD, G. N.: Pvt. Co. C. Date and place of enlistment not stated. D. at Camp McCausland, WV 7/7/62.

WOOD, GRANDISON: Pvt. Co. C. Enl. 1862 at Dublin, Va. or 11/24/63 at White Sulphur Springs, Greenbrier Co., WV. Listed as a prisoner 11/10/61. Det. government shop at Dublin, Va. on 12/31/63 roll and 4/8/64 roll. Present on 10/31/64 roll. POW paroled at Lewisburg, WV 4/25/65. Age: 27. Farmer. 5'10". Fair complexion, brown hair, blue eyes. Res. Fayette Co., WV 1860.

WOOD, J. M.: 2nd Lt. Co. C. Enl. 6/6/61. Place of enlistment not stated. Elected 2nd Lt. 6/6/61. Not re-elected 5/1/62. Dropped. 5/2/62. Successor: A. C. Rowan. This is possibly James Nelson Wood. Millwright, Fayette Co., WV. 1860. Age: 27.

WOOD, MARSHALL: Pvt./Musician. Co. C. Res. Fayette Co., WV. Enl. 5/1/62 at White Sulphur Springs, Greenbrier Co., WV. Inventor of an improved arm. Considered a genius inventor, wanted by U.S. troops who put up a $1000 reward for his capture. On 3/6/62 Lt. Col. William A. Jackson said Wood had nearly completed and efficient repeating rifle. Det. government work at Dublin, Va. on rolls for Sept.-Oct. 1862 and 12/31/62. Det. gunsmith at Dublin, Va. (and Wytheville, Va.) on rolls for 3/4/63 and 6/11/63. Det. government shop at Lynchburg, Va. on 12/31/63 roll. Deserter not entitled to bounty on 4/18/64 roll. Deserted 10/26/64. Paroled at Manchester, Va. 4/21/65 and given permission to go to his home in Greenbrier Co., WV. Mechanic/machinist. PWR says enl. 5/10/61. Artist, Fayette Co., WV 1860. Age: 30.

WOOD, OLIVER: Corp./Pvt. Co. C. Enl. either 1861 at Charleston, Kanawha Co., WV or 5/1/62 at White Sulphur Springs, Greenbrier Co., WV. Det. to work in government shop on Sept.-Oct. 1862 roll. Det. government shop at Dublin, Va. on 12/31/62 roll. Det. gunsmith in government shop at Dublin, Va. on rolls from 3/4/63 through 4/18/64. Present on 10/31/64 roll. POW paroled at Lewisburg, WV 4/27/65. Age: 25. Farmer. 6'. Light complexion, brown hair, blue eyes.

WOOD, THOMAS D.: Pvt. Co. D. B. Nelson Co., Va. Res. Clay Co., WV. Enl. 5/30/61 at Summersville, Nicholas Co., WV. Sick at Dublin, Va. July 1862. Present on 10/31/62 roll. Left sick at Beverly, WV 3/25/63. Capt. either at Bulltown, WV or Nicholas Co., WV by 17th WV Inf. 2/13 or 15/65. Another card says surrendered at Huttonsville, WV. Forwarded to Clarksburg, WV 2/23/65. Took oath and sent north 2/24/65. Present at Clarksburg, WV 2/28/65. Another card says deserted and surrendered Mar. 1865. Age: 23. Farmer.

WOODRUM, GREEN: Pvt. Co. I. Res. Boone Co., WV, 1860, age 20. Enl. 6/29/61 at Charleston, Kanawha Co., WV. Deserted 9/18/62. Brother of Robert C. Woodrum.

WOODRUM, ROBERT C.: Pvt. Co. I. Res. Boone Co., WV 1860, age 18. Enl. 6/29/61 at Charleston, Kanawha Co., WV. AWOL 7/26/61. Brother of Green Woodrum.

WOODS, M. G.: Corp./2nd Lt. Co. E. Enl. 6/18/61 at Charleston, Kanawha Co., WV. Corp. at organization. On mounted guard at White Sulphur Springs, WV 8/13-14/61. Left sick at White Sulphur Springs, WV 8/20/61. Elected 2nd Lt. 5/1/62. Apptd. 2nd Lt. 5/2/62. WIA at New Market, Va. 5/15/64. Permanently disabled by wd. and recommended for retirement by commdg. officer 10/20/64. Absent wded. on 10/31/64 roll.

WOODS, SAMUEL BERNARD: 1st Lt. Co. K. B. 12/12/28. Col., 142 Regt. Va. Militia. Enl. 6/6/61 at Charleston, Kanawha Co., WV. Elected 1st Lt. 6/6/61. Tendered resignation 8/2/61. Resgnd. Aug. 1861. Afterward on militia service in Fayette Co., WV. PWR says also served in 8th Va. Cav. D. 12/26/97. Bur. Old Methodist Cem., Oak Hill, WV. Blacksmith, Fayette Co., WV 1860. Age: 30. Bahlmann says resgnd. and became farrier in 8th Va. Cav.

WOODWARD, LUKE TOBIAS: Pvt. Co. E. B. in Augusta Co., Va. Enl. Mar. 1863 at Staunton, Va. Served in Co. D. 7th Battn. Art. then Co. E, 22nd Va. Inf. Fought at Piedmont, Winchester, Cold Harbor, and Spotsylvania. Capt. near Richmond, Va. Feb. 1865. POW Ft. Delaware. Relsd. June 1865. Res. South River District, Augusta Co., Va. 1910 at age 63. Res. Waynesboro, Va. 1912. Retired carpenter at Waynesboro, Va. when joined Stonewall Jackson CV Camp, Staunton, Va. D. 3/2/1929 Waynesboro, Va. Another PWR says served in Co. B, was res. of Jackson Co., WV and musteed 1863 for 2 years.

WOODWARD, WILLIAM L.: Pvt. Taylor's Co. Enl. 6/19/61 at Frankford, Greenbrier Co., WV. Age: 22. Co. transf. to Wise Brigade 7/10/61.

WOOLWINE, WILLIAM: Pvt. Co. A. Date and place of enlistment not stated. On a clothing list 4 QR 1864. POW paroled at Charleston, WV 5/11/65. Age: 44. 5'8". Light complexion, dark hair, gray eyes, dark whiskers.

WRIGHT, GEORGE A.: Pvt. Co. H. Enl. 5/8/61 at Charleston, Kanawha Co., WV. Absent cause unknown on 8/31/61 roll. Det. A.C.S. Dept. on rolls from Sept.-Oct. 1862 through 2/28/63. Det. Commissary Sgt. at New Bern, Va. on rolls from 4/30/63 through 6/30/63. A.C.S. Dept. at Dublin, Va. on rolls from 12/3/63 through 4/1/64. Present on 4/30/64 roll. Det. A.C.S. Dept. on 12/25/64 roll. POW paroled at Charleston, WV 5/2/65. Age: 23. 5'8". Dark complexion, dark hair, gray eyes, dark whiskers.

WRIGHT, HUGH CALDWELL: Pvt. Co. K. B. ca. 1828 Craig Co., Va. Enl. 4/24/62 at White Sulphur Springs, Greenbrier Co., WV. Det. government shoe shop on rolls from 12/30/63 through 5/1/64. Exchd. 3/15/65. At Chimborazo Hosp. for chronic diarrhea 3/19-24/65. Res. Craig Co., Va. in 1898 at age 69. Bur. Crawford Cem., Rt. 615 on Whitlow farm, across from U.S. Forest Service Office, Craig Co., Va.

WRIGHT, GEORGE A.: Pvt. Co. H. Enl. 5/8/61 at Charleston, Kanawha Co., WV. Absent cause unknown on 8/31/61 roll. Det. A.C.S. Dept. on rolls from Sept.-Oct. 1862 through 2/28/63. Det. Commissary Sgt. at New Bern, Va. on rolls from 4/30/63 through 6/30/63. A.C.S. Dept. at Dublin, Va. on rolls from 12/3/63 through 4/1/64. Present on 4/30/64 roll. Det. A.C.S. Dept. on 12/25/64 roll. POW paroled at Charleston, WV 5/2/65. Age: 23. 5'8". Dark complexion, dark hair, gray eyes, dark whiskers.

WRIGHT, HUGH CALDWELL: Pvt. Co. K. B. ca. 1828 Craig Co., Va. Enl. 4/24/62 at White Sulphur Springs, Greenbrier Co., WV. Det. government shoe shop on rolls from 12/30/63 through 5/1/64. Exchd. 3/15/65. At Chimborazo Hosp. for chronic diarrhea 3/19-24/65. Res. Craig Co., Va. in 1898 at age 69. Bur. Crawford Cem., Rt. 615 on Whitlow farm, across from U.S. Forest Service Office, Craig Co., Va.

WRIGHT, ROBERT K.: Pvt. Co. E. Enl. 7/18/61 at Porter Shoals, WV. Left sick at White Sulphur Springs, WV 8/20/61. NFR.

WRIGHT, SAMUEL E.: Pvt. Co. K. B. Craig Co., Va. ca. 1833. Enl. 4/24/62 at White Sulphur Springs, Greenbrier Co., WV. Deserted at Packs Ferry, WV 10/17/62. Under arrest for being AWOL on 2/28/63 roll. Said to be a deserter but since returned as a paroled prisoner on Mar.-Apr. 1863 roll. Absent sick since 3/15/64 on 10/31/64 roll. Deserted and surrendered Mar. 1865. Took oath at Charleston, WV and sent north 4/9/65. Age: 33. Blacksmith. 5'6". Dark complexion, dark hair, hazel eyes. Res. Craig Co., Va. Res. Craig Co., Va. in 1898 at age 65. Bur. Via Cem., Rt. 643 off Rt. 615, Craig Co., Va. PWR says wded. in left arm.

WRIGHT, THOMAS A.: Pvt. Co. A. Res. Mason Co., WV. Enl. 5/26/61 at Buffalo, Putnam Co., WV. Deserted 2/1/64. Capt. at Winchester, Va. 9/19/64. POW Pt. Lookout, MD 9/24/64. Exchd. 3/15/65. Deserted and surrendered Apr. 1865. Took oath at Charleston, WV and sent north 4/10/65. Age: 22. Farmer. 6'11½". Fair complexion, fair hair, hazel eyes.

WYATT, WILLIAM H.: Pvt. Co. A. Enl. 5/1/62 at White Sulphur Springs, Greenbrier Co., WV. Capt. at Lewisburg, WV 5/23/62. POW Camp Chase, OH 5/30/62. Transf. to Johnson's Island, OH 9/6/62. Exchd. 11/22/62. At Chimborazo Hosp. for very serious (old?) 5/25/64 to 7/3/64. Paroled at Charleston, WV 6/10/65. Age: 24. 5'7". Sandy complexion, light hair, blue eyes, sandy whiskers.

WYLIE, ROBERT J.: Co. F. Res. Monroe Co., WV. D. sometime prior to 1914.

YOUNG, ANDREW: Corp. Swann's Co. Enl. 5/25/61 at Charleston, Kanawha Co., WV. Age: 19. Co. transf. to Co. K, 59th Va. Inf. 7/25/61.

YOUNG, ARCHIBALD P.: Pvt. Co. H. B. 1/2/42 Kanawha Co., WV. Laborer, Charleston, WV 1860. Enl. 5/8/61 at Charlseton, Kanawha Co., WV. Joined co. 3/21/63. Present on 4/30/63 roll. WIA at Droop Mountain, WV 11/6/63. Absent wded. on rolls from 4/1/64 through 4/30/64. Recvd. at medical director's office, Richmond, Va. from Lynchburg, Va. 9/22/64. Absent wded. on 12/25/64 roll. POW paroled at both Harpers Ferry, WV and Charleston, WV 4/27/65. Age: 23. 5'4". Dark complexion, dark hair, dark eyes. Postwar res. Mill Creek, Kanawha Co., WV. PWR says served in Swann's Co.

YOUNG, AUGUSTUS: Pvt. Co. D. Enl. 11/18/63 at Camp Bailey. Present on 12/31/63 roll. NFR.

YOUNG, JAMES: Rank unknown. Co. D. Enl. 11/?/63. Place of enlistment not stated. Deserted at Camp Gauley, WV Jan. 1864.

YOUNG, SAM: Pvt. Co. H. Enl. 5/8/61 at Charleston, Kanawha Co., WV. Capt. by enemy on 4/1/64 roll. Capt. Jan. 1864 head of Big Sycamore Creek, Clay Co., WV. Escaped and returned to regt. Still alive in 1901.

YOUNG, SAMUEL L.: Pvt. Swann's Co. B. 12/22/39 Kanawha Co., WV. Enl. 5/25/61 at Charleston, Kanawha Co., WV. Age: 21. Co. transf. to Co. K, 59th Va. Inf. 7/25/61. Postwar res. Aaron's Fork of Little Sandy, WV. PWR says middle initial is M.

ZIRKLE, JAMES L.: Pvt. Co. A. Enl. 6/20/61 at Poca, Putnam Co., WV. Unaccounted for as of 7/24/61. Bur. Claypoole Cem., Logan Co., WV.

BIBLIOGRAPHY

Manuscripts

Autobiography of William F. Bahlmann, courtesy of Larry Legge, Barboursville, West Virginia. Copy in possession of the author.

W. Leigh Barksdale Letter, Headquarters First Brigade, Wharton's Division, Army of the Valley, November 9, 1864. Copy in possession of the author.

Carnifex Ferry Battlefield State Park Files, courtesy Randy Rice, Superintendent, Carnifex Ferry Battlefield State Park, Summersville, West Virginia.

John P. Donaldson Papers, War Library and Museum, Philadelphia, Pennsylvania.

Samuel D. Edmunds Memoir, "The Battle of Droop Mountain," courtesy Mike Smith, Superintendent, Droop Mountain Battlefield State Park, Hillsboro, West Virginia.

General John B. Floyd Papers, Duke University, Durham, North Carolina.

Hollywood Cemetery Burial Records, Hollywood Cemetery, Richmond, Virginia.

Huntington Gallery and Library, San Marino, California. "A Brief and Imperfect Account of the Action at Scary Creek," by George S. Patton.

Correspondence of the Patton Family Genealogy.

List of deserters from the 22nd Virginia Infantry, Confederate broadside, Salt Sulphur Springs, Virginia, August 20, 1862, W. W. Loring commanding. West Virginia University Library, Morgantown, West Virginia.

Museum of the Confederacy, Richmond, Virginia.

George S. Patton Papers.

Christopher Q. Tompkins Papers.

National Archives, Washington, D.C.

Andrew R. Barbee, Lt. Col., 22nd Virginia Infantry, Misc. 1864 Letters Received 1863-65, Department of West Va. (Grant), Entry 5691 (Inventory Vol. I), Record Group 393, U.S. Army Continental Commands 1821-1920, Location Stack Area 10W2, Row 11, Bottom Shelf.

Papers of General John B. Floyd.

Population Schedules of the Eighth Census of the United States, 1860.

Population Schedules of the Thirteenth Census of the United States, 1910.

National Archives Record Group No. 109, Compiled Service Records of Confederate Soldiers Who Served In Organizations From The State Of Virginia. 22nd Virginia Infantry, Microcopy M324, rolls 647-655.

George S. Patton Papers, courtesy Ruth Ellen Patton Totten, South Hamilton, Massachusetts.

Records of Confederate Military Units, 22nd Virginia Infantry, Virginia State Library, Richmond, Virginia.

Sampson Smith Letter, Camp near Fishersville, Virginia, December 29, 1864. Original in the possession of Col. Harvey E. Sheppard, Maitland, Florida.

Southern Historical Society Collection, University of North Carolina, Chapel Hill, North Carolina.
Catherine B. Broun Diary, (No. 2389).
Tucker Family Papers, (No. 2605).

Tompkins Family Papers 1800-71 and 1792-1869. Collection also includes a handwritten account of the war by Col. C. Q. Tompkins called "Record of the Revolution." Virginia Historical Society, Richmond, Virginia.

Diary of Andrew Stiarwalt, Sr., 23rd Ohio Volunteer Infantry. Rutherford B. Hayes Presidential Center, Fremont, Ohio.

United Daughters of the Confederacy Interviews with Craig County Soldiers, courtesy Brenda Williams, Newport, Virginia.

Veterans Grave Registration In West Virginia. A WPA Project. Card File located at the West Virginia State Archives, Charleston, West Virginia.

Virginia Military Institute Archives, Lexington, Virginia. Includes papers of John McCausland, George S. Patton, John K. Thompson, and students who served in the 22nd Virginia Infantry.

West Virginia State Archives, Charleston, West Virginia.
"A list of those persons who have served in the Kanawha Rifle Company."
Chilton Family Papers.
"Kanawha Riflemen."
Noyes Rand Papers.
James Clark Welch Papers.

West Virginia University, Morgantown, West Virginia.
Roy Bird Cook Collection:
Record Book, Company H, 22nd Virginia Infantry, 1862.
Diary of John Maston, Co. B, 22nd Virginia Infantry, 1864.

Published Primary Sources

Abbott, Joel H. "A Civil War Narrative," in *History of Fayette County*, by J. T. Peters and H. B. Carden, Charleston, West Virginia, 1926.

"Bahlmann Gives History of Men Serving in '61," Charleston *Gazette,* Sept. 17, 1922.

Bahlmann, William F. "Down In The Ranks," *The Journal of the Greenbrier Historial Society,* Vol. II, #2, Oct. 1970.

Broun, Thomas. "Reminiscences of Major Thomas L. Broun," *Confederate Veteran,* IX (1901).

Brown, Joseph Alliene. *Memoirs of a Confederate Soldier, Joseph Alliene Brown, As Told to his Grandson, Samuel Hunter Austin,* Abingdon, Virginia, 1940.

Clark, W. C. "The Journal of a Soldier of 1861," *West Virginia Review,* Nov. 1930. Based on the diary of Joab Smith, Co. B., 22nd Virginia Infantry.

Cowles, Calvin. *The Official Military Atlas of the Civil War.* Washington, D.C., 1891-1895.

Cox, Jacob D. *Military Reminiscences.* 2 vol. New York, 1900.

Early, Jubal A. "Early's March To Washington In 1864," *Battles and Leaders,* Vol. IV, pg. 492-499.

________ "Winchester, Fisher's Hill and Cedar Creek" *Battles and Leaders,* Vol. IV, pg. 522-530.

Faulkner, Virginia (compiler). *Dear Annie.* Parsons, West Virginia, 1969.

Hassler, Frederick W. B. "A Military View of Passing Events From Inside The Confederacy: The Campaign In West Virginia," *The Historical Magazine,* Vol. VI, New York, 1869.

Heth, Henry. "Memoirs of Henry Heth," *Civil War History,* VIII (1962).

Imboden, John D. "The Battle of New Market, Virginia May 15, 1864," *Battles and Leaders,* Vol. IV, pg. 480-486.

Mays, James H. *Four Years For Old Virginia,* edited by Lee Mays. Los Angeles, California, 1970.

Morgan, John. *The Last Dollar.* Kansas City, ca. 1895.

Reynolds, William C. *Diary of William Clark Reynolds: A Private In The Kanawha Riflemen,* Union, KY (?), 1956 (?).

Ruffner, Joseph. "The Tables Turned," *West Virginia Historical Quarterly, Vol. 1,* #4, Oct. 1901, pg. 63-66.

Smith, Isaac Noyes. "A Virginian's Dilemma," edited by William C. Childers. *West Virginia History,* Vol. 27, #3, Apr. 1966. The original Smith diary is in the Jacob D. Cox Papers, Oberlin College.

Thompson, Cameron L. "The Battle of Scary Creek," *Confederate Veteran,* XXVI (1918).

United States War Department. *The War of the Rebellion: A Compilation of the Official Records of the Union and Confederate Armies.* 128 volumes. Washington, D.C., 1880-1881.

Welch, Levi. "Battle of Scary." *West Virginia Historical Quarterly* Vol. 1, #1, Jan. 1901.

Newspapers and Periodicals

Kanawha Valley Star. 1860-1861.
Charleston *Gazette.*
Charleston *Daily Mail.*
West Virginia History.
West Virginia Review.
Confederate Veteran, 1893-1932.
Civil War Times Illustrated.
Blue and Gray.
Greenbrier Independent.
Monroe Watchman.
Upper Vandalia Historical Society Journal.
Southern Historical Society Papers, 1876-1959.
Virginia Country's Civil War.
National Guard Magazine, Sept. 1984.

Published Sources

Atkinson, George A. and Alvaro F. Gibbons. *Prominent Men of West Virginia.* Wheeling, West Virginia, 1890.

Bowman, Mary Keller. *Reference Book of Wyoming County History.* Parsons, West Virginia, 1965.

Cohen, Stan. *The Civil War In West Virginia: A Pictorial History.* Charleston, West Virginia, 1976.

________. "Col. George S. Patton and the 22nd Virginia Infantry," *West Virginia History,* 1961.

Comstock, Jim. *West Virginia Heritage Encyclopedia and Supplemental Volumes.* Richwood, West Virginia, 1968.

Cullum, George. *Biographical Register for West Point of the Officers and Graduates.* New York, n.d.

Cutchins, John A. *A Famous Command: The Richmond Light Infantry Blues.* Richmond, Virginia, 1934.

Davis, William C. *The Battle of New Market.* New York, 1975.

Dickinson, Jack. *8th Virginia Cavalry.* Lynchburg, Virginia, 1986.

Dyer, Frederick H. *Compendium of the Rebellion.* 2 vol. Des Moines, 1908.

Evans, Clement A. *Confederate Military History.* 13 vol. Atlanta, Georgia, 1899.

Hale, John P. *History of the Great Kanawha.* Madison, Wisconsin, 1891.

Hardesty, H. H. *Hardesty's Historical and Geographical Encyclopedia,* New York, 1884.

Hite, Delmer and Jackson County Historical Society. *Jackson Countians In America's Wars 1775-1920.* Ripley, West Virginia, 1978.

Hull, Forrest. "George S. Patton, Rebel," *Holland's Magazine of the South,* May 1945.

Johnson, Robert U., and C. C. Buel, editors. *Battles and Leaders of the Civil War.* 4 vol. New York, 1887-88.

Johnston, Jane Echols, and Brenda Lynn Williams. *Hard Times Vol. I* (1986) and *Hard Times Vol. III* (1987). Both Craig County Historical Society.

Laidley, W. S. *History of Charleston and Kanawha County, West Virginia.* Chicago, 1911.

Lewis, Thomas A. *The Shenandoah In Flames: The Valley Campaign of 1864.* Alexandria, Virginia, 1987.

Lowry, Terry. *The Battle of Scary Creek: Military Operations In The Kanawha Valley, April-July 1861.* Charleston, West Virginia, 1982.

________ *September Blood: The Battle of Carnifex Ferry.* Charleston, West Virginia, 1985.

Mickle, William E. *Well Known Confederate Veterans.* New Orleans, Louisiana, 1915.

Moore, Frank. *The Rebellion Record.* 12 vol. New York, 1861-71.

Moore, George E. *A Banner In The Hills.* New York, 1963.

Morton, Oren F. *History of Monroe County, West Virginia.* Staunton, Virginia, 1916.

Peters, J. T. and H. B. Cardens. *History of Fayette County, West Virginia.* Charleston, West Virginia, 1926.

Rice, Otis. *History of Greenbrier County.* Parsons, West Virginia, 1986.

Scott, J. L. *36th Virginia Infantry.* Lynchburg, Virginia, 1987.

________ *36th and 37th Battalions Virginia Cavalry.* Lynchburg, Virginia, 1986.

Shetler, Charles. *West Virginia Civil War Literature.* Morgantown, West Virginia, 1963.

Sons of Confederate Veterans Album. Houston, 1986.

Summers, George W. *Pages From The Past.* Charleston, West Virginia, 1935.

Swain, G. T. *History of Logan County.* Logan, West Virginia, 1927.

Walker, Gary C. *The War In Southwest Virginia 1861-1865.* Roanoke, Virginia, 1985.

Wallace, G. S. *Cabell County Annals and Families,* 1935.

Wallace, Lee A. *A Guide To Virginia Military Organizations 1861-65.* Richmond, Virginia, 1964.

Warner, Ezra. *Generals In Gray.* Baton Rouge, Louisiana, 1959.

Wise, Barton H. *The Life of Henry A. Wise of Virginia.* New York, 1899.

22ND VIRGINIA INFANTRY

2nd Edition

ADDENDUM AND ERRATA

Since publication of the *22nd Virginia Infantry* for the Virginia Regimental Histories Series in 1988 much new material has been located and contributed, particularly from Bob Driver, Brownsburg, Virginia; Scott Hutchison, Powhatan, Virginia; Tim McKinney, Kanawha Falls, West Virginia, and a number of others, including those with ancestors having served in the 22nd Virginia Infantry. As well as many new names to include in the roster, can be found much additional material for names already in the roster. Following that is a list of errors and the proper corrections, some are minor typographical and/or printing errors, and some are historical errors which need to be corrected, although they are usually minor. Included next is a list of corrections for the various maps and charts, then some new photos of members, and a list of material submitted after the bulk of this supplement was completed as well as some new data on the 22nd Virginia. As with all such projects of this nature, more new material will continue to be discovered and I hope all authors in the series will continue to update and correct their respective volumes.

—Terry Lowry
1991

ADDITIONAL NAMES FOR THE ROSTER

ALEXANDER, JOHN W.: Co. F. Postwar roster. D. 9/9/1907 at age 73. Bur. Mount Pleasant Cem., Sinks Grove, WV. "A brave Confederate soldier" inscribed on headstone.

ALLEN, G. A.: Co. G. Res. Wyoming Co., WV. Date and place of enl. not stated. D. 10/18/64 at Baltimore, MD. Bur. London Park Nat. Cem. See also entry in roster under Amon Allen.

ALLEN, HENRY: Co. K. Bur. Confederate Section, Woodbine Cem., Harrisonburg, VA.

BAILEY, A. CALDER: Some postwar sources indicate he may have served in 22nd Va. Inf. Adj. 8th Va. Cav. Date of commission 5/15/62. KIA at Chambersburg, PA. 7/30/64. Claim filed by Maria C. Bailey 1/14/65. Postwar sources state he was beaten to death by the citizens of Chambersburg, PA. Also spelled Caulder. PWR says served in Co. K, 22nd Va. Inf.

BAILEY, DANIEL P.: Co. F. Postwar res. Summers Co., WV. D. 1904. Bur. Mount Pleasant Cem., Sinks Grove, WV. (Possibly same as D. Bailey on roster).

BAILEY, E. H. C.: Surgeon. Postwar record only.

BLANKENSHIP, JOSEPH: Co. F. Res. Greenbrier Co., WV. D. 4/28/1911 at Falling Springs. (Possibly same as J. H. Blankenship on roster).

BOWEN, SAMUEL A.: Co. F. Postwar roster.

BOYD, MADISON: Co. F. Brother of Robert and William Boyd. WIA (severe head) at Cold Harbor, Va. June 1864. Never fully recovered. D. 4/15/99 his body being found outside an open window at Hinton Hosp., Hinton, WV where he had been receiving treatment for cancer. Bur. Mount Pleasant Cem., Sinks Grove, WV. (Possibly same as Mat Boyd on roster).

BROUN, JOSEPH McADAM: Co. H. B. 12/23/35 in Loudon Co., Va. Pre-war res. Kanawha Co., WV. Kanawha Riflemen 1861. Capt. and Q.M., 3rd Regt., Wise Legion at Sewell Mt., WV. 1861. Postwar lawyer, Charleston, WV. Brother of Thomas Lee Broun. In *Confederate Veteran* 1908, page 139.

BURDETTE, JOHN I.: Co. H. D. 12/8/1909 near Wolf Creek, WV from poor health brought on by stroke and paralysis. Age: 62. Bur. Mount Pleasant Presbyterian Church Cem.

BURDETTE, WASHINGTON: Co. F. Capt. at Lewisburg, WV. 5/23/62. Postwar res. Wolf Creek, WV. D. 6/18/1907 at age 67. Bur. Longacre Cem., near Fort Spring, WV. (Possibly same as George W. Burdette on roster.

CALDWELL, GEORGE W.: D. 3/3/65 of "billous colic" in Craig Co., Va. on wife's pension application.

CAMPBELL, WILLIAM: B. in Rockbridge Co., Va. ca. 1825. Farmer, age 36, 6th Dist., Rockbridge Co., 1860 census. Enl. in Rockbridge Rangers at Lexington, Va. June 1861. Enl. in Co. C, 14th Va. Cav. in Monroe Co., WV. 1863. Res. Amherst Co., Va. Transf. to 22nd Va. Inf. 1863.

CARTER, N. R.: Res. near Sinks Grove, WV for many years. Later res. Junta, Summers Co., WV. Then res. Pence Springs, WV. D. 6/10/1916 at Pence Springs, WV at age 75. PWR only.

CRINER, LEWIS B.: Co. F. KIA at Frazier's Farm, VA 1862. PWR only.

CRONE, JAMES: Co. A. D. 9/5/64 at Elmira, NY. Bur. Woodlawn Nat. Cem., grave 236. (This may be Pvt. James Crone, Co. A, 26th Battn. Va. Inf. who enl. in that unit as a conscript 6/23/63 at Lewisburg, Greenbrier Co., WV. Capt. at Cold Harbor (Gaines Mill), Va. 6/3/64. POW Pt. Lookout, MD 6/11/64. Transf. to Elmira, NY 7/12/64. D. of chronic diarrhea 9/5/64 at Elmira, NY. Bur. section 172, Woodlawn Nat. Cem., Elmira, NY).

DIEHL, J. R.: Pvt. Co. D. PWR only.

DUNN, JAMES P.: Co. F. Killed in 1861. PWR only.

GILPIN, JOHN: Co. F. D. 10/15/64 at Baltimore, MD. Bur. London Park Nat. Cem.

HAYS, JAMES F.: Pvt. Co. F. PWR only.

HONOKER, JOHN WILLIAM: Co. F. "Served with honor as a member of the Rocky Point Grays." D. 8/10/1928 at age 80. Bur. Sarton Cem. PWR only.

HYPES, JOHN H.: Co. C. D. 8/22/64 at Elmira, NY. Bur. Woodlawn Nat. Cem., Elmira, NY, grave #32.

INGLES, GEORGE: Co. F. D. 6/1/1902. PWR only.

KENNEDY, A. J.: Co. I. D. 3/6/65 at Elmira, NY. Bur. Woodlawn Nat. Cem., Elmira, NY. grave #2394.

LEMONS, WILLIAM: Co. F. D. 10/4/1914 at age 78. Bur. Mount Pleasant Cem., Sinks Grove, WV. (Possibly same William Lemons who served in 26th Battn. Va. Inf.).

LIONBERGER, JOHN HENRY: 1st Lt. Co. D. V.M.I. 1863. Also served in Co. D, 7th Va. Cav.

McLAUGHLIN, SAMUEL: No information except that he may have served.

McMILLION, NATHAN: B. ca. 1843 in Greenbrier Co., WV. Farmhand, age 17, Lewisburg, WV P.O., 1860 census. Enl. as a Pvt. in Co. A (2nd), 14th Va. Cav. in Greenbrier Co., WV. 4/21/62. Transf. to 22nd Va. Inf. in exchange for William W. Wolfenbarger 10/31/62. Farmhand, age 27, Falling Springs Dist., Greenbrier Co., WV 1870 census. Farmer. Brother of Robert M. McMillion. D. 12/24/93 in Greenbrier Co., WV.

MILLER, HENRY: Co. I. Postwar roster.

MYERS, C. C.: Co. I. D. 8/2/64 at Elmira, NY. Bur. Woodlawn Nat. Cem., Elmira, NY, grave #142.

O'RORK, JOHN: Co. I. D. 12/14/64 at Elmira, NY. Bur. Woodlawn Nat. Cem., Elmira, NY grave #1120.

PACK, JOHN: Co. F. Son of Matthew Pack. D. during war from being shot in leg by accident.

PARISH, W. J.: Pvt. Co. C. Enl. 5/1/62 at White Sulphur Springs, Greenbrier Co., WV. Deserted on Sept.-Oct. 1862 roll.

PARK, ABRAM (ABRAHAM) S.: 3rd Lt. Co. B. B. 2/27/44. Res. Ripley, WV, age 16. Enl. 6/1/61 at Ripley, Jackson Co., WV. Elected 3rd Lt. 6/27/62. Promo. 7/22/63. Absent wded. on Nov.-Dec. 1863 roll. Furlo. at Montgomery White Sulphur Springs, WV 1/7/64. Present on Apr. 1864 roll. Deserted July 1864. Admitted to U.S. Gen. Hosp. at Frederick, MD 7/25/64. Deserted at Waynesboro, Va. 9/29/64. Paroled at Charleston, WV 5/15/65. Age: 21. 5'6". Fair complexion, dark hair, hazel eyes. D. 9/22/88.

PARK, ALBERT GALLATIN JR.: Pvt. Co. B. Res. Loglick Run, WV 1860, age 25. KIA.

PARKER, SAM: Co. K. William Bahlmann says KIA at Lewisburg, WV 5/23/62.

PARKS, BUSHROD W.: Pvt. Co. H. Res. Charleston, WV 1860, age 14. Res. Cinn., OH. Enl. 9/15/62 in Kanawha Co., WV. Capt. at Cold Harbor, VA 6/3/64. POW Pt. Lookout, MD. Transf. to Elmira, NY 7/12/64. Took oath at Elmira, NY and relsd. 6/15/65. 5'10". Fair complexion, light hair, hazel eyes.

PARKS, J. CREED: Pvt. Co. H. Res. OH 1860, age 16. Enl. 5/8/61 at Charleston, Kanawha Co., WV. Capt. at Cold Harbor, VA 6/3/64. POW Pt. Lookout, MD. Transf. to Elmira, NY 7/12/64. Killed in a train wreck at Shohola, PA 7/15/64 en route to Elmira, NY. Listed as recvd. at Elmira, NY. 7/17/64. Bur. in unmarked grave at Woodlawn Nat. Cem., Elmira, NY.

PARRISH, WILLIAM I.: Co. C. Enl. 5/10/61. Postwar roster. (This may be the man listed as W. J. Parish on roster).

PARSONS, JO: Pvt. Co. B. Enl. 9/22/62 at Charleston, Kanawha Co., WV. Present on 11/16/62 roll. NFR.

PARSONS, WILLIAM: Pvt. Co. B. Enl. 9/22/62 at Charleston, Kanawha Co., WV. Capt. at Winchester, Va. 9/19/64. POW Pt. Lookout, Md. 9/23/64. Exchd. 3/15/65.

PERRY, CALVIN M.: Co. F. D. 7/19/1925 at age 78. Bur. Riverview Cem., Lowell. PWR only.

POWERS, J. L.: Quartermaster. PWR only. (This may actually be J. A. Powers).

QUENSENBERRY, JAMES G.: D. 8/13/64 at Baltimore, MD. Bur. Woodlawn Nat. Cem., Elmira, NY.

RADER, DANIEL K.: Co. H. B. 8/18/23 in Botetourt Co., Va. D. 8/29/89. Bur. Mt. Union Church Cem.

REASER, PHILIP: Co. D. Killed in a train wreck at Shohola, PA 7/15/64 carrying POW's to Elmira, NY. Bur. in an unmarked grave Woodlawn Nat. Cem., Elmira, NY. Listed on prison records as belonging to 26th Battn. Va. Inf.

RIFE, STEWART S.: B. in Monroe Co., WV. Enl. in 22nd Va. Inf. at Camp McCausland, WV 6/20/62. Transf. to Co. F, 26th Battn. Va. Inf. and dischd. from that unit 12/27/62 for chronic rheumatism and debility. Age: 27. 5'7". Light complexion, blue eyes, dark hair. Farmer.

RIPLEY, MACK: Co. F. D. 9/5/1919 at age 78. PWR only.

ROBINSON, M. F.: Co. K. Alive in Rockbridge Co., Va. in 1927. Possibly William S. Robinson.

SHIRES, CHRISTOPHER G.: Co. F. D. 8/18/1918 at age 83. Bur. Oak Grove Cem., Gates. PWR only.

SIERS, ISAAC: Res. Nicholas Co., WV. On a list of civilian prisoners from Fayette and adjacent counties held by Confederate forces, March 1862. Claimed he was in Col. Tompkins' regt. and was wded. in a skirmish near Charleston, WV. Evidence at time suggested him to be a deserter and recommended he be turned over to military dept. *History of Summers County* lists an Isaac Siers who was "once a member of Co. G, 6th WV (Federal) Cav. but deserted and joined the rebel army," and says he was a postwar res. Summers Co., WV.

SMITHSON, WILLIAM BARTLEY: Pvt. Co. F. Transf. to Co. C, 18th Va. Cav.

SOOKLOVE, ________: Co. K. KIA at Winchester, Va. 9/19/64. Bur. Stonewall Jackson Cem., Winchester, Va.

SURBER, THOMAS OLIVER: Pvt., Co. H. B. 5/2/46. Enl. 3/31/64 or 4/1/64 at Lewisburg, Greenbrier Co., WV. Age: 18. On clothing list 5/13/64. At Staunton, Va. hosp. for measles 5/28/64-6/6/64. Left on account of Staunton, Va. being evacuated by C.S. troops. A short distance from town met Maj. Monroe who gave him a pass to stay in the county until fit for travel. Met the 22nd Va. coming from Cold Harbor, Va. and joined in. Fought at Lynchburg, Va. but still having measles was sent to hosp. at Harrisonburg, Va. Rejoined command at Fisher's Hill, Va. 10/18/64. Fought at Cedar Creek, Va. the next day where he was hit by an enemy rifle ball, breaking his gun and knocking him down as the bullet lodged in his haversack. Was on the march from near the TN state line to meet Gen. Lee when heard the news of Lee's surrender. Was near Salem, Va. when the regt. disbanded on 4/9/65, according to him. Postwar res. Craig Co., Va. Married Eliza Windle 12/28/69. Res. Craig Co., Va. at age 52. D. 9/12/1934 at age 88 at home of grand-daughter. Bur. West View Cem., Old Rt. 42, Newcastle, Va. He believed himself to be the last of Co. H, being only one of three surviving Craig Co., Va. veterans out of over 700 who served.

SUTER, MORTIMER W.: Sgt./Pvt. Co. B. Res. Brooke Co., Va. Enl. 6/1/61 at Ripley, Jackson Co., WV. Capt. at Lewisburg, WV 5/23/62. Age: 21. POW Camp Chase, CH 5 or 6/30/62. Exchd. 8/25/62. Det. at Richmond, Va. hosp. on 12/31/62 roll. Transf. on 2/28/63 roll. 5'5". Light complexion, black hair, blue eyes, brown whiskers.

SWANN, SAMUEL R.: Asst. Surgeon. Field and Staff. Appears on a list of field and staff officers but no entry card in CSR. B. 1835 in Powhatan Co., Va. Univ. of Va. Commissioned Asst. Surgeon, U.S. Navy. Three years duty off coast of Africa. Health impaired disabling him from active duty. Moved to Charleston, WV with his two brothers, John and Thomas Swann. At outbreak of war resgnd. commission from U.S. Navy. Apptd. surgeon for Wise Legion. D. March 1862 at Norfolk, Va.

SWEENEY, JAMES: Pvt. Co. H. WIA (arm and leg) at Scarcy Creek, WV 7/17/61. Later Maj. in 60th Va. Inf.

TAYLOR, JOHN R.: Lt. Swann's Co. On a list of officers in Swann's Co. for 1861 but not in CSR.

TOMLINSON, JOSEPH: Co. D. D. 11/3/64 at Elmira, NY. Bur. Woodlawn Nat. Cem., Elmira, NY grave #751.

TOMPKINS, CHRISTOPHER QUARLES JR.: Son of Col. Christopher Quarles Tompkins and Ellen Wilkins. Originally a 2nd Lt. in Co. K. Then apptd. Adjutant Spring 1864. Present at New Market, Cold Harbor, Lynchburg, Winchester, Fisher's Hill, and Cedar Creek.

UFFERMAN, CHARLES: Lt. Swann's Co. On a list of officers in Swann's Co. in 1861 but not in CSR.

WOLFE, HUGH CALDWELL: D. June 1908. There is a man by the same name bur. at Lancaster Baptist Church who was born 11/19/13 and died 7/2/1905. PWR only.

ADDITIONAL MATERIAL FOR NAMES ALREADY IN THE ROSTER:

ACORD, JOHN: Mustered 1st Corp. Thurmond's Co. 9/19/62 **(not** 9/12/62).

ALFORD, JOSEPH A.: B. 6/10/33. D. 7/27/1910. Bur. Alford Cem. on Alford Farm, 30 ft. from fence, N.E. corner. Upright marker, Putnam Co., WV. (Possibly same as Joseph A. Alfred who served in Co. G, 8th Va. Cav.)

ALLEN, WILLIAM H.: WIA at Totopotomoy Creek, Va. 1864. "Served until end of hostilities with undaunted resolution and courage. As an example of his gallant conduct as a soldier, his comrades related that at Cold Harbor, Va. in June 1864 a Federal shell with the fuse sputtering fire, fell into the entrenchments held by the 22nd Va. Men ran and dodged in every direction, but Mr. Allen sprang forward and hurled it over embankments where an instant later it exploded without danger to the Confederates." Postwar res. Monroe Co., WV. D. 12/24/1914 at age 77. Bur. Green Hill Cem., Union, WV.

AMICK, JAMES A.: B. 11/4/35. D. 1/3/1915. Bur. Amick Cem., N. on Rt. 60, 10 mi. from Rainelle, WV to 10 on 10 N. past Ravenseye, WV, almost to Rt. 41. Cem. on right or E. of 40. Fayette Co., WV. Upright marker to grave.

ANKRUM, WILLIAM R.: Deserted 5/12/64.

ARMISTEAD, JAMES A.: Full name is Alford Armistead. B. 1835. Son of James Miller Armistead. Brother of Josiah, John Washington, Matthew S., and Thomas Jefferson Armistead. Res. mouth of George's Creek on Elk River 1840. Owned 1200 acres land, Roane Co., WV 1846. Farmer before and after war. D. 7/1/1908. Bur. Armistead Cem., Emory Nichols farm, Roane Co., WV.

ARMISTEAD, J. H.: Believed to be Josiah ("Little Joe") Armistead. B. 1843. Son of James Miller Armistead. Brother of James Alford, John Washington, Matthew S., and Thomas Jefferson Armistead. D. at Camp Chase, OH. Family account says capt. by home guard and made to dig his own grave. Bur. Armistead Cem., Sarah Armistead Wright farm, Hurricane Creek Rd., Rt. 23, Clio, Roane Co., WV.

ARMISTEAD, JAMES M.: Full name is James Miller Armistead. B. 1814. Father of James Alford, Josiah, John Washington, Matthew S., and Thomas Jefferson Armistead. D. ca. July-Aug. 1900. Bur. Armistead Cem., Sarah Armistead Wright farm, Hurricane Creek Rd., Rt. 23, Clio, Roane Co., WV.

ARMISTEAD, JOSIAH: Same as J. H. Armistead.

ARMISTEAD, J. W.: Full name is John Washington Armistead. B. 1845. Son of James Miller Armistead. Brother of James Alford, Josiah, Matthew S., and Thomas Jefferson Armistead. D. during the war. Bur. Armistead Cem., Sarah Armistead Wright farm, Hurricane Creek Rd., Rt. 23, Clio, Roane Co., WV.

ARMISTEAD, M. S.: Full name is Matthew S. Armistead. B. 1840. Son of James Miller Armistead. Brother of James Alford, Josiah, John Washington, and Thomas Jefferson Armistead. D. during the war. Bur. Armistead Cem., Sarah Armistead Wright farm, Hurricane Creek Rd., Rt. 23, Clio, Roane Co., WV.

ARMISTEAD, THOMAS: Full name is Thomas Jefferson Armistead. B. 1847. Son of James Miller Armistead. Brother of James Alford, Josiah, John Washington, and Matthew S. Armistead. D. 4/10/1918. Bur. Armistead Cem., Sarah Armistead Wright farm, Hurricane Creek Rd., Rt. 23, Clio, Roane Co., WV.

ARMSTEAD, ALFRED A.: This is the same man as James Alford Armistead according to descendants.

ARMSTRONG, VIRGIL S.: B. 1836. D. 1917. Bur. Pine Hill Cem., Ripley, WV. No marker.

AYERS, JAMES MADISON: D. 7/31/1909 at age 66. Bur. Haynes graveyard near Packs Ferry, WV.

BAILEY, JAMES LAWRENCE: Brother of Maj. Robert Augustus Bailey.

BAILEY, THEODORE F.: Possibly same man as T. F. Bailey who enl. as a Pvt. in Co. H, 8th Va. Cav. 8/12/64 in Tazewell Co., Va. POW paroled at Charleston, WV 5/9/65 at age 26.

BAKER, GEORGE: Middle initial is W. not A.

BALDWIN, JOEL: Probably same man who enl. in Co. B, 36th Va. Inf. 8/1/62 in Monroe Co., WV. Present with that regt. until KIA at Strasburg, Va. 8/17/64. Bur. Mt. Jackson, Va.

BALDWIN, JOHN: Probably same man who enl. in Co. B, 36th Va. Inf. 8/1/62 in Monroe Co., WV. Res. Boone Co., WV. KIA at Fayetteville, WV 9/10/62.

BARLEY, ALEXANDER: Appears to be same as A. B. Barley who enl. as a Pvt. in Co. F (1st), 14th Va. Cav. 9/15/62 in Putnam Co., WV. Deserted from that regt. 9/23/62.

BARR, JOHN W.: D. 6/30/91 at age 60. Bur. Spruce Cem., 1½ mi. E. from Gunville, Mason Co., WV on Rt. 5 to Rt. 52, continue 1½ mile on Rt. 52 to cem., 100 ft. W. of rd. Upright marker.

BARRETT, ANDREW J.: Almost undoubtedly same man who enl. in Co. D, 36th Va. Inf. 5/27/61 at Peytona, WV. Res. Boone Co., WV. Promo. Sgt. Nov. 1861. Present through Dec. 1863 roll. Apparently transf. to 22nd Va. Inf.

BARTON, NORMAN P.: B. in New York City, NY. Hollywood Cem. records state d. 12/23/23.

BEAN, WILLIAM M.: B. 4/10/32 Dropping Lick Creek, Monroe Co., WV. Postwar farmer, Waitesville, Monroe Co., WV.

BECKETT, JAMES DANIEL: Enl. at age 15 and served last two years of war. "Always cherished his services to the Confederate army as one of the proudest memories of his life." Postwar attended school at Parkersburg, WV. Later attended Norwood Academy at Union, WV. Introduced the game of baseball at Union, WV by bringing the rules of the game from Parkersburg, WV. Spent remainder of life in Monroe Co., WV. D. 2/13/1933 at age 86. Bur. Green Hill Cem., Union, WV.

BECKETT, WILBUR F.: Native of Monroe Co., WV. After war moved to Roanoke, Va. D. 1/12/1929 at age 83. Bur. Evergreen Cem., Roanoke, Va.

BECKLEY, ISAAC C.: Enl. in Co. D, 36th Va. Inf. 5/27/62 in Logan Co., WV. Age: 20. Present until transf. to 23rd Battn. Va. Inf. 2/1/63.

BEIRNE, ANDREW: PWR says also served as Maj. D. 1/26/88 at Lewisburg, WV. Bur. Old Stone Church Cem., Lewisburg, WV.

BELCHER, WILLIAM H.: Probably same man as William Harrison Belcher who enl. in Co. B, 36th Va. Inf. 5/1/62 in Giles Co., Va. Present with that regt. until listed sick in Pearisburg, Va. hosp. on Aug. 1863 roll. Present with 36th Va. from Oct. 1862 until wded. in Pearisburg, Va. hosp. on Oct. 1864 roll.

BENSON, WILLIAM E.: D. 9/4/1916 at age 87 at Augusta, GA. Body brought to Lewisburg, WV and bur. at Old Stone Church Cem., Lewisburg, WV. PWR says spelled Bensen.

BESS, JOHN L.: Middle name is Lee and was b. March 1837.

BIBBY, ROBERT: B. 2/29/39 at Charleston, WV. D. 5/27/1926 of peritonitis at Charleston, WV. Bur. Spring Hill Cem., lot 5, section 8, Charleston, WV. Upright marker and CSA emblem.

BOITNOTT, DAVID WILLIAM: B. in Franklin Co., Va. Some postwar sources state he may be William David Boitnott and possibly spelled Boitnolt.

BOOTHE, JOHN T.: Farmer, age 21, Roxalana, Roane Co., WV 1860 census.

BOWDEN, J.: Probably John H. Bowden. After war returned to Monroe Co., WV. His wife d. prior to 1922 and he had four sons and three daughters. D. 6/1/1922 at age 78. Bur. Sinks Grove Baptist Cem., Sinks Grove, WV.

BOYD, ROBERT: PWR says MWIA at White Sulphur Springs, WV 8/26-27/63. "Bullet severed aciatic nerve and he died within the year." Brother of Madison and William Boyd.

BOYD, WILLIAM: WIA (shot through heel) at Cedar Creek, Va. 10/19/64. D. 2/23/1917 at age 78 at home. Cause of death was blood poisoning developing from his Cedar Creek wound. Bur. Mount Pleasant Cem., Sinks Grove, WV. Brother of Madison and Robert Boyd.

BRAUGHMAN, WILLIAM: Postwar source says name is William Lyle Broughman (pronounced Bruffman). D. 10/20 or 26/98. Widow filed for pension in Alleghany Co., Va. (Probably same man as William Bruffman on roster).

BRIDGETT, JACOB H.: Probably same man who furnished a substitute in Chapman's Battery 8/1/62.

BROWN, JAMES M.: Enl. 3/13/62 in Co. K, 14th Va. Cav. at Blue Sulphur Springs, Greenbrier Co., WV. Absent sick through 4/62. Present 5-6/62. Dischd. for disability 7/9/62. Enl. in Co. G, 22nd Va. Inf. 1863 for two years. D. prior to 6/1906.

BROWN, THOMAS: B. 1844 Boone Co., WV. D. 9/?/75 at Peytona, WV in a mine accident. Bur. Brown Cem., 3½ mi. S. of Gordon P.O., Rt. 26, Boone Co., WV. Upright marker.

BROWN, WILLIAM W.: Full name is William Washington Brown. Amputated limb refused to heal and gave him much trouble from time to time. Blood poisoning at last set in and he d. 6/2/97 from the effects, at age 67 at his home on the Knobs, Wolf Creek, WV area. Bur. Leach graveyard.

BRUFFMAN, WILLIAM: Apparently same as William Braughman.

BRYAN, THOMAS A.: Full name is Thomas Andrew Bryan. B. 1844 Rappahannock Co., Va. Moved to Baltimore, MD at age 12. Remained with Bryan's Battery throughout war after transferring from 22nd Va. Inf. Postwar farmer and supplied material for the railroad. Moved to Baltimore, MD and operated a candy business and later a stove-making business. D. 12/30/1909 at Baltimore, MD. Bur. Woodbine Cem., Harrisonburg, Va. His uncle was Lt. Col. Barbee of the 22nd Va. Inf.

BURKE, W. H.: Full name is William H. Burke. D. 12/29/80 from brain damage suffered in an accident.

BURNS, D. H. M.: May have also served in McClanahan's Battery.

CABELL, HENRY CLAY: Possibly same Henry Clay Cabell who was B. 9/22/40 and D. 7/8/1909. Bur. Old Stone Church Cem., Lewisburg, WV.

CAMPBELL, JAMES: Transf. to cav. Never married. Dentist. Elected to clerk of Monroe Co., WV Court. D. 11/17/99 at age 68 at home. Bur. Mount Pleasant Cem., Sinks Grove, WV.

CAMPBELL, JOHN H.: B. at Pickaway, WV. In 1880 moved to McClede Co., Missouri. D. 7/22/1922. Bur. from White Oak Pond Church near late home.

CAMPBELL, L. M.: Full name is Lafayette M. Campbell. B. ca. 1834. Laborer, age 26, 1st Dist., Augusta Co., Va. 1860 census. Hired as a substitute for Samuel P. Campbell 4/1/62 to serve as a Pvt. in Co. H, 14th Va. Cav. Transf. to 22nd Va. Inf. 7/18/62.

CARR, CLAUDIUS: B. 11/13/46. Also served in Co. E, 60th Va. Inf. D. 3/6/1921 in Clay Co., WV of hardening of the arteries. Bur. Spring Hill Cem., lot 15, section B, Charleston, WV.

CARR, SAMUEL BILL: Middle name should be spelled Bell. Res. Alleghany Co., Va. 1905.

CARTER, ARCHIBALD: "A good soldier and cherished with pride his service in the Confederate Army." D. 1/19/1916 at age 80 at home. Bur. Sinks Grove Baptist Cem., Sinks Grove, WV.

CARTMILL, JAMES H.: Full name is James Henry Cartmill, Jr. Son of James Henry Cartmill and Lucinda Stephens. Brother of John Gilbert Cartmill. B. 1/29/39. D. May 1923.

CARTMILL, JOHN G.: Full name is John Gilbert Cartmill. B. 2/7/44 in Putnam Co., WV. Son of James Henry Cartmill and Lucinda Stephens. Brother of James Henry Cartmill, Jr. "Fought at Scary Creek with John K. Thompson, Tom Harmon, and Frank Counts." Married 2/17/67 to Virginia Helen McGuire. Member Missionary Baptist Church. D. 6/25/1927 (Saturday 3:00 P.M.) at age 83 yrs., 4 mos., 18 days. Bur. Spring Hill Cem., Huntington, WV.

CHANDLER, WILLIAM S.: Later served as Capt., Co. K (Capt. William S. Chandler's Co.), 2nd Regt., Va. State Line (1862-63).

CHURCH, ANDREW: Middle initial is T. B. ca. 1839. Son of Woodard Church and Mary Gabbart of Lewisburg, WV. Blacksmith, age 21, Lewisburg P.O., Greenbrier Co., WV 1860 census. Enl. as a Pvt. in Co. A (1st), 14th Va. Cav. at Lewisburg, WV 5/23/61. Disbanded 11/22/61. Enl. in Co. K, 14th Va. Cav. 1/22/62 at Lewisburg, WV. Det. service 4-8/62. Present 9/62-2/63. Absent sick 3-4/63. Ab. on detail with enrolling officer at Lewisburg, WV 7-9/63. "Froze to death on horseback on the mountain between White Sulphur Springs, WV and Lewisburg, WV one winter day while going to Lewisburg, WV to visit a sister."

CLARKSON, ALEXANDER Q.: Probably same Alexander Clarkson who served as a Pvt. in Co. I, 8th Va. Cav. Captured in Mason Co., WV 3/30/63. Age: 18. POW Camp Chase, Oh. Exchd. Res. Kanawha Co., WV

COLE, CHARLES M.: B. 9/16/25. D. 6/30/97. Bur. Marvel Cem., Marvel, Fayette Co., WV Four mi. N. of Ansted, WV. Upright marker.

COLE, G. W.: Full name is George W. Cole. B. 10/13/25. D. 10/16/1914. Bur. Marvel Cem., Marvel, Fayette Co., WV. Four miles N. of Ansted, WV. Upright marker.

COLE, WILLIAM H.: Bur. Stonewall Cem., Winchester, Va.

COOPER, JAMES: Also served in Co. H, 60th Va. Inf. D. 9/29/1907. Bur. Spring Hill Cem., Confederate lot, Charleston, WV. Upright marker.

CORLEY, WILLIAM M.: Bur. Mt. Rupert Cem., 3 mi. E. of Reger, Upshur Co., WV on Rt. 5. Cem. 40 ft. N. of rd. No marker.

COUNTS, SAMUEL FRANKLIN: D. 6/3/1903 at Scott Depot, WV.

CRAWFORD, J. D.: Bur. Confederate Cem., Univ. of Va., Charlottesville, Va.

CROW, WILLIAM: Bur. Pine Hill Cem., Ripley, WV. No marker.

DAME, ROBERT C.: Possibly same Robert C. Dame who was B. 1847 and D. 1917. Bur. Low Moor Cem., Alleghany Co., Va.

DAMRON, HENRY C.: Possibly same as H. C. Damron who served as a Pvt. in Co. H, 8th Va. Cav. and AWOL from that regt. July 1862.

DAVIS, HIRAM G.: Possibly same man who enl. in Co. E, 36th Va. Inf. 5/25/62 in Giles Co., Va. Present with that regt. until AWOL on Oct. 1862 roll. Again present on Dec. 1862 roll until trnasf. to 23rd Battn. Va. Inf. 3/13/63.

DAVIS, J. H.: Possibly same J. H. Davis who served as a Pvt. in Co. B, 14th Va. Cav. but does not appear on muster rolls of that regt. WIA at Spotsylvania C.H., Va. 1864. Receiving a pension in Nelson Co., Va. 1902.

DEWS, SAMUEL STONE: D. 1919. Bur. Dews Cem., near Ansted, WV.

DILWORTH, JOSEPH W.: Bur. Independence Cem., W. of Sandyville, Cabell Co., WV on rd. to Ravenswood, WV. Take gravel rd. to church. No marker.

DUNN, JOHN H.: B. 1/28/38 in Giles Co., Va. Brother of William A. Dunn. Married 5/12/60. Postwar farmer, Brush Creek, Red Sulphur District, Monroe Co., WV.

DUNN, RUFUS K.: Enl. in Bryan's Battery 4/13/63 at Thorne Spring, Va. Present with that unit through final roll. Paroled at Charleston, WV. 5/25/65. Age: 28. 6'. Gray eyes, sandy hair, sandy whiskers.

DUNN, W. L.: Full name is Wesley L. Dunn. D. 7/25/1909 at age 79 at home. Bur. Peterstown, WV.

DYE, CORNELIUS H.: Middle name is believed to be Harrison.

EAGLE, JOHN: D. 2/6/1924 at Ronceverte, WV. Bur. Bethel Burying Ground.

EDMONDS, SAMUEL D.: B. Halifax Co., Va. D. 1/4/1926 at Old Soldiers Home, Richmond, Va. Bur. Hollywood Cem., Richmond, Va.

ELKINS, DANIEL: Full name is Daniel P. Elkins. B. 2/9/(?)3. Veteran Grave Registration Card says served as a Capt. in 36th Va. Inf. but does not appear on records of that regt. Does appear as Capt. of Co. B, 1st Va. State Line 1862. D. 4/15/1906. Bur. Elkins Cem., Costa, Boone Co., WV. Upright marker.

EVANS, EDWARD R.: D. 4/3/1909 in Allegheny Co., VA. May have also served in 60th Va. Inf.

FINK, RICHARD P.: B. 4/22/23 in Monroe Co., WV. Son of Jasper Fink and Catherine Miller. Res. of Greenbrier Co., WV at age 26. Married Rebecca F. McCoy 8/26/50. Postwar res. Ronceverte, WV.

FISHER, HENRY: Shot for desertion 8/22/64 near Winchester, Va.

FLESHMAN, THOMAS J.: Son of Simeon Fleshman and Sarah Thompson. Married Leah H. Lewis 12/14/65. Postwar overseer of roads and school trustee of sub. dist. 17, Greenbrier Co., WV. Stonemason and farmer raising grain. Res. Blue Sulphur Springs, WV. Card says name is Thomas H. Fleshman.

FLOWERS, J. K.: WIA (right thumb, splintering the bone — ball lodged in palm of hand severing the leaders and 3 fingers broken) 5/23/64. Carpenter, Roanoke, Va 9/14/94 at age 48 on pension application. Still alive 1904.

FORREN, JOHN C.: PWR says enl. at age 16. B. 3/14/45 Monroe Co., WV. Postwar farmer and fisherman. Married twice. D. 11/15/1922 at age 77 at his son's home in Fayette Co., WV. Bur. Longanacre Cem., Fort Spring, WV.

FOUTS, W. M.: PWR says name is William W. Fouts. Enl. Apr. 1862 for three years. Res. Bedford Co., Va. 1899. Age: 65.

FOX, ISAAC L.: Probably same Isaac L. Fox who was in Bryan's Battery and was sick at home Dec. 1862. Recvd. clothing Mar. 1863. Then apparently joined 22nd Va. Inf.

FRIDLEY, SAMPSON: B. in Allegheny Co., Va. Widow received a pension in Allegheny Co., Va. and said he was killed by someone in Staunton, Va. in Nov. 1917.

FRIDLEY, WILLIAM H.: Typographical error — Should be Allegheny Co., Va. — not WV.

FRIDLEY, JOHN: Postwar source says correct spelling is Fridley.

GADD, ANDREW P.: B. 9/25/28 in Franklin Co., Va. PWR says originally in 23rd Battn. Va. Inf. and transf. to 22nd Va. Inf. in exchange for Isaac French McKinney. PWR says served as 2nd Lt. at New Market, Fredericksburg (?), and Cedar Creek (not likely as he was capt. at Winchester 9/19/64). D. 3/30/1909 at Flat Top, WV. Bur. Gadd Cem., on old Gadd farm, 0.9 mi. on private rd. off Rt. 21, near Flat Top, WV. Upright marker.

GAYLOR, JEREMIAH: Nicknamed "Jerry". Bur. Gaylor and Asbury Cem., 2 mi. E. of Rock Branch school, ¼ mi. N. of Rock Branch Creek on hill. Putnam Co., WV. No marker.

GILLISPIE, GEORGE: Deserted 5/12/64 at or near Staunton, Va.

GLOVER, JOHN H.: Possibly same John H. Glover who enl. in Bryan's Battery 3/27/62 at Lewisburg, WV. Present with that unit until deserted 7/24/64. Sick at home on Aug. 1864 roll. Paroled at Charleston, WV 12/10/64. Age: 32. 5'4". Gray eyes and dark hair. Farmer.

GRAY, JAMES: Middle name is Medley.

GREGORY, HENRY: B. 4/29/39.

GREGORY, JOHN: Middle initial is W. Also served in Co. E, 60th Va. Inf. D. 5/18/1908. Bur. Spring Hill Cem., Confederate lot, Charleston, WV.

GREYBILL, A.: Possibly same man as Abram Graybill who served in Co. E, 36th Va. Inf. and deserted from that regt. 6/22/64 near Salem, Va.

GUNN, WILLIAM R.: Possibly William Richmond Gunn who served as Capt. of Co. D(2nd), 8th Va. Cav. B. Ohio Co., Va. 1832. Apptd. 9/18/62. Capt. at Union Co., TN 12/6/63. Escaped from cars en route from Louisville, KY to Camp Chase, OH 1/18/64. Under Gen. Early's orders to bring in absentees 2/18/65 at New Bridge Church. Founder of Gunn's Rangers. Postwar WV Legislature. D. 5/19/1908. Bur. Lone Oak Cem., Pt. Pleasant, WV.

GWINN, OLIVER W.: D. 1906 at age 73. Bur. Riverview Presbyterian Cem., Alderson-Creamery Rd. Brother of William A. Gwinn.

GWINN, WILLIAM A.: PWR says D. 2/2/1915 at age 79. Bur. Riverview Presbyterian Cem., Alderson-Creamery Rd. Brother of Oliver W. Gwinn. "Served with courage and fidelity."

HALL, LEDFORD ALLEN: B. 6/27/45. Age: 20 at enl. and res. Allegheny Co., Va. Bur. Crown Hill Cem., Clifton Forge, Va.

HANSFORD, FELIX G.: Middle name is Gilbert.

HARFORD, HARVEY: D. at Lynchburg, Va. during the war. Bur. Old City Cem., Lynchburg, Va.

HARLOW, HENRY JACKSON: Typographical error — Louis Co. should be Louisa Co.

HAWKINS, CHARLES M.: PWR says Co. F. Postwar editor *Danville Daily Register*. Mercantile business, Union, WV 1874. D. 6/30/1906 at age 60.

HELMS, MASON V.: Typographical error — year of death should be 1863 — not 1862.

HENDRICK, OBEDIAH: B. 1830. D. 1890. Bur. Hendrick Cem., 2 mi. W. of Ansted, WV on Rt. 60. Cem. 75 yds. to right of hill. Family marker.

HENDRICK, T. P.: PWR says first name is Taulbert. Bur. Hendrick Cem., 2 mi. W. of Ansted, WV. Cem. 75 yds. to right of hill. No marker. Bur. to right of Laura L. Forlaine.

HENLY, CHARLES: Middle initial is W. on postwar record.

HEPPENSTALL, SAMUEL: Middle initial is C. Bur. Old City Cem., Lynchburg, Va.

HINTON, PETER: B. near Centreville (Greenville), Monroe Co., WV. D. 4/27/97. Bur. near Broad Run Church. "A faithful soldier."

HOWARD, ROBERT: D. 2/24/1926 at age 81 at his home at Hollywood, WV. Bur. in a nearby cem.

HUDDLESTON, JACKSON BURWELL: D. 12/25/1916 at Fayetteville, WV. Bur. Poteet Cem., 1.7 mi. S. of Oak Hill, WV E. of rd. on Rts. 19-21 at Hill Top, WV. Upright marker. CSA emblem.

HUDDLESTON, JOB: Res. Fayette Co., WV 1860, day laborer, age 22. Bur. Odd Fellows Cem., ¼ mi. W. of Rts. 19-21, Oak Hill, WV. Marker could not be located.

HULL, JOHN McKEE: Served first in 5th Va. Inf.

HULL, WILLIAM CRAWFORD: Taylor's Co. B. 10/5/30. Son of Jesse Hull of Lowry's Mills, WV. Brother of John McKee Hull. Postwar res. Greenbrier Co., WV.

HUMPHREYS, C. L.: B. 11/17/31. D. 4/29/1917. Bur. Humphries familly cem., Rt. 616, Allegheny Co., Va. Spelled Humphries on grave.

HUMPHREYS, CHARLES K.: Middle name is King. PWR spells Humphries.

HYPES, ASBURY: Bur. Hypes Cem., 0.7 mi. from Littlesburg, Mercer Co., WV on Rt. 23 to cem. 300 ft. E. of rd. Upright marker family stone.

HYPES, JACOB L.: Middle name is Lewis. B. 10/21/34 in Craig Co., Va. D. 12/28/1922 at Bluefield, WV of pneumonia. Bur. Hypes Cem., 0.4 mi. S. of Princeton, WV.

INGLES, JAMES H.: Bur. Confederate Cem., Univ. of Va., Charlottesville, Va.

INGLES, JOHN P.: "Served bravely and faithful until end of war." Crippled after the war by a gunshot wd. D. 5/10/1925 at age 81 at home. Bur. Sinks Grove Baptist Cem., Sinks Grove, WV.

INGRAHAM, GRANVILLE P.: Transf. to 60th Va. Inf. 4/1/62.

JACKSON, C. L.: Transf. to Capt. Thomas E. Jackson's Artillery 2/4/64. Capt. with that unit at Fisher's Hill, Va. 9/2 or 23/64. POW Pt. Lookout, MD 9/30/64. Exchd. 3/17/65. Paroled at Charleston, WV 5/11/65. Age: 19. 6'. Dark complexion, gray eyes, black hair.

JOHNSON, HENRY: (of Co. K). Possibly same Henry Johnson who was B. 1840 and D. 1/7/95. Bur. Chestnut Grove Cem., 4 mi. S. of Fayetteville, WV on Rt. 19-21. Upright marker.

JOHNSON, ISRAEL: Bur. Johnson family cem., #1-1 mi. on old Summerlee Rd. S.E. to Oak Hill, WV. No marker to grave.

JOHNSON, JOHN: WIA (shot in breast) at New Market, Va. 5/15/64. PWR says d. from wd.

JOHNSON, JOHN M.: Printing error — delete information about being wounded at New Market and d. from effect. This information applies to John Johnson.

KINCAID, WILLIAM R.: PWR says middle initial is H. D. 6/8/1922.

KNIGHTSTEP, WILLIAM C.: Probably same man who enl. in Bryan's Battery 6/15/62 at Salt Sulphur Springs, WV. Det. Aug. 1863. Present with that unit Oct. 1863 until wded. on Aug. 1864 roll. Paroled at Lewisburg, WV 4/26/65. Age: 24. 5'9". Hazel eyes and dark hair. Res. Jackson Co., WV 1916.

LAFAN, JOHN C.: Possibly same man as John Cain Lafan who was B. 1/26/44. Enl. in Co. H, 36th Va. Inf. 6/22/62 in Giles Co., Va. Present with that regt. until AWOL on Oct. 1862 roll. Present again on Dec. 1863 roll until deserted 5/27/64. D. 7/11/1906.

LEACH, EDMUND K.: From Wolf Creek, WV area. D. 4/9/1902 at age 65. Bur. Mount Pleasant Cem., Sinks Grove, WV.

LEACH, JAMES HARVEY: "Took part in 31 battles, proudest memory of his life." Postwar res. Keenan, WV. D. 11/1/1924 at age 82. Bur. Oak Grove Church Cem., Gates.

LEE, LEANDER S.: Possibly same L. S. Lee who was B. 1841 and D. 1909. Bur. Teays Hill Cem., St. Albans, WV. Upright marker to grave. CSA emblem.

LEMONS, JOHN ALEXANDER: PWR says D. 10/3/1927 at age 90. "Cherished memories of his war years as the proudest part of his life."

LEMONS, JOHN ANDERSON: Bur. Old City Cem., Lynchburg, Va.

LEMONS, WILLIAM HOUSTON: D. at home of his daughter, Mrs. Porter Jarvis, at Sinks Grove, WV.

LEWIS, CALEB V.: B. 1/16/44 in Va. Son of George and Susan Lewis. Brother of George Lewis. Student, age 13, 2nd Dist., Greenbrier Co., WV 1860 census. Records of 14th Va. Cav. say he enl. in Co. K of that regt. 10/11/63 at Lewisburg, WV and say his middle initial is G. Transf. to Co. A, 14th Va. Cav. date unknown. Records of 14th Va. Cav. say he was KIA 6/26/64 in Greenbrier Co., WV while taking the weapons from Union stragglers that had surrendered, was shot by one of them. PWR says he was assinated by U.S. stragglers while another PWR says he was killed 6/27/04 while "boldly pursuing Hunter on his retreat from Lynchburg." Bur. Muddy Creek Presbyterian Church Cem., near Blaker's Mills, Greenbrier Co., WV.

LEWIS, GEORGE: Middle initial is W. B. 7/11/184(?) or 1836 in Greenbrier Co., WV. Son of George and Susan Lewis. Brother of Caleb V. (or G.) Lewis. Transf. to Co. K, 14th Va. Cav. although not on muster rolls of that regt. KIA at Droop Mountain, WV 11/6/63. Bur. Muddy Creek Presbyterian Church Cem., near Blaker's Mills, Greenbrier Co., WV.

LEWIS, JOEL S.: Enl. as a Pvt. in Co. I, 8th Va. Cav. 9/1/63 at Abbs Valley, Tazewell Co., Va. Capt. at Lynchburg, Va. 6/64. POW Camp Chase, OH 7/2/64. Paroled at Charleston, WV 7/5/64. Age: 34. Res. Kanawha Co., WV.

LIPPS, ARCH: Also known as Archibald Lipps. Res. Alderson, WV. Photo appears in Lewisburg, WV Civil War monument book. D. 2/1/1908 at age 71.

LIPSCOMB, WILLIAM H.: Appears on a roster of 36th Va. Inf. as having enl. 8/8/61 as Capt. of his own co. and transf. out of regt. shortly thereafter.

LIVELY, CHRISTOPHER: B. 5/5/29 at Lindside, Monroe Co., WV. Son of Joseph Lively and Mary Pack. Married Matilda Margaret Thrasher 12/18/55. D. 4/8/1907 at Lindside, WV. Bur. Thrasher-Lively Cem. near Lindside.

LIVESAY, GEORGE W.: Middle name is Washington. B. 12/16/24 in Greenbrier Co., WV. Appears on a postwar roster as a Pvt. in Co. K, 14th Va. Cav.

LOONEY, ALEXANDER C.: WIA (badly) at Droop Mountain, WV 11/6/63. Res. Sweet Springs Valley, Monroe Co., WV. D. 4/30/1917. Bur. Maple Grove Cem., Gates.

LOVEJOY, MILTON A.: B. 1839. D. 2/28/1926 of old age at Liberty, WV. Bur. Coleman farm, ½ mi. W. of Sanhill School, Putnam Co., WV. No. marker to grave.

LUCAS, EDWARD: Middle initial is H. or M. Bur. Old City Cem., Lynchburg, Va.

LYPES, ALLEN LEFFEL: Bur. Old City Cem., Lynchburg, Va. Listed as Lipes.

MARTIN, JOSEPH: B. 1/10/38 at Sandstone, WV. D. 1/3/1923 at Red Sulpher Springs Mountain, WV of old age. Bur. Halidon Cem., near Green Sulphur Springs, WV.

MAYES, BENJAMIN F.: WIA at Cold Harbor, Va. 6/3/64.

McCLOUD, AUGUSTUS: Probably same Augustus McCloud who enl. in Beckett's Co., 36th Va. Inf. 5/16/61 at Hurricane, Putnam Co., WV. Probably then transf. to Swann's Co., 22nd Va. Inf. 5/22/61 at Charleston, WV.

McCLUNG, MICHAEL B.: Probably Michael Bright McClung who was B. 1/1/39 in Greenbrier Co., WV. Enl. as a Pvt. in Co. K, 14th Va. Cav. at Lewisburg, WV 1/22/62. Present 3-10/62. Ab. on leave 11-12/62. Present 1-4/63. Presence or ab. not stated 4/1/64. Ab. on det. duty to get a horse 9-10/64. Present 11-12/64. Paroled at Charleston, WV 5/10/65. 5'8". Fair complexion, blue eyes, dark hair, dark whiskers. Gd. Keokuk College, Iowa M.D. D. Burden, KS prior to 1903.

McCLUNG, W. A.: Served as a Pvt.

McCUTCHEON, GRANVILLE S.: B. 2/28/18 in Greenbrier Co., WV. Also served as 3rd Sgt., Crawford's Co., 14th Va. Cav. Also in Swann's Co., 3rd Va. State Line. Capt. 1/62. POW Wheeling, WV 1/18/62. 5'9½". Fair complexion, blue eyes, sandy hair. Apparently same man on roster as G. S. McCutcheon.

McDERMOTT, ADDISON: D. 7/6/64. Bur. Old City Cem., Lynchburg, Va.

McKINEY, ISAAC FRENCH: Transf. to Co. B, 23rd Battn. Va. Inf. in exchange for Andrew P. Gadd. Was a musician with 23rd Battn. and drew one month pay as Sgt. Present on July-Aug. roll with that unit and present as musician on 10/31/63, 12/31/63, and 4/1/64 rolls. Capt. at Winchester, Va. 9/19/64. POW Pt. Lookout, MD 9/23/64. Exchd. 3/15/65. Paroled at Charleston, WV 6/12/65. Age: 25. 5'11". Light complexion, light hair, blue eyes.

McNEILL, JAMES M.: Middle name is Monroe. B. 5/9/23 at Buckeye, WV. Son of William and Nancy Griffey McNeill. Musically inclined and many of his songs were sung around the campfires of the 22nd Va. Inf. After his release as a POW took train to Staunton, Va. then walked home to Nicholas Co., WV. Later moved back to Buckeye, WV. Married for second time 1873. D. 3/26/1911.

McNUTT, JAMES R.: Bur. at Stonewall Cem., Winchester, Va.

McQUEEN, A.: First name is Archibald.

MERIDETH, L. D.: Possibly same L. D. Merideth listed on a postwar roster as having served in Bryan's Battery.

MILLER, JAMES O.: From Sweet Springs, WV. Postwar farmer and Justice of Peace, Sweet Springs, WV. D. 3/17/1909. Bur. Carmel Presbyterian Cem., Gap Mills, WV.

NEIL, WILLIAM H. H.: D. 1/21/1929 at Lockwood, WV at age 88.

NELSON, JAMES M.: Middle name is Marion. D. 1/8/1923 at age 81 at his home at Alderson, WV. Bur. Riverview Presbyterian Cem., Alderson-Creamery Rd. "A faithful soldier."

NEWHOUSE, SIMON: Bur. Newhouse Cem., 2½ mi. N.E. of Guthrie, Kanawha Co., WV on Rt. 27. No marker.

NICKELL, ROBERT M.: B. ca. 1845 in Va. Farmhand, age 15, Peterstown, Monroe Co., WV 1860 census. Enl. in Co. A (2nd), 14th Va. Cav. as a Pvt. 7/12/62 at Union, WV. Present with that regt. 9-10/62. Ab. on scout in Botetourt Co., Va. after deserters 2/28/63. Present 3-4/63. Ab. on leave 3/22-4/1/64. Promo. 2nd Sgt. 4/1/64. Present through 10/31/64. Promo. 1st Sgt. 10/31/64. Ab. sick 11-12/64. Deserted to the enemy at Charleston, WV 4/4/65. Age: 20. 5'6". Fair complexion, hazel eyes, fair hair. Farmer, Monroe Co., WV. PWR says "fought at Chancellorsville, Antietam, Gettysburg, and Richmond."

NICOL, GEORGE T.: Probably same man as George T. Nickell who was B. 12/13/41 in Monroe Co., WV. Farmer, Union, Monroe Co., WV 1860 census. Enl. in Bryan's Battery 3/27/62 at Lewisburg, WV. AWOL 7/1/63. Enl. as a Pvt. in Co. E, 14th Va. Cav. 10/9/63 at Lewisburg, WV. Present or ab. not stated 4/1/64. Elected 3rd Lt. 5/27/64. Present 9-10/64. Cmdg. co. 11-12/64. Paroled at Charleston, WV 6/12/65. Age: 22. 5'10". Fair complexion, blue eyes, sandy hair, sandy whiskers. Lumberman, Gilmer Co., WV 1868-70. Justice of the Peace, Union, WV 1872-75. Deputy U.S. Marshall 1879. Farmer, Union, WV. Justice of the Peace, U.S. Marshall 1884.

NOUNNAN, JAMES H.: Records of 8th Va. Cav. say enl. as a Pvt. in 8th Va. Cav. 5/8/61 at Buffalo, Putnam Co., WV. Transf. from Co. H, 22nd Va. Inf. to 8th Va. Cav. 8/4/61 (Contradictory information). D. at Lee Camp Old Soldiers Home, Richmond, Va. 10/1/1900 at age 66.

NUTTER, T. A.: Possibly Theodore Nutter who served in Co. E, 62nd Va. Inf. although service record fails to correspond. If this is Theodore Nutter he D. at Chestnut Ridge near Phillipi, WV. Bur. Elk City (Hall) Cem., just W. of Elk City, Barbour Co., WV.

OGBURN, MOSES M.: Probably same as M. M. Ogburn who enl. as a Pvt. in Co. I, 8th Va. Cav. 4/? in Mercer Co., WV. Re-enl. 1/1/64. Absent foraging horses 4/63. Returned from desertion 12/31/63. Paroled at Charleston, WV 5/10/65. Age: 33.

OILER, GEORGE W.: REs. Lewisburg, WV. Conscript in Co. A, 26th Battn. Va. Inf. at Monroe Draft, WV - n.d. Capt. at Cold Harbor, Va. 5/31/64. POW Pt. Lookout, MD 6/8/64. Transf. to Elmira, NY 7/9/64. Took oath at Elmira, NY and relsd. 7/11/65. 5'7". Dark complexion, gray eyes, dark hair.

OTT, JEWIT: Probably same man as Jewitt G. Ott who enl. in Beckett's Co., 36th Va. Inf. 5/16/61 at Hurricane, Putnam Co., WV. Age: 19.

PATTON, GEORGE S.: Actuals bur. Stonewall Cem. [not Mt. Hebron], Winchester, Va.

PAXTON, CEPHAS S.: Son of Abner and Hester Elmore Paxton. Another PWR says D. 1/1/1913. Bur. Witcher's Creek Cem., 1 mi. up Witcher's Creek Rd. from Rt. 60. Upright marker.

PAYNTER, HIRAM HILL: Possibly Hill Painter who is bur. in Poteet Cem., Oak Hill, WV. No marker. CSA emblem. PWR says WIA at Charleston, WV 1862.

PERRY, WILLIAM GEORGE: After serving his one year service in 22nd Va. Inf. he enl in Co. A, 36th Battn. Va. Cav. 8/7/62 at Union, WV. Sick on June 1863 roll of that unit. Then present until capt. at Fisher's Hill, Va. 9/23/64. Age: 26. POW Pt. Lookout, MD. Paroled 5/11/65. 5'11". Hazel eyes and dark hair. D. 4/15/1913 at age 74 at his home at Fort Spring, WV.

PRICE, JAMES POLK: D. 10/25/91.

PURSLEY, WILLIAM E.: B. 1842. D. 1926. Bur. Short Cem., 4 mi. E. of Spencer, WV on Rt. 119, then ¼ mi. S. of Rt. 119 on Divide between Slate and Boggs Fork Creeks. Upright marker.

QUARRIER, MONROE: Another postwar source says D. at age 82.

RADER, A. S.: First name is Adam.

RAY, CHESLEY: Middle initial is A.

RIFFLE, MARTIN V. B.: B. ca. 1834 in Va. Farmer, age 26, Clay Co., WV 1860 census. Enl. as a Pvt. in Co. I, 14th Va. Cav. at Little Birch, WV 10/4/62. Capt. in Greenbrier Co., WV 11/26/62. POW Camp Chase, CH and Alton, IL. Exchd. 4/8/63. Transf. to Co. I, 17th Va. Cav. 1/28/63. Transf. to 22nd Va. Inf. 10/1/63. D. before 1919.

RIPPETOE, LAFAYETTE: Bur. Red Sulphur Springs, WV.

ROBERTS, THOMAS ALGERNON: Married 1865. Res. Salem, Va. 1886.

ROBINSON, JOHN R.: WIA (left leg) at New Market, Va. 5/15/64. Res. Reba, Bedford Co., Va. 4/3/88, age 55 on pension application. Spelled Robertson.

ROLLYSON, JOHN: Two burial sites indicated — Mt. Jackson, Va. cem. and Soldiers Cem., New Market, Va. Also spelled Rollson.

RORER, PETER F.: B. 10/27/37 in Patrick Co., Va. "First battle was at Gauley Bridge, WV where he was capt. and taken to Richmond, Va. where he joined Lee's army at battle of Chancellorsville and continued through entire war until final conflict at Gettysburg." D. 3/28/1937 at Princeton, WV of flu. Bur. Bolin Cem., ½ mi. W. from Athens, WV. No marker. Spelled Rorrer on Veterans Grave Registration card.

ROSS, JOHN T.: B. 11/11/40. Enl. 5/2/62 in Capt. W. D. Thurmond's Partisan Rangers and elected 3rd Lt. D. 6/15/96.

RUFFNER, MEREDITH PAYSON: PWR says B. at Red House, Putnam Co., WV. Also says transf. 8/?/63 to Swann's Battn. Six months as a N.C.O. then disability discharge. Says middle name is Peyton.

RULE, ANTHONY L.: Another postwar source says D. 1/19/1912.

SAMMONS, CHARLES B.: Middle name is Baxter. Enl. 6/3/61 in Co. F, 135th Va. Militia. Capt. cmdg. co. on 6/3/61 to 10/31/61 roll.

SAMS, A. A.: Possibly Andrew A. Sams who enl. in Chapman's Battery 8/12/62 at Narrows, Va. Present through final roll with that unit. Age: 20. 5'11". Blue eyes and brown hair.

SARVER, GEORGE W.: Probably George W. Sarver who enl. in Co. H, 36th Va. Inf. 10/21/62 in Giles Co., Va. Present with that regt. until June 1863 then NFR.

SARVER, J. G.: Possibly Jeremiah G. Sarver who enl. as a Pvt. in Co. H, 36th Va. Inf. 4/18/62 in Giles Co., Va. Present with that regt. until capt. at Winchester, Va. 9/19/64. POW Pt. Lookout, MD. Exchd. 3/15/65. Paroled 6/23/65. Age: 35. 5'11". Blue eyes and black hair.

SAWYERS, WELLINGTON: Pvt. Enl. at Lewisburg, Greenbrier Co., WV. Res. Allegheny Co., Va. Age: 30. Surrendered Apr. 1865.

SELDOMRIDGE, PETER F.: B. 10/22/39 in Greenbrier Co., WV. D. 3/12/1919 at Forest Hill, WV of old age. Bur. Forest Hill Cem., 100 ft. E. of Forest Hill, WV on Rt. 21. Upright marker. PWR says also served in Capt. Black's Co. (?).

SHANNON, NEWSON: D. 12/4/62. Bur. Red Sulphur Springs, WV. Also listed as Newton Shannon.

SHEILDS, WILLIAM H.: Possibly same man as William Hamilton Shields who enl. in Bryan's Battery 2/9/64 at Lewisburg, WV. Present with that unit through final roll. Paroled at Charleston, WV 7/19/65. Age: 30. 6'. Brown eyes, dark hair, dark whiskers. D. 1902.

SHIRES, J. T.: B. in Monroe Co., WV. Son of Richard and Millie Neil Shires. Married 6/3/51 to Aurelia R. Lynch. PWR says enl. in Co. F, 22nd Va. Inf. in 1864. Postwar mercantile business. Postwar res. Ronceverte, WV.

SHUMATE, RUFUS A.: Enl. in Bryan's Battery 4/1/63 at Throne Spring, Va. On leave Aug. 1863. Present Oct. 1863 through final roll. Paroled at Charleston, WV 6/12/65. Age: 27. 5'11". Blue eyes and dark hair. Res. Monroe Co., WV 1916. B. 3/12/38. D. 11/24/1920. Bur. Shumate Cem., 2.3 mi. S. from Oakvale on Rt. 219/6. Cem. 200 ft. E. of rd. Upright marker.

SIZEMORE, LEVI: Middle initial is J.

SKAGGS, HENRY H.: B. 1/17/39. D. 2/24/1914. Bur. Marvel Cem., 4 mi. N. of Ansted, WV. Upright double marker to grave.

SKAGGS, HIRAM: D. 12/?/1901. Bur. Marvel Cem., 4 mi. N. of Ansted, WV. Upright double marker to grave.

SKAGGS, THOMAS A.: Middle name is Anderson. Bur. Marvel Cem., 4 mi. N. of Ansted, WV. Upright double marker to grave.

SMITH, A. H.: Possibly Augustus H. F. Smith who enl. as a Pvt. in Co. G, 36th Va. Inf. 5/3/62 in Bland Co., Va. Present and promo. to Corp. Deserted 6/29/64.

SMITH, BALLARD J.: B. 1830. D. 1914. Bur. Falling Springs Presbyterian Church Cem., Allegheny Co., Va.

SMOOTH, GABRIEL: Printing error — name should be Gabriel **Smoot.**

SPESSARD, J. P.: Name is apparently Jacob F. Spessard. Res. Union, WV 1874. Business. Res. Fincastle, Va. 1882. Store. Returned to Union, WV. Mercantile. D. 9/8/1908 of kidney and heart failure. Age: 72.

STARCHER, EDWARD: Bur. Cherry Grove Cem., 0.8 mi. S. of Alto, Jackson Co., WV. on Rt. 22/6 then 1.8 mi. S.E. on rd. to cem., 200 ft. N. of rd. No marker.

STEEL, WALDEN M.: B. 7/17/36. D. 4/25/1900. Bur. Ward Cem., 2.4 mi. E. from Flat Rock, Mason Co., WV on Rt. 5 to Rt. 20, 1½ mi. N. on Rt. 20 to cem. 25 ft. W. of rd. Upright marker. PWR says name is Walden M. Steele.

STEWART, JOHN ANDERSON: B. in Va. D. in Mingo, WV of senility. Bur. in Mingo Cem., 0.2 mi. N. of Mingo, WV on Rt. 62/1. Cem. 1000 ft. N. of rd. Upright marker.

SUMMERS, WILLIAM SYDNEY: Same as William S. Summers who enl. as a Pvt. in Co. I, 8th Va. Cav. 11/1/62. Det. to Commissary Dept. 8/63. On receipt rolls to 12/30/63.

SURBAUGH, G. WASHINGTON C.: Same as George Washington C. Surbaugh who was B. 5/19/34 in Greenbrier Co., WV. Enl. as a Pvt. in Co. A (2nd), 14th Va. Cav. 5/5/62 at White Sulphur Springs, WV. Apptd. in 22nd Va. Inf. 4/10/63. Farmer. D. 6/12/1912 at Rainelle, WV. Bur. Sewell Baptist Church Cem., Sewell Mountain, WV.

SWANN, JOHN STERLING: Son of Richard Swann and Miss Belt. V.M.I. class of 1842 (non-graduate). Resgnd. from V.M.I. 1/18/41 to take effect 1/21/41. Matriculated from Powhatan Co., Va. Guardian: Benjamin L. Belt. Came to Charleston, WV in Spring of 1849. PWR says formed a pre-war company in Logan Co., WV. Elected Capt. of Swann's Co., 22nd Va. Inf. 5/8/61. Present as Capt., Co. A, 26th Battn. Va. Inf. on 4/30/63 roll. WIA (slight flesh/head/ear) at Gardner's Farm, Va. 5/30-31/64. Admitted to Chimborazo Hosp. 6/1/64. At Chimborazo Hosp. with gunshot wd. to right ear and scalp 6/2-3/64. At Wayside Hosp., Richmond, Va. for very serious head wd. 7/6-27/64. At Charlottesville, Va. Hosp. 7/27/64 — resumed furlo. 7/27/64 —

was stopping at Wayside in transf. and left during the week. Capt. at Winchester, Va. 9/19/64. POW Ft. Delaware 9/27/64. Took oath at Ft. Delaware and relsd. 6/17/65. 5'10". Fair complexion, light hair, blue eyes. Res. Kanawha Co., WV. Never married. "A brave and gallant man at all times." In later years was "an old cranky bachelor & when he died was practically out of practice and almost a recluse." D. at Charleston, WV.

TAYLOR, ELI: Possibly same Eli Taylor who served in 36th Battn. Va. Cav. Bur. Taylor #1 Cem., 1 mi. S. of Frametown, WV on Rt. 4 to junction of Rt. 30/3 then 1 mi. on Rt. 30/3 to cem. Braxton Co., WV. No marker to grave.

TEUBERT, JOHN: Middle initial is G. D. 2/3/1904. Bur. Teubert Cem., ½ mi. from Chimney Corner up Big Creek off Rt. 19/21 & 60, Fayette Co., WV. Upright marker.

THOMAS, WILLIAM B.: D. in Lynchburg, Va. hosp. Bur. Old City Cem., Lynchburg, Va. [Co. D?]

THURMOND, ROBERT GIVEN: Bur. Thurmond Cem., 5½ mi. S. on Rt. 25 from Thurmond, Fayette Co., WV. Upright marker.

TINGLER, GEORGE W.: Middle name is Washington.

TOMLINSON, ANDREW N.: Probable same Andrew N. Tomlinson who enl. in Bryan's Battery 3/27/62 at Lewisburg, WV. Present through final roll with that unit. D. before 1917.

TONKIN, L. E.: B. 3/23/36. D. 10/28/96. Bur. Tonkin Cem., 1.4 mi. N. of Corley P.O., WV on Rt. 22, 200 yds. W. of rd. Braxton Co., WV. Upright marker.

TRIPLETT, E. BAINS: Bur. Stonewall Cem., Winchester, Va.

TRIPLETT, MARSHALL: PWR says in Co. #3, Capt. Marshall Triplett's Co. of Rangers. Apptd. Capt. 3/18/62. PWR also says Capt. in Greenbrier Co., WV.

TROUT, HENRY S.: Was in hosp. for 'febris bilius.'

TUCKWILLER, M.: Middle initial is J. Bur. Confederate Cem., Univ. of Va., Charlottesville, Va.

TYREE, ALEXANDER N.: D. 3/2/1913 at age 75. Bur. Riverview Cem., Ronceverte, WV.

TYREE, JAMES: Middle initial is H. PWR says Co. H. Bur. Falling Springs Presbyterian Church, Allegheny Co., Va.

TYREE, JOSEPH M.: Middle name is McClung. B. 1840. D. 1920. Bur. Westlake Cem., Ansted, WV. Upright double marker to grave.

VERSER, L. J.: WIA at Cold Harbor, Va. 6/3/64. PWR says name is I. J. Verser.

VEST, JOHN M.: Possibly same John M. Vest who serve as a Pvt. in Co. K, 60th Va. Inf. and was B. 9/17/18 at Jumping Branch, WV and D. 7/6/99 at Jumping Branch, WV of old age. Bur. Farley Cem., 2½ mi. W. of Jumping Branch, WV on Rt. 3. Cem. ½ mi. N.E. of rd. Upright marker.

VICKERS, JOHN H.: Possibly same John H. Vickers who enl. as a Sgt. in Co. G, 8th Va. Cav. in Monroe Co., WV (n.d.). Capt. near Knoxville, TN 12/18/63. POW Louisville, KY. Forwarded to Rock Island, IL 1/23/64. Transf. for exchange 3/20/65. Paroled at Charleston, WV 5/3/65 at age 24.

WATTS, LEWIS FRANKLIN: Also served as a Pvt. in Co. D, 14th Va. Cav. at Lewisburg, WV 3/1/64. Absent on detail Tailor at Rocky Point (Sinks Grove), WV 6/31/64. Present 9-10/64. Capt. at Cedarville, Va. 11/12/64. POW Pt. Lookout, MD. Exchd. 2/15/65. Paroled at Lewisburg, WV 4/27/65. 5'10½". Light complexion, light hair, blue eyes, dark whiskers.

WEEKS, CORNELIUS: Deserted. D. 5/21/64 of Lynchburg, Va. hosp. Bur. Old City Cem., Lynchburg, Va.

WETZELL, GEORGE W.: Probably same George Washington Watzell who enl. as a bugler in Bryan's Battery(?) 4/10/62 at Lewisburg, Greenbrier Co., WV. Capt. at Lewisburg, WV 5/23/62 (as either a member of the 22nd Va. or Bryan's Battery). POW Camp Chase, OH. Exchd. Present with Bryan's Battery until on leave Dec. 1863. Sick with typhoid Aug. 1864. Paroled at Charleston, WV 6/12/65. Age: 36, 5'7". Blue eyes and light hair. D. before 1917.

WHITT, THOMAS: Enl. in Co. C, 36th Va. Inf. 11/15/63 at Narrows, Va. Present until transf. to 17th Va. Cav. Jan. 1864. Transf. to 22nd Va. Inf. 2/28/64.

WHITTEN, LEWIS: Postwar roster says "remembered by his comrades in Co. F" and says middle name is Austin. Res. Greenville, WV. Enl. in Co. K early in war. Postwar farmer. D. 2/26/1920. Age: 88. Bur. Keaton Cem.

WICKLINE, BERNARD PAYTON: Ord. Sgt. Co. F. Enl. at age 18 and served until end of war. Not in CSR but probably same Bernard P. Wickline who enl. in Co. A, 26th Battn. Va. Inf. 11/23/63 at Lewisburg, Greenbrier Co., WV. Conscript. On clothing receipt 4/4/64. Capt. in Cold Harbor, Va. campaign. POW in hands of the enemy on 10/31/64 roll. Paroled at Charleston, WV 6/12/65. Age: 20. 6'. Dark complexion, brown hair, hazel eyes. Postwar res. Gap Mills, WV. D. 1/15/1916 at age 71. Bur. Cokesbury Church Cem., Sweet Spring Valley.

WILKINSON, AMBROSE B.: D. at Lanham, Putnam Co., WV. Bur. Melton Cem. #2, two mi. W. of Lanham P.O. on Vince Lanham farm near the river bank. No marker to grave.

WILLIAMS. R. L. D.: D. while in prison.

WILSON, JOHN HENRY: 2nd Sgt. WIA at Cold Harbor, Va. 6/3/64. D. 1/10/1912.

WILSON, WILLIAM L.: WIA (shot in arm) at Cold Harbor, Va. 6/3/64.

WOODRUM, GREEN: B. 12/20/39 in Boone Co., WV. D. 3/4/1912 at Lory, Boone Co., WV. Bur. Woodrum Cem. 1½ mi. E. from Lory, WV on Rt. 6 to cem. which is 1200 ft. N. of road. Upright marker. Middle initial is B.

YOUNG, ARCHIBALD P.: Also called Archer. Enl. 5/8/61 in Swann's Co. at Charleston, Kanawha Co., WV. Then Co. A, 26th Battn. Va. Inf. until transf. to Co. H, 22nd Va. Inf. 3/20/63.

ERRATA AND CORRECTIONS:

Page 15 - paragraph 3 - line 8: should read - The full-scale retreat, beginning on July 24, went relatively well until July 25 when some 700 men under -

Page 16 - paragraph 1 - line 3: Many accounts say Wise reached Lewisburg July 31.
Page 16 - paragraph 4 - line 2: Should say that Floyd arrived and took charge August 11.
Page 16 - paragraph 4 - line 6: Should read - but in mid-August the two quarrelsome generals -
Page 19 - paragraph 4 - line 1: Should be September 12 (*not* 13).
Page 19 - paragraph 4 - line 3: Should be 14th (*not* 17th).
Page 19 - paragraph 4 - lines 5 and 6: Should be - reached there on the 17th (*delete* same day).
Page 20 - paragraph 2 - lines 12 and 16: Change to - On September 24 Lee came to Sewell Mountain with the 22nd and 36th Virginia regiments, the 13th Georgia and 14th North Carolina and Floyd returned to the mountain with his troops October 1st. On September 25 the 22nd Virginia built breastworks at Camp Defiance — etc.
Page 20 - paragraph 3 — line 6: Should be September 26 (*not* 29).
Page 20 - paragraph 4 - line 1: Should be October 12 (*not* 11).
Page 20 - paragraph 5 - line 2: *Delete* on October 23 he discovered - replace with - *realized.*
Page 21 - paragraph 1 - line 2: Begin sentence with - On October 21 Floyd immediately set up camp - etc.
Page 28 - paragraph 2 - line 12: Singleton was in Co. K (*not* Co. H).
Page 31 - paragraph 2 - line 1: Should be eastern (*not* western).
Page 31 - paragraph 2 - line 6: Spelled Colonel (printing error).
Page 47 - paragraph 3 - line 15: Singleton was in Co. K (*not* Co. H).
Page 47 - paragraph 3 - line 18: Should be - pick (*not* picked).
Page 51 - paragraph 2 - line 2: *Delete* Mounted.
Page 53 - paragraph 2 - line 12: Capt. Thompson of Co. A was not captured at Droop Mountain.
Page 53 - paragraph 2 - line 17 and 19: Should read - Averell reaching Beverly on November 12 and Duffie at Charleston on the 17th.
Page 62 - paragraph 3 - line 1: Printing error - should say - While still in the trenches - etc.
Page 63 - paragraph 3 - line 13: Should read - had seen "Strong", the Negro cook replied - etc.
Page 70 - paragraph 3 - line 5: Should be Charlestown (*not* Charleston).
Page 71 - paragraph 1 - lines 2 and 12: Both times should be Charlestown (*not* Charleston).
Page 71 - paragraph 2 - line 8: Should be Charlestown (*not* Charleston).
Page 87 - paragraph 1 - line 3: Should be - killing a Yankee (*not* at Yankee).
Page 103 - paragraph 2 - line 2: *Delete* Kanawha County and replace with Putnam County.
Page 114 - paragraph 5 - line 6: Delete *was* at end of sentence.
Page 118 - paragraph 2 - line 2: Change necessary to necessarily.

Page 118 - paragraph 2 - line 4: Change surround to surrounding.
Page 119 - paragraph 2 - line 10: Typographical error - tie should be time.
Page 119 - paragraph 3 - line 7: Change approximately to approximate.
Page 119 - paragraph 7 - line 2: Change other to others.

ERRATA ON MAPS AND CHARTS

Page 27 - Giles C.H. - 2nd WVa. Cav. of Paxton and Gilmore should be white to distinguish them as Federal units.

Page 45 - White Sulphur Springs - Oley and two units south of creek should be white to distinguish them as Federal units.

Page 49 - Droop Mountain - 6th Battalion should be 26th Battalion and should not be on map at all.

Page 59 - New Market - "Field of Last Shots" should be "Field of Lost Shoes."

Page 88 - Second note in middle of page should be 12/30/63 roll, *not* 12/30/64 roll.

LATE ADDITIONS:

ALLEN, JAMES H.: PWR says middle name is Henry and also served in Co. A, 45th Battn. Va. Inf. and enl. in that unit 4/2/63 in Boone Co., WV. WIA (shoulder by sabre) at Droop Mountain, WV 11/6/63. Later in Co. A, 27th Va. Cav. Later capt. and POW at Ft. Delaware. Took the oath and relsd. 6/12/65. Was elected 2nd Lt. in Co. A, 45th Battn. Va. Inf. Brother of William Perry Allen.

ALLEN, WILLIAM: Middle name is Perry. Age 18 when KIA. Brother of James Henry Allen. A. "Mr. Powell from Greenview was with him when he died."

McMILLION, NATHAN: B. in Greenbrier Co., WV circa 1843. Farmhand, age 17, Lewisburg, WV 1860 census. Enl. as a Pvt. in Co. A (2nd), 14th Va. Cav. 4/21/62 in Greenbrier Co., WV. Transf. to 22nd Va. Inf. 10/31/62 in exchange for William W. Wolfenbarger. Farmhand, age 27, Falling Springs Dist., Greenbrier Co., WV 1870 census. Farmer. D. 12/24/93 in Greenbrier Co., WV. Brother of Robert M. McMillion. Name does not appear on 22nd Va. Inf. roster but is mentioned in record of Willam W. Wolfenbarger and called Nathan McMullen.

McMILLION, ROBERT M.: B. circa. 1841. Farmhand, age 19, Lewisburg, WV 1860 census. Enl. (date unknown) as a Pvt. in Co. A (2nd), 14th Va. Cav. Transf. to 22nd Va. Inf. in exchange for John A. Myles 10/20/62. Brother of Nathan McMillion.

RICHMAN, A. T. C.: Pvt. Dischd. on surg. certificate 11/19/61 at Camp Piney, WV. This name is not on the roster.

ETC. — Additional Information of Interest:

Capt. John K. Thompson, Co. A, 22nd Va. Inf. was in command of the regiment at Cold Harbor and Col. Patton commanded the brigade. This is confirmed in a letter written by Thompson to George M. Edgar found in the George M. Edgar found in the George M. Edgar Papers, Southern Historical Collection.

The Rocky Point Grays were from Rocky Point, Monroe Co., WV which today is known as Sinks Grove, WV.

Col. William A. Jackson was assigned command of the 22nd Va. Inf. 11/10/61 at Camp Dickerson, WV.

Capt. Thomas Belt Swann, Co. E, had a court-martial trial set 10/15/61 at Camp on New River but no charge was specified but there was a list of officers who were to hear his case at Camp Dickerson, WV 10/16/61. He was released and returned to duty on the day of the hearing.

Maj. Isaac Noyes Smith was granted a leave of absence "until such time as he hears from his resignation" 10/29/61 at Camp Dickerson, WV.

Col. Christopher Q. Tompkins was granted a leave of absence "until he hears about his resignation" 11/5/61 at Camp Dickerson, WV.

Maj. Thomas Smith of the 36th Va. Inf. was assigned to duty with the 22nd Va. Inf. and ordered to report to Col. Tompkins 10/29/61 at Camp Dickerson, WV.

Courtesy: Sandra S. White, Lewisburg, WV

Pvt. John C. Forren, Co. F. Photo taken about 1915.

Pvt. Richard Lafayette Deal, Co. C

1st Sgt. Samuel F. Tyree, Co. C

Courtesy: Wayne Casto, Nitro, WV

Pvt. John Gilbert Cartmill, Co. A

Lt. Samuel Hunter Austin, Co. B

M. N. Humphreys, Co. K

Pvt. Joseph H. Lipps, Co. B

ABOUT THE AUTHOR

Photo by Patty Tyler

Born in Charleston, West Virginia, in 1949, ***Terry Lowry*** is a 1967 graduate of South Charleston High School where he worked on the school newspaper. He graduated in 1974 with a B. A. in History from West Virginia State College (now University) and studied Civil War History at Marshall University Graduate School.

A professional musician since 1965, he was contributing music editor for *The Charleston Gazette* 1970-75 and music editor 1977-78. Lowry spent two years with the Circulation Department of *The Atlanta Journal* and was employed for 20 years with the Circulation Department of Charleston Newspapers, Inc.

He published his first book, *The Battle of Scary Creek: Military Operations in the Kanawha Valley, April-July 1861,* in July of 1982. He has also published *September Blood: The Battle of Carnifex Ferry* (1985); two volumes in the Virginia Regimental Series, *22nd Virginia Infantry* (1988) and 26th (Edgar's) *Battalion Virginia Infantry* (1991); and *Last Sleep: The Battle of Droop Mountain November 6, 1863* (1996).

He co-authored with Stan Cohen *Images of the Civil War in West Virginia* (2000). Additionally, his Civil War articles have been published in *North South Trader, Wonderful West Virginia,* and *Confederate Veteran* magazines, as well as in the *West Virginia Hillbilly.* He has also been a contributor to the Time-Life series of books and remains an avid collector of Civil War memorabilia.

He was historian/curator of the Craik-Patton House, Charleston, West Virginia, for two years and in 2001 has been a historian at the West Virginia State Archives, Charleston, West Virginia.

In 2009 he published *Bastard Battalion: A History of the 83rd Chemical Mortar Battalion in World War II,* which critics have called "one of the best World War II unit histories ever written." The surviving members of the 83rd appointed him Official Historian of the unit.

He wrote *The Battle of Charleston: The 1862 Kanawha Valley Campaign* in 2016 and retired from the West Virginia State Archives the following year.

The author can be reached by e-mail: ***TLSnoop@aol.com***

Made in the USA
Monee, IL
05 November 2026

67815704R00136